# Synthese Library

Studies in Epistemology, Logic, Methodology, and Philosophy of Science

Volume 484

The aim of *Synthese Library* is to provide a forum for the best current work in the methodology and philosophy of science and in epistemology, all broadly understood. A wide variety of different approaches have traditionally been represented in the Library, and every effort is made to maintain this variety, not for its own sake, but because we believe that there are many fruitful and illuminating approaches to the philosophy of science and related disciplines.

Special attention is paid to methodological studies which illustrate the interplay of empirical and philosophical viewpoints and to contributions to the formal (logical, set-theoretical, mathematical, information-theoretical, decision-theoretical, etc.) methodology of empirical sciences. Likewise, the applications of logical methods to epistemology as well as philosophically and methodologically relevant studies in logic are strongly encouraged. The emphasis on logic will be tempered by interest in the psychological, historical, and sociological aspects of science. In addition to monographs *Synthese Library* publishes thematically unified anthologies and edited volumes with a well-defined topical focus inside the aim and scope of the book series. The contributions in the volumes are expected to be focused and structurally organized in accordance with the central theme(s), and should be tied together by an extensive editorial introduction or set of introductions if the volume is divided into parts. An extensive bibliography and index are mandatory.

Adam Rieger • Stephan Leuenberger
Editors

# Themes from Weir: A Celebration of the Philosophy of Alan Weir

Springer

*Editors*
Adam Rieger
School of Humanities
University of Glasgow
Glasgow, UK

Stephan Leuenberger
School of Humanities
University of Glasgow
Glasgow, UK

ISSN 0166-6991 ISSN 2542-8292 (electronic)
Synthese Library
ISBN 978-3-031-54556-6 ISBN 978-3-031-54557-3 (eBook)
https://doi.org/10.1007/978-3-031-54557-3



This Springer imprint is published by the registered company Springer Nature Switzerland AG
The registered company address is: Gewerbestrasse 11, 6330 Cham, Switzerland

Photograph by Neil McDonnell

# Introduction: Themes from Alan Weir

**Stephan Leuenberger and Adam Rieger**

## I

After a distinguished career at Queen's University Belfast and the University of Glasgow, Alan Weir was forced into early retirement in 2015 due to illness. When he was in better health, a three-day workshop 'Themes from Alan Weir' was held in Glasgow in December 2018. The present *Festschrift* consists partly of contributions to the workshop, and partly of papers subsequently commissioned. Its aim is to celebrate Weir's contributions to philosophy.

In this Introduction, we wish to give the reader a sense of Weir's intellectual character, and sketch a few of his ideas.

Most of Weir's published work has been in the philosophy of logic and the philosophy of mathematics. However, his interests range widely, and he has also published in the philosophy of language, the philosophy of perception, and the history of analytic philosophy. He has always been a systematic philosopher, making sure that his views in different areas of philosophy fit together. In recent years, he has become more explicitly a system-builder in his work on metaphysics. In that extensive work—not yet published—he aims to formulate a mereological theory that furnishes a complete ontology, to which everything can be reduced.

In his choice of topics to work on and views to defend, Weir shows a healthy disregard for philosophical fashion. His trademark move is to rehabilitate views that are natural and have powerful intuitive support, but have been discredited to the extent they are no longer considered by other researchers. We will sketch a few such views, as well as Weir's relaunched versions, in the next section.

Weir's work displays a rare combination of striking originality, philosophical sensitivity, and technical ingenuity. He cares deeply about the issues he writes about, and does not embrace novelty for novelty's sake. In his writings, he leaves no stone unturned in his attempt at solving the problem at hand. Difficulties are confronted head-on when they crop up, rather than finessed, or saved up for consideration in future work. Packaging his contributions into small publishable units, made acceptable to as wide a range of background commitments as possible, has never been Weir's aim. As a result, his writings are densely packed with ideas and arguments.

Due to these characteristics, as well as the relative unfamiliarity of some of the views he advocates, reading Weir can be a challenge. It is both taxing and rewarding. But his work is not only challenging but, in places, funny, thanks to the humour with which it is sprinkled.

Weir's voracious mind has interests beyond philosophy. He is a fervent advocate of Scottish independence, and, characteristically, offers his own justification for it, distinct from the one offered by his party, the Scottish National Party. His subtle reasoning is laid out in three substantial essays, which are available on his website.[1]

## II

In this section, we wish to characterise briefly Weir's overall philosophical position, as well as to highlight three of his most striking contributions: a neo-formalist account of mathematics, and two closely related 'neo-naive' theories, of truth and of sets, respectively. Given the subtlety of his views, any short summary has to simplify them, and elide distinctions that Weir has carefully drawn.

Weir's basic philosophical outlook is a form of physicalism. Notably, he rejects the existence of mathematical entities in a Platonic realm, outside space and time. He himself prefers the term 'naturalism'. However, given that he tends to emphasise the ontological rather than the methodological elements of that multi-faceted position, 'physicalism' may be more apt. He may think that label has lost its usefulness, having been associated with a mere supervenience claim. Weir insists that any physicalism worth having is a form of the identity theory, according to which all entities are physical. He has articulated his ontological commitments in passing in a number of places, and is working out the details in the unpublished metaphysical work that we mentioned in the previous section.

The first of the distinctive views which we will summarise, his neo-formalist philosophy of mathematics, is developed in his monograph *Truth Through Proof.* The view revives an idea that is intuitively appealing if one tries to account for mathematics in a physicalist framework: that the truth of a sentence of the language of mathematics does not consist in its correspondence to some extra-linguistic reality, but rather in the existence of a proof. Here, a proof is understood to be a sequence of symbols satisfying certain conditions. A physicalist ontology may be expected to be able to accommodate syntactic entities like proofs.

However, formalism is widely held to be untenable—to be one of the theories whose decisive refutation is a clear example of progress in philosophy. The usual reason given is that results due to Gödel and Tarski tell us that no system will generate proofs for all truths.

Weir's key move is to liberalise the notion of a proof, in such a way that these results do not apply. Specifically, he allow proofs of infinitary length. This involves an idealisation of our ordinary notion of proof, for whose legitimacy Weir makes

[1] https://sites.google.com/site/weirysite/scottish-independence-referendum-1.

an impressive case. His argument can be broken down into two claims: first, that idealisation to finite proofs of arbitary length is legitimate; and second, that if idealisation to arbitrary finite proofs is legitimate, so is idealisation to infinite proofs. It is the second claim that is distinctive to Weir. Given the naturalistic background metaphysics, the central distinction is between those proofs which can be taken to be inscribed by some physical system in the universe and those that are not—and that distinction divides short finite from long finite proofs, not finite from infinite ones. Of course, this defence of the second claim puts pressure on the first one, which Weir resourcefully deflects. Weir's philosophy of mathematics is discussed in Chap. 8, by Mary Leng.

The naive theories of truth and of sets that Weir revives consist of intuitively compelling principles: the disquotation principle that '$p$ is true' is equivalent—in some sense—to $p$, and the comprehension principle that for every condition, there is a set of things meeting that condition. These theories are widely held to be refuted by the liar paradox and by Russell's paradox, respectively.

The standard response to Russell's paradox involves denying the comprehension principle, while typical responses to the liar paradox either deny that the paradox-generating liar sentences are well-formed or restrict the disquotation principle in some way. What is attractive about Weir's neo-naive theories is that they do none of these things: both the disquotation principle and the comprehension principle hold in full generality. Neo-naive theories share this virtue with paraconsistent ones; but unlike the latter, they are consistent.

Weir's neo-naive theories achieve this feat by weakening the background logic. Specifically, derivability in Weir's neo-classical logic fails to satisfy a certain generalised version of transitivity. His approach thus belongs to what has come to be the called 'substructural' family. Weir's neo-classical logic is discussed in detail in Chap. 4, by Elia Zardini.

Weir emphasises that if our theories are going to be consistent, we face a choice between giving up natural introduction and elimination rules for the truth predicate and for set abstracts, and giving up structural principles of derivability. He makes a strong case for the former having stronger epistemological credentials than the latter. In doing so, Weir is turning the tables on the strategy of appealing to structural rules to explain why certain pairs of introduction and elimination rules do not correspond to any connective—a strategy first deployed by Belnap in his diagnosis of what is wrong with Prior's 'tonk' connective.

In a series of papers, Weir has discussed semantics for his neo-classical logic. The basic move is to abandon bivalence, and to require a certain version of falsity-preservation from the conclusion to the premisses in addition to the usual requirements of truth-preservation from the premisses to the conclusion. He has explored various valuation spaces for this logic, 3-valued, continuum-valued, and even surreal-number valued. He aims to account for the legitimacy of using full classical logic in certain contexts by arguing that all sentences have one of the classical truth values in those contexts.

Our sketch of Weir's ideas in this section has barely scratched the surface. For a detailed exposition, we refer the reader to his published work (a full bibliography of which can be found at the end of this volume), and, once more, to Chaps. 4 and 8.

## III

After this Introduction and Alan Weir's philosophical Memoir, the volume features six contributions to the philosophy of logic. As a long-standing defender of classical logic, Timothy Williamson challenges logical revisionists to account for the applicability of mathematics (Chap. 2). In his contribution to the volume (Chap. 3), Weir takes up the challenge on behalf of his neo-classical logic, and argues that the constraints on the application of certain modes of reasoning may provide a welcome safety net. Neo-classical logic is also the subject of Chap. 4, which we already mentioned: Elia Zardini provides a detailed analysis and evaluation of Weir's logic, with its restriction on transitivity. After that, Neil Tennant discusses a dilemma for a cricket umpire, and how a proponent of intuitionistic logic may respond to it (Chap. 5). Roy T. Cook studies some conditions on abstraction principles that Weir had suggested in his work on neo-Fregeanism (Chap. 6), and Eric Snyder and Stewart Shapiro examine how plural logic might deal with mass nouns (Chap. 7).

The next block of papers deals with topics in the philosophy of mathematics. As we mentioned, Mary Leng continues a debate she has had with Weir over a number of years, concerning the respective merits of his formalism and her own fictionalism about mathematics (Chap. 8). Marianna Antonutti Marfori examines whether mathematical naturalism and revisionism about mathematics—two key commitments of Weir's—are compatible (Chap. 9). The papers by Graham Priest and Marcus Giaquinto engage with topics in the philosophy of set theory (Chaps. 10 and 11).

The rest of the papers in the volume engage cover a variety of topics. They span the philosophies of mind and language, metaphysics, epistemology, and political philosophy—a fitting tribute to the range of Weir's philosophical interests and contributions. Graham Peebles denies that introspection presents us with, or directly causes us to conceive of, qualia (Chap. 12), while Alexander Miller offers a quasi-realist interpretation of Kripke's 'sceptical solution' to the rule-following paradox (Chap. 13). The paper by Stephan Krämer (Chap. 14) uses truthmaker semantics to investigate various notions of equivalence that are tighter than logical equivalence. The volume comes to a conclusion with contributions by two of Weir's Glasgow colleagues. Gary Kemp discusses the two-envelope paradox (Chap. 15), and Adam Rieger (Chap. 16) argues in favour of adopting a novel voting system.

## IV

Finally, we would like to close with some remarks of a more personal nature. Alan is a superb philosophical interlocutor, full of insightful comments on others' work. His live presentations of his own work are in a class of their own: the philosophical material, often of a highly technical nature, is lightened with Alan's unique humour, typically involving pictures, cartoons, and outrageous jokes which bring the house down. He is fantastically good company, in the pub or elsewhere, with an unending supply of anecdotes, many hilarious and some quite hair-raising (fortunately a good sample of these is available in Alan's memoir, included in this volume). He led the department through a particularly difficult period, when the university was being 'restructured', with sensitivity and wisdom. He is a man of enormous intellectual energy, but also great kindness, warmth, and generosity. The whole department was thrilled to have him back in our community after his illness. Alan has been a wonderful colleague and friend in Glasgow, and we feel privileged to have worked alongside him.

# Contents

# Chapter 1
# A Memoir

**Alan Weir**

**Abstract** In this Memoir I describe the trajectory of my philosophical career, highlighting the themes which I have pursued. I endeavour to set it all in the context of changes to the philosophical and academic environment over the course of my adult life, interspersing a few vignettes which may be of interest to the younger generation.

I must start by expressing my enormous gratitude to Stephan Leuenberger and Adam Rieger for organising the conference in December 2018 on which this volume is based and then going on to bring the book to fruition. In addition, of course, I must thank the speakers and participants at the conference, and the contributors to the volume (overlapping but distinct groups of people). I will start by outlining a little bit of the context in order, I hope, to make the nature and structure of this Festschrift clear.

The conference might well have emerged as a memorial conference; needless to say I am very glad it did not. For one thing all the available evidence suggests that the subjects of memorial conferences cannot go out for a drink in the evening with the participants, dispute their conclusions, try desperately to turn the conversation away from the defects of their philosophy on to trivia and gossip; and so on. To explain: following a diagnosis in 2013 of the incurable disease Idiopathic Pulmonary Fibrosis – mean life span from diagnosis 3 years – I had to take medical retirement from Glasgow University in 2015 when the symptoms worsened, rendering work impossible. The conference in 2018 took place because I was very fortunate, in 2016 and weeks from death, to receive the only get out of jail card for this condition, a lung transplant, bilateral in my case. Of course this means that my escape from death is the result of someone else's tragedy, a very sombre thought.

A. Weir (✉)
University of Glasgow, Glasgow, UK
e-mail: Alan.Weir@glasgow.ac.uk



A. Rieger, S. Leuenberger (eds.), *Themes from Weir: A Celebration of the Philosophy of Alan Weir*, Synthese Library 484,
https://doi.org/10.1007/978-3-031-54557-3_1

On my initial diagnosis I thought of putting together a collection of my papers, writing some interlinking material and a sketch of the overall philosophical position I'd hoped to set out in the three or so volumes I had ludicrously imagined I would produce after the publication of my 2010 monograph. The referees for the project were very kind in their report but gently suggested that there weren't too many top ten hits among my published essays: a top 100 in the specialist philosophy of mathematics chart, perhaps, but that was about it. Why didn't I just write a monograph defending the rather ambitious and wide-ranging position the published papers were partly working towards?

So that is what I have been doing, alas even more slowly than was my wont. But these circumstances partly explain the format of the conference and the volume: contributors were told they need not focus on ripping apart (or even endorsing!) a publication of mine but were to feel free to write on any of the themes I have published on, such as philosophy of mathematics, naïve set theory, liar paradoxes, philosophical logic more generally, Quine, metaphysical realism and some topics in philosophy of mind and epistemology.

So I have not written a set of replies to critics. I have, however, written a piece, Weir (2024), prompted by a challenge from Tim Williamson in his contribution. In this introductory autobiographical sketch my aim is to outline the trajectory of the emergence of the above-mentioned themes in my career. But I also hope to provide a brief account which may be of interest to younger philosophers of what it was like to enter the profession in the late 1970s, as a white, male philosopher (so not a very unusual profile!) and give an indication of how, and to what extent, things have changed. All this with the important provisos (a) that my memory is pretty hazy and of course (b) even white, male philosophers differ quite a lot in philosophical views, backgrounds, education, academic employment and so on. So generalisations from my experience, even if I've captured it fairly accurately, should be made very tentatively.

I start with my entry into what might be considered the trial phase of a professional philosophical career: full-time post-graduate studies.

## 1.1 The Dreaming Spires

I arrived in Oxford in 1977 at the beginning of Michaelmas term, as it is known in those parts, to commence the B.Phil degree, the (excellent) master's degree at Oxford (which, with its customary arrogance, it refuses to call a master's degree). I missed the initial formal admission to Oxford, being otherwise occupied at the time in a sheriff court in Scotland answering charges of assault and breach of the peace; plea bargaining got me off the former charge but I was fined for the latter. This wasn't particularly problematic for me at Oxford, as the charge of assault related to throwing rotting vegetables, gladly donated by a nearby Asian shop, at the leader of the then very active neo-Nazi National Front, who was up from England being interviewed by a local radio station. To the amazement of those who knew of my somewhat erratic throw-outs as a goalkeeper, I actually scored a direct hit. Seeing the boys in blue come after me I headed towards the serried ranks of IS – International Socialist – and IMG – International Marxist Group – comrades confident they would hold off the agents of capitalist oppression and enable my escape. But this confidence proved to be entirely misplaced, they melted like the driven snow and I was duly huckled by the polis, to put it in Scots. Initially I feared getting a beating from the police in the cells, not unheard of in those days. But there was more of a danger from internecine warfare among some of the other imprisoned comrades: the question of whether the Soviet Union was a degenerate worker's state, or merely a deformed one (or was the 'merely' the other way round?) for example, used to excite more passion among members of the Trotskyite groupuscules than defeating Tories.

At any rate, having arrived late I had to present myself by special arrangement to P.F. Strawson at his rooms in Magdalen college, to be officially and legally admitted. But I was late for that too, having celebrated, a bit too exuberantly, my arrival the night before with some fellow philosophers I had newly met. Still, very hoarse, near blinded by blazing, as it seemed to my aching brain, autumn daylight and incapable of uttering anything much more complex than Quinean observation sentences – 'Sunny'; 'Lo: Deer' – I did manage to convey to the great Strawson who I was and duly got the required forms signed. He may well have been tempted at that point to abandon his purely descriptive metaphysics in favour of revisionary metaphysics, a creature of a metaphysical category hitherto unknown to human speculation having appeared before him.

I had no direct contact with Strawson, *père*, thereafter, apart from a very amiable conversation with him and Hao Wang at a Quine conference in St. Louis more than a decade later (of which conference more anon). A pity, as he was a bit of a hero of mine at the time. I'd heard him on the radio in Bryan Magee's *Conversations with Philosophers*, the aural precursor of his televised *Men of Ideas* series. The radio programme was important to me as a schoolboy, no doubt feeding my desire to become a philosopher (though I must have been inclining that way to have listened to it). Moreover I'd already read, before going to Oxford, Strawson's book *Individuals* and had this notion that I might find Strawson blind-folded at his desk, listening to a sequence of bleeps from an enormous antique radio, a broadcast of a Stockhausen piece perhaps, while emitting a steady, though varying in pitch, droning sound (the radio that is). Readers of Strawson's book will doubtless have themselves performed a similar experiment in order to test the coherence of the idea he considers in Chap. 2, namely that a purely aural world would be non-spatial and so, it might be thought – though Strawson does not conclude this – not a world of objectively existing bodies. Alas I did not find him in that state, pioneering X-Phi, at least as far as I remember; but of course memory is fallible.

Initial formalities negotiated, albeit in a somewhat unorthodox fashion, I discovered that I had (crash) landed right in the middle of the 'Davidsonic boom' at Oxford. John McDowell and Gareth Evans were all the rage, their highly influential collection *Truth and Meaning* having been published the previous year. The burning questions of the day included such things as: whether, inverting Tarski, the concept of truth could be taken as a primitive in order to shed light on linguistic meaning; whether a 'homophonic' theory of truth, given in the same language as the target language – except for a strengthened truth concept – could explain what it is to understand a language; the extent to which formal truth theories could be applied to natural language; if so, whether to just the 'empirical' fragments of the language or more generally, for example to moral or aesthetic discourse; and so on.

Of course Oxford had, and has, a very large philosophy faculty especially if one includes all the college lecturers, temporary positions, visiting philosophers and so on; so those engaged in wrestling with Davidson's programme were no doubt a minority among the permanent staff. Still there is no doubt that among the young, the postgraduate community, the junior research fellows and so on, the work of the likes of Davidson, McDowell, Evans, Christopher Peacocke and others associated, to varying degrees, with that programme were viewed by a great many as 'where it is at'. Not so much, however, by me, as will become evident.

The B.Phil consisted of three examinable papers and a dissertation which could be used as a component part of a future D.Phil. (as Oxford call Ph.D.s), something forbidden in most universities that I am aware of. A common pattern was to complete the B.Phil. in the regulation 2 years then, for those of us on state postgraduate grants which lasted 3 years, attempt to expand the thesis to a finished D.Phil. in our third year when we had some money to live on. Like very many I didn't manage to do this and worked on finishing the D.Phil. during a further 4 years of temporary jobs, mostly teaching jobs but for a period as a barman.

My chosen B.Phil. subjects were (i) Philosophical Logic, (ii) Original Authorities in the Rise of Mathematical Logic and (iii) Hegel and Marx. Charles Taylor, who was my supervisor in Mathematical Logic – woops a typing slip there – in Hegel and Marx rather, was astonished at the combination. This greatly pleased me:- I smugly congratulated myself on my singularity. So I was very intrigued to see, when reading Crispin Wright's account of his philosophical education (Wright, 2009) that he chose exactly the same B.Phil. papers. Crispin, however, completed the B.Phil. in just 1 year! Stakhanovite!

My dissertation (and subsequent D.Phil.) was on 'The Philosophy of Language of W.V. Quine'. So at that time I viewed myself as a philosopher of language/logic (undoubtedly the fashionable specialisms at the time), though with a particular interest in philosophy of mathematics. The interest in Marx was motivated in great part by my political views and activities but I always have had a more general interest in political philosophy. Nor did I view studying Hegel (under, of course, a world-renowned Hegel scholar) as merely background for studying Marx; I was independently interested in Hegel. Explaining how it came to pass that I arrived in Oxford with this constellation of interests would require travelling back a little in time to my undergraduate education, I'll provide a small snippet later.

In 3 years in Oxford 1977–1980 I did spend a lot of time in political activity, mostly the toytown world of student politics. When I went to university aged seventeen, to a university whose name escapes me, such is my memory, but it is somewhere East of Glasgow, I was actually an anarcho-communist, but like many I had travelled far to the right by the time I had reached the advanced age of twenty two or so. However, since I had started out on the political pitch not so much on the left wing as outside the stadium among the programme sellers, this landed me on the left wing of the Labour Party, in a (very dull) group known as the Clause Four group (after the socialist clause in the Labour Party constitution of 1918). It was around then I met Marcus Giaquinto and had many a discussion with him ranging from the legitimacy (or lack of it in my view) of the Bolshevik smashing of the Constituent Assembly in January 1918 to abstruse questions in mathematical logic and philosophy of maths. At this stage I very naively thought that the Labour Party believed in self-determination for all nations, including Scotland, and that after devolution of some (minimal) powers to Scotland (which we voted for in 1979 but were denied by the Labour Party) one could work inside that party for equality in status and powers for the Scottish people and our parliament with those of other small European nations. I was a bit like a Jesuit thinking of entering the Taliban in hopes of converting them.

The Oxford University Labour Club however, especially from my vantage point – I organised the visiting speaker meetings for some time – was a very unusual one as many of the Labour great and good came to speak, including serving Cabinet ministers of the 74–79 government and many a tale I have to tell about what happened in the wining and dining after the talks. But my lips will forever remain sealed (as long as those crates of top class malts keep coming). Here I will try to stay focussed on philosophical matters.

## 1.2 Postgrad Studies

As well as supervisors for one's chosen subjects – in addition to Taylor I had David Bostock for Mathematical Logic and Christopher Peacocke for Philosophical Logic – one had an overall supervisor who typically guided one through the dissertation. I was very fortunate in having L. Jonathan Cohen, Jon Cohen to many, as my supervisor for the B.Phil. dissertation. Alas I think he is not very well-known among young philosophers now but he was an extremely interesting philosopher with an unusual range. His biggest influence on me was the work he did on the distinction between acceptance of propositions – voluntary – and belief in them – not directly under the control of the will he held – and its implications for rationality, although this did not percolate through to me until later.

At any rate, we hit it off very well. He was an excellent supervisor and also very important for my future career since, somewhat scandalously one could (indeed should) argue, Oxford, and to a lesser extent Cambridge, were overwhelmingly dominant in those days in the matter of jobs placement; letters of recommendation (always entirely confidential) from supervisors, often to ex-students or colleagues from Oxbridge on 'colonial postings' teaching out in the 'provinces' (as Jon Cohen had been, in Scotland) were vital. I came to know I was very well served by him in that regard.

One other person who would feature prominently in any account of Oxford at the time is Michael Dummett, the Wykeham Professor of Logic, with his massive book *Frege's Philosophy of Language* not long published and very influential. Dummett never supervised me and I had little direct contact with him but he certainly was an important influence, positively and negatively. I did attend quite a few seminars of his, or ones in which he was a major contributor (at an extremely high philosophical level of course); he also made significant contributions to the thick clouds of cigarette smoke which enveloped seminars in those carcinogenic days, clouds which often obscured fellow participants to the extent that the seminar table was sometimes like a mountain range with only the tallest philosophers emerging from the clouds.

Seminars were a little different then not only in the dangers from smoke inhalation: the ethos of the question and answer section differed too. In those far off times the instant the Q&A[1] starting gun was fired the hitherto silent seminar audience[2] transmutated into hyenas ravenously tearing the philosophical prey to pieces and leaving barely a tattered remnant of the thesis behind (even if half of the hyenas actually believed it and had in fact defended it in print). I will say the following on this matter. I have some sympathy with colleagues who regret the passing (if it actually has passed) of these, shall we say, vigorous times. My own instincts are always to attack a philosophical view presented to me and seek out its weaknesses even if, indeed perhaps especially if, I am attracted to it. One doesn't have to be a Popperian (please don't be a Popperian) to think that this sort of critical attitude can lead to the rooting out of philosophical error. And I do believe, firmly, there is such a thing as philosophical error (but would attack the whole concept in any seminar in which someone else defended that self-same belief). Moreover the ritualistic praise which nowadays sometimes prefaces interventions ('loved your talk so much; but ...') can get a bit sickly; we don't want our discussion periods to become, as it seems they are in some other disciplines, occasions for lavish flattery of the speaker liberally peppered with references to the interlocutor's brilliant, in print or online, contributions to the same topic (or, in the worst cases, just plugs for the interlocutor's musings on largely unrelated matters).

On the other hand, some of the more heated, supposedly critical, clashes of the past tended to involve alpha males, at least as they fondly believed themselves to be, playing out personal rivalries which had nothing to do with philosophy. That aside, it is perfectly possible to raise problems and possible weaknesses politely, and whilst acknowledging strong points. It is certainly possible to try to do this, without being patronising and in a sensitive way, when in debate with someone much less senior than oneself. So I don't regret it at all if the rutting stag days, the humiliation of nervous junior staff days, have gone for good; and of course I wholeheartedly

---

[1] As the 'discussion period' is now called.

[2] Unless there happened to be mathematical logicians or especially computer scientists in the audience. They have an accepted practice of intervening throughout a talk, often forcing one to respond, with increasing irritation, with comments such as: 'this objection/request for clarification I will be addressing later' then, further on in the talk, 'will be addressing shortly', eventually: 'will be addressing in the next five seconds if only you'd shut the fuck up'.

welcome the prohibition of cigarettes and the promotion of a physically healthy atmosphere in seminar rooms and lecture theatres.

## 1.3 Undergraduate Days in the Murmansk of the South

My two deepest interests in my postgraduate studies, abiding interests as it turned out, were philosophy of mathematics and Quine. Both of these had arisen in my undergraduate studies at Edinburgh, the soi-disant Athens of the North (or as some of us from Glasgow and the West say, the Murmansk of the South[3]). So I'll say a little here on my undergraduate preparation, though I will have to pass on how someone raised in a tin shack in a shanty town on the outskirts of Glasgow ended up at university.[4]

At Edinburgh I received an excellent, broad-ranging philosophical education, broad-ranging as regards 'Western' or European philosophy that is (though the Scholastic period was almost entirely invisible, a result of the Reformation no doubt). And I value this broad range, being interested in most areas of philosophy, and indeed had hoped to contribute to a much wider range than I have.

Political philosophy, for example, was a keen interest both from an (intermittently) activist and from a theoretical standpoint and at Edinburgh I had an excellent grounding in political philosophy from Vinit Haksar, one of the few persons of colour working in philosophy then (has much changed?). Vinit's main interest was in perfectionism in moral and political philosophy but the big battle of the day in analytic political philosophy was between the liberal (which at that time meant roughly 'social democratic') John Rawls and the extreme right wing 'anarcho-capitalist' Robert Nozick. I think I still hadn't reached my fairly short lived Marxist phase at this stage, but Rawls I wrote off as an apologist for capitalism. As for Nozick, he was completely inside the pale.[5] Nonetheless I found many of his negative arguments against 'statist' socialism very powerful and the emphasis on individual autonomy and against treating all of humanity as one big container to be filled with the maximum amount of 'utility-fluid' struck a chord. The positive argument in favour of the impermissibility of going beyond the minimal state I found pathetically thin. Still do.

---

[3] Actually when I ventured out East I don't think the ancient rivalry figured very prominently at all: I worked myself into a lather about it when I got a job in Glasgow later. Then again, I met hardly anybody actually from Edinburgh when an undergraduate, certainly very few working class locals, and hung about in an almost entirely (mutually comprehensible) Glaswegian clique.

[4] Those familiar with the *At Last the 1948 Show* 'Four Yorkshireman' sketch, later made famous by *Monty Python*, will recognise what is going on here. Alas I have been a frequent loser in 'Four Yorkshireman' battles as to who has the most proletarian background, though going to university certainly upped my win rate.

[5] For me the good guys are beyond the pale, at least if it is the Irish Pale we are talking about, the baddies inside it!

But the biggest influence on me at Edinburgh was a young, very energetic logic lecturer, Neil Tennant, an evangelist for Gentzen/Prawitz tree-form natural deduction and, in his research work, for intuitionist (and later relevant) logic as against classical logic. He sold me on natural deduction, not on intuitionist relevantism. He also introduced me to the work of Quine, the dominant influence on me. My systematic study of him, and of his masterpiece *Word and Object* in particular, began in the summer of 1976 when I participated in a regular exchange Edinburgh has with the Ivy League Dartmouth College New Hampshire, Neil being the accompanying lecturer. Neil has recounted my misadventures in New York getting there (Tennant, 2024). I had a few other misadventures over the summer, including an encounter with a giant, drunken, American "footballer" in a fraternity, the original for Animal House I reckon, who shoved a fellow exchange student off the juke box whereupon I flung an entire pint of beer over the monster and was swiftly felled in one blow. Nevertheless the monster ended up pummelled to the ground.[6]

Being the poor relations, the Edinburgh exchange students were given food vouchers for the canteen. I discovered that in this canteen it was permissible to go round as many times as one liked and help oneself to as much food as one wanted. The giant college "football"[7] guys, monstrous or not, all seemed to go round at least twice. So I sold off half my food vouchers cut price to the other students in my block, who were all African-American guys on some sort of affirmative action programme, to give myself some extra beer money, eating giant meals only every other day. The result was a sort of reptilian existence. The day after a food day would start slowly and ponderously. I'd lie about in the sun immobile most of the morning, moving little, dull of thought; by late afternoon I was razor sharp, by evening ravenous.

A year later came finals. In those days all one's eggs were in one basket, your degree depended entirely on how you did on a fortnight or so's worth of examinations right at the end. What if you were a woman at the wrong point of your menstrual cycle? Tough luck women, was the attitude back then. Though in favour of having substantive examination components in assessment, I am very glad that sort of arrangement has gone – there has been some progress to report over the years!

The only exam I remember is the Ethics exam. Having attended a Labour Party fund-raiser the night before and failed to adhere to my vow to be moderate (in drink if not politics) I slept-in and arrived late. I sat down flustered, looked at the exam paper, and one question leapt out the page:

Is weakness of the will possible?

[6] Note that I have made no claim here as to who was responsible for the subjugation of the brute!

[7] I have to scare-quote that. Genuine *football* (I can't bring myself to use the S word for it), is played mostly with the feet, not the hands!

'Too fucking right it is', I wrote, 'why only last night ...'. Ok, I didn't actually write that. But akrasia has been an interesting theoretical and practical question for me ever since.

I have no idea how I did on that exam, or on any other, in particular, but overall it went well for me, I secured a first and was ready to head off to Oxford, where you found me way back in Sect. 1.1.

## 1.4 Quine and Neo-Logicism

My interest in philosophy of mathematics, sparked by Neil was deepened by working with David Bostock on the philosophy of maths B.Phil. paper. Bostock was an excellent supervisor and was carrying the torch for logicism, in his *Logic and Arithmetic* volumes (Bostock, 1974, 1979). In my last year at Oxford I attempted to prove the Peano-Dedekind axioms from what is known as Hume's Principle (HP).[8] I still have the handwritten effort in a notebook, but can't be bothered trying to figure out where I failed, as I no doubt did, to anticipate the proof in the Appendix to Wright, 1983.

But Quine was my main preoccupation. My views on Quine, and the lessons we should learn from his work, have changed a little over the decades but the fundamentals have not changed much since my D. Phil. dissertation. I am highly sympathetic to Quine's physicalistic naturalism, and much more sympathetic to his behaviourism about language than current orthodoxies. However in the dissertation I criticised Quine's 'atomistic' behaviourism as far too crude, and in conflict with his semantic holism about theoretical language, setting out the objection going back to Geach in the 1950s that mental states issue in behaviour only in combination with one another. The semantic holism I also criticised, arguing for a *molecular behaviourism*: grasp of language consists in a well-founded hierarchy of verbal dispositions winding down to atomistic dispositions at the bottom. Here there is influence from Dummett, but I argued one can espouse molecularism in language without lapsing into verificationism and thus 'anti-realism', as Dummett preferred to call his version of idealism.

Rejecting atomistic behaviourism I found Quine's scepticism about the existence of a fine-grained synonymy relation wanting. So, though convinced that the methodology of looking for necessary and sufficient conditions for the application of interesting concepts was a dead-end, especially in its ordinary language version,[9] I did hold then, and for a long time after, to the very non-Quinean idea of philosophy as a distinctive non-empirical discipline of conceptual analysis, but charting highly

[8] Informally: the number of $\varphi$s is identical with the number of $\psi$s iff there is a one:one correspondence between the $\varphi$s and the $\psi$s. Dummett (1998) objected to the association of Hume with the principle but such honorific namings are commonplace in maths.

[9] But alas it seems to be still alive in some areas of the discipline.

complex webs of interaction rather than simple quasi-definitions. Since synonymy is kosher, there remained a place for me for a realm of analytic truths, among them HP, whose analyticity I thought could provide a route to a non-platonist validation of classical mathematics. My commitment to physicalistic naturalism was never total. In the dissertation I argued for something like attribute dualism in order to avoid the absurdity of ontological relativity. That has gone by the wayside but physicalist theories of consciousness have never struck me as plausible.

As was not uncommon, I did not manage to complete the D.Phil. in the year after the end of my master's degree, and, so by, 1980, with my 3 year grant from Scotland over, was on the look out for job.

## 1.5 The Precariat: 80–83

This I found back in Edinburgh, where I landed a 1 year position and found a very supportive environment amongst my former teachers. My first task was to give a lecture course on Descartes. I remember well the very first lecture. I told the class who I was, set out the prospectus for the course and the reading list and then ended by saying that if anyone found my accent hard to understand, just raise their hand and I would take them outside and give them a good kicking. Total silence! Then a few titters after a moment or two; but I decided to leave off the jokes until I was more into the swing of lecturing.

Being back amongst my old comrades, but now with a bit more money to spend, did have its dangers. As will be clear from this memoir, I have been no stranger to the 'Devil's Buttermilk' over the years and have had many a convivial session in the pub discussing everything from football to football (and sometimes a bit more). But, like everyone from my neck of the woods, I know quite a few people whose lives and those around them have been destroyed by alcohol. What causes these people to go on a downward spiral they never pull out of, whilst others do and become masters (most of the time) rather than servants of the demon drink? A lot of factors perhaps – some genetic, some environmental, maybe 'strength of will' (a very interesting and puzzling concept) contribute in part. Anyway that first year was probably the time when I was most in danger of going on a downward spiral but, thankfully, and maybe good luck is another element here, I did not.

I was also in danger of unemployment or falling out of the profession. I next secured a 6 month job at the University of Birmingham, which I enjoyed a lot, though Birmingham was still not fully recovered from the dreadful IRA bombings in the city centre way back in 1974. There followed a couple of temporary jobs, their duration asymptotically approaching zero, a period of unemployment and some bar jobs over a 6 month period, in one of which I featured in a centre spread (as it were) in the tabloid *Daily Record* under the headline 'Brains behind the Bar'. Toe-curdlingly embarrassing of course, but I couldn't resist the fame. So it was looking as if, like many highly qualified philosophers, I was about to fall out the profession when salvation came from down South, from the 'Scottish' college in Oxford, Balliol.

## 1.6 Balliol 83–5

I remember quite well the interview at Balliol, in particular all the candidates being invited for a drink in the buttery by the Master, Anthony Kenny, who asked me first what I would like. A half pint of bitter I said (a quantity which I never imbibed before or after), thinking this would go down well. Next came Marcus Giaquinto, whom I used to regularly drink under the table in the 70s. But to my horror Marcus ordered a pint, everyone else following suit, leaving me feeling like a complete eejit. Given the drinking habits of the 'champagne socialist' section of the Fellowship, I doubt if it was my faux moderation which got me the 2-year Tutorial Fellowship.

Balliol was an enormously exciting place to teach, more exciting than 'Teddy Hall', that is St. Edmund Hall, where I was a postgraduate, though I am, of course, for ever grateful to that institution. One might expect there are lots of tales to tell about my time at Balliol, and indeed there are; but once again those crates of malt keep me mostly schtum.

Focussing on the academic side, I would say first of all that the job was hard work. Though I was a Fellow with the same rights as all others,[10] I was essentially a college tutor, doing very little lecturing, nothing outside college. But even then the work rate (albeit only during the rather short Oxonian terms) was high: it included twelve one:one tutorials a week in which one heard and discussed a pre-prepared essay. One of these was Ian Rumfitt's first philosophy tutorial, which he informed me was going to be on Hume. As I settled into my chair preparing to go over the usual stuff on ideas and impressions, or maybe induction, I was startled to hear the opener, promising to challenge Kemp Smith's naturalist interpretation of Hume and defend the older, sceptical Dugald Stewart reading. I climbed back onto my seat and picked up pen and paper to take notes and try to keep up. At the end rather than set him a topic I gingerly asked what are you thinking of for next week? Sluga's critique of Dummett's interpretation of Frege was the answer. Not your average undergraduate!

I was hired primarily to cover for Bill Newton-Smith[11] during his stint as Senior Proctor and was helped greatly by him, and indeed all the philosophers at Balliol. During my time there I completed and was awarded the D.Phil., to the great pride of my parents, my supervisor being Michael Woods, an extremely careful and attentive

[10] Which enabled me to vote down the honorary degree proposed for Margaret Thatcher and take part in a very merry dinner at Balliol to celebrate Gerry Cohen's appointment to the Chichele chair, a gathering of 'Hampstead champagne socialists' (but also including Donald Davidson) which would have had the Tory chatterati turning purple in outrage.

[11] Who sadly has died since I submitted the initial draft of this memoir.

reader of drafts. Something of an anachronism even then, for example by being a don living in the college, he probably would not fit into the university system nowadays, and sadly died relatively young before he could see the full fruits of his work published. I also did the groundwork for a number of papers, some (1985b, 1986b) setting out my molecular behaviourism and opposition to semantic holism but also one on 'grue' which never saw the light of day until much later (Weir, 1995). This arose from my interest in Hume, further nourished by regular discussions over a pint (or twa) with Simon Blackburn then editor of *Mind;* he never published the paper – should have bought him more drink! In the paper as finally published I tried to give a version of grue which was epistemologically and semantically[12] no more complex than green – so not like the usual 'green if observed before time *t*, blue otherwise' definitions – a version which I argued was much harder for the anti-Humean non-sceptic to deal with. I still think this is right; and also wonder whether the grue/green distinction as I developed it has implications for the differing concepts of 'woman', gender on one side versus biological sex on the other, though alas such topics are getting more difficult to discuss in an open fashion today.

The Oxbridge collegiate system is often vaunted as nurturing inter-disciplinary fertility, as dons from widely disparate domains meet and compare and contrast their specialisms over lunch or dinner. While some no doubt greatly exaggerate this, I did find there to be a lot of truth in it, possibly because, as a young single academic, I spent a lot of time in post-prandial discussion! The college held regular work-in-progress seminars where I learnt, for example, about chaos theory from a visiting physicist. Donald Davidson, who was a visiting Eastman professor when I was there, turned up for my, wildly over-ambitious and over-long, talk and was helpful, if a little bemused. I didn't have much contact with him though; I remember most his turning up wearing the Davidson clan tie for the annual Holywell Manor Burns Supper and telling me about his Scottish ancestry.[13]

On the non-academic side I did regularly escape to a very different Oxford, to a wonderful Irish pub, not a faux Irish one but one almost entirely frequented by actual Irish people and which claimed diplomatic immunity from English drink laws with regular after-hours 'lock-ins' where there was some great singing and music-making (well it all seemed great at the time). But I did also make lots of academic friends, including a rather unlikely one in art historian David Ekserdjian who tried, with only limited success at the time, to introduce me to opera. He also introduced me on occasion to some of his high Tory friends who, though later very well-known, will remain nameless for I don't like to name-drop, as I was saying to Barack Obama only the other day.

---

[12] Though the focus was epistemological; a semantic take, as found in Kripke's *Wittgenstein*, features a bit more in Weir, 2003a, where I take a rather different view from my fellow symposiast Alex Miller (2003, see also 2024).

[13] This wasn't quite as wild as one I went to in the trade union college, Ruskin College, in the 1970s which ended up in a wild west brawl between Chuckies and Stickies, that is supporters, it was rumoured in fact members, of the Provisional IRA and the Official IRA. Not your usual Burns Supper!

That said, one occasion on which there were a lot of names to drop was the 90th birthday party of alumnus Harold 'You've never had it so good' Macmillan, Prime Minister of the UK 1957–1963. After an easy-to-chew meal in hall with all the students, the Fellows plus various Balliol alumni guests, including former Prime Minister Edward Heath and former Chancellors Denis Healey and Roy Jenkins, repaired to the Senior Common Room for some refreshments, there being a scarcely finite supply of whisky, brandy, gin and so forth lying around. Come the wee sma oors, only a few were left but this select band included Macmillan himself, sipping, very slowly, on his drink, a very watered down whisky I think. But 'Respect Man!', he was 90!

He was surrounded by a little coterie of admirers, including a Balliol classics student named Boris Johnson who I think had gatecrashed; even given the amount of drink taken I don't think anybody there would have suspected that here we had a future Prime Minister seated at the feet of Macmillan. Abandoning my group of well-oiled Fellows and wandering unsteadily across the Common Room to hear what gems of wisdom they were imbibing, I hadn't listened long when Macmillan come out with the incendiary 'when England stood alone in 1939' or some such. Immediately I was seized with a burning urge to lift the aged Tory out his seat and administer a 'Glasgow Kiss', that is a head butt. Fortunately, a glimmer of foresight caused a vision of the next day's shrieking red-top headlines to rise up in my imagination: 'GLASGOW MAN BEATEN UP BY 90 YEAR OLD EX-PREMIER' (for, even at 90 versus my 30, Harold was undoubtedly in better shape than me at this stage of the proceedings) and I staggered back to my comrades muttering fearsome imprecations against English imperialism.

My imprecations led to one of our number, later a pillar of the UK establishment (again keep those crates of malt coming) embarking on a lusty rendition of 'Sean South of Garryowen', not, I suspect, standard fare at birthday parties of UK Prime Ministers, being an IRA song; a typical one, that is to say a celebration of defeat and martyrdom, in this case of the virulently reactionary, probably fascist, South in the IRA's 'border campaign' of the 1950s. Being rather too knowledgeable in the matter of both Republican and Loyalist ditties, I joined in. However in the interests of balance, I immediately followed with an, initially solo, rendition of 'The Ould Orange Flute', one of the few funny Loyalist airs (at least if you can find mirth in a celebration of ethnic cleansing). This put me in some danger from Sean South's serenader but assistance arrived in the unlikely form of Maurice Keen, a very sweet medieval historian who I had assumed was a classic English gent but who, it turned out, also knew every last verse. In fact he was Anglo-Irish, he had served in the Royal Ulster Rifles in the 1950s where, I assume, he learned the song (but it was not he who did for South!).

## 1.7 Bees in Bunnet

Now those who know me will recognise, in my reaction to the conflation of 'England', 'Britain' and 'United Kingdom', one of the most insistently buzzing of the many bees in my bunnet, particularly as regards the question of the UK head of state. For one of the many respects in which the English are a remarkable people is that none of them seem to have a clue who their head of state is. I suppose the only comparable cases are uncontacted indigenous peoples such as still exist in the Amazon or the Andaman Islands. At any rate, I have never met an English person, no matter how well-educated, who has been other than bamboozled when I point out that[14] there is no more a Queen of England than a Queen of Sheba, a Wicked Witch of the West, nor, (I regularly direct this one at English philosophers of language) a King of France.[15]

Getting the semantics of "The Queen of England", "The King of England", "The Prime Minister of England" (or indeed "The Prime Minister of Britain", non-denoting since 1/1/1801) right is surely the key test of a theory of singular definite descriptions. There really ought to be vast outlays of money devoted to setting up centres, programmes, research institutes dedicated to studying the question and comparing Meinongian versus Schmeinongian[16] and other answers. Bertrand Russell – the 'real Meinong', it has been argued[17] – is an interesting case, here as in other matters. Not only was he under the delusion for most of his life (though not the later stages, as Jim Levine once demonstrated to me) that there were English monarchs after the 1st May 1707 (Russell, 1903, p. 739) he was also under the wholly mistaken belief that there have been Prime Ministers of England:

> [T]he proposition is certainly about the present Prime Minister of England; and the present Prime Minister of England is Mr. Arthur Balfour. The *is* here expresses absolute identity; and if *a* and *b* are identical, whatever holds of *a* holds of *b*. (Russell, 1994, **4**, p. 317)

This is interesting because Arthur Balfour, he of 'Balfour Declaration' fame/infamy, Prime Minister of the United Kingdom of Great Britain and Ireland 1902–1905, was in fact not only a philosopher he was a Scottish philosopher, born

[14] The main draft of this piece dates from March, 2022, prior to the change of monarch. I have left the example of the definite description 'The Queen of England' as it is, since ignorance about its semantic status is what I have railed against for decades now! So context of utterance: March 2022; first-order circumstance of evaluation: March 2022; second-order circumstance of evaluation: "now"!

[15] Nor is she [again we are in March 2022], in addition to being the Queen of the UK (and 14 other realms, at least at the moment) also the Queen of England, anymore than she is the Queen of the car park at St. Mirren stadium, or of New South Wales.

[16] Oliver & Smiley, 2016, pp. 7–8.

[17] See Alex Oliver (1999) where he argues (p. 263) that Russell when attacking Meinong is actually attacking his earlier self, and not the very real Alexius Meinong Ritter von Handschuchsheim, identical, he argues, pace David Lewis (and indeed my earlier self) with Lewis's 'Bogus Meinong'.

in East Lothian, though not really a Common Sense one.[18] And he was far from the only "Prime Minister of England' who was Scottish or of entirely Scottish ancestry.

Pedantry, pedantry you may say. And in a sense it is: in fact I find the, until recently, almost universal practice in England of using 'England', 'Britain' and (to a lesser extent) 'United Kingdom' as synonymous in most political contexts understandable and mildly amusing rather than irritating. Understandable because, though most Scottish nationalists don't like to admit this, Scotland was extinguished as a political entity on the dark day of the 1st May 1707: proposals for a federative union were rejected by England in favour of incorporation – the "incorporating union" – with invasion threatened on failure to agree, followed by violent suppression of the insurrections which did occur. The idea that England also vanished on that day wasn't even entertained by anyone then (not by Scottish unionists such as Hume for example[19]) any more than anyone believed Scotland disappeared from the political realm when it acquired the Northern Isles or believes that Canada became a political dodo when Newfoundland joined in 1949. Still I do find phrases like 'England stood alone against Germany in 1939' highly offensive, phrases I have come across a number of times from English people of my generation. I had difficulty restraining myself when hearing Bernard Williams come away with this in a talk on methodological individualism at the Oxford Philosophical Club in the 1970s. But I promised Stephan and Adam I would restrain my tendency to fulminate on this particular bumbee in ma bunnet ('bonnet', as it is scriven in thon southron dialect o Scots kent as Inglis).[20]

I do, however, have to admit to having a lot of Bees in the Bunnet, they accumulate with age. A fairly randomly ordered list would include in addition to the English monarchs who do not exist even among the shades (as Russell might put it):

> ⓪ The phrase "in principle possible"; ① the phrase "nothing over and above"; ② fixation with definitions; ③ misuse of confidence intervals in statistics; ④ "reductive" as a lazy term of abuse; ⑤ so-called "identity criteria"; ⑥ "extensionality" versus "intensionality"; ⑦ conflation of predicate senses/meanings with properties (attributes/universals); ⑧ confusion of impredicativity with higher levels/types; ⑨ people who put their bags on the seat next to them in public transport, as if they are worth twice the rest of us.

[18] He defended an anti-naturalist stance but one which was not a crude opposition to the science of his day but appealed to a form of epistemic holism. See Root (1980). Root views him as an 'English man of letters' (p. 141) though, since Balfour was Eton-educated like so many of the Scottish ruling class, this is perhaps not wholly inaccurate. On Balfour and Russell's philosophical exchanges see Monk (1996), pp. 338–9.

[19] Hume identifies himself as English in the preface to the Treatise (1739/1888), p. xvii. It was only later in life, in the face of rising anti-Scots bigotry in England, that he seems to have developed a strident Anglophobia, though, bad though this is, it is nowhere near as objectionable as his racism.

[20] For any interested in these matters see my defence of Scottish Nationalism in *The Philosophers Magazine*, Weir, 2014, and the pieces written during the independence referendum campaign of 2014 at https://sites.google.com/site/weirysite/scottish-independence-referendum-1.

And there's more:

⓿ misuse of the type/token distinction; ❶ the very ideas of grounding/fundamentality; ❷ Edinburgh; ❸ the supposed determinate versus determinables distinction; ❹ the Little Austrian Bombardier (aka Ludwig Wittgenstein); ❺ "numerical" identity versus some supposed other type; ❻ people who audibly type on their laptops during seminars, particularly those who are emailing or on social media and not playing the blindest bit of attention to the seminar; ❼ 'binary' as a term of abuse; ❽ describing Scots as 'slang'; ❾ people with lots of bees in their bunnets.

And there's yet more:[21] but that will give a rough idea of the kind of grumpy old man territory I currently occupy.

## 1.8 Belfast

At any rate, to leave the bees behind and return to the chronology, when my fellowship came to an end I was touching thirty and very much in need of a permanent job. These were very scarce in those days too. A tempting opportunity presented itself: a tenured position in Belfast, still in the middle of deep civil strife – 'The Troubles'. I had never been there; but violent sectarianism: I know all about that I thought. What's more, surely nobody else will be crazy enough to apply? Indeed there weren't many applications but there were a few very high quality ones. But I managed to secure the position: perhaps being from the 'Celtic fringe' helped.[22]

Expecting not to be there for all that long, I stayed in Belfast for over 30 years. Obviously there is a lot that can be said about the non-philosophical aspects of life there but, turning to the philosophical side, Tim Williamson, then at Trinity Dublin, has referred (Williamson, 2024) to the dual nature of Ireland, in philosophy as in all else. There were two departments at Queens' University, Belfast: the Department of Scholastic Philosophy and the Department of Philosophy. It looked to some like a divide between a Catholic department and a Protestant one. It wasn't quite that, of course, but everyone in the Scholastic department was from the nationalist community, nobody in the latter was. Although the head of the former was a Catholic priest, James McEvoy, an expert on Grosseteste, and Tim Lynch worked on the neo-Thomist Bernard Lonergan, the divide had elements of the continental/analytic division. Some resemblance, perhaps, to the divided departments at Sydney in the late 1960s. But things were much darker here. Not

[21] Including the use of scare-quoting as a substitute for genuine intellectual argumentation. I plead innocent to the charge of hypocrisy and inconsistency notwithstanding the flurry of scare quotes above, and not by dint of utilising a non-classical logic. It's all right if one provides reasons for rejecting or querying the scare-quoted term. And I will, in the above cases, given enough time!

[22] I was asked at the end of the interview, in the 'personal interests' part: 'what football team do you support?'! Unthinkable now. I think some of the non-locals on the panel were puzzled by the answer 'St. Mirren': "sounds Catholic but it isn't Celtic: what can it all mean?"

only was the Scholastic department bombed by loyalists, the wife of one colleague was murdered by them in front of their children; he would have been too if he had been at home at the time. Most of my nationalist friends believed that the British security services were behind incidents like this. At the time I thought they were being paranoid. But later revelations (cf. the reports by former Metropolitan police chief Lord Stevens) about the role of the British intelligence services in 'running' loyalist paramilitary serial murderers leads me to think this is far from far-fetched.

My own attitude to the departmental divide was somewhat complicated. I arrived in Belfast a committed Scottish nationalist highly sympathetic to Irish nationalism (a position which might seem odd to those who favour 'geographical nationalism'). One might think that the terrible death toll (at least by recent European standards) and civil divisions might turn one against 'nationalism' but, though I came to appreciate the complexities of the situation more, it only confirmed me in my belief that civic nationalism was not a variant of, but in fact antithetical to, the best answer to, the ethnic nationalism of which British nationalism – unionism – has historically been a vicious example.[23] On the other hand, I am pretty sectarian on the analytic/non-analytic divide. I did have discussions with the neo-Scholastics in Belfast about why Neo-Scholasticism was not on the analytic side of this divide. Certainly the scholastics of the medieval period attracted derogatory remarks from early modern philosophers – pedantic, nit-picking, irrelevant etc – which mirror criticisms of analytic philosophy.[24] As for non-analytic or 'continental' philosophy, my very limited grasp of what comes under this heading leads me to think it is a very mixed bag.[25] For example, at Edinburgh we were exposed to quite a lot of neo-Hegelianism which seems to have survived longer in Scotland than elsewhere;[26] to Bradley in particular. On Bradley, Geach wrote:

> Reading a few pages of *Appearance and Reality* now makes me feel as if I had drunk several pints of beer. (Geach, 1991, pp. 2–3)

Your point being, Big Pete? Is this not a recommendation?

Sure Bradley can be pretty woolly, but the masters of analytic philosophy, such as Russell and Quine, can often be quite obscure, when you penetrate beneath the surface of their polished prose. And even the best logicians make some tremendous howlers in natural language reasoning. Moreover philosophy is not just B follows

---

[23] It is very important not to see such conflicts as occurring between nationalists and non-nationalists. It is part of the legacy of formerly imperial nations to think of their nationalism as not really nationalism at all, as 'patriotism' perhaps, and obviously rational, moderate, divinely ordained or whatever, unlike the 'separatist' nationalism of smaller nations which have had the temerity to break free or want to break free.

[24] Consider also the non-existent debate over how many angels can dance on the head of a pin, which seems to have been invented by an eighteenth century Swiss Protestant. See Clark, 1993; Mcdonald Ross, 1985.

[25] In recent decades the historians have also emphasised the interchanges between philosophers supposedly on opposite sides of the divide, Husserl and Frege, Meinong and Bradley contra Russell, for example, who all seem to have thought they were working in the same area.

[26] See Candlish (2007) p. 3, p. 189, fn 4.

from A. We need people to come up with new concepts and ways of thinking, and perhaps some people are good at this creative business but not very rigorous.

That being said the influence, on people writing in English, of the once fashionable French philosophies of the swinging sixties and a little later seems to have been very bad. A worrying theme which surfaces in the famous/infamous Sokal hoax is an apparent intellectual dishonesty in the parading of scientific terms which the authors seem to have no grasp of, a bit like innocent children laughing at dirty jokes they do not know the meaning of. I say 'apparent' because I have not done the work which would be needed in making a firmer assessment, namely reading substantively in the key figures, ideally in French. Nor have I any intention of doing so, any more than I intend reading the works of L. Ron Hubbard to confirm my belief, based on testimony, that there is nothing of worth there. Life is too short, we have to make choices about whose testimony to trust in regard to what to read, what to work on, choices which we cannot be certain are the right ones.

At any rate Belfast in the mid 1980s was an unusual environment for an analytic philosopher, in a number of respects. How did my philosophical work develop in these circumstances? I don't think it was influenced a great deal; some may see this as a defect. I was pretty much working a lone furrow there, though Christopher McKnight, who worked both in ancient Greek philosophy and in analytic philosophy was a great sounding board, and there were the Dubliners including Jim Levine, Maria Baghramian, Jim O'Shea and, for a time, Tim Williamson, to meet with at the Irish Philosophical Club.

My main goal initially was to complete a monograph in philosophy of mathematics. But right at the start of my time at Queen's I had what I took to be something of an epiphany regarding the right way to respond to an unanswered philosophical problem going back (like so many) to the Ancient Greeks, the *Modes of Agrippa* for example (Jonathan Barnes at Balliol may have put me on to these). As I presented it, when my ideas eventually made it into print,[27] the problem is a dilemma posed by a *Philosophical Irrationalist*, someone who reasons and argues just like the rest of us (indeed just is us, in highly self-critical mode) and attempts to destroy from within (to deconstruct even!) our standard notion of rationality. Contrast the *Rabid Irrationalist*, who makes no attempt at argumentation but, for example, simply puts words targeted for attack, such as "Truth", "Objectivity", "Rationality", in scare-quotes (or italics). This tactic, which is not confined solely to non-analytic philosophers, is effective in the way that repeating what someone says in a high-pitched whine is effective as a debunking strategy amongst children. The Rabid Irrationalist should simply be ignored. Paul Feyerabend, at least in some of his moods, is a modern candidate for a Philosophical Irrationalist and should not be ignored.

A key premiss in my argument against the Irrationalist ('Philosophical' understood from now on) is a voluntariness constraint: for any action or mental state to be evaluable as rational or irrational it must be under our direct control. The Irrationalist

[27] Weir, 1996b, 1997, 2000a.

concludes from that premiss that there is complete symmetry between someone who acts or judges on the basis of nothing more than groundless capricious acts of will and someone who does so rationally according to any of the usual methodologies, whether foundationalist or holist. The only difference between the those we usually class as rational and the capriciously irrational is long-windedness, not a criterion of rationality by our own account.

The voluntariness constraint, of course, is not universally accepted. Quite the contrary. I presented counter-arguments to the standard arguments against the constraint for the latter seem to me to be weak. But I think that a main reason people flinch from that constraint is that they see, at least dimly, that it seems to lead to irrationalism. So my positive argument (Weir, 2000a, 2004b, pp. 403ff) is that this is not the case, though it does lead, I claim to a highly restrictive 'neo-Humean' account of theoretical (but not moral or practical) rationality (see Weir, 2022a, §XI).

I extended my general account of rationality to moral rationality, arguing that it leads to a theory in the 'Golden Rule' tradition. My motivation for this venture into moral philosophy was actually financial! I entered a thesis on the topic for the Green Moral Philosophy Prize at Oxford for which there was a prize of a couple of thousand pounds which would have come in very handy. I did come second, which is creditable for a non-specialist.[28]

Actually the question of irrationalism is one where Belfast did have an influence on me. I remember being appalled on hearing the reply by a Sinn Fein politician (and QUB graduate) when asked by a journalist if there wasn't a contradiction between the IRA's campaign, with its disincentive effect on investment, and Sinn Fein's holding the UK government responsible for the high levels of unemployment. Yes, that's a contradiction, he said, no doubt thinking he was being disarmingly frank, but sure contradiction is the motor which drives progress forward. Or some such. I think he was influenced by what he may have taken to be Marxism (though Engels and Plekhanov are the sources), that is by 'dialectical materialism'; definitely not dialetheism. Similar themes emerged from the opposite side, namely Loyalist killers, some of whom appealed to antinomian Calvinism. I once stood Graham Priest in front of a mural glorifying the loyalist paramilitary Ulster Volunteer Force and which included a quotation from Deuteronomy, only a couple of chapters after a reprise of the Ten Commandments, 'thou shalt not kill'[29] and all:

> thou shalt smite them, *and* utterly destroy them, thou shalt make no covenant with them, nor shew mercy unto them. (Deuteronomy, **7** 2, (King James VI version))

It was perfectly clear who these Loyalists had lined up for destruction: not the Canaanites nor Amorites of the text. So is it perfectly ok for a dialetheist terrorist or religious extremist to affirm both that thou shalt not kill and also that one

[28] You should mention there were only two entrants (*ed.*).

[29] Of course many interpret the texts as having implicit qualifications, e.g. do not kill anyone inside your own community, or exception clauses, self-defence etc. But consider, in evaluating dialetheism, the simpler case of a direct contradiction: 'never kill under any circumstances and also sometimes kill in such and such ones'.

can, perhaps even should, kill people from such and such community? Graham's answer was that not all contradictions are equal (Priest, 1998, pp. 420, 423). To put it in terms of probability theory (and he develops a dialetheist theory thereof) with rational credibility interpreted as a type of probability, some contradictions have non-zero probability, in some cases indeed probability one; others, such as $(0 = 0 \wedge 0 \neq 0)$ or the antinomian or terrorist or biblical literalists' contradictions, they have credibility zero. There is no simple algorithm which enables us to determine which level of credence to give.

This answer does not work for me, as a neo-Humean who accepts very deep limits on theoretical rationality and only, at best, a subjectivist account of probability in the sense of rational credence. But Graham is not, in my book, a Prince of Darkness like Feyerabend; he upholds a distinction between rationality and irrationality and tries to stay on the former path. My claim is that his dialetheism does lead one (by accepted methods of reasoning) to irrationalism, despite best intentions.

I criticised Priest's dialetheism in 'There are no True Contradictions', (Weir, 2004b) though the above neo-Humean objection was excised from the shorter printed version. It started with a classical proof of $\neg(P \wedge \neg P)$. Presenting the paper in talks, I used to pause right there and say, 'Ok that's it, let's go to the pub'. Except when I gave the talk at Leeds when, before I could utter this, John Divers piped up to say exactly that. It took me a long time to forgive him for this (at least six pints and a hauf or twa of whisky).

Graham was taking part in one of my 'morbid tours' which I put on as visiting speaker organiser, usually in car, sometimes on foot. I am happy to report no philosophers wimped out on coming over to give us talks. Some, indeed, were disappointed on the occasions when, on tour, we did not get stopped by armed patrols. Stewart Shapiro was a bit taken aback, right enough, but not unduly fazed when, as we walked down the Falls Road, British soldiers on foot patrol pointed their rifles straight at him and then me from a few feet away, as was their wont to do with everyone in Republican areas.

But what became of my goal, the submission of a monograph on philosophy of mathematics? Actually my goal was to produce at least three volumes, covering logic, maths, philosophy of mind and moral and political philosophy. I have always aimed at being a systematic philosopher covering a wide range: it's just that I got rather bogged down at my starting point! I did submit the first draft of what became *Truth through Proof* to Clarendon Press Oxford, Jon Cohen general editor, in 1989. I got a 'revise and resubmit', which I successfully completed: about two decades later!

Since I haven't managed to find either print or electronic copies of that material, I rely on my memory as regards how much changed in the intervening time. And my memory, as so often, is fairly blank on this. I do remember being annoyed by a referee's comments on my use of HP, thinking she or he had a mistaken view of the import of Azriel Levy's proof (Levy, 1969) of the non-conservativeness of a set-theoretic version of HP in ZF minus the Axiom of Foundation. So I think at this stage I was still a pretty loyal logicist: the (supposed) analyticity of HP provided an anti-platonist answer to the epistemological problems for naturalism raised by mathematics.

By 1992 however, I was certainly quite well advanced towards my settled 'neo-formalist' position on mathematics. I gave a talk on 'Putnam, Gödel and Mathematical Realism' at a conference on Putnam in Dublin,[30] and published the ideas from it the next year (Weir, 1993). Although I didn't describe my position as 'neo-formalist', some key ideas were there. These included the distinction between, in my later terminology, informational content, along the lines of Fregean *Sinn*, and *metaphysical content* which specifies what makes-true or make-false the sentence whose content it gives in a way which helps explain how speakers use the sentence to relate its components to reality. One example I gave, of a type familiar from Kaplan's character/content distinction, was sentences like 'yesterday that table was in the storeroom'. In many contexts that sentence itself is best suited to give its own informational content homophonically. But we may conjecture that its metaphysical content is given by something more like 'the currently most salient table was in the storeroom yesterday', which is not synonymous with it, indeed speakers of the object language may not even have the concept of salience, though metatheorists can use it to explain their verbal behaviour.

I then extended this idea to projectivist accounts of, for example aesthetic judgements, whose metaphysical content, but not informational content, is given, for the projectivist, by propositions involving attitudes of the speaker. The influence of Hume and Blackburn is obvious here but unlike Blackburn, at least in his best-known *non-cognitivist* phase, I, as a deflationist about truth, saw no need not to ascribe truth and falsity to aesthetic judgements. For they can be correct or incorrect relative to considered judgements of one's peers, or some such. The Frege/Geach problem, is therefore not a worry (Weir, 1993, p. 263; 2010, pp. 44*n*, 83).

Finally mathematics is treated in a different way again. The metaphysical content of a mathematical sentence is that a concrete proof of the sentence exists (even if only as a concrete structure which could feasibly be interpreted by speakers as a proof). The biggest problem I saw for this view at that time was Gödel's first incompleteness theorem on the negation-incompleteness of arithmetic. By Weir, (2010) I saw Gödelian speed-up considerations[31] – comprehensible sentences may have no comprehensible proofs or disproofs – as the bigger worry. I noted already in 1993 that Gödel's first incompleteness theorem only applies to formal theories which are recursively axiomatisable and argued that the widespread belief that proof cannot play its epistemic role unless proofs are finite and proofhood effectively decidable is without foundation, pointing out that nearly all formal proofs in mathematics and metamathematics have more symbols than the estimated number of elementary particles in the universe. This is a position I have maintained

[30] It was at this talk that I first came across legendary Glaswegian philosophers John Divers and Alex Miller. In the discussion period John started his question by complementing me on my comprehensible diction. But nobody else bar Alex and myself understood a word he said.

[31] Weir, 2010, pp. 179–80; 2015, p. 34; Gödel, 1936.

steadfastly ever since (Weir, 2010, Chapter Six §II, Chapter Seven; 2015) and underlies one of the bees in my bunnet, my hostility to appeals to 'in principle possible'.

The paper also reveals the increasing influence, positively and negatively, of Graham Priest's dialetheism. Negatively in that I reject it; positively in that I accept his critique of hierarchical solutions to the semantic and set-theoretic paradoxes. I employed a form of revenge paradox argument and embraced naïve truth theory and naïve set theory. These, I still maintain are consistent, my most extended defence of this position being the *Mind* paper 'Naïve Set Theory is innocent', (Weir, 1998c). How can that be, given inconsistency, indeed triviality, is provable in quite weak logics? In the 1993 paper I reference Fitch's restricted logic but commend a stronger logic, the first appearance of what I later called 'neo-classical logic'.[32] The key idea is to restrict classical rules by imposing *determinacy* constraints. Apparent proofs can fail to be proofs if certain premisses or conclusions are not determinately true or false. In that article I embraced a very radical notion of proof as non-formal indeed as dependent on what is actually the case.

I soon moved back towards a more conventional notion of proofhood as entirely formal but retained two elements from that paper. Proofhood is not finitary and consequence is non-transitive. So in the latter case I found myself forming part of a philosophical lineage stretching back through my teacher Neil Tennant to Smiley (and Geach), though I remember being baffled by the notion when Neil first talked to me about it. It is good to see that investigation of non-transitive logics is now in the hands of brilliant young logicians such as Elia Zardini (this volume). On the other hand my restriction on transitivity is less strict than Neil's and the operation of transforming classical into the corresponding non-classical proofs by the addition of determinacy claims, when needed, produces only a linear increase in size, not exponential as in his case (see Weir, 2024, §I).

That was in the future though, as was a move away from logicism. During the next few years I became one of the neo-Fregean loyal opposition, posing difficulties for logicism in the existence of pairwise inconsistent abstraction principles – the *Embarrassment of Riches* objection (Weir, 2003b)[33] – and in the strength of the impredicative second-order logic, the existential theorems which follow from it in particular, needed to carry through the logicist programme.

Here I was heavily influenced by Stewart Shapiro whom I met in one of my frequent trips to St. Andrews. As I recollect, I came across him watching a rugby game on television and keen to know what rules, if any, the players were following. I wasn't much help there, rugby having been written off by me, after experience of rugby clubs at universities, as a game for fat fascists. But though he learned little from me on rugby I learned a great deal from him about mathematical logic,

---

[32] Pretentious? Moi? Certainement pas. I do seem to have a penchant for neo-Xs though.

[33] I think my most cited paper though I have been somewhat out-gunned by work by Fine (2002), Roy Cook (2009, this volume) and Øystein Linnebo, cf. the joint paper Cook and Linnebo (2018). For a critique of Kit Fine's philosophy of maths, see Weir (2005c, 2007a).

philosophy of maths, and indeed malt whiskies! We collaborated on two papers on the neo-logicist programme, sympathetic but critical (Shapiro & Weir, 1999, 2000). The first, which was mostly written by Stewart (and has a lot more citations than the second, to which I contributed a lot more!) led to me having an honorary naming experience, (cf. Hume's Principle, Sect. 1.4 above) at a great conference in Munich celebrating 100 years of Russell's paradox. After having failed to persuade the conference that the proposition 'Russell was Welsh' bears investigating as possibly indeterminate in truth value, I went along to a talk by a young philosopher of mathematics, Gabriel Uzquiano, whom Stewart had told me to look out for. I was enjoying the talk until Gabriel put up on the board 'The Shapiro-Weir Theorem'. What a shock: what the hell does that mean I thought. After a few moments, I remembered what it meant but couldn't for the life of me remember how we (i.e. Stewart) had proved it and was terrified someone in the audience would say 'well look, that's Weir there, ask him about the proof'. Fortunately Gabriel moved on pretty quickly.

Over the time I taught in Belfast one change in the profession I should note was a significant increase in the number of women in philosophy, though a painfully slow one, and still with a long way to go. I think I was a bit naïve then as regards the sexist nature of our profession. I was well aware, from the testimony of just about every woman I knew, that sexual harassment, assault and worse is pretty much universal experience for women. But I thought our profession was a better than average environment, not because I thought philosophical education would wean men away from sexism but because I thought prudential self-interest and the threat of losing a middle class job would be a deterrent. I did not realise that sexist philosophers realised it was very easy for them to get away with it, very difficult for women to report abuse. That does seem to be changing in the right direction. One example I was almost directly made aware of was the plight of a philosopher who had to resort to the desperate expedient of listening to my haverings[34] at the bar in order to escape the persistent attentions of a very famous, much older, philosopher who had been following her around and pestering her at a conference. Maybe I should have dug him up there and then.

Whilst there has been some improvement for women, it doesn't look to me as if the situation for people of colour has improved much at all in philosophy. Nor has, at least in proportional terms, the situation for working class philosophers as far as I can tell. It isn't very easy to find data on these cases which perhaps tells a story in itself. Guessing on the basis of accents, philosophers with family backgrounds outside the professional classes or upper middle class are still very thin on the ground in the UK. I doubt much can be done about that without a radical socialist transformation, in particular without workers control[35] of the economy, so over the

[34] Don't confuse this with 'swithering'. On pain of death.

[35] I do not mean here the Lenin/Trotsky/Stalin idea of workers control, meaning ruthless control exercised by the 'advanced workers', i.e. themselves and those in close agreement with them, over everyone else.

pay of executive managers and the distribution of wealth and power of individual or small groups of shareholders. So, just as I dream of finally taming naïve set theory, I still dream of my vision of market socialism (I like to call it 'Hayekian Socialism', it winds up lots of different types of people) coming to pass. I'm not holding my breath though.

(Though in favour of a form of 'market socialism', my settled view is not one of those rightly disparaged both by Marxists and Nozickians, of workers' cooperatives or shareholding workers. No, if I ruled the world, not only would every day be the first day of spring, not only would a large chunk of independent Scotland's GNP go to St. Mirren football club, I'd also abolish equity or voting shares and put every economic enterprise, above a minimum size, under the control of the workforce.)

## 1.9 Small Worlds

On the subject of philosophy conferences, younger philosophers may wonder what life was like back in the days before universities built fancy en suite accommodation for rich students which could also be used during conferences. Grim folks: it were grim up north, grim down south – grim everywhere. Not at all the glamorous jet-setting world of Morris Zapp and Persse McGarrigle wittily portrayed in David Lodge's campus novels such as *Changing Places* and *Small World*. As a matter of fact I read the latter flying from Dublin to one of the more important conferences in my career, an 80th birthday celebration of Quine's work at St. Louis. By chance I was allocated a spacious seat opposite a flight attendant and she had seen the television version and wondered if the conference I was going to would match up to a Lodge novel (for which see ahead).

In UK conferences in those far-off times accommodation was usually in student halls, in grossly uncomfortable beds with mattresses covered in plastic sheets to cover up the stains. At one, in London, I received an unexpected wake-up call about six am when water came through the roof onto me as a result of a leak from the shower rooms above. (It must have been an American, to be taking a shower at that unearthly time.) Accommodation aside, there was great camaraderie at the conferences, and the circuit of giving talks at other departments (where one was often put up by colleagues in their own homes) was a great way to get to know the profession and catch up with new ideas.

The trip to America for the Quine conference was, of course, a bit special. One highlight was Lord Anthony Quinton plummeting from a raised platform during Quine's talk, to the great amusement of Burton Dreben. Up until my direct perceptual acquaintance with him there I had assumed Dreben was a Nabokovian character who only appeared in Quine's prefaces but had no concrete existence, not even as a Williamsonian semi-concrete dot.

I will pass over some other interesting aspects of the week I spent there but I must relate the exchange I had with Quine. At eighty, he had nothing very new to say at the conference. But his stamina, in replying to each talk, was mightily impressive.

During one of these sessions I plucked up the courage to ask him a question about his substitutional account of logical truth in *Philosophy of Logic*.[36] Was it not a problem that it relied on the notion of (first-order) arithmetic truth, a much more complex notion, indefinable in the language of first-order arithmetic (by Tarski's theorem) than the notion of logical truth, which is not only definable but low in the arithmetical hierarchy namely $\Sigma_1^0$ – recursively enumerable?[37] I asked this in my best Scottish Standard English, the dialect I was taught at school to use in the classroom, on formal occasions and so on. 'I'm sorry I can't understand a word you are saying', Quine replied, to much mirth, the first known example of truly radical non-translatability: between Glaswegian English and Ohio English! Actually that bit isn't quite true since 'a chain of interpreters of a sort can be recruited of marginal persons across the darkest archipelago'[38] (Quine, 1960, p. 28) and sure enough a marginal person surfaced in the form of Charles Parsons, who translated the question to be put to Quine! I can't actually remember what Quine's response then was. Later, at a conference in Reading in 1996, by which time Quine's mental powers were quite clearly fading, he had no trouble understanding me at all; we had a chat about his Manx ancestry. I took this as a sign that my accent had been corrupted by years of speaking in standard Scottish English!

Tim Williamson also attended the 1988 conference. After it finished we both headed back to Dublin from the airport at St. Louis at the same time but by different routes. Rather like a commutative diagram in category theory, we met up again next morning at exactly the same time at adjacent baggage carrels at Dublin airport. I was blootered, having been plied with free drink by the Small World flight attendant all the way back across the Atlantic, despite her disappointment that nothing Lodgean had happened to me, topping up with a couple of pints of Guinness at Shannon airport which, in those days, was an enforced stop for Aer Lingus flights from JFK to Dublin. Blootered, but in full possession of my luggage, if not my faculties. Tim, by contrast, was entirely sober but his luggage hadn't turned up. Make of that what you will, as regards strategies for not losing one's luggage when flying!

I will mention one more memorable conference: the Joint Session of the Aristotelian Society at Liverpool in 1995. I missed the start of the talk given by my friend Tim Crane because I'd persuaded some postgraduates (some later to become

---

[36] Quine, 1970, pp. 49–56.

[37] I presupposed that Quine thought of his account as some sort of explication of the notion of logical truth. In fact he gives a number of definitions in that chapter. One might interpret him, therefore, as holding only that the substitutional account is extensionally equivalent to an explanatory account, if there is such a thing, but with the virtue of requiring only a modest ontology. The passage is a good example of some unfortunate aspects of Quine's thought. For example, his tendency to conflate language and theory, here the language with the theory of arithmetic. There also occurs a semantic variant of his 'theorem-fixation', as Neil Tennant has put it (Tennant, 1997, p. 4), his strange tendency to ignore logical consequence and prioritise logical truth.

[38] Some of Quine's language in discussing radical translation – 'natives' being investigated by a 'jungle linguist' and so on – has worryingly colonialist overtones. From what I know of Quine's political views, perhaps it is for the best that I don't know more!

colleagues at Glasgow) to go up to Blackpool so we could test theories about the self over time (if there be such a thing) on the 'Big One' roller-coaster (at the time the tallest and steepest in the world). After Tim's talk I outlined the results. On buying the ticket at the bottom, Weir at time $t$ (the postgrads turned out to be too scared to go on the ride) predicted, correctly, that Weir at time $t'$, just about to hurtle down a 205 ft (62 m), 65% incline drop, would be filled with burning hatred for Weir at time $t$. Just then Geach approached our group and, after some initial philosophical chatter, launched into a political diatribe against the Chinese government, I suppose paramount leader Jiang Zemin must have been the target. Geach averred that he was the most evil leader since ... and here we all assumed Hitler or Stalin or Mao would appear. But to our surprise Shaka, King of the Zulu in the early nineteenth century, was accorded that position. Geach listed the terrible crimes he was said to have committed, as the audience gradually leaked away. Finally, with only Tim and myself present, he asked: did we know that Shaka had a very small penis? (We didn't.) When he become King, we were told, he had all those who had teased him as a youth impaled on stakes. At this point I made the first interjection: 'Well ye cannae really blame him for that, can ye?'. Geach fixed a fearsome glare on me and then stormed off, leaving Tim and myself helpless with laughter.[39]

(To balance up that anecdote, I heard Geach give an extempore talk at a wonderful conference on Lukasiewicz in Dublin after the Joint Session there in 1996. The Poles in attendance adored Geach of course. Though he didn't really say anything philosophically new, his prose in an off-the-cuff speech was almost as flawless as his written prose, which I greatly admire. His moral, religious and political views: not so much! He is an interesting philosopher of language and logic though.)

## 1.10 Sharman

In 1999 by far the greatest catastrophe in my own personal life occurred when my younger sister Sharman, my only sibling, died in childbirth (fortunately her daughter survived). The personal effect on her family and friends and colleagues (the wonderful Citizens' Theatre in Glasgow, where she was general manager, immediately cancelled their performances) can be imagined and I will not dwell on those matters here. Obviously these moments of grief and loss raise in the most acute form philosophical and religious issues which everyone grapples with whilst the sort of philosophy I specialise in was not uppermost in my mind in the days and weeks that followed; but, surprising as it may seem, some such issues did arise.

[39] Though Shaka seems to have been pretty appalling, Geach's account was perhaps based on the racist, colonialist historiography according to which the Zulu *Mfecane* devastated large parts of Southern Africa supposedly leaving virtually uninhabited 'empty lands' conveniently at hand to be occupied by Boer settlers expanding out of the Cape. For a more balanced view, see Shillington (2019) pp. 552–556.

For not long after her death I became aware, through a tangle of personal and professional connections in the legal-medical world, that this was not just a terrible tragedy. I managed eventually to institute a Fatal Accident Inquiry[40] (one of the longest held up until then) into her death. A lot emerged from that, for example that after having been in hospital for a number of days with suspected pre-eclampsia, she had agreed to be a subject for an examination of medical students and became alarmed when the examiners informed the students the diagnosis was pre-eclampsia and the correct course – she was 37 weeks pregnant – delivery. As a result she had threatened to discharge herself and go to another hospital if not delivered, which they then agreed to do the next day. But before that happened she collapsed with severe epigastric pain and severe hypertension and died from a brain haemorrhage.

The FAI had some positive effects, I believe, on the practice in the hospital but its findings were limited because of a conflict of opinion among the four expert witnesses. The result of that was that I reported the two who had not found that Sharman, in the careful language of these inquiries, had not received the standard of care she might reasonably have expected and which had she received would probably have saved her life, to the General Medical Council which has the power to debar medical practitioners. However this path was then blocked after a judge overturned a GMC decision to strike off the medical register a well-known expert witness in cot death cases, Professor Sir Roy Meadows. This had led to the GMC suspending cases involving alleged malpractice by expert witnesses such as the one I was pursuing. *Quis custodiet ipsos custodes* issues to the fore here of course. I do still feel guilty in not taking up this case again; perhaps I will, but the whole process was emotionally very exhausting.[41]

The specialist philosophical matters which arose included current controversies in the philosophy of statistics such as the misuse of significance values, the blind belief that the number 0.05 has some special mystical value being only one of them. (Incidentally Meadows had been accused of vastly underestimating the likelihood of multiple cot deaths in one family by multiplying the probability of non-independent events together.) Another issue was the role of double blind clinical trials. The hospital's practice in place of not, as a matter of course, delivering women with severe pre-eclampsia at 37 weeks gestation was justified on the basis that there were no double blind trials investigating the benefits or harms of this practice. Neither, we can be sure, have there been any such studies of getting pre-eclamptic women to drink bottles of bleach; the most senior of the experts criticised this defence along those lines, albeit in more measured tones.

The main defence was that Sharman did not have severe pre-eclampsia, indeed it was maintained that she did not have pre-eclampsia until her collapse. Looking into what the medical textbooks said about this was a bit of an eye-opener. Doctors differentiate between signs and symptoms, the latter roughly speaking

[40] Roughly equivalent to a Coroner's Inquest in England.

[41] And exhausting in other ways too: I sent in a 60,000 word complaint to the medical authorities. Very likely nobody read more than the first few lines of it.

being unquantified matters of judgement, often patient experiences, the former objectively measurable. Sharman had plenty of symptoms of severe pre-eclampsia, seeing repeated flashing lights for example. Signs are different. But I discovered that criteria for conditions in textbooks are often given as simple lists, with no explicit indication of whether they are to be taken disjunctively or conjunctively. Rather alarming for a logician though my impression from talking to medical people is that a disjunctive interpretation is implicitly understood: one hopes universally so by all medical students.

Even more alarming was the treatment of consistency, though not in the logical sense. The key signs of pre-eclampsia are proteinuria and hypertension, and she quite definitely met the textbook criteria required for severe pre-eclampsia as I interpreted the textbooks. The precise systolic and diastolic boundaries for hypertension and severe hypertension are, of course, arbitrary to a degree, the sort of thing which can be used to generate 'slippery slope' Sorites paradoxes.[42] Moreover since blood pressure is labile, especially in pre-eclampsia, it is important not to rely on one measurement but to look at a number of them separated in time, 4 h is often given as a minimum. But the medical staff in Sharman's hospital were, we were told, taught that consistency requires *all* or *nearly all* measurements to be above the boundary. So, for example, if your blood pressure oscillated every 4 h from 185/125 to 115/75 and back again, you would not even be classed as having hypertension, never mind severe hypertension by these doctors! And in fact Sharman's BP did oscillate in this sort of way until the crisis point. This practice seemed to me to diverge radically from elsewhere so it was very alarming to see such wide divergence of interpretation.

So the debates on statistical practice: classical versus Bayesian, base rate fallacies, the inversion fallacy, misuse of significance values and confidence intervals, these all became much less abstract for me. It is rare for me to quote Ludwig Wittgenstein with approval, but Norman Malcolm's recollection of him saying:

> [What] is the use of studying philosophy if all that it does for you is to enable you to talk with some plausibility about some abstruse questions of logic, etc., if it does not improve your thinking about the important questions of everyday life? (Malcolm, 2001, p. 93)

certainly has resonance.

## 1.11 Final Years at QUB

It was some time after the events, of course, before I was able to think about the philosophical implications properly, although when I returned immediately after the Inquiry to normal working patterns I had reached the part of my philosophy of science course when I discussed the experiments of Semmelweis on maternal death

[42] Cf. Edgington, 1997, pp. 295–6, Weir, 2022a, §XII.

rates at his hospital in Vienna, a famous example of the use of control variables; this was rather difficult to teach. I should record here my deep gratitude to the philosophy department at Queen's, and in particular its head at the time Jonathan Gorman, for their support during this period, enabling me to attend most of the Inquiry.

I never turned, however, to publishing on the philosophy of statistics. The main thrust of my work in the next period was roughly in three areas. Firstly the application of my neo-classical logic to problems such as dialetheism and the liar paradox, (Weir, 2004b, 2005b). Here I worked in a classical metatheory. Tim Williamson (2024) objects to non-classicists using classical logic in the metatheory. With respect to giving a soundness or consistency proof for naïve set theory (NST) I agree that only a bootstrapping use of NST and non-classical logic is legitimate, or indeed possible. Contrary to appearances, it is not at all trivial, if indeed NST is consistent relative to the logic, and my attempts at such consistency proofs have thus far not been, I now believe, successful. That classical logic is not legitimate flows from my anti-realist position on mathematics. However I see no objection to using classical metatheory to give the semantics for my neo-classical logic as applied to vagueness in ordinary or scientific language. This is because I am a *metaphysical bivalentist*: I affirm, for example, $\forall x(x = x)$ and $\forall x \forall y[x \,\varepsilon\, y \vee \neg(x \,\varepsilon\, y)]$, where $\varepsilon$ is a primitive of a first-order theory of properties, $x \,\varepsilon\, y$ to be read as $\langle x$ instantiates $y\rangle$. I do not conclude from these metaphysical theses that:

Santa Claus = Santa Claus;
that cloud = that cloud;
Hamish is bald or isn't bald;

even assuming context is fixed from one part of each sentence to any another. The first fails because the term is empty, the second two may fail as a result of indeterminacy induced by vagueness. So I have to restrict ($\forall$E) and ($\exists$I) which I do in a way formally parallel to common restrictions in free logic (Weir, 2022a, §V, 2024, §IV). The world is sharp, but language links up to it in messy ways invalidating classical logic.

A second cluster of papers concerned themselves largely with Quinean exegesis, arguing, contra Gary Kemp and Peter Hylton, that Quine's overall position is incoherent, there being an irreconcilable inconsistency between his holism, epistemic and semantic, and the reductionism of his atomistic behaviourism (Weir, 2005a, 2006b, 2013c). Quine never resolved the inconsistency because he never fully shook off the 'semantaphobia' prevalent in the Vienna Circle prior to Tarski's rehabilitation of the notion of truth and never fully shook off his nominalism.

Finally there were papers developing my molecular behaviourist approach to language and mind, arguing against holist views, and also functionalism in 'More Trouble for Functionalism' (Weir, 2001). Thus in my joust with Alex Miller (Weir, 2003a; Miller, 2003) at the Aristotelian Society Joint Session in Belfast, I continued to argue that the cognitive structure of the mind is a well-founded hierarchy of dispositions winding down to atomistic dispositions of the Quinean sort at the bottom. At that time, I still thought this left a distinctive place for philosophy as charting out the 'conceptual geography' revealed by this hierarchy.

One outlier among those papers is the paper on an 'Ultra realist' theory of perception (Weir, 2004a), my entry into the lists on the side of the naïve realist. If a theory is naïve, I'm for it! The 'ultra' was to signify that I viewed my version as more externalist than competitors, as close to a 'no-self' position as one could coherently go. It is certainly very different from the dominant naïve position *disjunctivism*, both in its content and in the fact that it attracted far less attention! A basic concept was that of situation or state of affairs, but that does not mean I espoused a metaphysics of objective facts or states of affairs. I have long rejected that, using the (to me) conclusive argument that since Wittgenstein (in the positive body of the *Tractatus*) has an ontology of states of affairs, there cannot be any! The idea, rather, was that perceptions, and more generally experiences such as direct experiential memory, illusions and hallucinations, have an organic structure similar to that posited by situation theorists, with actual objects, properties and relations as their constituents. So, I argued, the only situations which exist are subjective, mind-dependent entities but with wholly objective constituents 'bound' into situations. Illusion and hallucination, the fatal rocks on which disjunctivism founders, are explicated as collages, as it were, of real objects and properties and relations, though they might not be bound together in reality. Borrowing a long-running joke from the great English comic Eric Morecambe, I say that in an illusion or hallucination only real actual entities occur, but not necessarily in the same order as in reality.

The biggest difficulty I have found in convincing people of this position[43] is the feeling that it entails some crazy anti-naturalist views on causation; that in remembering talking to someone long dead one somehow moves that person and the event through time to the time of the memory. There is no such implication. True the view is not readily characterised as physicalist but my hard line naturalism has always been tempered by the feeling that the 'hard problem' cannot be answered physicalistically.

At this time there also appeared symptoms of rising respectability. I took over from David Evans as Head of Department at Queen's from 1998 to 1999. The tenure was brief because at this time the university decided to merge the philosophy and scholastic philosophy departments. It made sense from their point of view (all the staff would be entered together in the UK Research Exercises for example) but not at all from the perspective of the Scholastics who would find themselves outnumbered, rather like nationalists in the Six Counties of Northern Ireland. They feared, with good reason, the eventual disappearance of their tradition at the university. I favoured giving them as much autonomy as possible and so did not find favour with the appointing committee, much to my relief actually.

Later still I was appointed to the panel for the Research Assessment Exercise (RAE) for 2008. But before that exercise happened, I had moved to the department at Glasgow.

---

[43] My graduate student in philosophy of mind Graham Peebles was completely unconvinced, and remains so to this day! More recently I have dropped entirely the notion of states of affairs, even subjective ones, from my naïve realism, see ahead footnote 50.

## 1.12 Dear Green Place

I was delighted to be offered a chair of philosophy at Glasgow, succeeding Bob Hale, alas no longer with us. As always, I had to rely on the support of my wife Ann, who had her own social work career in Belfast, where our three children were also still at school. The solution was a typical Scottish male one: while she worked and looked after the children I stayed in Glasgow 4 days a week in term time with my mother, at that time still alive and well, and far too willing to cook, iron and so on for me, the only cost to me being frequent warnings to watch myself coming home from our Tuesday evening post-seminar sessions in the pub. (I was over fifty by this stage!)

Initially things were hard going. I took over as head of department from Gary Kemp in 2007, after Gary suffered a near-death illness which thankfully he recovered from. There were the sort of staff problems which crop up in lots of departments. On top of this the RAE imposed a tremendous work load on me because the staffing situation meant I could not take a reduced workload, in fact mine was above average, and the RAE methodology requires an enormous amount of reading.

I do still think that we need research quality assessments, rather than handing out research money randomly or according to historical precedent (i.e. to Oxbridge). You cannot assess quality other than by reading the material, and I tried to read everything assigned to me, though not all the books were read closely. Perhaps some other method might work better, with less of a workload. But the aspect I disapproved of most was the introduction of impact measures into a research assessment. Almost the whole panel signed a letter to the *Times Higher* pointing out the historical ignorance in thinking that practical fruits of blue skies research can be reasonably sought over a 10–15 year framework. We pointed to the gap between the pioneering work on formal languages by Frege/Peirce until the emergence of the modern IT revolution.[44] Needless to say this is not to disparage applied research; but funds for such research should be allocated for specific projects with tendered bids. Amalgamating impact (including 'outreach' success which has no obvious correlation with research quality) into the framework of research exercises has led to an enormous waste of time and money and perhaps a threat to the integrity of academics who have to conjure up impact statements.[45]

---

[44] I may have been behind this example. Philosophers perhaps over-egg it a little bit but I think it is overall a fair point. The gap between the invention (or discovery for platonists) of complex numbers and their practical utilisation in alternating currents and quantum theory is an even more glaring example of the dangers in trying to justify funding blue skies research to the taxpayer (and I do not deny justification is needed) in terms of likely immediate results.

[45] I included a rant against impact, and also against the belief that there is something good about interdisciplinarity per se, in the preface to my monograph. I got so worked up I forgot to acknowledge the Arts and Humanities Research Council grant which helped me write the book. *Mea Culpa* and belated thanks now!

Still there were many good aspects to the experience, the camaraderie with the other panellists forged as we were locked away in none too luxurious hotels (to avoid any suspicion of corruption) and in particular getting to know the state of many areas of UK philosophy much better. This was heartening – I came away with a very high opinion of the intellectual quality of our profession.

Once I got through that, I settled down to work in a wonderfully collegiate department in the world's greatest city: what's not to like! My monograph on neo-formalism eventually appeared in 2010. It didn't trouble the scorers as much as I would have liked but I was heartened that no review found (in my objective opinion!) any serious flaws. I initially thought I had failed to explain adequately my position on the relation between concrete proof, which is what makes true or false mathematical claims and formal proof, namely that the equation of truth and proof at the heart of the position must take place, like any systematic theory in this area, as an equation between truth and proof *formally specified*. And that this is perfectly legitimate as long as there are concrete feasible proofs of the equation, e.g. of truth in arithmetic and provability in ω logic. But now I am inclined to think the view was just too novel. I made another attempt to get the idea across in my investigation of the relationship between informal proof and formal proof in Weir (2015).[46]

The most fruitful interaction I have had in this domain is with Mary Leng. I attacked fictionalism in Chap. 5 of the book on a number of fronts, including the unavailability, if one took Field's original position literally, of the conservative extension results necessary for explaining the applicability of maths. I acknowledged that some versions of the position, not best classed as 'fictionalist', can do better but claimed they came close to neo-formalism which those anti-platonists would do better embracing. I have not managed to convince Mary of this though! (Leng, 2015, this volume.) A crucial issue at dispute is whether modal notions can be taken as primitive. I am too close to Quine's position to accept that. Nor do I seem to have converted any platonists. It is not that I have anything against abstract objects, by the way, it's just that I don't number them [ouch] among my acquaintances: the naturalised epistemology of platonism is not in good shape.[47]

Philosophy of mathematics has been an abiding interest, but I always intended developing to full fruition my, as I think, systematic and coherent philosophical vision stretching from maths, logic and science through philosophy of mind and including moral and political philosophy. I'm just a slow developer! One area I tended to bypass for most of my career was metaphysics. The theories and ideas which came to dominate it, Kripke on modality and rigid designators, Lewis's metaphysics, supervenience, fundamentality, grounding: they all passed me by, writing them off as either as not very interesting or confused. Even the rigorous

[46] Taking off from anti-formalist work by Rav (1999). On informal rigour, see Marianna Antonnuti Marfori (2010).

[47] Though I do like to tell my platonist colleagues that, unlike them, I am directly acquainted with Numbers. The historian of science Ronald L. Numbers, to be exact, who was perhaps rather bemused at the vigour with which I shook hands on meeting him after a talk he gave at QUB.

clarificatory work by the two Stephans, Leuenberger and Krämer, has failed to move me.[48] Grounding, for example, seems to me a classic case of projection[49] of features of the mind and our conceptual apparatus onto the world; I take a similar view with what I see as the projection of logico-grammatical structure, such as being negative or disjunctive, from expressions and concepts onto properties and relations. Of course this stance stands in need of justification, which I think can only be broadly abductive, since an alternative explanation of the match between conceptual and metaphysical structure is available: the former is representing the latter, rather than imposing structure where there is none.

I did start to become interested in non-modal metaphysics around 2010 when I first presented a paper with the title ‘Properties as Fusions’, later ‘Properties and Relations as Fusions’. I think there was a Quinean stimulus here. His rejection of properties or attributes has never seemed to me to be backed up with a plausible argument and an ontology of attributes is essential (in a metaphysically light use of the term!) in any scientific endeavour, behaviouristic linguistics as much as any. Quine considers identifying properties with mereological aggregates but points out apparently fatal problems. I tried to get round them by weakening standard mereology and incorporating relations by using a trick from Hazen. I was still pursuing the identification of properties with fusions during a wonderful spell in the beautiful South Island, New Zealand organised by Alex Miller in 2014. But some time later I became enamoured of multigrade relations and reworked the whole metaphysics in those terms, with the ordinary mereological part relation, and also the instantiation relation, falling out as a special case binary case of a multigrade constitution relation. See Weir (2022b) for a theory of absolutely everything![50]

My scepticism about the role of modality is something I do still take from Quine, bolstered by the ‘quasi-Quineans’ at Glasgow, Gary Kemp and Adam Rieger and Glaswegian exile John Divers (see, for example, his 2017). Like Adam I am doubtful that counterfactuals can have fully objective truth conditions (Rieger, 2017). However given the close (though not neatly formalisable) links between counterfactuals and disposition terms, that does raise problems for my combination of realism about everyday and scientific discourse together with my dispositionalist molecular behaviourism. How can scientific claims be objectively true if two conflicting accounts of the dispositions which constitute grasp of them can each be correct, as my views on counterfactuals permit? A global anti-realism threatens. I think realists can surmount this problem, but it is a difficult one.

Another area where I have become even more Quinean in my anecdotage is the philosophy of philosophy and also, with some qualifications, the old terrain of the synonymy relation; the influence of Shapiro and Williamson has been marked here.

---

[48] For example, Leuenberger, 2014, 2018; Krämer, 2018.

[49] Here I am thinking of projection more on the lines of the ‘Left Hegelians’ such as Feuerbach than Hume.

[50] I now drop the idea of mental experiences as being like states of affairs in favour of a naïve realism whose metaphysics is based on my ‘multigrade mereology’ (Weir, 2020b).

Whereas I had thought of philosophy as distinctive and a priori, its task to map out conceptual geography, more recently I have come to see this as a hangover from the era of conceptual analysis and assigned this task to a future empirical semantics, perhaps working in tandem with AI in building robots which can grasp human language.[51]

On synonymy, I have remained resolutely anti-Quinean in that I think his scepticism is based on a crude verificationist behaviourism, his ersatz notion of synonymy far too thin and sparse, collapsing most theoretical sentences together. The truth lies in the opposite direction. Meaning is a really rich but also really messy notion. Not only is the meaning of utterances, as well as their truth conditions, highly context-relative. As Shapiro says (Shapiro, 2014, Chapters Four and Five) whether two expressions are synonymous is itself a context-dependent claim. But this does not lead me to nihilism about semantics, such as we find in Wittgenstein. The details of semantic interrelationships may be unsurveyable (Williamson, 1994, p. 209) like the tangle of neural connections, but they are just as objective and determinate. Semantic theory cannot take the same shape as a physical theory expressible in a few relatively simple equations, but that does not rule out empirical semantics of the sort mentioned in the previous paragraph. Indeed there have to be these determinate semantic links otherwise there are no objective logical consequence relations[52] and the Irrationalist deconstruction of rationality will go through.

What of the philosophy of philosophy? Here I have now come round to the rather Quinean perspective of Tim Williamson. Philosophy is best seen as an armchair 'soft science' like mathematics (automated theorem-proving notwithstanding). The truth value of mathematical claims is determined by formal methods, more specifically proofs, concretely realised. What makes true, or false, philosophical claims? THE WORLD: I have nothing more informative to say, in short order, as to what this world is. That is a central matter of philosophical dispute. What methodology should we adopt in searching for truth in philosophy? I now find the broadly abductive methodology of Williamson attractive. Philosophy is a bit like cosmology, only even more distantly related to empirical confirmation/disconfirmation.

[51] Here is another frighteningly quick change which has occurred in the relatively short time between first submitted draft and final version. I never thought anything like ChatGPT would emerge in my lifetime. I am inclined to reject the idea that a 'disembodied' neural network, however complex, can think, never mind have conscious experience. But wired-up to a sensori-motor apparatus – that is a different matter. I also think we have reason to be fearful about such a development. Perhaps we should try to prevent all-purpose intelligent robots emerging, for example by restricting them to 'modular' entities limited to very specific cognitive and practical domains.

[52] On this point I seem to be in august company: Priest (1979), Dummett (1981, p. 596), Wright (1986), Shapiro (2000), Tennant (2003, p. 692) all make related points.

## 1.13 Conclusion

This brings me back to my starting point. Just as with the death of a loved one, being diagnosed with a terminal illness with a short expected lifespan does cause one to ponder now and again on the Big Questions which analytic philosophers are accused of turning away from. I pretty much ignored them until Idiopathic Pulmonary Fibrosis began to hit the swift downturn phase which is usually the endgame. In this condition an unknown cause triggers a mechanism which leads to scarring of the lung sacs, a mechanism without a stop button, so that one suffocates to death by stages.

The only escape route requires being accepted onto a lung transplantation programme. Most people with terminal lung conditions are not even considered, and those who are have to go through a rigorous assessment procedure. Awaiting the verdict of my four-day assessment was much more stressful than awaiting exam results. Fortunately the wonderful Institute of Transplantation at Freeman Hospital, Newcastle took me on. So here I saw the other, and far more representative, side of the NHS than Sharman saw. But this is only the beginning of the journey. Given the shortage of donor organs one has to get very ill before going 'active' on the list. That was no problem in my case; the danger lay in becoming too weak to be operated on and I got very close to that point, if not perhaps a bit beyond it. Without the support of Ann and the rest of my family, I would never have remained active enough to have survived.

Not that I was very active. I was linked up to oxygen concentrators 24/7, even when sitting in the shower, until I could no longer do that, the sense of drowning being too much to bear. I did connect strongly with numbers then, or token numerals anyway, as counting out breathing patterns was one way I used to stop panic attacks; that was more effective than morphine.

Six weeks after being listed the call came at night, and I was blue-lighted by ambulance the 180 miles to Newcastle and scrubbed and made ready for the operation. But it did not go ahead because the donor lungs were found to be unsuitable. Not unusual, we were warned this might well happen, but still pretty devastating to be driven back up the road, car loaded up with oxygen tanks, to wait again. Then nothing. Though I never gave up, I did know that I would likely be dead in a few weeks. Did philosophy help me here?

Well I was very touched by the messages of support from religious people, of lots of different faiths and denominations (joking that I wanted to get them from as many different ones as possible to 'cover all the bases'). But though it did cause some reflection, I couldn't see why on earth God would listen to prayers to save me having permitted the Holocaust (or, harder for the free will defence, countless terrible deaths in earthquakes and from other natural causes). Given my neo-Humean views on the limited powers of reason, though, I do not rule out theism as irrational.

However I abandoned theism in primary school. For pretty poor reasons: if I had been brought up in a Hindu region of India I would have been a Hindu: that sort of argument. But, to put it in Humean/Calvinist terms, I have never been caused to change my beliefs on this. I have, however, long being agnostic on the after-life, something I think one can reasonably both hope for, and fear, even if one is a physicalist.

As to fear: dark thoughts did increasingly come to me before I got the second, successful, call another 6 weeks later, thoughts perhaps owing to my national background. I began to worry that there was a Deity of an even more terrible aspect than the one worshiped by the 'predestinarian doctors' as characterised by Ramsay.[53] They believe there is an eternal paradisical afterlife awaiting a tiny minority, randomly selected, at least as far as human evaluations of moral worth can tell. The even more fearful notion is that our world was indeed created by a mind even more complex than the world itself; but an infinitely malevolent one who will torture us all for eternity. The good things of life don't tell against this idea for perhaps the diabolical Deity put them there to make the eternal and unending torment even worse, much as some optimistic theists give a 'hills and valleys' response to the problem of evil.

Thinking on this prospect, as I did when not simply counting out my breaths and making sure each one had a successor, put an entirely different, darker colouration on the world. Perhaps this is how Schopenhauerians see things. At the time I thought to myself that if I actually manage to pull off a last-ditch escape I must not forget how the world appeared to me in my worst moments speeding ever closer towards asphyxiation. But I predicted I would forget. And so I have, pretty much. Nature, at least my nature, has proved too strong, the memory of the experience is dim. Intellectually I still accept the evil Creator hypothesis as a possibility, but a possibility too terrible to dwell on for any length of time without risking insanity. And fortunately for me I'm not built thus to dwell. One can be favoured with good psychological health as well as good physical health without necessarily doing anything to deserve it, and I think I am so favoured, through a type of stupidity or blindness. I do not dwell on dark possibilities, at least if they are not imminent and pressing, and am easily led away (too easily my family would say) from concentrating on mundane, often troublesome, reality to speculation about abstruse matters of philosophy.

I have been very lucky in being able to do so for most of my life. Think of how many potentially brilliant philosophers were never able to achieve their potential because they were born into a society where this is not possible, or because of their sex or economic status. Even had I died from pulmonary fibrosis at sixty, I would have been a very fortunate individual, compared to the general run of humanity. Primarily because of my good fortune in the most important thing, namely family. But also because I have been able to get acquainted with the thoughts and arguments of great philosophers and discuss them with, almost without exception, marvellous

[53] 'The grosser pagans contented themselves with divinizing lust, incest, and adultery; but the predestinarian doctors have divinized cruelty, wrath, fury, vengeance, and all the blackest vices' saith the Chevalier Ramsay, quoted in Hume (1757), p. 66.

colleagues. We live in a period in which it is not absurd to worry about the very survival of civilisation. But to young philosophers struggling to get a foot in the academic ladder in these Hard Times my advice is: go for it and the very best of luck. The philosophical life is very much worth living.

## References


Antonutti Marfori, M. (2010). Informal proofs and mathematical rigour. *Studia Logica, 96*, 261–272.

Bostock, D. (1974). *Logic and arithmetic. Vol. 1: Natural numbers*. Clarendon Press.

Bostock, D. (1979). *Logic and arithmetic. Vol. 2: Rational and irrational numbers*. Clarendon Press.

Candlish, S. (2007). *The Russell/Bradley dispute*. Palgrave Macmillan.

Clark, S. (1993). Where have all the Angels gone? *Religious Studies, 28*, 221–234.

Cook, R. (2009). Hume's Big Brother: Counting concepts and the bad company objection. *Synthese, 170*, 349–369.

Cook, R. (this volume). Abstraction and modest reflection. In S. Leuenberger, & A. Rieger (Eds.), *Themes from Weir*. Springer.

Cook, R., & Linnebo, Ø. (2018). Cardinality and acceptable abstraction. *Notre Dame Journal of Formal Logic, 59*, 61–74.

Divers, J. (2017). How skeptical is Quine's "Modal skepticism"? *The Monist, 100*, 194–210.

Dummett, M. (1981). *Frege: Philosophy of language* (2nd ed.). Duckworth.

Dummett, M. (1998). Appendix on the term 'Hume's Principle. In M. Schirn (Ed.), *The philosophy of mathematics today* (pp. 386–387). Clarendon Press.

Edgington, D. (1997). Vagueness by degrees. In R. Keefe & P. Smith (Eds.), *Vagueness: A reader* (pp. 294–316). MIT Press.

Fine, K. (2002). *The limits of abstraction*. Clarendon Press.

Geach, P. (1991). A philosophical autobiography. In H. A. Lewis (Ed.), *Peter Geach: Philosophical encounters*. Kluwer.

Gödel, K. (1936). Über die Länge von Beweisen. *Ergebnisse eines mathematischen Kolloquiums, 7*, 23–24.

Hume, D. (1739/1888). *Treatise of human nature* (L. A. Selby-Bigge, Ed.). Clarendon.

Hume, D. (1757). *The natural history of religion*. Bradlaugh Bonner.

Krämer, S. (2018). Towards a theory of ground-theoretic content. *Synthese, 195*, 785–814.

Leng, M. (2015). Review of *Truth through proof. The Philosophical Quarterly, 65*, 288–292.

Leng, M. (this volume). Formal proof, informal proof, formalism, and fictionalism. In S. Leuenberger, & A. Rieger (Eds.), *Themes from Weir*. Springer.

Leuenberger, S. (2014). From grounding to supervenience? *Erkenntnis, 79*(1), 227–240.

Leuenberger, S. (2018). Global supervenience without reducibility'. *Journal of Philosophy, 115*(8), 389–422.

Leuenberger, S., & Rieger, A. (Eds.). (2024). *Themes from Weir*. Springer.

Levy, A. (1969). The definability of cardinal numbers. In J. J. Bulloff et al. (Eds.), *Foundations of mathematics (Symposium papers commemorating the sixtieth birthday of Kurt Gödel)*. Springer.

Malcolm, N. (2001). *Ludwig Wittgenstein: A memoir* (2nd ed.). Clarendon Press.

Mcdonald Ross, G. (1985). Angels. *Philosophy, 60*, 495–511.

Miller, A. (2003). Objective content II. *Supplementary Proceedings of the Aristotelian Society, LXXVII*, 73–90.

Miller, A. (2024). What is the sceptical solution? In S. Leuenberger, & A. Rieger (Eds.), *Themes from Weir*. Springer.

Monk, R. (1996/7). *Bertrand Russell: Spirit of solitude*. Vintage.
Oliver, A. (1999). A few more remarks on logical form. *Proceedings of the Aristotelian Society, 99*, 247–272.
Oliver, A., & Smiley, T. (2016). *Plural logic* (2nd ed.). Oxford University Press.
Priest, G. (1979). Two dogmas of Quineanism. *The Philosophical Quarterly, 29*, 289–301.
Priest, G. (1998). What's so bad about contradictions? *The Journal of Philosophy, 9*, 410–426.
Quine, W. V. (1960). *Word and object*. MIT Press.
Quine, W. V. (1970). *Philosophy of logic* (1st ed.). Prentice-Hall.
Rav, Y. (1999). Why do we prove theorems? *Philosophia Mathematica*, 5–41.
Rieger, A. (2017). Was Quine right about subjunctive conditionals? *The Monist, 100*(2), 180–193.
Root, J. D. (1980). The philosophical and religious thought of Arthur James Balfour. *Journal of British Studies, 19*(2), 120–141.
Russell, B. (1903). *The principles of mathematics* (Vol. I). Cambridge University Press.
Russell, B. (1994). In A. Urquhart (Ed.), *Collected papers*. Routledge.
Shapiro, S. (2000). The status of logic. In P. Boghossian & C. Peacocke (Eds.), *New essays on the a priori* (pp. 333–366). Oxford University Press.
Shapiro, S. (2014). *Varieties of logic*. Oxford University Press.
Shapiro, S., & Weir, A. (1999). New V, ZF and abstraction. *Philosophia Mathematica, 7*, 293–321.
Shapiro, S., & Weir, A. (2000). "Neo-logicist" logic is not epistemically innocent. *Philosophia Mathematica, 8*, 160–189.
Shillington, K. (2019). *History of Africa* (4th ed.). Red Globe Press.
Tennant, N. (1997). *The taming of the true*. Clarendon Press.
Tennant, N. (2003). Theory contraction is NP-complete. *Logic Journal of the IGPL, 11*, 675–693.
Tennant, N. (2024). The umpire's dilemma and the ashes of realism. In S. Leuenberger, & A. Rieger (Eds.), *Themes from Weir*. Springer.
Weir, A. (2022a). *Vague mathematics, indeterminacy and non-classical logic*. https://sites.google.com/site/weirysite/research-1
Weir, A. (2022b). *Multigrade mereology as a framework for ontology*. https://sites.google.com/site/weirysite/research-1
Weir, A. (2024). Indeterminacy and non-classical logic. In S. Leuenberger, & A. Rieger (Eds.), *Themes from Weir*. Springer.
For the references to Weir's published work, see the separate bibliography.
Williamson, T. (1994). *Vagueness*. Routledge.
Williamson, T. (2024). Can non-classical logic treat mathematics as exceptional? In S. Leuenberger, & A. Rieger (Eds.), *Themes from Weir*. Springer.
Wright, C. (1983). *Frege's theory of numbers as objects*. Aberdeen University Press.
Wright, C. (1986). Inventing logical necessity. In J. Butterfield (Ed.), *Language, mind and logic* (pp. 187–209). Cambridge University Press.
Wright, C. (2009). Foreword, on becoming a philosopher. *Synthese, 171*, 359–364.
Zardini, E. (this volume). Neo-classical logic, cloisonnist vagueness, non-objectual truth. In S. Leuenberger, & A. Rieger (Eds.), *Themes from Weir*. Springer.

# Chapter 2
# Can Non-classical Logic Treat Mathematics as Exceptional?

**Timothy Williamson**

**Abstract** The paper criticizes a 'lazy' strategy popular amongst contemporary advocates of non-classical logic motivated by non-mathematical phenomena (e.g. vagueness), which is to argue that, since the relevant phenomena do not occur in pure mathematics (e.g. because it is precise), classical logic (with e.g. the law of excluded middle) remains valid as restricted to the special case of pure mathematics, so no revision of classical mathematics is needed. The lazy strategy is shown to be inadequate because it fails for *applied* mathematics, when theorems of pure mathematics are instantiated with respect to physical situations as characterized in terms for which the phenomena motivating the non-classical logic still arise (e.g. imprecise physical descriptions). Thus, such non-classical logicians are under pressure to reconstruct as much mathematics as they can on the basis of their favoured non-classical logic. In most cases, the results will be too weak for purposes of the natural and social sciences. The chapter also includes a short memoir of the author's friendship with Alan Weir and their experience of philosophy and sectarian troubles when they both taught in Ireland.

**Keywords** Alan Weir · Classical logic · Non-classical logic · Law of excluded middle · Vagueness · Pure mathematics · Applied mathematics

Alan Weir's career and mine have been intertwined. Philosophical logic is central to our research. We were born within a couple of months of each other, in 1955. We both have doctorates in philosophy from Oxford, and connections with Balliol College, where Alan had a short-term Tutorial Fellowship in 1983–1985, and I had been an undergraduate in 1973–1976. Alan's first degree is from the University of Edinburgh, where he also held temporary lectureships before the Balliol position, and much later I held the chair of Logic and Metaphysics, though before his move

T. Williamson (✉)
University of Oxford, Oxford, UK
e-mail: Timothy.Williamson@philosophy.ox.ac.uk



A. Rieger, S. Leuenberger (eds.), *Themes from Weir: A Celebration of the Philosophy of Alan Weir*, Synthese Library 484,
https://doi.org/10.1007/978-3-031-54557-3_2

to a chair at Glasgow; Edinburgh was also home for part of my childhood, in the late 1950s and early 1960s. However, we first became acquainted not in Scotland or England, but in Ireland. For three years, 1985–1988, we were the two professional philosophical logicians on the island, both lecturers in philosophy, Alan at The Queen's University of Belfast (QUB) and I at Trinity College Dublin (TCD). We became good friends.

Two organizations served philosophy over the whole island: the Irish Philosophical Club and the Irish Philosophical Society. In those days, the distinction was clear: the Club was largely Protestant, the Society largely Roman Catholic. The difference was philosophically consequential, since it was roughly coextensive with that between 'analytic' and 'continental' philosophy. TCD was historically the Protestant 'West Brit' place; I started teaching there in 1980, soon after a Roman Catholic Archbishop of Dublin's unenforceable ban on his flock studying at TCD had been lifted. Things were starting to change, but the teaching staff of the School of Mental and Moral Science (in effect, the Philosophy Department) was still entirely non-Catholic. By contrast, the corresponding departments in the colleges of the National University of Ireland (University College Dublin, University College Galway, and so on) had a powerful Church presence, much more so than other departments in those colleges, often with priests in senior academic positions—philosophy was too close to theology to be left to chance—though the more junior lecturers might be phenomenologists, hermeneuticians, or Marxists rather than neo-Thomists. QUB compromised by having two philosophy departments, one on each side. Alan's lectureship was in the Department of Philosophy, *de facto* Protestant, but there was also a Department of Scholastic Philosophy, *de facto* and overtly Catholic. If there had been only one philosophy department at QUB, it would have been uniformly Protestant.

It was the time of the Troubles in Northern Ireland, while in the Republic the cultural prestige and political power of the Church had not yet collapsed in scandal. Many years would elapse before the phrase 'post-Catholic Ireland' came into use. Still, it was not difficult to be an atheist in Dublin, but the question remained: were you a Protestant atheist or a Catholic atheist? An atheist, brought up on the mainland by mildly Anglican-going-on-Quaker parents, I was obviously not a Catholic atheist, so by disjunctive syllogism had to be a Protestant one. Hoping to be independent, and with a big appetite for philosophy, I attended both the Club's meetings and the Society's, and was made welcome at both. Sectarian differences play far less of a role in Irish philosophy today, north and south of the border, but in the 1980s they were hard to ignore.

Predictably, the Club's main axis ran between the Protestant departments at TCD and QUB, though they did not exhaust its membership. Its annual meeting was held at the Ballymascanlon House Hotel, in County Louth, with the imposing remains of a megalithic tomb close. They stand just south of the border with Northern Ireland. Other regular guests were said to include the leadership of the Provisional Irish Republican Army (IRA), who could plan their campaigns of violence in comfort there, just out of reach of the Royal Irish Constabulary and the British Army. As far as I know, their meetings never overlapped with the Club's. Anyway, Alan and

I first got to know each other at the Club meeting in February 1986. He gave a talk in the cosy hotel lounge to an audience of philosophers literally in armchairs. Alan had and has more sympathy than I with Quine's behaviourism, though less than I with Quine's unwillingness (at least in practice) to give up classical logic. He used Quine's three-valued assent-dissent-abstain tables for the logical constants to argue fluently for a non-classical logic. Heresy! We locked horns in the discussion period and have been doing so ever since, though often at far too long intervals. For the next twenty years, Alan was the life and soul of Club meetings (despite the absence of souls from a Quinean ontology).

Obviously, Alan was much closer to the Troubles in Belfast than I was in Dublin, though I recall in 1981 watching through a window of the Arts Building in TCD as stragglers from a demonstration in support of the hunger-striking IRA Provisional Bobby Sands ran down Dawson Street, smashing the windows of parked cars as they passed. How Dublin car owners were to blame for a death by hunger strike in Northern Ireland was not entirely clear. Alan used to play for a non-sectarian soccer team in Belfast which, unusually, played both Protestant teams and Catholic ones. When they played Protestant teams, they had to pretend to be all-Protestant; when they played Catholic teams, they had to pretend to be all-Catholic. In Northern Ireland, Christian names tend to reveal affiliation. Thus each member of Alan's team needed both a Protestant name (as it were, 'Billy') and a Catholic one (as it were, 'Seumas'), to call safely for a pass. Naturally, in the heat of action, it was hard enough to remember one's own current name, let alone those of all one's teammates. Another of Alan's many anecdotes from that period concerned the Harland and Wolff shipyard, one of Belfast's major employers. For years, it had been a Protestant closed shop: no Catholic could get a job there. Eventually, it was forced to open its jobs to Catholics too. In petty retaliation, every Friday lunchtime, the Protestant workers would pack in early to the canteen and order fish, in an attempt to ensure that there would be none left for their Catholic co-workers to satisfy the principle of no meat on a Friday. My Oxford college serves fish and chips for lunch on most Fridays, pandemic permitting; it usually reminds me of Alan's story.

Disputes in logic may cut intellectually deeper than disputes in politics or religion, but tend to be fought out less vindictively, with fewer casualties. My friendship with Alan has easily survived our doctrinal differences. But it would be contrary to the spirit of our friendship to miss an opportunity to pursue our dispute over classical logic. Alan is of course a world leader in the application of one kind of non-classical logic to mathematics. I will therefore explain why I take going non-classical to have more serious consequences than is usually realized for mathematics. I have worked out the technical details of the argument in more detail elsewhere (Williamson, 2018). I will state it in general terms, leaving to Alan the task of determining exactly how the dialectic plays out for his own preferred approach—his formalist philosophy of mathematics makes his view distinctive in ways which a fuller discussion would of course need to address (see his *magnum opus*, Weir, 2010).

## 2.1 Mathematical and Non-mathematical Motivations for Logical Deviance

Mathematics constitutes by far the most sustained and successful deductive enterprise in human history. With only minor exceptions, mathematicians have freely relied on *classical logic*, including principles such as the law of excluded middle. They unquestioningly accept classical reasoning in proofs. When deviant logicians reject a classical principle, they face an obvious challenge: what does that mean for mathematics? Where does it leave theorems whose proofs rely on the principle? Absent a good response, the proposed alternative logic has not earned the right to be taken seriously.

How the dialectic goes from there depends on what motivates the rejection of the classical principle. Some motivations are *internal* to mathematics. The standard example is the intuitionistic rejection of excluded middle, motivated by a constructivist conception of mathematics in general and infinity in particular. The rejection of classical mathematics is the *point* of intuitionism, not a grudgingly accepted incidental cost. Intuitionists willingly accepted the burden of building mathematics anew using only intuitionistically acceptable reasoning. The result may be no adequate substitute for classical mathematics, but that is a further issue.

A more recent example of logical deviance with a motivation at least partly internal to mathematics is paraconsistent logic. Dialetheists endorse an unrestricted comprehension principle for sets, including the inconsistent instance that generates Russell's paradox, while revising the logic to avoid the total collapse classically implicit in the claim that the Russell set is self-membered if and only if it is not self-membered. Here the motivation is overtly to simplify and clarify the axioms of set theory: some rebuilding of mathematics is intended, not just accepted as a by-product.

Formalism in the philosophy of mathematics is another potential source of logical deviance. If the truth of a mathematical statement is equated with its provability, and its falsity with its refutability, then undecidable mathematical statements are neither true nor false. If such gaps are handled with three-valued logic, we can expect non-classical results.

Even in these internally motivated cases, the strategy favoured by non-classical logicians is not usually to construct a new mathematics from scratch, with no eye to the old. Often, it is to make the new as close as possible to the old, while starting with the deviant logic and respecting the motivation for reform. It is a project of *reconstruction*, not pure construction *ab initio*. That is hardly surprising, given the multitudinous successful applications of classical mathematics throughout science.

For other rejections of classical logic, the motivation is *external* to mathematics. Salient examples include various proposals motivated by phenomena of vagueness, such as sorites paradoxes. Typical cases involve non-mathematical words like 'heap', 'bald', 'red', and 'rich', which lack clear boundaries. On the (false) assumption that the unclarity does not merely manifest our characteristic ignorance of the boundary's location, vagueness is supposed to involve some sort of non-epistemic indeterminacy that undermines the bivalent dichotomy of truth and falsity. As for more obviously non-classical treatments of vagueness, there is no consensus as to which is most promising. Many reject excluded middle and postulate some form of many-valued logic, typically either a three-valued logic or a continuum-valued fuzzy logic. Another alternative is to reject Cut, the structural rule that allows a series of short arguments to be chained together as in a sorites paradox, in a generalization of transitivity for logical consequence. In any case, such proposals are usually not intended to disrupt classical mathematics. Core mathematical vocabulary is assumed to be precise, and so not to generate sorites paradoxes and other phenomena of vagueness. Thus classically valid reasoning is still supposed to preserve truth within a purely mathematical language.

One might contest the assumption that vagueness does not infect mathematics. For instance, even the language of pure set theory has been thought to be vague as to exactly what counts as a 'set', so that Cantor's continuum hypothesis comes out true on one legitimate sharpening but false on another (Feferman, 2011). However, even if such vagueness is granted, it is usually assumed to be irrelevant for purposes of the ordinary working mathematician. Admittedly, some standard notation of ordinary working mathematics is vague by intention: for instance, $x \approx y$ is read '$x$ is approximately equal to $y$', and $x << y$ as '$x$ is much smaller than $y$'. However, for the sake of argument we may concede that in principle such uses of vague notation in mathematical proofs can always be eliminated in favour of something more precise.

A further motivation external to mathematics for rejecting classical logic comes from the Liar and other semantic paradoxes. Metalinguistic terms like 'true' and 'false' do not figure in the core language of mathematical theories. Consequently, those who take the semantic paradoxes to motivate a retreat from classical logic to a non-classical logic usually assume that their logical reform leaves classical mathematics itself intact.

Yet another example of logical deviance motivated externally to mathematics is quantum logic, which rejects one of the classical distribution principle because it supposedly fails in the quantum world. This too is often assumed to do no damage to classical pure mathematics.

The idea that, for all externally motivated logical deviance shows, classical logic is fine for mathematics has been voiced by various theorists. For example, Hartry Field takes excluded middle to fail in the presence of semantic and sorites paradoxes, but suggests (Field, 2008: 101):

> Presumably excluded middle holds throughout ordinary mathematics, and indeed whenever vagueness or indeterminacy is not at issue.

Recently, Ole Thomassen Hjortland has articulated the idea in more detail (2017: 652–3):

> Mathematical proofs do contain an abundance of instances of classical principles: applications of classical *reductio ad absurdum*, conditional proof, disjunctive syllogism, the law of absorption, etc. The emphasis, however, should be on the fact that these are *instances* of classical principles. The mathematical proofs do not rely on any of these principles being unrestricted generalizations [...]. They do at most rely on the principles

> holding restrictedly for mathematical discourse, which does not entail that the principles of reasoning hold universally. Put differently, mathematical practice is consistent with these reasoning steps being instances of *mathematical* principles of reasoning, not generalizable to all other discourses. *A fortiori*, they may very well be principles of reasoning that are permissible for mathematics, but not for theorizing about truth.

I will argue that such claims are much too optimistic about the prospects of isolating mathematics from logical deviance in non-mathematical discourse. They overlook the capacity of pure mathematics to be *applied*.

## 2.2 The General Applicability of Mathematics

The simplest way of applying pure mathematics is just by substituting what may loosely be called 'empirical terms' for the variables in its theorems, when the latter are universal generalizations. Of course, many actual applications to science involve something less direct and more elaborate than that, but if pure mathematics cannot be kept free of externally motivated logical deviance in the simplest applications, then the Field-Hjortland strategy is already in serious trouble. Applications of mathematics to the natural and social sciences also make trouble for formalist philosophies of mathematics, since they are even harder than pure mathematics to make sense of as moves in a game like chess, and descriptions in mathematical terms of the natural or social world are typically not susceptible of formal proof.

We can start with a rather extreme external motivation for logical deviance: ontological vagueness. Consider the idea that there are vague objects, such as mountains, whose identity is vague in ways that make them counterexamples to the law of excluded middle for identity. A proponent of this view may hold that there is a mountain Everest and a mountain Chomolungma such that it is vague whether they are identical. The vagueness is not in the names but in the things named. Under the assignment of Everest and Chomolungma to the variables '$x$' and '$y$' respectively, the open formula $x = y$ is neither true nor false. Given the many-valued semantics which this theorist adopts, the disjunction of $x = y$ with $x \neq y$ is correspondingly neither true nor false under this assignment, so excluded middle for identity as a universal generalization is not true. Of course, this is a deeply contentious view, and very far from my own. Nevertheless, such views have been defended by serious, technically competent philosophers of vagueness (Parsons, 2000).

For present purposes, the point is that excluded middle for identity can be formulated as a sentence in the pure language of first-order logic with identity, without non-logical predicates or individual constants. It therefore seems to belong to the core language of mathematics. Although Everest and Chomolungma are not mathematical objects, they are still in the range of the first-order quantifiers. Thus the logical deviance, although motivated externally, still infects mathematics. One cannot rely on excluded middle even for formulas in its core language.

For proponents of the isolationist strategy, the problem may seem to have an obvious solution. They can restrict their defence of classical logic to *pure* mathematics, where by definition the quantifiers are restricted to *purely mathematical* objects. In particular, the quantifiers in set theory as a branch of mathematics are restricted to *pure* sets, sets to which only sets bear the ancestral of the membership relation. A set is pure if and only if all its members are. Thus Everest, Chomolungma, their singletons, pair set, and so on, are excluded from the quantifiers of pure mathematics. On this reinterpretation, excluded middle for identity is immune to the proposed counterexample.

But where does the proposed solution leave applied mathematics? At an elementary level, consider the statement that if there are exactly $m$ apples and $n$ oranges, and no apple is an orange, then there are exactly $m + n$ apples and oranges. It quantifies over non-mathematical objects, apples and oranges, and so falls outside pure mathematics, on the proposed restriction. It is not even clear how it might be derived from a theorem of pure mathematics, as just construed. Of course, Peano arithmetic itself quantifies only over natural numbers, but one might still expect pure mathematics as a whole to deliver the result that for any things and any other things, if there are exactly $m$ of the former and $n$ of the latter, then there are exactly $m + n$ of the former and latter together.

At a more sophisticated level, consider the application of group theory to the group of symmetries of a physical system. The symmetries are rotations, reflections, translations, or other mappings of the elements of the system onto each other. They are not purely mathematical objects. Thus group theory as a branch of pure mathematics on the proposed restriction does not deal with such symmetries. In some cases, the chief interest of a result in pure mathematics derives from its applications: Arrow's Theorem may be an example, with its application to voting rules.

At first sight, the problem of applied mathematics may in turn appear to have an easy solution. For it may be argued that any system of objects some of which are not purely mathematical is *isomorphic* to a system of purely mathematical objects. On this view, each set of non-sets is in a one-one correspondence with a set of pure sets, which induces a mapping of any properties, relations, or operations defined over the former to corresponding properties, relations, or operations defined over the latter. Thus one can prove a result in pure mathematics and then by isomorphism transfer it to impure applied mathematics.

The isomorphism argument appeals to standard habits of mathematical thinking. However, they are invalidated by the very restriction of mathematics at issue. For starters, the claim 'Every set of non-sets corresponds one-one with a set of pure sets' falls outside pure mathematics on the proposed restriction, since it quantifies over objects that are not purely mathematical: sets of non-sets. It would have to be derived from some impure theory, which on the envisaged view is not entitled to rely on classical logic. Indeed, the alleged counterexample to excluded middle for identity shows that the claim is quite dubious in this setting. For suppose that the set {Everest, Chomolungma} of non-sets corresponds one-one with a pure set. By hypothesis, pure set theory obeys classical logic, so either the pure set has at most one member or it has at least two members. Therefore, since it corresponds one-one with the pure set, either {Everest, Chomolungma} has at most one member, in which case Everest and Chomolungma are identical, or it has at least two members, in which case they are distinct. Thus the isomorphism argument conflicts with the assumption that excluded middle fails outside mathematics but not inside.

One moral of that argument is that if set theory is to provide a foundation for mathematics, it must be *impure* set theory, for example Zemelo-Fraenkel set theory with the axiom of choice and ur-elements (non-sets), rather than just pure Zermelo-Fraenkel set theory with choice. For a key feature of pure mathematics is that it can be applied outside itself, to natural and social science. Without impure sets, set theory fails to enable such applications. An account of isomorphisms between pure and impure sets is itself part of impure set theory, not of pure set theory.

In any case, even if we do find a sense in which every set of non-sets corresponds one-one with a set of pure sets, the strategy of transferring results about the latter to the former is undermined by the very hypothesis at issue, that classical logic holds when the quantifiers are restricted to purely mathematical objects but not when the restriction is lifted. For that hypothesis entails that what holds under the restriction is *not* structurally representative of what holds without the restriction. A parallel argument will apply to other putative foundational frameworks. Thus, given the hypothesis, the isomorphism strategy fails as a way of applying mathematics.

For all that has just been argued, some classical principles may hold under the restriction to purely mathematical objects but not more generally. But the point is that once you treat that as a live option, you cannot freely apply standard pure mathematics to science. Instead, you must develop an alternative mathematics for applications, based on your preferred non-classical logic. That is the hard work which proponents of the isolationist strategy were hoping to avoid.

Might the case of excluded middle for identity be somehow misleading? After all, the hypothesis of ontologically vague identity is notoriously problematic. We may therefore consider a deviant logician who nevertheless accepts excluded middle for identity as universal generalization. The next question is whether the standard, classical rule of universal instantiation can be applied to that universal generalization. If so, we can derive an instance of it for any closed singular terms, for instance, 'Everest' and 'Chomolungma'. But a non-classical logician who regards vagueness as semantic rather than ontological may still reject such an instance on the grounds that indeterminacy in how names refer can invalidate an instance of excluded middle.

The envisaged theorist accepts excluded middle for identity as a universal generalization but rejects a particular instance, and so must also restrict the standard rule of universal instantiation. Of course, that is not unprecedented: it happens in free logic. A similar restriction might be adopted here: to derive a particular instance from a universal generalization, one needs auxiliary existential premises for each name. One may then reject those premises when one regards a name as indeterminate in reference. The trouble with this strategy is that in effect it treats semantic indeterminacy as reference failure: vague names become virtually unusable in science, because they fall outside the reach of universal generalizations, including mathematical theorems. That applies not just to radically vague names but even to slightly vague ones, since the slightest vagueness in one name can make the instance fail, for a suitable choice of the other name, on the envisaged view of vagueness. In practice, virtually any term for a physical quantity is at least slightly vague.

Such problems for universal instantiation arise even when the quantifiers are restricted to purely mathematical objects. For instance, in a suitable context, one may want to instantiate the universal generalization with a definite description such as 'the velocity of that cloud in kilometres per hour', or 'the current number of oxygen molecules in the Pacific Ocean', or 'the largest small natural number'. One can talk vaguely about mathematical objects, just as one can about non-mathematical objects. On some precise, purely mathematical reading of the other term, the identity will then be vague. Nor is this problem specific to the identity predicate. It arises for other mathematical predicates too. The same examples show that, on a non-classical semantic view of vagueness, it is problematic to reason from the classical principle that real or natural numbers are totally ordered in the standard way to a vague instance. To avoid such problems on the semantic view, the relevant restriction is not to quantify only over purely mathematical objects but rather to use only purely mathematical vocabulary. However, that restriction merely exacerbates the problem of applied mathematics.

Is the difficulty specific to vague singular terms? If so, it should be resolved by the Quinean move of eliminating singular terms in favour of predicates. But, on reflection, such a move turns out to make little difference.

If we are allowed to quantify into predicate position, we have the second-order version of the problem of universal instantiation. For instance, on its intended interpretation the second-order universal generalization of excluded middle for predication is precise; informally, it says that for every way, everything either is or is not that way. On a suitably unrestricted understanding of the quantifiers, that generalization will be precise. Thus, if the problem for excluded middle is vagueness, then excluded middle for predication as a second-order universal generalization should hold. But if the second-order rule of universal instantiation is valid, then the universal generalization entails its instances. But when the predicate constant substituted for the predicate variable in the generalization is vague, those who take excluded middle to fail in case of vagueness will reject the instance. A similar problem arises if some classical principle other than excluded middle is taken to fail in case of vagueness. If mathematical theorems are formulated as precise higher-order generalizations, and classical logic holds for precise languages but not vague ones, then we cannot expect to instantiate those theorems with vague predicates, so their applications are blocked.

If one tries to approximate the generality of a second-order universal generalization with that of the corresponding first-order schema, the problem is more immediate. On the envisaged view of vagueness, the unrestricted first-order schema is invalid, because its instances with vague predicates fail.

The point generalizes to other motivations for non-classical logic external to mathematics, such as the semantic paradoxes. Instances where a truth predicate is substituted for the predicate variable are often alleged to fail, even though the universal generalization remains a classical theorem in the language of pure second-order mathematics. Similarly, for proponents of quantum logic, instances of the distributive law fail for predicates ascribing properties to a particle, even though the law itself may be formulated as a classical theorem in the language of pure second-order mathematics. In such cases, pure mathematics cannot be applied by means of second-order universal instantiation.

It will not do simply to claim that we can recover the instances of the second-order universal generalization when we need them in unproblematic cases. For the instantiating predicate may be problematic, generating a paradox, for some values of the individual variable and not for others, but to restrict second-order universal instantiation we must classify the predicate itself as problematic or unproblematic, since it is what gets substituted for the second-order predicate variable, not the corresponding open formula for some particular value of the individual variable.

Of course, in many circumstances (though not all) we can use first-order generalizations over sets instead of second-order generalizations into predicate position. Instead of instantiating the second-order predicate variable, we instantiate a first-order set variable with a set term. We could then formulate excluded middle as a universal generalization about set membership.

But that is no help. For in a particular instance the set will be characterized by a defining predicate, and set membership should reduce to predication, making the instance of excluded middle for membership of the relevant set reduce to an instance of excluded middle for the relevant predicate, which is just what the envisaged non-classical logician rejects, while accepting excluded middle for set membership as a first-order universal generalization, since it is a classical theorem in the language of pure mathematics. On the other hand, if the predicate fails to define a set, the application fails anyway. Either way, the envisaged non-classical logician still faces the first-order version of the obstacle to applying pure mathematics, already discussed above.

For a deviant logician who rejects bivalence, a more modest claim is that, even if not every theorem of classical mathematics is true, none of them is false. Thus one will never go too badly wrong if one relies on classical mathematics. Some initial results point that way. For instance, no valuation in the Kleene weak or strong three-valued semantics makes a classical theorem false. One might hope to extend that result to all theorems of classical mathematics.

However, the result depends on peculiarities of the Kleene semantics. Many forms of many-valued semantics allow some classical theorems to be false. Indeed, on such approaches, the slightest degree of vagueness can make some classical theorems come out perfectly false (see Williamson, 2018 for details).

That is far from the only case where a non-classical approach falsifies some classical theorems. For example, in intuitionistic mathematics, the theory of choice sequences has theorems which deny universally generalized principles of excluded middle for a given predicate. In intuitionistic logic, although it is inconsistent to deny any specific instance of excluded middle, it is not always inconsistent to deny a universal generalization over infinitely many instances of excluded middle.

Although not all non-classical logics diverge so dramatically from classical logic, claims of approximately classical behaviour must be established (using a non-classical metalogic!), not assumed.

## 2.3 Mathematical Reasoning

So far, applying pure mathematics has been treated simply as a process of instantiating its theorems. However, mathematics is used in science far more pervasively than that. For example, given a scientific law or hypothesis in the form of an equation, one uses ordinary mathematical reasoning to deduce its consequences, typically with statements of initial conditions or other auxiliary assumptions as extra premises. Rejections of classical logic often have the effect of calling such reasoning into question, even when motivated externally to mathematics.

To illustrate the variety of inference rules rejected by one non-classical logician or another, we need look no further than the problem of vagueness, which has been taken to motivate the rejection of almost any given rule:

Some fuzzy logicians define a valid argument as one such that on every interpretation on which every premise reaches a fixed threshold of degree of truth, say 0.9, so does the conclusion. By that standard, the rule of modus ponens comes out invalid: even if its premises are 'true enough', its conclusion may not be. By contrast, some paraconsistent accounts of vagueness invalidate the rule of disjunctive syllogism, on the grounds that it is not truth-preserving.

Others have rejected standard structural meta-rules, which are not specific to a particular connective, such as the Cut rule, a generalization of the transitivity of logical consequence, in order to block sorites paradoxes. For example, on the background assumption of the tolerance principle that the successor of any small natural number is small, where 'small' is vague, even if our other rules tell us that '1 is small' follows from '0 is small', and that '2 is small' follows from '1 is small', we typically need Cut to conclude that '2 is small' follows from '0 is small'.

The rejection of any of the rules just mentioned could also have been illustrated from non-classical treatments of the semantic paradoxes.

The difficulty of being classical inside mathematics but non-classical outside generalizes from theorems to rules and meta-rules. For suppose that a deviant logician accepts classical reasoning within mathematics but rejects a classical rule or meta-rule as invalid outside, and so rejects some classically valid reasoning because one of its steps applies the rule to a non-mathematical atomic expression such as a vague or semantic predicate. Now replace the vague expression by a new variable of the same grammatical type throughout the reasoning. Unlike the vague expression, the new variable lacks a distinctively non-mathematical meaning. Repeat the process until all the non-mathematical atomic expressions in the original reasoning have been replaced by distinct new variables. The result is some analogous reasoning which is still classically valid because it has the same relevant form as the original classically valid reasoning. Our deviant logician accepts the new reasoning because it is classically valid and contains no extra-mathematical elements. But we can recover the original reasoning simply by substituting back the vague expressions for the variables, or interpreting the latter in terms of the former.

The deviant logician will doubtless refuse to give up, and will instead reject the substitution of the vague expressions for the variables as illicit. This rejection is either sensitive or insensitive to the details of the reasoning. If it is insensitive, then it amounts to a blanket rejection of the application of all mathematics and logic to vague or semantic expressions, which puts the latter beyond the reach of reason. But terms such as 'heap' and 'small', 'true' and 'false', are *not* beyond the reach of reason. On the other hand, if the rejection is sensitive to the details of the reasoning, then the deviant logician is in effect already engaged in the project of reconstructing mathematics to determine, for purposes of applying it, how much can survive the retreat to the deviant logic.

Sometimes the substituted variable will be of a higher grammatical type, not usual in ordinary mathematical notation. The reasoning with it may then be more logical than mathematical in flavour. For present purposes, however, that is a superficial feature. Instead of the predicate variable, we could have used a variable for sets or characteristic functions. That would have given the substitute argument a more characteristically mathematical flavour, while raising the same key issues. If there is no set of all red things, no characteristic function for 'red', and so on, how can we apply standard set-theoretic or function-theoretic mathematical reasoning to the distinction between red and non-red at all?

For mathematical reasoning, the conclusion is the same as for mathematical theorems: if classical logic fails outside pure mathematics, and pure mathematics is fit to be applied outside itself, then classical logic fails inside pure mathematics too.

Another way to think of the background role of logic and mathematics in natural and social science is as a *closure operator*. We want to test scientific hypotheses by their logico-mathematical consequences, so a system of logico-mathematical reasoning should induce a mapping of each set of hypotheses to its set of consequences according to that system, its closure under the consequence relation. For simplicity, we can treat the hypotheses and the consequences as sentences of the same language. Indeed, for many purposes we can identify the system of reasoning with this closure operation. Such closure operations obey standard structural principles, including the Cut rule.

This is a problem for those who reject Cut in an attempt to solve semantic paradoxes or the paradoxes of vagueness. For example, suppose that we start by accepting that the successor of any small natural number is small and that 0 is small. We then deduce that 1 is small, and accept that too. Since we now accept that the successor of any small natural number is small and that 1 is small, we deduce that 2 is small, and accept that too, even though it does not follow from our original two postulates in a logic where Cut fails. We continue indefinitely, at each stage accepting only consequences in the Cut-free system of what we already accept. Evidently, we have not escaped the sorites paradox. In response to such difficulties,

David Ripley (2013, 2015), a friend of logics without Cut, admits that they are ill-suited to playing the closure role; he has other purposes in mind for them.

Another substructural strategy involves replacing sets of premises by *multisets*, in which the same sentence may occur many times. The idea is that using a premise many times may require many occurrences of that premise. This involves rejecting the rule of Contraction, on which one occurrence of a premise is as good as two. The paradoxes may then be blamed on illicit uses of Contraction. However, when we are investigating the consequences of a scientific hypothesis, 'How many occurrences of the hypothesis are we assuming?' is not a sensible question.

In general, substructural logics are ill-suited to acting as background logics for science.

## 2.4 Non-classical Mathematics

When non-classical logics are introduced to treat paradoxes of vagueness, semantic paradoxes, or the like, the aim is to provide a systematic framework for correct deductive reasoning with vague, semantic, or other vocabulary for which classical reasoning is supposed to break down. This should extend to a systematic framework for correct *mathematical* reasoning with the same vocabulary. As already argued, if the non-classical logic is needed at all, it is also needed for mathematical reasoning with those terms. If we could simply substitute them for variables in classical mathematical reasoning, there would have been no need for the non-classical logic in the first place. The lazy strategy does not work: mathematics must be redeveloped from scratch within the non-classical framework. Although there may be pockets within mathematics where everything behaves classically, on the non-classical view they cannot include mathematics for general applications.

What will non-classical mathematics be like? Obviously, that depends on the non-classical logic. As a case study, let us consider the prospects for the classical *least number principle* in a setting where excluded middle is supposed to fail for vague or semantic vocabulary.

A set-theoretic version of the least number principle says that every nonempty set of natural numbers has a least member. In practice, we often need to apply the principle to a complex predicate, to the effect that if the predicate applies to some natural number, there is a least natural number to which it applies. To obtain the predicate version from the set version, we need a set of exactly those things to which the predicate applies. For present purposes, we can ignore the set-theoretic digression, and focus directly on the predicate version. Classically, the least number principle can be simply proved from the principle of mathematical induction. But if excluded middle fails in case of vagueness, the least number principle is implausible for many vague predicates. For instance, it goes from the obvious premise that there is a large natural number to the highly problematic conclusion that there is a least large natural number—what is it?

In response to this problem, Hartry Field (2008: 100–1) proposes a watered-down version restricted to predicates which apply to some natural number and satisfy excluded middle for all smaller numbers. The point of the restriction is that in Field's framework it implies that the predicate behaves like a precise predicate from that number down, and since the predicate applies to that number, there is no need to look above it for the least satisfier.

The trouble with Field's restricted principle is that it is a one-off postulate in place of a mathematical *result*, proved from more basic principles. Consider any other theorem-schema of classical mathematics, formulated with predicate variables. Does the non-classical logician simply eyeball it, and make an educated guess at the minimal mutilation of it to escape counterexamples with vague or semantic predicates in place of the variables? Such an unsystematic, conjectural approach would fall far short of the standards of contemporary mathematics, which provides an enormous accumulating body of theorems ultimately derived from a very small group of first principles, such as the axioms of Zermelo-Fraenkel set theory with Choice and ur-elements. If mathematics were done in such an ad hoc spirit, one should have much less confidence in its results.

Alternatively, the non-classical logician might try to derive, within the favoured non-classical logic, some watering-down of all classical mathematics from a small group of first principles, presumably similar to those of classical mathematics but rendered invulnerable to counterexamples with vague or semantic predicates. That looks like an enormous undertaking, if done properly—and it is far from clear that it *can* be done properly.

The underivability of the least number principle from the rule of mathematical induction also demonstrates that the common practice of measuring the mathematical strength of theories of arithmetic by which instances of mathematical induction they prove is inadequate when the background logic is non-classical. For a theory may do very well by that standard, yet lack many classically trivial consequences of those instances, and so be near-useless for applications. In the non-classical setting, we must check that standard results over a much wider range are derivable before we can start to regard the theory as seriously applicable.

These difficulties with induction and the least number principle are just a foretaste of the challenges for the reconstruction of mathematics within a non-classical logic. One can expect the non-classical logician to respond wherever possible with local 'recovery' results to the effect that classical logic can be recovered for a specified class of formulas on condition that their constituents are 'well-behaved', for example by obeying excluded middle. For this strategy to be cogent, at least the initial recovery results must themselves be proved within the corresponding non-classical metalogic. For metalogic is itself a branch of mathematics; to use a classical metalogic at this stage would be to assume something of the very kind that was to be proved. That applies to the whole meta-logical background, including the numerical coding of formulas. It is doubtful whether the required non-classical proofs of recovery results will be available to non-classical logicians. More specifically, even if there is a classical proof that there is a proof in the relevant non-classical system of a result, that classical meta-theorem does not entitle non-classical logicians by their lights to rely on that result in their own inquiries, as opposed to using it *ad hominem* against their classical opponent. They need to display the non-classical proof itself, or at least to display a proof in their non-classical meta-logic that there is such a ground-level non-classical proof. That may be humanly impossible, for instance if the length of the corresponding non-classical proof grows exponentially with the length of the classical proof.

Suppose, however, that a recovery result can somehow be non-circularly obtained, and we use it to justify relying on classical mathematical reasoning in a given scientific explanation. This will involve adding an auxiliary premise to the explanation, to ensure that the relevant terms are well-behaved, for instance by obeying excluded middle. That might be assumed directly, just by listing the relevant instances of excluded middle on a case-by-case basis. However, the more extra premises one adds, the more the explanatory cost, just as *avoiding* reliance on a premise *adds* explanatory value, by making the explanation more elegant and economical. Alternatively, one might add both an extra major premise, to the effect that excluded middle holds for all precise, non-semantic terms, and an extra minor premise, to the effect that the terms at issue are precise and non-semantic. But such metalinguistic premises seem even more out of place in an ordinary natural scientific explanation of a non-linguistic state of affairs. They are not even partly why the state of affairs obtains. Thus the recovery strategy has a tendency to degrade ordinary explanations in natural science.

## 2.5 The Easy-Going Attitude

One sometimes encounters the easy-going attitude that classical logic is a good enough approximation for some purposes and not for others, just as physicists who accept Einsteinian special relativity theory may still treat Newtonian mechanics as sometimes but not always a good enough approximation. An advantage of this attitude is that it enables one to use classical logic when one judges it to be good enough, with no great pressure to specify the precise conditions under which it is good enough. One simply judges it to be good enough in ordinary mathematical contexts, but not good enough in the presence of paradoxes of vagueness or semantic paradoxes. That is compatible with several contrasting attitudes to the non-classical logic. Some may regard it as, unlike classical logic, correct without exceptions. Others may regard it as a better approximation than classical logic, but still only an approximation to genuine logical consequence. Still others may be undecided between those two attitudes.

The easy-going attitude is often used as an excuse for not developing an alternative mathematics on the basis of the preferred non-classical logic. But the excuse presupposes that the contexts in which one needs mathematics are disjoint from the contexts in which one needs non-classical logic. That presupposition is hard to motivate. Suppose, for example, that one takes the semantic paradoxes to require a non-classical logic. Versions of those paradoxes arise when one adds a truth predicate to Peano arithmetic, by the standard devices of Gödel numbering and diagonalization. Thus, in order to give a rigorous theory of truth in mathematics as a whole, the theorist's commitments require a development of mathematics on the basis of the preferred non-classical logic. Again, since quantum mechanics applies mathematics ubiquitously, those who propose non-distributive quantum logic as a serious alternative to classical logic are not excused from the need to reconstruct mathematics on the basis of their quantum logic.

Theorists who motivate a non-classical logic by appeal to paradoxes of vagueness may take themselves to escape the requirement, on the grounds that when applying mathematics in science one can *pretend* that the relevant non-mathematical terms are precise, because the gap between this pretence and reality is irrelevant for scientific purposes. But that argument is simplistic, for several reasons.

First, physics and other natural sciences are often in the business of making accurate quantitative predictions, sometimes with an expected error much less than one in a thousand. Vast sums of money may be spent on the basis of those predictions, as when a probe is sent to another planet. Theories are also tested by such predictions, as with a particle accelerator: then accuracy is needed for the value the quantity *would* take if the theory *were* true. Such predictions rest on complex mathematical reasoning and calculations, using standard classical mathematics. If there is vagueness in the relevant terms, and the true mathematics for vague terms is non-classical (because the correct logic for them is), then one would expect 'vagueness errors' in the predictions. Moreover, it would be unreasonable to expect the vagueness errors to take the form of random noise, normally distributed about the true value. Since the true non-classical mathematics will differ structurally from standard classical mathematics, the difference might easily result in errors of a more systematic kind, with some sort of structural bias. When scientists spend so much effort in meticulously controlling for other sources of error, should they not also make some attempt to control for vagueness errors? That would require a careful study of the nature of the divergences between classical mathematics and the true non-classical mathematics, which would in turn require the latter to have been developed. Here the easy-going attitude is not good enough.

Second, consider the application of probability theory to the beliefs and reasoning of uncertain agents who specify the relevant events in vague terms, as real agents do. Such vagueness is highly relevant to doxastic and evidential probability, because borderline cases result in characteristic uncertainty effects. If the true logic and mathematics of vagueness is non-classical, then any psychologically realistic account must take that into account. Probabilities are being assigned to 'vague events'. For reasons already explained, in such a setting non-classical theorists cannot simply help themselves to standard classical mathematics in the background. From their perspective, they must carefully justify any distinctive use of classical mathematics, on pain of distorting the very phenomena they are trying to understand. Here too the easy-going attitude is out of place.

Third, consider the use of mathematics in legal cases, for instance to analyse statistical evidence, as with DNA testing. In legal reasoning, such mathematical considerations interact with the imprecise terms in which the law is often cast and evidence is often given. If the true mathematics for such legal discourse is non-classical, that needs to be taken into account. When a verdict of innocence or guilt is at stake, the easy-going attitude is hardly appropriate.

Of course, in practice we seem to get by well enough in such cases without taking into account the possibility of vagueness errors in classical mathematics. But friends of a non-classical logic of vagueness should take no comfort from that fact, since it is some evidence that the logic of vagueness is classical after all (Williamson, 1994). It does not support the easy-going attitude.

Friends of non-classical logic sometimes assume that classical logic is simply refuted by paradoxes of one sort or another, and is no longer even a candidate for non-approximate correctness. But such crude falsificationism is methodologically naïve. Classical treatments are available for all the paradoxes used to motivate non-classical logics. Although some people find those treatments implausible, the issues are too subtle, abstract, and theoretical to be settled either way by such pre-theoretical judgments. Rather, in the long run they will be decided by an abductive comparison of the rival accounts (Williamson, 2013: 423–9, 2017). Since the accounts differ in their component logics, a major aspect of the comparison will naturally be the serviceability of those logics for mathematics and its applications in science, which is just where the easy-going attitude undermines non-classical logics. Since it invokes them only when the need is supposed to be most urgent, in the immediate presence of paradoxes, they lack the massive abductive support that accrues to classical logic in the vast range of non-paradoxical cases where explanations rely on standard classical mathematics. The effect of the easy-going attitude is to concede the abductive comparison to classical logic.

Pure mathematics is applicable to the world outside pure mathematics. As a result, in a non-classical world, pure mathematics is no safe haven for classical logic. Advocates of non-classical logic motivated by non-mathematical considerations have often underestimated the challenge of reproducing the success of classical mathematics within their framework. Over to you, Alan.

## References

Feferman, S. (2011). Is the continuum hypothesis a definite mathematical problem? In *Exploring the frontiers of incompleteness*. http://logic.harvard.edu/EFI_Feferman_IsCHdefinite.pdf

Field, H. (2008). *Saving truth from paradox*. Oxford University Press.

Hjortland, O. T. (2017). Anti-exceptionalism about logic. *Philosophical Studies, 174*, 631–658.

Parsons, T. (2000). *Indeterminate identity: Metaphysics and semantics*. Oxford University Press.

Ripley, D. (2013). Revising up: Strengthening classical logic in the face of paradox. *Philosophers' Imprint, 13*(5), 1–13.

Ripley, D. (2015). Contraction and closure. *Thought, 4*, 131–138.

Weir, A. (2010). *Truth through proof: A formalist foundation for mathematics*. Oxford University Press.

Williamson, T. (1994). *Vagueness*. Routledge.

Williamson, T. (2013). *Modal logic as metaphysics*. Oxford University Press.
Williamson, T. (2017). Semantic paradoxes and abductive methodology. In B. Armour-Garb (Ed.), *Reflections on the liar* (pp. 325–346). Oxford University Press.
Williamson, T. (2018). Alternative logics and applied mathematics. *Philosophical Issues, 28*, 399–424.

# Chapter 3
# Indeterminacy and Non-classical Logic

**Alan Weir**

**Abstract** This paper is a response to a challenge set by Timothy Williamson in his *Contribution* to this volume where he argues that, even leaving aside the difficulties non-classical logicians have if they wish to retain classical pure mathematics, these difficulties become well-nigh insuperable in the case of applied mathematics. I reply, on behalf of my own heterodox non-transitive logic. One novel aspect of the reply is letting assignments take *occurrences* of expressions as their domain, not expression types. On this basis, after presenting reasons for rejecting Williamson's epistemicism, I sketch a semantics for vague applied arithmetic including second-order Fregean arithmetic with Hume's Principle. I then argue that vagueness in at least some scientific contexts can be accommodated by treating it as a phenomenon which occurs 'inside' measurement uncertainty and that therefore the application of classical logic in these cases is legitimate, even if one espouses a non-classical theory of vagueness. Finally I argue that the non-classical account is to be preferred to the classical in cases such as the Sorites, particularly the special case of the Slippery Slope Sorites which does crop up in the social sciences.

**Keywords** Vagueness · Non-classical logic · Applied mathematics · Epistemicism

This volume is not intended to be one in which contributors discuss and criticise particular pieces of mine, and so there has been no need for a 'Replies to Critics' section. However at the end of his paper Tim Williamson writes (Williamson, this volume):

> Pure mathematics is applicable to the world outside pure mathematics. As a result, in a non-classical world, pure mathematics is no safe haven for classical logic. Advocates of non-classical logic motivated by non-mathematical considerations have often underestimated

---

A. Weir (✉)
University of Glasgow, Glasgow, UK
e-mail: Alan.Weir@glasgow.ac.uk

 
A. Rieger, S. Leuenberger (eds.), *Themes from Weir: A Celebration of the Philosophy of Alan Weir*, Synthese Library 484,
https://doi.org/10.1007/978-3-031-54557-3_3

the challenge of reproducing the success of classical mathematics within their framework. Over to you, Alan.

A philosophical gauntlet has been thrown down! This has fired me up, albeit with some trepidation in taking on such a powerful defender of classical logic, to attempt a reply on behalf of my version of non-classical logic. Since I take a holistic view of philosophical methodology I do not think it possible to defend adequately a view on, say, logic, without straying into some other philosophical compartments. But given my proclivity for straying into just about every other compartment, my response here will be conditional on the plausibility of a number of standpoints not directly on topic which will, for space reasons, be set out but not argued for. I attempt some defence of these views in a considerably longer version of this paper[1] which is online at: https://sites.google.com/site/weirysite/research-1

## 3.1 Logic

For a long time, starting in Weir, 1993, I have defended a non-classical logic I call 'neo-classical logic' (NCL); better, a family of logics since I have produced a few variants over the years. Here is the specification of logical consequence, for single conclusion logics (it is actually simpler and more symmetrical in the multiple conclusion case):

1. $\Delta, \varphi \vDash \psi$ iff for all models $\mathbb{M}$ and assignments $\sigma$

if $\mathbb{M}_\sigma[\theta] = 1$, for all $\theta \in \Delta$, then

if $\mathbb{M}_\sigma[\varphi] = 1$ then $\mathbb{M}_\sigma\ [\psi] = 1$ and if $\mathbb{M}_\sigma\ [\psi] = 0$ then $\mathbb{M}_\sigma[\varphi] = 0$[2]

A key idea here is that falsity-preservation upwards is as important as truth-preservation downwards.[3] Holding all premisses bar one true (value one), if it is true the conclusion is true, if the latter is false (value zero) so is that premiss. Of course in the context of classical bivalent semantics each direction entails the other. So the non-classical nature of NCL comes out in the specification of non-classical valuation spaces and the semantics for the connectives. I have long favoured the Lukasiewiczian or strong Kleene rules. Thus in Weir (2013*a*) I used the Lukasiewiczian framework, albeit using a different notion of logical consequence though one equivalent to the above in the trivalent case. In footnotes to that paper I acknowledged parallel ideas on consequence in Hartry Field's (2009) and in Dorothy Edgington's (1992) but applied, I said, to degrees of belief, not truth. I

[1] 'Vague Mathematics, Indeterminacy and Neo-Classical Logic', Weir, 2022a.

[2] $\psi$ is a logical truth, normally characterised by $\varnothing \vDash \psi$, is realised as $\top \vDash \psi$, where $\top$ is a constant of the language true in all models.

[3] Edgington (1993, p. 195) and Halbach and Horsten (2006, p. 690) concur.

should note, however, that I was mistaken about Edgington. In fact, in §III of her paper Edgington sought to generalise this to degrees of truth, contrary to what I said in footnote 14. This generalisation to truth was developed in greater detail in her important paper 'Vagueness by Degrees', Edgington (1997) which, alas, I failed to consider in my 2013*a* paper. *Mea Culpa*. This was a memory lapse. My memory resides, extended-mind style, mostly in emails and drafts stored on various hard drives and external drives. A rare burst of neural memory recently led to me locating an external drive and reviving notes I had made on 'Vagueness by Degrees', an email correspondence with Dorothy, and a draft book chapter on vagueness which included an attempt to respond to the objections she makes against truth-functional, degree-theoretic approaches. An unsuccessful response, I see, on revisiting it, perhaps a subconscious reason for the memory lapse!

My response now to Edgington[4] is:- capitulation! This is also my response to the challenges to multi-valued logic as a solution to vagueness to be found in Rosanna Keefe's observation (Keefe, 1998) that some vague concepts, such as 'fat', are 'multidimensional' in that intermediate truth grades are determined by a number of independent factors with no non-arbitrary way of imposing a linear order. My response is to move to a lattice semantics, the value space being a complete bounded distributive lattice $\mathbb{L}$ with a maximal element 1– true– and minimal 0 – false– though a compromise can be effected by including the unit interval [0,1] as part of the lattice. See Weir, 2022b, (Appendix to 2022*a*), §III for my reasons for this further relaxation and the full details of the lattice semantics.

A consequence of my account of consequence is that it is non-transitive. So, after initially giving Neil Tennant an incredulous stare when he tried to persuade me entailment is non-transitive, I have now come round to that position. However my restrictions on transitivity are less severe than his. They are implemented by putting determinacy constraints on certain natural deduction rules. For example, for ¬E the rule is:

| | | | |
|---|---|---|---|
| $\Gamma$ | (1) | $\varphi$ | Given |
| $\Delta$ | (2) | $\neg\varphi$ | Given |
| $\Sigma_{i\in I}$ | (3.*i*) | $D(\theta_i)$ | Given, $\forall\theta_i \in \Gamma \cap \Delta$ |
| $\Gamma, \Delta, \Sigma_{i,\, i\in I}$ | (4) | $\psi$ | 1,2 [3*i* $i \in I$], ¬E |

where, $\forall i$, $\theta_i \notin \Sigma_i$ and the operator D is used to express the *determinacy* of the embedded sentence.

[4] And also to Williamson, 1994, Chapter Four, §4.14.

This has the result that in NCL we have failure of Cut:

2. $A \wedge \neg A \vdash A$ and $A, A \wedge \neg A \vdash \bot$ but $A \wedge \neg A \nvdash \bot$.

though we can prove $D(A \wedge \neg A), A \wedge \neg A \vdash \bot$:

| | | | |
|---|---|---|---|
| 1 | (1) | $A \wedge \neg A$ | Given |
| 2 | (2) | $D(A \wedge \neg A)$ | Given |
| 1 | (3) | A | 1 $\wedge$E |
| 1 | (4) | $\neg A$ | 1 $\wedge$E |
| 1,2 | (5) | $\bot$ | 3,4 [2] $\neg$E |

The formal characterisation I give for determinacy is $D(\varphi) \equiv_{df.} \neg(\varphi \leftrightarrow \neg\varphi)$, with $\leftrightarrow$ the Lukasiewiczian biconditional (defined from the conditional in the usual way).[5] I'll call $D(\varphi)$ the *determination* of the wff $\varphi$. The result is that, with $\vdash_C$ being classical derivability, if $\Delta \vdash_C$ then $\Gamma, \Delta \vdash C$ for some $\Gamma$ all of whose members are determinacy claims of the form $\neg(\varphi \leftrightarrow \neg\varphi)$, $\varphi \in \Delta$; thus each member of $\Gamma$ is a classical theorem. Proofs can therefore be chained together, as required, albeit with the addition of some premisses. Unlike Tennant's *Core Logic* (Tennant, 2017), there is no exponential blow-up in converting a classical proof into a neo-classical proof from the same assumptions, plus determination claims. Again see Weir, 2022b, §IV for a full statement of the rules and for a soundness proof.

Now Stewart Shapiro will ask what do I think I am doing here (as he often does in many and varied contexts). In *Varieties of Logic* (Shapiro, 2014) he distinguishes context-relative versus more absolutist claims about the meanings of the logical operators, with a similar dichotomy with regards to the meaning of terms like 'valid' and 'logical consequence' (*op. cit.* Chap. 4.) In his terms, I have consistently argued for absolutism about the logical operators in the sense of there being a minimum core meaning to the terms across many 'logics' in special contexts such as mathematical reasoning.[6] On the other hand I accept that 'valid' has a lot of meanings and that even in the context of logic the idea of validity and consequence as relativised to a particular formal system– classical, intuitionistic, relevantist and so on, is a useful one. Nonetheless I think the above notion of consequence, in tandem with the absolutism about the operators, is particularly important in part because of my concern to block an ancient, but I believe still potent, argument for irrationalism (Weir, 1996b, 2000a, 2022b, §X). The rough dialectic of my counterargument is to accept the useful, if not entirely sharp or proof-architecture invariant, distinction between structural and operational rules. Among the operational rules some, I have argued, have the status of absolutely binding rules based on the minimal cross-framework meaning of the connectives;

[5] In the lattice semantics the 'semi-Lukasiewiczian' biconditional essentially has the same properties, for example the same consequence relation is sound under that semantics.

[6] Here I am close to the view Shapiro attributes to Greg Restall (Shapiro, 2014, p. 109); there is a minimal meaning logical particles have which is held in common among (most) rival logicians.

only if there are such rules can we defeat the Irrationalist. For the structural rules I advance the general principle that critical inquiry requires of us that we should adopt the strongest logic which never leads us from actually true premisses to a conclusion which is not actually true *nor* from an actually false conclusion to a non-false premiss. This package leads to the non-transitive NCL logic, assuming of course that bivalence fails.

## 3.2 Indeterminacy and Pure Mathematics

How could there be indeterminacy in pure mathematics? The intuitionists expound one way this can happen but I am unsympathetic to intuitionism. Another case is naïve set theory of which I am one of a small band of advocates, and one of the very few one who thinks it is not just non-trivial but also consistent. But let us also leave aside that rather extreme case. Indeterminacy and failure of bivalence can happen even in mainstream classical mathematics, if my 'neo-formalist' position, defended in *Truth Through Proof* (Weir, 2010) is right. On this account (at first pass) a mathematical sentence is true if there is a concrete proof of it, false if there is a concrete refutation. But won't it then be the case that Neil Tennant's example S:

$(2^{2^{2^{2^{2^{2}}}}} + 1)$ is prime.

(Tennant, 1997, p. 152)

will come out as lacking a truth value, since it is plausible there is neither a concrete, feasible proof nor disproof? Certainly for many sentences of $\Delta_0$ arithmetic that will be the case, and I would certainly take the failure of bivalence there as a reductio of any theory which led to it. My response, in the book (Chapter Six, §II) was to argue that the neo-formalist equation of truth with provability must be at the level of formal, *metamathematical* notions of proof and truth. From a neo-formalist perspective, metamathematics must be treated like any other part of mathematics. Tokens of metamathematical sentences, such as the claim that $\Delta_0$ arithmetic is negation-complete, are meaningful and true if concretely provable. And in fact we do have concrete proofs (or at least proof sketches) of this truth which enables the neo-formalist to affirm bivalence for that sector.

But this is a very restricted part of pure mathematics: what of the rest? To extend the account I rejected the common idea that formalised mathematical proofs must be finitary and subject to Gödelian constraints– that proofhood must be recursively decidable for example, if it is to formalise an epistemically acceptable concept of proof. This constraint is hard to justify given that almost all proofs and sentences of standard finitary PA have more symbols in them than there are elementary particles in the observable universe and so are utterly beyond human comprehension. Hence adding, for example, the $\omega$-rule and more generally expanding to infinitary languages with infinitely long expressions is an entirely legitimate idealisation of actual human practice. Not only do we have abbreviated concrete proof sketches of finitary proofs with e.g. $2^{1000}$ steps in them (cf. Weir, 2010, p. 243) we also have abbreviated concrete proof sketches of proofs, in ω-logic, of Gödel sentences for standard finitary logic: I gave an example which managed to fit into my finite book (*op. cit.* p. 251). The book then sketches an extension of this idea to set theory, although allowing that bivalence may well fail there, but harmlessly so. Thus for the neo-formalist, the General Continuum Hypothesis (GCH) is true in some frameworks, where it is an axiom for example, false in others just as Kant's '7+5 = 12' is true in ordinary arithmetic, false in arithmetic modulo 8. Where the GCH is neither provable nor refutable, e.g. in ZFC, it lacks a determinate truth value.[7]

## 3.3 Yet Another Semantics for Vagueness

But the cutting edge of Williamson's critique is the claim that non-classical logics cannot account for the success of classical mathematics in the sciences. In *Truth through Proof* my strategy for explaining the applicability of mathematics, a fatal problem for formalists according to Frege, was to appeal to conservative extension theorems, a strategy familiar from Field, though I argued fictionalists are not entitled to use these theorems. But in the book I considered only precise languages. Can the same strategy be applied to vague languages? First we need to set out a semantics for vagueness which fits in with the NCL approach, I'll call it *Neo-Classical Semantics for Vagueness*: NCSV.

I will start, as is common, by specifying what a model is. I will take a model $\mathbb{M}$ to be a fairly simple beast consisting, firstly, of a (non-empty) domain $\mathbb{D}$ of individuals. Secondly a model includes an interpretation $\mathbb{I}$, this being a set of assignments, each assignment $\sigma$ in $\mathbb{I}$ agreeing on constant terms. The language $\mathscr{L}$ I will take, largely for simplicity, to be a standard first-order one with identity and where its terms, individual and predicate, divide into constants and (singular) variables.

A distinctive aspect of NCSV is the characterisation of assignments. Assignments are *functions*, assigning a unique semantic value to each item in its domain, total over the variables but they can be partial over the constants. So, I depart from more standard views in taking each assignment $\sigma$ to assign members of $\mathbb{D}$ to constants as well as variables. If we stick to a single first-order domain of individuals then the obvious (though not inevitable) choice of semantic values for predicates are "extensions", that is, in the monadic case, members of the power set $\mathfrak{P}(\mathbb{D})$ of $\mathbb{D}$. A realist about properties and relations, among whose ranks I number, may want to work instead with a domain of relations or perhaps add a second order domain $\mathbb{D}^n_2$ of $n$-ary relations and so on up the higher-order ladder but again purely for simplicity and the utility of 'speaking to the Gentiles' I will stick with taking the values of the predicates to be sets of $n$-tuples of individuals.

I break with orthodoxy, however, by taking the domain of an assignment $\sigma$ (relative to a model $\mathbb{M}$ with domain $\mathbb{D}$) to include not only symbol types but also symbol *occurrences*. These I take to be pairs $\langle \tau, n \rangle$ consisting of a simple individual

[7] For a fuller defence see Weir, 2015 and 2022*a*.

symbol type τ– constant or variable–together with an occurrence index number $n$. The basic idea is as follows. For any atomic predication F^$s$ consisting of a predicate constant concatenated with a string $s$ of simple singular terms:

$$\mathbb{M}_\sigma \models \mathrm{F}{}^\wedge s \text{ iff } \langle \sigma(s_1, 1), \ldots, ..\sigma(s_n, n) \rangle \in \sigma(\mathrm{F})$$

$$\mathbb{M}_\sigma \Dashv \mathrm{F}{}^\wedge s \text{ otherwise.}$$

Read $\mathbb{M}_\sigma \models$ (F^$s$) as $\mathbb{M}_\sigma$ *satisfies* (F^$s$), $\mathbb{M}_\sigma \Dashv$ (F^$s$) as $\mathbb{M}_\sigma$ *dissatisfies* it.

Here $s_i$ is the $i$th term, $i \geq 1$, in the string $s$. For example, R*aba* is satisfied by σ iff $\langle\sigma(a,1), \sigma(b,2), \sigma(a,3)\rangle \in \sigma(\mathrm{R})$. Note that satisfaction of an atom by an assignment is entirely bivalent if we assume, as I will, a bivalent metalanguage. In particular, if $\sigma(s_i,i)$ is empty, for some $i$, then I stipulate that

$$\langle \sigma(s_1, 1), \ldots, ..\ \sigma(s_n, n) \rangle \in \sigma(\mathrm{F})$$

is dissatisfied by σ in the model.

Now we introduce the third and final element of our model, a two-placed function $\mathbb{V}_\sigma[\tau]$ whose inputs are pairs, first component an assignment σ, the second a simple symbol *type* τ subject to the following conditions:

(i) for every variable $x$, $\mathbb{V}_\sigma[x]$ is a unit set $\{\alpha\}$, $\alpha \in \mathbb{D}$;
(ii) for every individual constant $a$, $\mathbb{V}_\sigma[a] \subset \mathbb{D}$[8];
(iii) for every $n$-ary predicate constant F, $\mathbb{V}_\sigma[\mathrm{F}] \subseteq \mathrm{P}(\mathbb{D}^n)$, the power set of the $n$-ary product of $\mathbb{D}$[9];
(iv) for every assignment σ and symbol occurrence $\langle\tau,n\rangle$ on which σ is defined, $\sigma(\langle\tau,n\rangle) \in \mathbb{V}_\sigma[\tau]$;
(v) for every assignment σ, every individual constant $a$, every $n$ and for each $\alpha \in \mathbb{V}_\sigma[a]$ there is a *vague variant* $\sigma_v$ of σ with respect to $a$, which agrees everywhere with σ except (at most) that $\sigma_v(\langle a,n\rangle) = \alpha$.
(vi) for every assignment σ, every predicate constant F and for each $\mathrm{P} \in \mathbb{V}_\sigma[\mathrm{F}]$ there is a *vague variant* $\sigma_v$ of σ with respect to F, which agrees everywhere with σ except (at most) that $\sigma_v(\langle\mathrm{F}\rangle) = \mathrm{P}$.

A symbol type τ is *precise* relative to σ and $\mathbb{V}$ just in case its range $\mathbb{V}_\sigma[\tau]$ is a unit set, otherwise vague or empty. Note that the variables are all precise symbol types, precise terms. The $\mathbb{V}$ function gives us what we might call the *vague range* of each non-variable, the 'candidate' referents we might say of the singular constants, the

[8] Including the case $\mathbb{V}_\sigma[a] = \varnothing$ where $a$ is an empty term.

[9] Again we can have $\mathbb{V}_\sigma[\mathrm{F}] = \varnothing$, a case where F is an empty predicate term. The idea of an empty predicate is much more intuitive, however, when its denotation is distinguished from its extension, an empty predicate being one lacking denotation, and as a result thereof having an empty extension. But our eschewal of an ontology of properties and relations means we cannot represent this intuitive idea in this semantics.

candidate extensions of the predicates.[10] We need also to assume, as in standard models, that for every $\alpha \in \mathbb{D}$, every variable $x$ and every assignment $\sigma \in \mathbb{I}$, there is an assignment $\sigma[x/\alpha]$ which agrees with $\sigma$ except (at most) that it assigns, $\alpha$ to (each occurrence of) $x$. Standard set theory assures us of the existence of this supply of vague variants and $x$-variants of assignments enabling us to give the correct semantics for quantifiers.

A model $\mathbb{M}$, then, is a triple $\langle \mathbb{D}, \mathbb{I}, \mathbb{V} \rangle$. Indeterminacy is implemented via a supervaluationist account of truth relative to assignments where we supervalue *for atomic sentences only*. That is for atomic predication A:

$$\begin{aligned} \mathbb{M}_\sigma\,[\mathrm{A}] &= 1 \text{ if for all vague variants } \sigma_v \text{ of } \sigma, \mathbb{M}_{\sigma_v} \models \mathrm{A}; \\ &= 0 \text{ if for all vague variants } \sigma_v \text{ of } \sigma, \mathbb{M}_{\sigma_v} \mathrel{=\!\!|} \mathrm{A}; \\ &= i, i \in Int \text{ otherwise.} \end{aligned}$$

So the valuation space has a 'top' value, conventionally 1, corresponding to *being true*, a bottom value 0, *being false* and a set *Int* of intermediate values. In a simple trivalent case, $Int = \{i\}$, $i$ some value distinct from 0 and 1, a conventional choice being $i = ½$.[11]

The radical implications of NCSV can be seen in its treatment of identities of the form $a = a$. I will interpret the identity predicate as a precise one: $\mathbb{V}_\sigma[=] = \{\langle \alpha,\alpha \rangle : \alpha \in \mathbb{D}\}$ for every $\sigma$.[12] Nonetheless if $a$ is imprecise in $\mathbb{M}$, in particular if $\mathbb{V}[a]$ is neither a singleton nor $\varnothing$, the sentence (in the trivalent system) takes $\mathbb{M}$-value ½. There will be an assignment $\sigma$ such that $\sigma(a,1) = \sigma(a,2) = \alpha$ so that $\mathbb{M}_\sigma \models a = a$[13] and a vague variant $\sigma_v$ such that $\mathbb{M}_{\sigma_v} \mathrel{=\!\!|} a = a$, because $\sigma_v(a,1) = \alpha$, $\sigma_v(a,2) = \beta$, $\alpha \neq \beta$. Any sentence $x = x$, $x$ a variable will, however come out as true in each assignment $\sigma$ since it will be true in each vague variant simply because each occurrence of $x$ is assigned the same value.[14]

Thus where the better known supervaluationist accounts start from partial, gappy valuations and generate gappy supervaluations by generalising over all

---

[10] This heterodox move, of allowing different occurrences of the same expression to have different referents, may be criticised as a "mere technical trick". Certainly there is a high element of artefact about it; but all model-theoretic semantics has a strong element of artefactuality. One could note that the idea of not just different occurrences but different tokens of an expression having different referents may have a highly representational employment. But that is more obviously the case for context-dependence rather than vagueness, and I am not advocating an assimilation of the two. The role of this 'technical trick' here, rather, is to show the coherence of combining 'metaphysical bivalentism' (see ahead) with non-classical logic for vague expressions.

[11] For complications re empty terms, see again Weir, 2022a, §III.

[12] Here I am naughtily using the same sign '=' in the object language and metalanguage.

[13] In the official dialect, this will be $= aa$.

[14] A somewhat more cumbersome 'quasi-bivalent' semantics for identity sentences is given in 2022*b*, §IV; it is of interest because it delivers a compensating smoothness in the treatment of the quantifier rules in free logic settings.

completions,[15] here I am starting with complete bivalent valuations and generating gappiness via the idea of vague variants.

As regards the logical operators, an obvious choice given the NCSV theory, and one I have worked on in the past, is the continuum-valued Lukasiewiczian degree-theoretic extension of the strong Kleene tables to include the conditional I need for my determinacy concept. After capitulation to Edgington and Keefe this needs to be amended to 'a semi-Lukasiewiczian' account (Weir, 2022a, §IV) but, as shown in 2022*b*, §IV, this makes no important difference to the proof theory of the logic.

This semantics allows for the law of excluded middle (LEM) to fail, in that some instances are not determinately true.[16] If $Fa$ is neither satisfied nor dissatisfied by $\sigma$, if different vague variants of $\sigma$ come out differently, then both it and its negation get indeterminate values and the strong Kleene semantic clauses for $\vee$ and $\neg$ ensure that so too does $Fa \vee \neg Fa$.[17]

But how, then, to meet Williamson's challenge to a realist, *metaphysical bivalentist* such as myself who upholds classical laws such as the law of identity and bivalence with respect to property and relation instantiation:

$$\models \forall x\,(x = x) \quad \text{and} \quad \models \forall x \forall y\,(x \,\varepsilon\, y \vee \neg\,(x \,\varepsilon\, y))$$

where $\varepsilon$ is a primitive of a first-order theory of properties, $x \,\varepsilon\, y$ reading ‹$x$ instantiates $y$›? How can I do that whilst espousing a non-classical logic where theses such as excluded middle are not logical truths?

## 3.4 Quantifier Rules

Here the treatment of variables as precise whereas constants need not be is the solution to the problem, but one which of course requires non-classical quantifier rules, ∀E and ∃I in particular. Being much more favourable to free logic than Williamson (see his 2013, pp. 150ff) I restrict them in a fairly standard free logic manner, but with the addition of the same overlap restrictions needed in NCL for the

[15] But cf. Shapiro, 2006, pp. 79–80, who does not require completion, thus leading to a semantics which, consonant with his pluralism, is both more congenial to intuitionistic logic but allows for a 'classical recapture' for a restricted fragment of his language.

[16] Note, though, it is never determinately false. We should not say that some instances of LEM are neither true nor false, not if we reject dialetheism and uphold the De Morgan equivalences, as I do. This is a little similar to the intuitionists studied neutrality on instances of LEM with undecidable disjuncts, but disagreement is sharp when we come to schematic generalisations such as $\neg\forall x(Fx \vee \neg Fx)$ some instances of which (in analysis for example) are determinately true according to intuitionists, none according to me.

[17] A crucial clause here is one specifying that a disjunction is only true (value one) if at least one disjunct is. Edgington's semantics, based on probability theory in the framework of classical logic, lacks this clause. I defend my more restrictive semantics in 2022*b* §III.

rules $\vee$E, $\rightarrow$E, $\neg$E and $\rightarrow$I; thus a fairly uniform, not an ad hoc, treatment. The usual free logic restriction is to add a premiss specifying that the term for introduction or elimination is non-empty; in this case we need to add an extra condition specifying, in effect, that it is precise. A brute way to do this is to add a predicate constant E! subject to the (bivalent) semantic condition that in every admissible model

$$\mathbb{M}_\sigma\,[\mathrm{E!}(a)] = 1 \text{ iff } \exists\alpha\,(\mathbb{V}\,[a] = \{\alpha\}\,.$$

$$\mathbb{M}_\sigma\,[\mathrm{E!}(a)] = 0 \text{ otherwise.}$$

and I will do so here.[18] The modified rules are:

| | | | |
|---|---|---|---|
| $\Gamma$ | (1) | $\forall x\varphi x$ | Given |
| $\Delta$ | (2) | E! $(a)$ | Given |
| $\Sigma_i, i \in I$ | (3.$i$) | D($\psi_i$) | $\forall\psi_i \in \Gamma \cap \Delta$ |
| $\Gamma, \Delta, \Sigma_{i,\, i \in I}$ | (4) | $\varphi(x/a)$ | 1,2 [3.$i$] $\forall$E |

| | | | |
|---|---|---|---|
| $\Gamma$ | (1) | $\varphi(x/a)$ | Given |
| $\Delta$ | (2) | E! $(a)$ | Given |
| $\Sigma_i, i \in I$ | (3.$i$) | D($\psi_i$) | $\forall\psi_i \in \Gamma \cap \Delta$ |
| $\Gamma, \Delta, \Sigma_{i,\, i \in I}$ | (4) | $\exists x\varphi x$ | 1,2 [3.$i$] $\exists$I |

where, as usual, $\psi_i \notin \Sigma_i$. $\varphi(x/a)$ is a plural metalinguistic parameter. Substituends refer, possibly multiply, to any result of substituting occurrences of *a for one or more* occurrences of substituends for $x$ which are free for $x$ in $\varphi$.

These rules are neo-classically sound in the semantics modified to allow for empty terms, as are the following amended rules for identity:

=I

| | | | |
|---|---|---|---|
| $\Gamma$ | (1) | E! $(a)$ | Given |
| $\Gamma$ | (2) | $a = a$ | 1, =I |

[18] We can, however, also use the Williamsonian $\exists x(x = a)$ as a precise existence predicate if we utilise the 'quasi-bivalent' semantics for identity given in 2022*b*, §IV.

=E:

| | | | |
|---|---|---|---|
| $\Gamma$ | (1) | $\varphi(x := a)$ | Given |
| $\Delta_1$ | (2) | E! $(a)$ | Given |
| $\Delta_2$ | (3) | E! $(b)$ | Given |
| $\Lambda$ | (4) | $a = b$ | Given |
| $\Sigma_i, i \in I$ | (5.$i$) | $D(\psi_i)$ | $\forall\psi_i \in \cap[\Gamma/\Delta_1/\Delta_2/\Lambda]$ |
| $\Gamma, \Delta_1, \Delta_2, \Lambda, \Sigma_{i,\, i \in I}$ | (6) | $\varphi(x := b)$ | 1,2, 3, 4 [5.$i$] = E |

where $\psi_i \notin \Sigma_i$, $i \in I$ and ‹$\in \cap[\Gamma/\Delta_1/\Delta_2/\Lambda]$› abbreviates the disjunction of all pairs of intersection claims involving distinct occurrences of $\Gamma, \Delta_1$, $\Delta_2$ and $\Lambda$. Here $\varphi(x := a)$ is the result of substituting a term $a$ new to $\varphi$ for *all* free occurrences of $x$ in $\varphi$.

Finally to complete the NCL presentation of quantificational logic with identity we need an existence axiom schema:

E! Axiom:

| | | | |
|---|---|---|---|
| $\varnothing$ | (1) | E!$x$ | E!xistence |

We thus have $\vdash x = x$ (and so from $\forall$I, which is the standard classical rule in NCL,[19] $\vdash \forall x(x = x)$).

The upshot is that I can consistently maintain a realist, *metaphysically bivalent* stance, reasoning classically about reality when using precise language, whilst refusing to affirm classical theorems, for example excluded middle or reflexivity or transitivity of identity, when the instances contain imprecise constants, unless that is, appropriate additional determinacy assumptions are made.

## 3.5 Why I Am Not an Epistemicist

But why bother with all this semantic and proof-theoretic complexity, complexity which results in a significantly weaker logic than the classical logic we are familiar with and use (it is widely believed) in mathematics and the mathematical sciences? Why not just adopt epistemicism and classical logic? Isn't this the simpler, and more elegant, course? Moreover one key widespread argument in favour of an indeterminist response to the phenomenon of vagueness has come under severe challenge from Tim Williamson. This indeterminist argument appeals to the limitations of our cognitive faculties, at the most basic level our perceptual

[19] I am not countenancing universally free logic, for which see Tennant, 1978, pp. 167–8.

faculties. How could it be that the extension of 'hairy' includes someone with, say, 90,001 hairs but not someone with 90,000 when we could never tell the difference between someone in the first category and someone in the second (not without very elaborate methods which are no part of our normal practice with observational terms)? The words we utter have the meanings we finite, limited beings (wee, sleekit or otherwise) imbue them with in our linguistic practices.[20]

As Williamson points out, the same argument could be used against a trivalent semantics in which 'hairy' has a tripartite extension which partitions exclusively and exhaustively into a sharp extension, a sharp anti-extension and a sharp set of 'indeterminates' in between. We are incapable of distinguishing, by observation alone, between someone who is determinately hairy, just and no more, and someone who just misses that mark. The argument generalises to *n*-valued extensions and logics; continuum-valued logics fail even more spectacularly the test of dove-tailing neatly with linguistic practice.

This counter-argument I find compelling against the blunt form of pro-indeterminacy argument it is directed against. But it does not lead me to embrace epistemicism. We need also a positive account of how our grasp of language, of predicate terms for example, can be such that bivalence holds. Failure to do so is not a conclusive reason to reject epistemicism, to be sure. Our minds may be complex enough to enable us to have such grasp but not complex enough for us to figure out what it is or how it comes about. Or maybe the nihilist is right and there is simply nothing there for a systematic theory to capture. But I am temperamentally disinclined towards such defeatism. Wittgenstein is the arch-nihilist but I find worrying echoes of it in Williamson:

> The epistemic theory of vagueness makes the connection between meaning and use no harder to understand than it already is. At worst, there may be no account to be had, beyond a few vague salutary remarks. Meaning may supervene on use in an unsurveyably chaotic way. 1994, p. 209

The issue is not so much unsurveyability: even if we had a comprehensive record of the total corpus of a community's linguistic doings (a spooky scenario) we may well be incapable of formulating, let alone verifying, an explanatory account of how linguistic meaning is determined by these facts. The issue is rather whether we have reason to believe that there are any determinate semantic facts, knowable or unknowable, ('facts' in the lightweight sense of 'true propositions') and if so, how detailed they are, how fine-grained are the semantic inter-relations such as synonymy and conceptual-dependence.

I take Williamson to be a sceptic, there isn't much by way of such determinate semantic structure to discover or even wonder about. Quine, of course, argues against semantic determinacy, independently of considerations to do with vagueness. I differ from both. Of course we are very far from having an explanatory theory

[20] This argument also convinced Williamson initially, see his review of Roy Sorensen's *Blindspots*, Williamson, (1990), p. 138. But by the time of *Vagueness* (Williamson, 1994) he had gone over to the Epistemicist camp, see the preface p. xi.

of meaning (Weir, 2003*a*, sets out my vision of how any such should proceed). But one feature which will be present in any plausible formal semantics is a division between the *artefactual* elements of the semantics and the *representational*. So here I am very much in agreement with Shapiro, 2006, pp. 50ff. The distinction is fairly clear for physical theories. If, for example, we use a coordinate geometry then many aspects – choice of point of origin, orientation of axes, unit of measurement of the metric and so forth – are 'conventional', in the usual phrase, not intrinsic; they do not represent mind-independent aspects of reality. Other features, perhaps certain invariant features over transformations of the geometrical frameworks, are assumed to do so.[21]

What could the equivalent be for the theory of meaning? Here is where I am still heavily influenced by Quine: the representational aspects of the theory are those which denote or depict *verbal behaviour*. Though Quine focuses on quite a narrow range in the query and assent methodology which is part of his stimulus meaning theory, he is clear that a wide range of utterances, of any grammatical mood, count as verbal behaviour and a wide range responses to them too. I would also include 'internal' verbal behaviour in the 'mind's eye and ear'.

If these sorts of actual events were the only external reality constraining formal semantics a very radical indeterminacy looms, though a different type of indeterminacy from that which features in vagueness. In the first, radical, case we have many distinct *conflicting* interpretations equally compatible, it is claimed, with the 'intrinsic' facts which determine correctness. In the latter case, an interpretation can assign truth values other than (determinately) true and (determinately) false but there need be no claim that there is a plurality of determinately incompatible but equally correct interpretations assigning different values.[22]

Radical indeterminacy is inevitable if we restrict the factors which determine correctness of interpretation to actual utterances and responses to them. Appeal to counterfactuals greatly reduces semantic indeterminacy. In particular it demolishes ontological relativity in the extreme 'proxy function' form embraced by Davidson, Putnam and, at times, Quine. That this thesis was taken seriously is testament to the widespread tendency of philosophers when driving on certain terrain (here philosophy of language) to choose vehicles utterly lacking in either a brake or a reverse – aka modus tollens – gear. As a result they hurtle unstoppably over the Cliff of Absurdity unable to drop a hugely implausible assumption which is driving the argument (here that the extremely minimal constraints in the proxy function argument exhaust all the objective constraints on interpretation).

---

[21] Other approaches, such as Cartan's coordinate-free approaches, screen out many artefactual non-covariant aspects, but they can never be entirely eliminated.

[22] Supervaluational approaches to vagueness may seem to blur this distinction for here an interpretation consists in a multiplicity of 'sub-interpretations' – 'precisifications' – with truth values determined by generalisations over them all. But the precisifications are not competing with each other over what the correct truth value is but are components in a procedure for determining a single truth value (or semantic value more generally) for an expression. And the supervaluationist is at liberty to declare that exactly one supervaluational theory is correct for any given object language.

Let us leave radical indeterminacy for another day[23] and assume for the purposes of argument that the facts concerning actual verbal behaviour and the 'counterfacts' about what would be said or what reactions would ensue are the intrinsic facts to which any acceptable interpretation of a speaker, or community of speakers, must conform. What are the implications for the debate between epistemicist, classical approaches to vagueness versus non-classical indeterminist ones?

I note firstly the phenomenon of *epistemic diffidence* – widespread agnosticism, or absence of full-on belief or disbelief by many speakers with respect to many propositions. Not all cognitive agents are diffident.[24] One sometimes has the misfortune to be sitting on a bus or train across from a blowhard who is convinced that all in earshot are absolutely desperate to hear his (for in my experience blowhards are usually male), utterly definite opinions on the most various of subjects, few of which he has any knowledge of at all. However if we stipulate blowhards more narrowly to be people who entertain full-on belief or full-on disbelief about every proposition, never holding any to a grade or degree intermediate between zero or one, and who affirm or reject any proposition before them without any qualification – if that is what we mean by 'blowhard' – then there are very few, if any, blowhards about.

So the verbal behaviour of speakers concerning utterances which express those propositions will be tripartite, at least. In the Quinean picture, there will be both assent, dissent and neither – agnosticism. Further divisions are imaginable, perhaps we can make sense of degrees of assent, manifested, for example, as one is force-marched down a Sorites sequence of imperceptibly balder and balder humans and asked if they are hairy. To be sure assent/dissent reactions to utterances are just a tiny fragment of verbal behaviour even with regard to highly empirical terms. Moreover we must include not only *snapshot dispositions* (2010, pp. 49–53) to affirm or reject, assent or dissent, prompted by current stimulation but, most importantly, also *correctional* dispositions where, for example we revise our judgement that that was a rabbit, back there, just then (Weir, 2003*a*). The 'corrected verdict' to an observational sentence cued to a particular place and time is not a single response given in special ('ideal') circumstances but the distillation of many responses, both snapshot and revisionary– 'no that wasn't a rabbit after all'– with the latter (counterfactually) given at places and times around the location to which the snapshot response was prompted.

Even taking 'verbal behaviour' thus widely will not eliminate indeterminacy of the type associated with vagueness for observational predicates (and here I leave aside ontological relativity). There will be borderline cases where we would hesitate to give any settled verdict, even in optimal circumstances and even after reflecting on snapshot judgements (or the lack thereof) from other perspectives and angles. Moreover even a idealised meta-theorist able to 'survey' all the counterfacts about

---

[23] Weir, 2006*b* and 2013*a* provides more material for gluttons for punishment.

[24] It is an interesting question, which I leave for future work, whether one can write off the non-diffident as irrational, on the basis for example of standard theories of rationality, such as decision theory based on maximising expected utility.

speakers correctional dispositions would not, in all cases, find a single verdict emerging as the dominant one. The limitations of our perceptual powers which so many philosophers appealed to in support of indeterminacy of truth value in vagueness do certainly show that non-blowhards do not have dispositions which would fix on settled, corrected assent or settled, corrected dissent in every case.

But since I have acknowledged there is bound to be an element of artefactuality in formal semantics why not settle such indeterminacy by an act of bivalent, as it were, stipulation, acknowledging that the boundary thereby decreed is an artefact, is in a certain sense conventional, with the boundary placed in a central position in the 'no-man's land' in between corrected assent and corrected dissent? The reason is that the element of artefactuality can only legitimately occur in regions where there is freedom of manoeuvre, where the objective facts to which the semantics must answer do not force only one choice. Any semantics which does not place someone with 150,000 head hairs in the positive extension of 'hairy', or does place someone with 500 hairs in that extension, *is incorrect*, clashes with the behavioural facts which determine meaning. But any formal semantics which places an individual for which the corrected verdict would settle on agnosticism in the positive or the negative extension i*s also objectively incorrect*. The areas where there is leeway are the two boundaries, for example individuals, in a Sorites line from the hirsute to the severely follicly challenged, who are borderline in the sense that, to take the example of the positive borderline, the speakers' dispositions are 'mixed', the responses to their being hairy being assent in some circumstances, agnosticism in others, or perhaps there is no clear disposition. The existence of these kinds of dispositions might be tested by repeated questioning with respect to individuals in the line who are unchanged in relevant respects, and getting a mix of assent and dissent (borderline) or fairly steady agnosticism (indeterminate). Similarly there will be borderline cases at the other, negative, dissent, edge. There will be a number of individuals in each such border region. Where *exactly* we settle the border between the positive extension and the rest is an arbitrary matter, not fixed by the behavioural facts, likewise for the border with the negative extension. If there are finer grades of assent or affirmation or similar linguistic acts then more than two boundary points may emerge, and so an $n$-ary semantics $n > 2$ (with continuum semantics an idealisation, used for smoothness as the continuum is in some cases in physics, e.g. fluid mechanics). But away from their borderline regions it is objectively incorrect to draw a boundary where the corrected verdicts are uniformly agnostic.

What of higher-order indeterminacy? Certainly the concepts of 'correct interpretation', 'provoking a mix of assent and dissent' and 'generating steady agnosticism' are themselves vague meta-theoretic concepts. If we feel the need, we can investigate these in a meta-meta-theory where we would be looking at semantic theorists' linguistic behaviour with respect to phrases such as 'mix of assent and dissent', and 'steady agnosticism'. We might also find there, while looking to 'distilled' corrected verdicts (where we wean out some verdicts because they occur too infrequently, or in far from optimal situations and so on) cases of steady agnosticism with respect to semantic verdicts, including steady agnosticism with respect to the phrase 'steady agnosticism'. Here again an element of artefact will come in with respect to setting precise higher-order boundaries in the formal semantics. The regress which emerges is not a vicious one. Metaphysical bivalentists have compelling reason to believe that we can ascend up a chain of every more precise meta$^n$languages with a Citadel of Precision as the asymptotic ideal limit.[25]

It might be objected that this whole picture smacks of verificationism and Dummett-inspired anti-realism.[26] Heaven forfend. Quine's behaviouristic approach to language does indeed to lead to very similar anti-realist conclusions. But only, I have long argued, if one adopts *atomistic* behaviourism. So here *is* a Dummettian idea I accept, the *molecularity* of language (Dummett, 1976, p. 79), something Quine does partially accept, in his verdict matrix theory, but alas only to a very limited degree. Grasp of language consists in a well-founded hierarchy of linguistic dispositions winding down to dispositions specifiable in purely behavioural terms at the bottom. The whole structure of correctional dispositions presupposes such molecularity. But it does *not* entail 'luminosity'. Even for highly observational predicates like 'hairy', or 'round' or 'crimson', the account does not entail that in every circumstance it is possible for speakers to know whether the predicate applies or not. For example, if the correct application of F at *p*, *t* in speaker S's idiolect is determined in large part by what S would say about (F at *p,t*) at a range of nearby places and times (verdicts weighted by such factors as proximity to *p*, *t*), it will not be possible to know what that is by, *per impossible*, simultaneously occupying all these positions and then somehow summing them all up, at *p*, *t*. (For more on molecularity and realism, see Weir, 1985*b*, 1986*b*, 2003*a*, 2022*a*).

## 3.6 The Application of Mathematics

It is generally assumed that the mathematical sciences (perhaps with some exceptions in computer science) use classical logic as their logical basis. Williamson thus challenges non-classicists to explain how the mathematics used in science can be applied when the logical framework is non-classical. This seems to be an especially difficult challenge for a formalist such as myself, the inability to account for the application of mathematics to science being one of Frege's major charges against formalism.

> it is applicability alone which elevates arithmetic from a game to the rank of a science. (§91 in Frege, 1903/1980 p. 167.)

The account, from a neo-formalism perspective, of the application of mathematics in Weir, (2010) Chapter Five, pays no attention to vagueness. However I believe the account presented there can be carried forward essentially unchanged, using the

---

[25] For the Citadel, see Williamson, 1994, p. 169. He criticises the very idea but this is in the context of an attack on nihilists regarding the concept of vagueness, among whose ranks I certainly do not belong.

[26] See again Williamson, 1994, p. xi.

NCSV model theory presented above rather than, as in the book, standard model theory.

I refer to that chapter for a detailed defence of my position. I appealed to conservative extension theorems which show, roughly speaking, that one cannot prove anything 'empirical' about an 'empirical' sub-domain using mathematics which could not be proven from empirical premisses alone. Thus the following bridge principle, more exactly set of *numerical equivalence* schemata, can be shown to be conservative, under the interpretation of n$x$ as a numerical operator. That is, n$x\varphi x$ is assigned, in every admissible model, an element representing the cardinality of the extension of $\varphi$, which may include non-mathematical instances.

$$\begin{aligned} &\exists_0 x \varphi x \leftrightarrow \mathrm{n}x\,(\varphi x) = 0 \\ &\exists_1 x \varphi x \leftrightarrow \mathrm{n}x\,(\varphi x) = \mathrm{S}0 \\ &\ldots \quad \ldots \quad \ldots \\ &\exists_n x \varphi x \leftrightarrow \mathrm{n}x\,(\varphi x) = \mathrm{S}^n 0. \end{aligned}$$

$\exists_0 x\varphi x$ abbreviates $\neg\exists x\varphi x$ and $\exists_n x\varphi x$ can be schematically defined by:[27]

$$\begin{aligned} &\exists x_1, \ldots \exists x_n \Big[ (\varphi x_1 \wedge \ldots \varphi x_n) \wedge \\ &(x_1 \neq x_2 \wedge \ldots x_1 \neq x_n \wedge x_2 \neq x_3 \wedge \ldots x_{n-1} \neq x_n) \wedge \\ &\forall y\,(\varphi y \rightarrow (y = x_1 \vee \ldots y = x_n)) \Big] \end{aligned}$$

Two further bridge principles which can also be shown to be conservative – adding them to a non-mathematical theory T never leads to non-mathematical conclusions not provable from T alone – are:

Addition:

$$\frac{\neg\exists x\,(\varphi x \wedge \psi x)\,;\ \mathrm{n}x\varphi x = \mathrm{S}^n 0;\ \mathrm{n}x\psi x = \mathrm{S}^m 0}{\mathrm{n}x\,(\varphi x \vee \psi x) = \mathrm{S}^n 0 + \mathrm{S}^m 0}$$

Multiplication:[28]

$$\frac{\mathrm{n}x\varphi x = \mathrm{S}^n 0;\ \forall x\,(\varphi x \rightarrow \mathrm{n}y\,(\mathrm{R}xy) = \mathrm{S}^m 0)\,;\ \mathrm{Func}^{-1}\,(\mathrm{R})}{\mathrm{n}y\,(\exists x\,(\varphi x \wedge \mathrm{R}xy) = \mathrm{S}^n 0 \times \mathrm{S}^m 0)}$$

[27] In the book I defined $\exists_{n+1} z\varphi z$ more succinctly by $\exists z(\varphi z \wedge \exists_n y(\varphi y \wedge y \neq z))$ with $z$ new in $\varphi$. The long-winded formulation makes the conservativeness proof easier. Here 'S' denotes the successor operator of Peano arithmetic, ‹$\mathrm{S}^n 0$› the zero term prefaced by $n$ occurrences of S.

[28] In this schema, substituends for R behave like second-order free variables: 'de-schematicised' in second-order logic, we would generalise over R. So it has to be treated here, in its first-order truncation, as a precise predicate. See Weir, 2022b, §VII.

The neo-formalist, I argued, can account for the applicability of some, at least, elementary parts of arithmetic by appeal to conservative extension theorems. But that was working with the idealisation of a fully precise language. What if we move to a vague language, interpreted by the NCSV semantics: do the same bridge principles come out as true in these models; are they still conservative over empirical language? Bluntly put: yes! See 2022*a* for the details. But some detail here, in particular on the interpretation of the numerical operator will, I hope, help clarify how the NSCV semantics can be extended to cover mathematics.

When we add term-forming operators such as $\mathrm{n}x$ to our language the object language now is a union $\mathscr{L}_\omega = \bigcup_{i\in\omega} \mathscr{L}_i$ with $\mathscr{L}_0$ a base language $\mathscr{L}$, in our case containing the language of pure arithmetic with primitives 0, S, + or $\times$, and with $\mathscr{L}_{i+1}$ generated from $\mathscr{L}_i$ by expanding the stock of singular terms by adding $\mathrm{n}x\varphi x$, for each open sentence $\varphi x \in \mathscr{L}_i$. Each purely arithmetic term occurring in $\mathscr{L}$ gets a precise interpretation. The extension function of a predicate $\varphi x$ (relative to an assignment which fixes the value of all the other variables) at language $\mathscr{L}_i$ – $\mathrm{Ext}^i_\sigma[\varphi x](\ldots)$ – is divided into three components (non-exclusive): i) the truth range, the objects in $\mathbb{D}$ mapped to true, to value one; ii) the indeterminate range, those objects mapped to a value in *Int*, the values other than 1 or 0. And the positive range, the union of the previous two ranges.

Where the $\mathrm{Ext}^i_\sigma[\varphi x](\ldots)$ function is binary, mapping into {0,1}, and of finite size $k$, then $\mathbb{V}_\sigma[\mathrm{n}x(\varphi x)] = \{k\}$; where the truth range is infinite $\mathbb{V}_\sigma[\mathrm{n}x(\varphi x)] = \{\infty\}$, $\infty$ some element of $\mathbb{D}$ chosen to index any extension of infinite size. For non-binary $\mathrm{Ext}^i_\sigma[\varphi x](\ldots)$ with finite truth range, the value range for $nx(\varphi x)$ at $\mathscr{L}_{i+1}$ is a consecutive set of numbers starting from a lower bound $k$, the size of the truth range, upper bound the size of the positive range, $\infty$ if that is infinite. The general principles of NCSV give us that $\mathbb{M}_\sigma[\mathrm{n}x\varphi x = \mathrm{S}^k0]$ is 1 if $\mathbb{V}_\sigma[\mathrm{n}x(\varphi x)] = \{k\}$; is 0 if $k \notin \mathbb{V}_\sigma[\mathrm{n}x(\varphi x)]$ and is indeterminate otherwise. We can fix the precise indeterminate value at $\mathscr{L}_{i+1}$ by setting it equal to $\mathbb{M}^i_\sigma[\exists! n\varphi]$ so that

$$\mathbb{M}^{i+1}_\sigma\left[\mathrm{n}x\varphi x = \mathrm{S}^n0\right] = \mathbb{M}^i_\sigma[\exists_n x\varphi]$$

takes value one.[29] With this semantics all three bridge principles come out as true in the vague language (2022*b*, §§V–VII).

What of identities of the form $\mathrm{n}x\varphi x = \mathrm{n}x\psi x$? The obvious principle governing these is Hume's Principle, in its usual second-order form:

$$\mathrm{n}x\varphi x = \mathrm{n}x\psi x \leftrightarrow \exists R\,[1-1\,(R, \varphi, \psi)]$$

[29] This semantics also fixes the value of each identity $\mathrm{n}x\varphi x = t$ for purely arithmetic $t$ since for each $t$ there is some $k$ such that $t = \mathrm{S}^k0$ is true in each (admissible) model; and indeed true identities among pure arithmetic terms are all provable, false ones disprovable, in very weak arithmetic theories such as Q.

where the right hand side expresses the existence of a one:one correspondence between the $\varphi$s and the $\psi$s, something which can be formalised in a number of well-known ways.

We can extend the semantics for n$x$ further to ensure the truth of all instances of Hume's principle. As well as extending our semantics to second-order logic, a complication arises in setting the value range $\mathbb{V}[\mathrm{n}x\varphi x]$ where the positive range is infinite. For example, if both the truth and indeterminate ranges are of cardinality $\aleph_\alpha$ then the cardinality of the positive range will also be $\aleph_\alpha$. Without modification of our policy we would get $\mathbb{V}[\mathrm{n}x\varphi x] = \{\aleph_\alpha\}$ even though $\varphi$ is vague, and we need $\mathbb{V}[\mathrm{n}x\varphi x]$ to be a plural set in that circumstance. The solution is to use ordinal, rather than cardinal numbers (of course this comes to the same thing in the finite case) in the specification of value ranges for numerical terms to secure the validity of Hume's Principle. (See 2022*b*, Appendix, §VIII.)

Having provided this semantics for the bridge principles, the conservativeness result for the application of the arithmetic theory can proceed in essentially the same way as in the bivalent case. This on its own does not settle how chains of inferences (I include calculations as inferences) will fare even if those bridge principles can be appealed to, since the neo-classicist has, it seems, to apply the determinacy restrictions, for example on the identity rules, whenever using vague terms such as 'the number of grains of gold in that heap'. One strategy of response features in the next section where I look a little beyond arithmetic.

## 3.7 Magnitudes

For so far we have dealt with rather simple mathematical applications in vague languages. As I noted in Weir, 2010, Chapter Five, there will be no general recipe. Each case of application needs to be looked at in detail, with particular attention to what the plausible philosophical account of the relevant branch of science is. But Williamson challenges non-classicists to account for the use of mathematics to make detailed predictions (and explanatory retrodictions) in the sciences. What can the neo-classicist, in particular, say in response?

To make progress, we can look at *magnitudes* of physical qualities, for example of qualities such as mass, charge, distance and energy. These I take to be physical entities though often represented mathematically by scalars such as natural, real or complex numbers.[30] Magnitudes are not first-order properties which in turn are

[30] Some use 'magnitude' and 'quality' interchangeably, and use 'degree' or 'intensity' or, in non-continuous cases, 'grade' for what I am calling magnitude. And some use 'quantity' for what I am calling 'quality'. My terminology is influenced by the use of 'magnitude' in many contexts with the connotation of size or extent. So I see *having mass* as a quality possessed by all massive objects equally but the magnitude (or quantity, or degree if you like) of the quality differs from one instance to another.

instances of *qualities*; for qualities are not second-order properties.[31] A dumbbell is both an instance of the quality *having mass* and an instance of a particular magnitude of that quality, 4 kg say. So the metaphysics of magnitudes and qualities, their structural interrelationships for example, is quite a complex matter.[32] But the exact nature of magnitudes does not matter for present purposes, what matters is the *precision* of expressions representing magnitudes. (However thinking of them as first-order properties, not extensions, at least for the sake of the argument, probably makes the line of argument here easier to follow.)

Developments in technology and physics have led not only to ever more accurate measurement of magnitudes but to ever more precise definitions/explications. No longer are units of measurement always defined in terms of prototype objects, such as the Parisian standard metre, a platinum bar, or the International Prototype Kilogram, aka the 'Le Grand K', a ball of platinum and iridium in *Le Bureau International des Poids et Mesures* in Parc Saint-Cloud, Sèvres, Paris, although the latter remained the standard until surprisingly recently, 2018 in fact. Mass is now defined in terms of the Planck constant, the metre as $299{,}792{,}458^{-1}$ of the distance light travels in a vacuum in a second; the latter is defined in terms of another 'invariant of nature', the caesium frequency $\Delta\nu_{Cs}$ and so on. If the background theories here are correct (which of course we cannot be certain of) then terms built up from these constant terms using precise mathematical terms are themselves precise expressions designating physical magnitudes. Absolute precision is not confined to mathematics.

Of course even if the scientific theory is right, our measurements in any given case will be subject to measurement uncertainty which one might think is a separate matter from vagueness; but see ahead. At any rate vagueness, even if greatly reduced in fundamental physics, is still present not only in the historical language of previous scientists, which an adequate theory of vagueness must account for; it is present in other sciences and indeed in practical applications of fairly fundamental sciences. 'The mass in kg of asteroid 52 Europa' or 'the distance in metres between 52 Europa and 511 Davida' (both relative to a time) inherit some vagueness from the vagueness of the physical boundaries, centres of mass etc., of the asteroids in question.

Let us consider the simple case of a quality such as distance or mass characterised in the historic fashion via prototypes; let us say the terms express *prototype concepts* in these cases. One way of approaching the concept of distance or mass is via congruence relations, e.g. $(x,y \text{ Cong } z,w)$. In the case of spatial distance this might be read along the lines of: ‹rod with end points $x$ and $y$ is spatially congruent with rod with end points $w$ and $z$›. An empiricist explanation of those relations will advert to such things as placing rods end to end and judging whether the ends coincide, concatenating congruent rods and placing the result against another, placing two objects in scales and judging whether they balance and so forth. This is the tradition

[31] Here I mean 'second-order' in the logicians' sense, not that common in philosophy of mind in which it means 'impredicative first-order property'.

[32] On which I expatiate in 'Multigrade Metaphysics', Weir, 2024.

of the representational theory of measurement, as developed by philosophers such as Patrick Suppes (Suppes, 1957, Chapter Twelve). One looks for axioms governing the empirical operations and a formal applied mathematical theory in which one can prove the existence of an interesting isomorphism between mathematical structures and empirical structures satisfying the axioms, one which generates mathematical values for magnitudes.

Now I do not want to embrace this sort of representationalism *tout court*. Critics have pointed out that a great deal of measurement is much more indirect and sophisticated than the sort which features in the cruder empiricist versions of representationalism; nor does the explanation of the success of a measuring system always require postulating isomorphisms, or even homomorphisms between mathematical and physical structures. And I certainly do not want to embrace the empiricist idea that all qualities and magnitudes are in some sense directly empirical observable.

For it is clearly nonsense to claim that 'this iron ball has mass 0.5 kg' ever *meant* anything like:

> were we to transport it to Sèvres and place it on a pan along with an exactly similar ball in a (well-calibrated) balance with Le Grand K in the other pan then the scales would balance

no matter how many qualifications we add, to do with the mass not changing in transport etc. etc. Not to mention the fact that these qualifications and concepts such as 'well-calibrated' would introduce a vicious circularity into any supposed analysis anyway.[33] If mass is an objectively existing quality then, it is safe to conclude, ‹*x* is the same mass as *y*› stands for on objectively existing relation and two objects are correctly said to be the same mass just when they bear that relation to each other, whether or not they can be brought into close contact with one another without changing mass, and whether or not their bearing that relation can be accurately measured by us. I reject operationalism about theoretical qualities and magnitudes in other words.

Nonetheless I do believe the representationalist picture yields at least a sizeable part of the true story about measurement and more generally the application of mathematics to physical reality. And it gives us a good grip on how a plausible, formal, non-bivalentist semantics can be given for some magnitudes as conceptualised in early times when more primitive theoretical and practical methods were all that was available – and when, as a result, there was more vagueness in scientific language.

Indeed I claim any account of how human beings can come to grasp concepts which pick out objective congruence and equivalence relations, applying as they do to a huge variety of pairs of objects across space and time, must nonetheless start from a grasp of primitive observational concepts, largely subterranean concepts buried at the base of the complex molecular hierarchy of more theoretical concepts. Our grasp of the primitive concepts, down in the deeper strata of semantic

[33] I doubt there is any better account of what the sentence *means*, what its informational content is, other than 'this iron ball has mass 0.5 kg'.

geology, is constituted *in part* by the sort of behaviour which verificationists and operationalists appeal to.

How does vagueness come in? Let's take mass as our example, and imagine a nineteenth century context, of balances and spring scales and so forth. This is certainly not a theory-free zone: evaluation of measurement results will be corrected in the light of then current theory: heat expansion, variable effects of the gravitational field and so forth. Testing of scales will involve taking repeated measurements in presumed constant environments in order to look at the spread of results and calculate variance; it will include testing weights against standard weights, perhaps even an all expenses paid trip to Paris for a confrontation with Le Grand K. For instruments believed to be well-calibrated, the spread of results may well approximate a well-known distribution, most notably a normal distribution.[34] This will partly confirm the conjecture that the instrument is well-calibrated and suggest the statistic to use on the data to generate the measured value, often the mean. These results can also be used to back up counterfactual claims about what the pattern of results would be in the circumstances featuring in correctional dispositions – more epistemically optimal situations, a larger number of tests and so forth. Plausibly the approximation to a normal distribution will have smaller standard deviation in such situations.

This suggests the following semantics for magnitudes of mass, as the latter was understood then. The semantics is advanced from our twenty-first century citadel of partial precision; I will use italics for the more precise concepts of the metalanguage – *mass*, *kilogram*, *kg* – which the metatheorist is to use to give a semantics for the older vaguer, unitalicised concepts expressed by 'mass', 'kilogram' and so on.

To motivate the semantics we engage in a thought experiment, imagining how linguists venturing out from the citadel would go about investigating the cruder concept at large in the community under study. We imagine that they investigate speakers' responses to attributions of mass to standardised sample objects and weighing apparatuses, focussing in particular on what they take to be best practice, winnowing out competence failures and sub-optimal conditions. More than that (and in contrast, typically, with community members seeking to estimate measurement error) they take into consideration *correctional dispositions*, deselecting from the corpus those positive responses which are later revised when judging from other standpoints or after reviewing in the light of peer commentary; perhaps switching round positive and negative responses or changing agnostic ones to full affirmation or rejection, also in the light of correctional behaviour. They end up with a chart displaying the range of *masses* of standard test objects to which subjects give a positive response to 'is of mass 1kg', weighted by the proportion of times

[34] Though not always: a number of distributions are important in measurement theory, including the uniform distribution, bimodal and certain types of skewed distributions. But this simple widespread case serves to illustrate the problem and develop the solution in a way which I do not think is biased at the outset in favour of my neo-classical approach.

community members respond to that mass positively, and perhaps also by some measure of the degree of assurance of response, filtering out responses which do not stand up after correction.

We further imagine that, if they do a thorough and extensive study, they find a distribution of results which looks roughly normally distributed around *mass* $1+\delta$ *kg* (where $|\delta|$ is small relative to 1). So the highest number (perhaps weighted number) of corrected affirmations that a sample $x$ satisfies 'mass = 1kg' is given to objects of mass $1+\delta$, with proportions dropping off roughly normally on either side. To simplify let us suppose that the response level hits zero the same distance on either side.

The next step, as usual in theories involving messy phenomena such as human behaviour, is to go for a smooth mathematical idealisation: our imagined Metatheorists of the Citadel extrapolate as smoothly as seems reasonable from their chart to a normal probability distribution function $f(x) =$

$$\frac{1}{\sqrt{2\pi\sigma}} e^{-[x-(1+\delta)/\sigma]^2/2}$$

mean $1+\delta$, standard deviation $\sigma$, $x$ the *mass*, $f$ yielding the probability of a (corrected) positive response to 'its mass is 1kg' (or ‹the $\text{mass}_{\text{kg}}$ of $x = 1$› more formally) in standardised circumstances. The smaller the standard deviation the linguists extract from best practice, the less vague the object language concept is, at least at the 'determinately true' end; the smaller the absolute value $|\delta|$ is, the closer their older prototype concept of 1 kg is to our precise (or more precise) concept of 1 *kg*.

If this thought experiment is a plausible (highly idealised) sketch of how an adequate interpretation of the prototype mass term should proceed, a reasonable semantics which emerges is:

$$\begin{aligned}&\mathbb{V}\left[\langle\text{the mass}_{\text{kg}}\text{ of } x = 1\rangle\right] = \\ &\left\{m : \exists y\left[y \in (1+\delta+\varepsilon, 1+\delta-\varepsilon) \wedge \forall x^{35}\left(\text{mass}_{\text{kg}}(x) = y \text{ iff } x\ \varepsilon\ m^{36}\right)\right]\right\}\end{aligned}$$

[35] But the existential quantifier $\exists x$ would do just as well here because when a magnitude is indexed by a number, relative to a particular measuring system, all instances of the magnitude are indexed by the same number.

[36] ‹$x\ \varepsilon\ m$› reads ‹$x$ is an instance of magnitude $m$›.

where [1+δ+ε, 1+δ–ε] is a 100γ% confidence interval[37] for distribution *f*, that is:

$$\int_{1+\delta-\varepsilon}^{1+\delta+\varepsilon} \frac{1}{\sqrt{2\pi}\,\sigma} e^{-[x-(1+\delta)/\sigma]^2/2} dx = \gamma.$$

The expression ‹the mass$_{\mathrm{kg}}$ of $x = 1$›, then, like any vague expression, has a spread of candidate values, in this case these are *magnitudes*, the values of the variable '*m*' above, a subspecies of the quality *mass*, as captured in our more precise metalanguage.

If γ is 0.95, 95% of corrected positive responses, in optimal situations etc., would be given to magnitudes within a range of $\pm\varepsilon$ of the central value 1+δ according to the linguists' idealised counterfactual extrapolation from their data. The concept 'having mass 1kg' in the object language thus picks out not a single magnitude, but a range, just as other imprecise terms do according to NCSV theory.

What particular degrees (or grades) of truth should we assign objects, given these data? One suggestion: for any objects with mass above or below the $\gamma = 0.95$ $\delta\pm\varepsilon$ range, the semantics assigns value zero to ‹the mass$_{\mathrm{kg}}$ of $x = 1$› whilst the condition for $\mathbb{M}_\sigma$[the mass$_{\mathrm{kg}}$ of $x = 1] = 1$ is that σ(*x*) have *mass* 1+δ *kg*. But on reflection, since the concept is not fully precise and the choice of δ itself just an approximation from limited data, we would want to allow masses very close to 1+δ to count as determinately of mass 1 kg. We can achieve that by relaxing a little our third clause in the semantics for atomic sentences by allowing that where the satisfaction status – satisfied/dissatisfied – varies across vague variants we let the precise truth value be *any* member of the valuation space $\mathbb{L}$, and do not restrict always to *Int*, that is $\mathbb{L}\backslash\{0,1\}$. In some cases, an extremal value can occur even when not determined by supervaluation. Furthermore in the case of magnitude terms, I suggest all the truth values will occur in [0,1], even in a wider lattice-theoretic valuation space.

Our clause for assigning degrees of truth in atomic ascriptions of magnitude value might then, I suggest, invoke a monotone decreasing function *i*:

$\mathbb{M}_\sigma$[the mass$_{\mathrm{kg}}$ of $x = 1] \geq i(\xi)$ where $mass(x) \in (1+\delta) \pm \zeta$, for $\zeta \leq \varepsilon$ and

$$\int_{1+\delta-\varsigma}^{1+\delta+\varsigma} \frac{1}{\sqrt{2\pi}\,\sigma} e^{-[x-(1+\delta)/\sigma]^2/2} dx = \xi$$

[37] The use, or rather abuse, of confidence intervals is currently a hot topic in statistical methodology. I am very far from thinking we can make uncritical appeal to them. Indeed in 2022*b*, §X, I use this debate as an example of philosophy usefully intervening in scientific practice: philosophical critiques or revisions of the logic used in mathematical and scientific practice are on a par with current critiques in the philosophy of statistics of widespread statistical (mal)practice. Such critiques are not to be ruled out by a principle of extreme deference to scientists but evaluated in the light of detailed and informed consideration and debate.

Thus suppose we choose a 95% confidence interval for the value $\gamma$ which generates upper and lower bounds of $\pm\varepsilon$ round our empirically determined central value $1 + \delta$ beyond which ‹the mass$_{\text{kg}}$ of $x = 1$› is determinately false. And suppose bound $\pm\zeta$ yields $\xi = 0.90$ and so gives us a narrower 90% confidence interval. Then the degree of truth of ‹the mass$_{\text{kg}}$ of $x = 1$› for objects with *mass* in the range $(1 + \delta) \pm \zeta$ will be $\geq i(0.90)$. Setting $i(\xi)$ to 0 for $\xi \geq 0.95$, means taking $\varepsilon$ to be the (sharp!) cut-off where determinate falsehood starts; $i(0.90)$ will be somewhere in (0,1). If we set $i(\xi)$ to 1, $\xi \leq 0.1$ we allow ourselves to make determinately true claims about mass other than in the very rare, possibly vanishingly rare, circumstances of an object whose mass is right on the mark.

But where are our Metatheorists of the Citadel to get the value $\gamma$ from? How do they determine the shape of $i$? There is clearly a fair amount of arbitrariness about this procedure. As noted earlier in agreement with Shapiro (2006, pp. 50ff.) semantic theory, like theory in general, is a mix of the artefactual – the conventional and stipulative – and representation (accurate or inaccurate) of external objective reality. The sharpness of the various boundaries, between objects definitely lacking mass 1 kg and the rest, between objects having mass 1 kg to degree $\leq \frac{1}{\sqrt{2}}$ and the rest etc., all this is artifice. No such boundaries exist in mind-independent reality, they only come into existence as and when stipulated by metatheorists. Indeed since the Metatheorists of the Citadel are in fact pure fiction, the sharp boundaries are thus, in a sense, doubly unreal. It is easy, then, on NCSV semantics, to explain how at least some vague boundaries are unknowable: there is nothing there to know. But when laid down, they are not completely arbitrary. Some ways of drawing the boundaries will be incorrect others (more than one though) correct. 'Correct' is itself indeterminate in the metatheory, but the regress here is not at all vicious.

The idea of using a normal distribution of semantic values for magnitude terms is at the heart of my response to the final objection from Williamson which I will now attempt to address.

## 3.8 Vague Predictions

Williamson writes:

> physics and other natural sciences are often in the business of making accurate quantitative predictions, sometimes with an expected error much less than one in a thousand. Vast sums of money may be spent on the basis of those predictions, as when a probe is sent to another planet. Theories are also tested by such predictions, as with a particle accelerator: then accuracy is needed for the value the quantity *would* take if the theory *were* true. Such predictions rest on complex mathematical reasoning and calculations, using standard classical mathematics. If there is vagueness in the relevant terms, and the true mathematics for vague terms is non-classical (because the correct logic for them is), then one would expect 'vagueness errors' in the predictions. Moreover, it would be unreasonable to expect the vagueness errors to take the form of random noise, normally distributed about the true value. Since the true non-classical mathematics will differ structurally from standard classical mathematics, the difference might easily result in errors of a more systematic kind,

> with some sort of structural bias. When scientists spend so much effort in meticulously controlling for other sources of error, should they not also make some attempt to control for vagueness errors? That would require a careful study of the nature of the divergences between classical mathematics and the true non-classical mathematics, which would in turn require the latter to have been developed. Here the easy-going attitude is not good enough. (Williamson, this volume p. 54.)

My strategy of response here will be obvious from the vague semantics for magnitude terms I set out in the previous section. Firstly I take the structures of pure mathematics to be just the classical ones (naïve set theory aside). Secondly I take measurement uncertainty in these cases properly to include vagueness spread. Vagueness errors are encompassed in measurement errors because the correct theory of vagueness of magnitude terms will take the values of the magnitudes which determine intermediate values for the truth of an ascription of a magnitude to be normally distributed around the magnitude which determines value one. If NCSV theory is correct, it is entirely reasonable, pace Williamson, to expect the vagueness errors to take the form of random noise, normally distributed about the true value.

Consider the example of predicting asteroid positions in the future from current data. This will not be possible if the system is currently in a chaotic state. But if not, then the astronomer will plug in values for the position (of the centre of mass, perhaps) of an asteroid and its momentum together with other relevant bodies – the sun and planets, nearby large asteroids – plug these into, let us assume, a deterministic theory. This might take the form of a function yielding future position and momentum, the theory perhaps an approximation, given the unsolvability in general of the $n$-body problem, $n > 2$, from the approximately true Newtonian mechanics. But the scientist will also allow for measurement error. With a powerful enough computer to hand, the scientist can plug in a range of values for the solar bodies all inside the estimated measurement errors and get the computer to predict for each selection the position of the asteroid. These initial values might give us a convex subspace of a phase space. If the system is non-chaotic the output range for the asteroid, the measurement error for the asteroid's future position and momentum, will not be much larger than the input over the desired time period, the sub-space of final values not much greater in volume.

What if the scientist has been persuaded to adopt neo-classical logic? She should reason classically, using a classically programmed computer for the task, if she thinks all potential determinacy assumptions which might be needed in the neo-classical rules are true. But while, as noted, there are empirical claims which are as precise as mathematical ones, particularly as science moves beyond prototype concepts, this is not the case in the current example.

However on the NCSV semantics our scientist, at least if she is a metaphysical bivalentist, can step back and reason *classically*[38] in the metalanguage about the reasoning, whether by paper and pencil or by the computer, involved in predicting

[38] Or better, neo-classically but adding the needed determination claims to yield the same conclusions as classical logic.

future values for the asteroid from historic ones. If the semantics for vagueness sketched above for mass applies to location and momentum terms each will be presumed to have (in a specific context) a spread of candidate values. If this spread is a sub-space inside the measurement error subspace then reasoning classically in the metatheory, as a metaphysical bivalentist is entitled to do, the neo-classicist will conclude that the final state will be not a single point inside the measurement error sub-space, but a final 'vague' sub-space also inside that same measurement sub-space. The same practical conclusions can be drawn.

Here, though, I have assumed that the vagueness space determined by corrected verdicts is inside the measurement space. But corrected verdicts not only delete some snapshot verdicts, they can also add or reverse verdicts, turn negative into positive for example. Might we not find that, looking at a table of actual estimates of the value of a magnitude, when subjects come to revise and correct their estimates many of the corrected ones fall well outside the chosen confidence interval, perhaps well into the tails of the distribution as determined by theorists looking only to estimate measurement error and not paying much attention to correction and revision? If so, the vagueness sub-space may not nest inside the error space.

This is indeed possible. Take a dramatic case. We have a non-prototype magnitude concept based on some supposed physical constant. But our theory is wrong, the constant actually varies periodically but, by chance, all our measurements of the magnitude so far have occurred during periods in which the "constant" on which we partly base the measurement takes the same fixed value. Later measurements outside those periods confront us with the choice of either assuming the magnitude varies drastically, doubles or halves periodically say, or else dropping the assumption of a constant parameter. The latter may be the smoothest course. In that case it is not so much that the corrected verdicts determine a vagueness space which is not inside the error space; rather the whole concept has to be abandoned.

This is an extreme case; there may well be intermediate cases in between the unthreatening one – vagueness a subspecies of measurement error – and the dramatic one. Firmer conclusions could only be drawn after detailed work in non-classical measurement theory. For now, my claim is that, for at least some cases of predictions in vague languages, the neo-classicist is entitled to draw exactly the same conclusions as the classicist.

## 3.9 Conclusion

This raises a different objection: many of my attempts to explain how neo-classical logic and semantics for vague terms can account for the application of maths in a vague language might seem to take an overly conciliatory line. Neo-classical logic collapses into classical logic by the addition of determinacy constraints; or we simply reason classically in the metatheory, appealing to metaphysical bivalence. Are there any areas (outside naïve set theory and naïve semantics) in which classical logic is too strong, where the needed determinacy constraints do not apply, so that to reason validly we need to restrict ourselves to neo-classical logic?

I think there are such regions of discourse, those where the Sorites makes its appearance for example. That might seem an unpromising example – there is no science of 'Heapology'. But consider some 'softer' sciences such as history and jurisprudence, or more practical aspects of law and medicine:- specifying test criteria for when a condition is present for example, and consider a special type of Sorites, Slippery Slope Sorites, which can crop up when reasoning about such things as legal age boundaries or determining rules of thumb for interpreting key elements of legislation (cf. Edgington, 1997). For those reasoning neo-classically, we must distinguish the 'Binary Sorites', where all the conditional premisses are wrapped up in a single conjunction or universal generalisation, from the 'Plural One' where we have a plurality of conditional premisses. Sorites reasoning is valid in the plural case in neo-classical just as in classical logic. But for the former, without the assumption that the 'inductive premiss' is determinately true, the Sorites reasoning is neo-classically invalid. Will, then, neo-classicists reach different conclusions from the classicist, will the two sides reason differently and as a result draw different practical conclusions? I tentatively suggest (Weir, 2022a §XII, XIII) that they will, using a thought experiment concerning a Twin Earth where there are two nations, the Logicean People's Republic (LPR), where the educated are taught classical logic, and the Democratic Republic of Logicea (DRL), which brings students up in a neo-classical environment. (All Hail the DRL!)[39]

What about reasoning in philosophical papers such as this? Here too I think neo-classical logic ought to be applied, as per the DRL, although the sense in which 'ought' is used here requires careful elucidation. How would we check for this? Could the idea of automated proof-checking be exported from mathematics, where there are widely agreed (though hard to specify) standards of informal rigour, to philosophy? Possibly, in limited areas: see my 2022*a* §XIII.

But all this takes us far from the immediate task of answering the Williamsonian challenge. That is actually an entire project, in the intersection of philosophy of science, logic and mathematics, which may reach different conclusions in different areas. Here I have only touched on a few basic topics – the neo-classical coherence of metaphysical bivalence, elementary applied arithmetic and some speculations on logic for vague magnitude terms. I hope I have done enough to convince some that the prospects for the neo-classicist are not as bleak as Tim believes.

[39] As it happens, the two countries are neighbours with a tense DeLogicised zone in between them in which people in transit are only permitted to use a very restricted relevant logic, currently FDE (first-degree entailment) though fierce battles have broken out in the past as to which should be used. To this day, there is still no agreement on which logic to use in order to settle that dispute.

## Bibliography

Dummett, M. (1976). What is a theory of meaning II. In G. Evans & J. McDowell (Eds.), *Truth and meaning*. Oxford University Press.
Edgington, D. (1992). Validity, uncertainty and vagueness. *Analysis, 52*, 193–204.
Edgington, D. (1993). Wright and Sainsbury on higher-order vagueness. *Analysis, 53*, 193–200.
Edgington, D. (1997). Vagueness by degrees. In R. Keefe & P. Smith (Eds.), *Vagueness: A reader* (pp. 294–316). MIT Press.
Field, H. (2009). *What is the normative role of logic* (pp. 251–268). Proceedings of the Aristotelian Society.
Frege, G. (1903/1980). Frege against the formalists: Vol II of Frege 1903, §§86–137. In M. Black & P. Geach (Eds.), *Translations from the philosophical writings of Gottlob Frege* (3rd ed.). Blackwell.
Halbach, V., & Horsten, L. (2006). Axiomatizing Kripke's theory of truth. *Journal of Symbolic Logic, 12*, 97–122.
Keefe, R. (1998). Vagueness by numbers. *Mind, 107*, 565–580.
Shapiro, S. (2006). *Vagueness in context*. Oxford University Press.
Shapiro, S. (2014). *Varieties of logic*. Oxford University Press.
Suppes, P. (1957). *Introduction to logic*. Van Nostrand.
Tennant, N. (1978). *Natural Logic*. Edinburgh University Press.
Tennant, N. (1997). *The taming of the true*. Clarendon Press.
Tennant, N. (2017). *Core Logic*. Oxford University Press.
Weir, A. (2022a). *Vague mathematics, Indeterminacy and Neo-Classical Logic*. https://sites.google.com/site/weirysite/research-1
Weir, A. (2022b). *Appendix to 2022a*. https://sites.google.com/site/weirysite/research-1
Weir, A. (2024). *Multigrade metaphysics*. https://sites.google.com/site/weirysite/research-1. For the references to Weir's published work, see the separate bibliography.
Williamson, T. (1990). Review of *Blindspots* by Roy Sorensen. *Mind, XCIX*, 137–140.
Williamson, T. (1994). *Vagueness*. Routledge.
Williamson, T. (2013). *Modal logic as metaphysics*. Oxford University Press.
Williamson, T. (this volume). Chapter 2: Can non-classical logic treat mathematics as exceptional? In A. Rieger & S. Leuenberger (Eds.), *Themes from Weir: A celebration of the philosophy of Alan Weir*. Springer.

# Chapter 4
# Neoclassical Logic, Cloisonnist Vagueness, Nonobjectual Truth

**Elia Zardini**

**Abstract** In a series of works, Alan Weir has suggested that the paradoxes are due not so much to the operational principles of classical logic, but to some of its structural principles. Weir's insightful suggestion has opened up the programme of developing a system that, while validating the principles of naive truth (or of naive vagueness), preserves the classical essence of the logical operations and

---

It is for me a huge pleasure and honour to contribute to a volume celebrating Alan's life and work. I first came across Alan when I was a PhD student in St Andrews: he came over from Glasgow to give a talk on his formalist theory of mathematics and I remember being blown away (not only by his jokes but also) by his bold acceptance of infinitary methods (whose influence is evident e.g. in my Zardini 2015c). In the same period, I read his Weir (1998a) and I remember being blown away (not only by his style but also) by his bold denial of structural principles (whose influence is evident e.g. in my Zardini 2008a). However, our real friendship began some years later, when I contacted him asking how he would deal with (a) a certain version of Curry's paradox featuring as consequent (b) the claim that Real Madrid will win the next Champions League. That query spurred a very long and entertaining exchange that developed in many directions, not all of them related to item (a). Through the vicissitudes of our lives, that exchange is still ongoing. Earlier versions of the material in this paper were presented in 2018 at the Workshop *Themes from Alan Weir* (University of Glasgow) and in 2022 at the Workshop *Indeterminacy and Paradoxes* (Complutense University of Madrid). I'd like to thank both audiences for very stimulating comments and discussions. Special thanks go to Marianna Antonutti, Víctor Aranda, Ludovica Conti, John Divers, Shimpei Endo, Stephan Krämer, Mary Leng, James Levine, Alex Miller, Sergi Oms, Stewart Shapiro, Timothy Williamson and an anonymous referee. Especially special thanks to Alan for illuminating conversations throughout all these years that have helped me to understand better both the philosophical motivations and the technical details of his system, as well as for extraordinarily generous and helpful comments on earlier versions of the material in the paper. Needless to say, I bear the sole responsibility for any remaining misrepresentation of the system in the paper (and for the tortuous title). (That said, should the reader detect any Scottish-nationalist tendencies in the paper, she will know their source.) I'm also grateful to the editors Stephan Leuenberger and Adam Rieger for inviting me to contribute to this volume and for their support and patience throughout the process. Work on the paper was supported by the Ramón y Cajal Research Fellowship RYC-2017-22883. Additionally, support from the Basic Research Program of the National Research University Higher School of Economics is gratefully acknowledged. I also benefited from the FCT Project PTDC/FER-FIL/28442/2017 *Companion to Analytic Philosophy 2*, from the Project PID2019-105746GB-I00 of the Spanish Ministry of Science and Innovation *Linguistic Relativity and Experimental Philosophy*, from the Project 2019PIDPID-107667GB-I00 of the Spanish

 
A. Rieger, S. Leuenberger (eds.), *Themes from Weir: A Celebration of the Philosophy of Alan Weir*, Synthese Library 484,
https://doi.org/10.1007/978-3-031-54557-3_4

remains coherent by denying instead, in a philosophically motivated way, some of the classical structural principles. This paper first presents Weir's logical system and argues that it does not adequately preserve operational classicism. The paper then presents Weir's nontransitive system of vagueness and argues that it does not adequately preserve naive vagueness. The paper finally presents Weir's nontransitive system of truth and argues that it does not adequately preserve naive truth. As a contrast, the paper mentions how these shortcomings are overcome by my favoured nontransitive system of vagueness and by my favoured noncontractive system of truth.

**Keywords** Classical logic · Contraction of logical consequence · Naive truth · Naive vagueness · Transitivity of logical consequence

## 4.1 Paradoxes and Structural Principles

Alan Weir's pioneering work on *nontransitive* logics (Weir 1993; 1996; 1998a; 1998b; 1999a; 1999b; 2004; 2005; 2015)[1] has mainly revolved around the paradoxes of "*selfreference*" (for want of a better term, see Yablo 1985, p. 340). Among these, historically and theoretically prominent are the *semantic* paradoxes; to fix ideas, without loss of generality, I'll focus on them (as well as, to a somewhat lesser extent, on another family of paradoxes to which Weir devotes much less attention during the sustained formal development of his system, but which he very tellingly mentions when it comes to providing some introductory philosophical justification for it: the

Ministry of Science and Innovation *Worlds and Truth Values: Challenges to Formal Semantics*, from the BBVA-Foundation Project Grants for Scientific-Research Projects 2021 *Unstable Metaphysics* and from the FCT Project 2022.03194.PTDC *New Perspectives on the Objects and Grounds of Structural Rules*.

[1] Some time after the bulk of this paper had been written, I became aware of exciting new work in progress by Weir that further develops his work on this topic in promising directions. Given that such developments are still ongoing, an adequate discussion of them will have to wait for another occasion; however, as far as I can tell, while they certainly represent an improvement in several respects, *mutatis mutandis* all the essential points made in the paper apply to them just as well.

E. Zardini (✉)
Department of Logic and Theoretical Philosophy, Faculty of Philosophy, Complutense University of Madrid, Madrid, Spain

International Laboratory for Logic, Linguistics and Formal Philosophy, School of Philosophy and Cultural Studies, National Research University Higher School of Economics, Moscow, Russia

LanCog, School of Arts and Humanities, University of Lisbon, Lisbon, Portugal paradoxes of *vagueness*).[2] The semantic paradoxes can be understood as forcing a dilemma between *classical logic* on the one hand and *naive semantic notions* on the other hand.[3] Among these, historically and theoretically prominent is *naive truth*; to fix ideas, without loss of generality, I'll focus on it. Weir's take on the dilemma is to stick to (naive) truth and consequently revise classical logic (AYE). So far, that's on the right track but is also nothing very exciting—something even Englishmen could already come up with (e.g. Priest 2006). We know Weir can do better.

Indeed. Weir observes that classical logic can be analysed into two components (AYE): one consists of the principles governing *logical operations* (for example, the entailment of *simplification* $\varphi \,\&\, \psi \vdash \varphi$, which governs conjunction), the other one consists of the principles governing premises and conclusions regarded as (as is put by Zardini 2018a, p. 241) a *structure of unstructured objects* (for example, the entailment of *reflexivity* $\varphi \vdash \varphi$, which only depends on the premise and conclusion's being the same object). Notice that traditional philosophically motivated revisions of classical logic have accepted the *structural principles* of classical logic and have instead denied only some of its *operational principles* (usually—*intuitionism* (Brouwer 1908), *many-valued* logics (Łukasiewicz 1920), *relevant* logics (Orlov 1928)...—but not always—*quantum* logics (Birkhoff and von Neumann 1936)—those concerning negation and implication). Weir then makes what I regard as a profoundly insightful suggestion: that *the paradoxes are due not so much to the operational principles of classical logic, but to some of its structural principles*—that the real agent pulling the strings of paradox hides below the *Überbau* of the operational principles in the *Bau* of the structural principles (AYE). I think Weir's suggestion is entirely spot-on (at least under the untraditional conceptualisation of structural principles offered in Zardini 2021b, pp. 507–519; 2023b), and it has opened up the following programme: *develop a system that, while validating the principles of naive truth (or of naive vagueness), preserves the classical essence of the logical operations and remains coherent by denying instead, in a philosophically motivated way, some of the classical structural principles*. *Classicism* as a whole

---

[2] There is a long-standing tradition in philosophy of logic of regarding the semantic paradoxes (if not all the paradoxes of selfreference) and the paradoxes of vagueness to be essentially the *same kind* of paradox (see Priest 2010 for a recent revival of that tradition and Oms and Zardini 2021 for its criticism, see also Zardini 2021b, pp. 494–506 for some background discussion). I don't read Weir as signing up to that tradition (AYE) (I'll henceforth so mark points on which I emphatically agree with—and, often, concerning which I've substantially been influenced by—Weir; for good measure, I'll also mark with 'NAE' tangential points on which I emphatically disagree with Weir): as I understand him, to put it in familiar terms (derived from Chihara 1979), while he isn't committal about what the right kind of *diagnosis* is in either case, what he argues for is that the same kind of *treatment* is appropriate in both cases (NAE).

[3] I don't take *denial of "selfreference"* to be a serious option (footnote 27). Moreover, I take naive semantic notions not to discriminate between occurrences of expressions that, by normal standards, do not differ in any semantically relevant respect (they refer to the same object—e.g. a Liar occurrence—and predicate of it the same property—e.g. being false), so that I consider *occurrence-based* approaches as willy-nilly taking the first horn of the dilemma. Weir (2000); 2002 persuasively argues against some of those approaches (AYE); Zardini (2008c) against some others.

must be knocked down, but *operational classicism* in particular must be left to stand. I myself have tried to implement that programme in the case of the semantic paradoxes and in the case of the paradoxes of vagueness, developing systems that exhibit substantial similarities with, but also important dissimilarities from, the one developed by Weir (I'll give some details in footnotes 50 and 26 respectively). As for the rest of this paper, in Sect. 4.2, I'll present Weir's logical system and assess how well it preserves operational classicism; in Sect. 4.3, I'll present Weir's system of vagueness and assess how well it preserves naive vagueness; in Sect. 4.4, I'll present Weir's system of truth and assess how well it preserves naive truth.

## 4.2 Transitivity and Neoclassical Logic

In Weir's hands, the programme is implemented essentially by providing a semantics that allows for sentences to have a *nonclassical status* and then define the relation of logical consequence in a philosophically motivated way out of such semantics, with the result that that relation invalidates an important structural principle of classical logic. By way of spoiler, letting $\vdash$ be the relation of *logical consequence*,[4] such a principle is the metaentailment of *transitivity with side premises and side conclusions*:

($\text{TRANS}_{\vdash}^{\text{SPSC}}$) If $\Gamma_0 \vdash \Delta_0, \varphi$ and $\Gamma_1, \varphi \vdash \Delta_1$ are valid, $\Gamma_1, \Gamma_0 \vdash \Delta_0, \Delta_1$ is valid.[5]

Notice that the most direct application of the notion of transitivity to the case of the relation of logical consequence seems rather the metaentailment of *transitivity with multiple premises and multiple conclusions*:

($\text{TRANS}_{\vdash}^{\text{MPMC}}$) If $\Gamma \vdash \varphi$ and $\varphi \vdash \Delta$ are valid, $\Gamma \vdash \Delta$ is valid,

[4] I cannot emphasise this enough: $\vdash$ is the relation of *logical consequence* (which may or may not coincide with a relation of *derivability*), and consequently principles about $\vdash$ are principles about logical consequence (which may or may not coincide with principles about a relation of derivability). Therefore, if the right conclusion to draw from the discussion to follow is that certain principles concerning $\vdash$ (e.g. contraction, to be introduced in footnote 17) are invalid and certain others (e.g. transitivity, to be introduced in the current sentence in the main text) are valid, that is the conclusion that *certain principles about logical consequence (not (simply) principles about a relation of derivability or principles of any other sort of "proof architecture") are invalid*. (Not that I'm assuming that logical consequence can be reduced to *semantic* facts either, see Moruzzi and Zardini 2007, pp. 161–170 for an overview of the main options on this question.)

[5] For clarity, I'll officially follow Weir in assuming a format where *premises and conclusions are each combined into sets* (although I think that, at the end of the day, that assumption is fatal, see especially footnotes 17, 50 for discussion), but it should be clear how the relevant passages should be modified if one is working with a different mode of combination of premises and of conclusions.

since that seems a direct instance of the general notion of *transitivity of a relation* $R$:

(TRANS$_R$) If $Rxy$ and $Ryz$ hold, $Rxz$ holds[6]

(obviously, (TRANS$_{\vdash}^{\text{SPSC}}$) entails (TRANS$_{\vdash}^{\text{MPMC}}$) but not *vice versa*).

Partly by way of terminological digression, for the reason emerging in the last paragraph, it is sometimes preferred to call (TRANS$_{\vdash}^{\text{SPSC}}$) '*cut*' rather than anything having to do with 'transitivity'. However, I regard such a preference as multiply misleading. I'll give my reasons in what I consider increasing order of importance. Firstly, 'cut' (*Schnitt*) is originally and primarily the term for the metarule corresponding to (TRANS$_{\vdash}^{\text{SPSC}}$) in the sequent calculus (Gentzen 1934): as such, *it suggests some sort of proof-theoretic setting that need not be in place at all* (footnote 4), since one can fruitfully investigate whether (TRANS$_{\vdash}^{\text{SPSC}}$) is valid independently of any proof-theoretic system (that is, as long as one has a grip on logical consequence independent of proof theory—as most of us do). Secondly, if one wants to get really anal about use of the word 'transitivity', one should read footnote 6 and stick to (TRANS$_{\vdash}$) as the *only principle deserving to be called 'transitivity'*; that is however such an impotent principle in many run-of-the-mill cases where one would traditionally appeal to "transitivity" (e.g. cases with multiple premises) that it would make the resulting use of the word 'transitivity' fade into insignificance. Thirdly, as far as I can see, the most natural reason for thinking that (TRANS$_{\vdash}^{\text{SPSC}}$) is valid is given by the idea that logical consequence is transitive (in the supertight sense of (TRANS$_{\vdash}$)) *and that it continues to be so if the* $\gamma \vdash \varphi$-*entailment is valid only under certain conditions* (truth of all its side premises $\Gamma_0 \setminus \{\gamma\}$ and falsity of all its side conclusions $\Delta_0$) *and the* $\varphi \vdash \delta$-*entailment is valid only under certain conditions* (truth of all its side premises $\Gamma_1$ and falsity of all its side conclusions $\Delta_1 \setminus \{\delta\}$), *by requiring that the* $\gamma \vdash \delta$-*entailment be valid under the combination of all those conditions*—indeed, it's hard to come up with any decent

---

[6] Well, (TRANS$_R$) is naturally understood with '$x$', '$y$' and '$z$' ranging over *objects of the same kind*, whereas their counterparts in (TRANS$_{\vdash}^{\text{MPMC}}$) aren't naturally so understood (on pain of postulating invisible lassos around $\varphi$ in (TRANS$_{\vdash}^{\text{MPMC}}$)). Even worse, (TRANS$_R$) is naturally understood with '$x$', '$y$' and '$z$' ranging *in the same way* over *a single domain of objects*, whereas their counterparts in (TRANS$_{\vdash}^{\text{MPMC}}$) can't be so understood (try to replace '$\varphi$' with a set variable!). Shifting from *singular quantification over sets* to *plural quantification over sentences* (as generally advocated by Oliver and Smiley 2016, pp. 43–44) would only address the first point, and would bring in additional difficulties, since it would not by itself cover the case of entailments with *no premises* (or with *no conclusions*) and (contrary to the way in which a set-based format can be modified into a sequence-based format) it would not allow modifications *able to represent the invalidity of structural principles* such as contraction or commutativity (unless one is prepared to envision something like "ordered pluralities"). All this indicates that not even (TRANS$_{\vdash}^{\text{MPMC}}$) is after all such a great candidate for being a "direct instance of the general notion of transitivity of a relation $R$". A certainly better candidate is the metaentailment of *transitivity with single premise and single conclusion*:

(TRANS$_{\vdash}$) If $\gamma \vdash \varphi$ and $\varphi \vdash \delta$ are valid, $\gamma \vdash \delta$ is valid,

where '$\gamma$', '$\varphi$' and '$\psi$' are naturally understood as ranging over the single domain of sentences.

reason for thinking that ($\text{TRANS}_{\vdash}^{\text{SPSC}}$) is valid that is not a variation on the one just rehearsed. It is therefore extremely natural (though I'm not saying absolutely necessary) to understand ($\text{TRANS}_{\vdash}^{\text{SPSC}}$) as a principle of *extended* transitivity (in the supertight sense of ($\text{TRANS}_{\vdash}$)), and therefore totally appropriate to label it as such. Fourthly, the most direct application of the notion of "cut" to the case of the relation of logical consequence seems rather the principle represented by ($\text{TRANS}_{\vdash}^{\text{MPMC}}$). When there is *something else combined with the cut formula* in either of its occurrences (as is the case with $\Delta_0$ and $\Gamma_1$ in ($\text{TRANS}_{\vdash}^{\text{SPSC}}$)), the notion of "cut"—with its simple idea of *"deleting" a conclusion occurrence and a premise occurrence of the same formula in two different entailments and combining the rest in a single entailment by "fusing" the two occurrences of the turnstile—does not give us any solid guidance as to whether and how such an extra material should occur in the resulting entailment.* To regard ($\text{TRANS}_{\vdash}^{\text{SPSC}}$) as an instance of "cut" requires an *extension* of the notion of "cut" *similar* to the one required of the notion of "transitivity" for regarding ($\text{TRANS}_{\vdash}^{\text{SPSC}}$) as an instance of "transitivity".

Beyond the terminological issue, I've been delving into these distinctions between different versions of transitivity since, as we'll see, it is a fundamental feature of Weir's system that it invalidates only ($\text{TRANS}_{\vdash}^{\text{SPSC}}$) but not ($\text{TRANS}_{\vdash}^{\text{MPMC}}$), whereas, as I'll argue, a nontransitive approach to naive vagueness is viable only if ($\text{TRANS}_{\vdash}^{\text{MPMC}}$) and indeed ($\text{TRANS}_{\vdash}$) themselves (over and above ($\text{TRANS}_{\vdash}^{\text{SPSC}}$)) are invalidated. (I'll also argue that a nontransitive approach to naive truth is consistent with certain desirable intersubstitutability principles only if ($\text{TRANS}_{\vdash}^{\text{MPMC}}$) and indeed ($\text{TRANS}_{\vdash}$) themselves (over and above ($\text{TRANS}_{\vdash}^{\text{SPSC}}$)) are invalidated, but that the resulting system is bound to have nonclassical features concerning implication and negation that are rebarbative to lovers of operational classicism, see footnote 45.)

In more detail, Weir assumes the very natural *Łukasiewicz 3-valued valuation scheme* (Łukasiewicz 1920),[7] but deviates from the definitions of logical consequence that are usually adopted in that framework. Weir is a staunch defender of the philosophical importance of a (multiple-premise and) *multiple-conclusion* setup for logical consequence (AYE),[8] and, in understanding his position here, it will indeed help to assume such a setup.[9] It will also help to deploy a fairly neutral

[7] More precisely, Weir oscillates a bit about *implication* (e.g. Weir 2005, pp. 226–227). For the purposes of this paper, to fix ideas, we can however assume the Łukasiewicz 3-valued valuation scheme for implication too—nothing crucial in our discussion will rely on this detail. (I should also note that Weir 2015 works with more complex value spaces, but that again nothing crucial in our discussion will rely on working with the simpler 3-valued space (see footnote 23 for an exemplification of this point).)

[8] Actually, after recent conversations with Weir, I realise that might be a bit of an overstatement (NAE).

[9] Importantly, throughout, I'll understand the setup so that premises are combined *conjunctively* and conclusions are combined *disjunctively* (see Zardini 2021b, pp. 507–516; Zardini 2023b for elaborations on how exactly to understand those adverbial modifications). Such is the standard understanding, and there are indeed good reasons in its favour, even abstracting away from the

understanding of the values of the Łukasiewicz 3-valued valuation scheme: 1 is something like *best value*, 1/2 is something like *average value* and 0 is something like *worst value*.[10]

---

fact—which will become apparent in this section and is actually already by itself decisive given this paper's focus on operational classicism—that *it is this understanding that makes it possible to express many features of the classical essence of the logical operations in the first place*. Firstly, *inference from premises to conclusions* is a transition from some acceptance-related attitude towards the premises taken together to some acceptance-related attitude towards the conclusions taken together, and the most natural ways for sentences to be accepted together are either as facts or as alternatives. (This *psychological* point admits of a *truth-theoretic* variation: *preservation of semantic status from premises to conclusions* is an implication from some truth-related status of the premises taken together to some truth-related status of the conclusions taken together, and the most natural ways for sentences to be true together is either as facts or as alternatives.) Secondly, just as the natural idea of combining $\varphi$ and $\psi$ as premises is the one of having premises that are (typically, jointly) *stronger* in that they yield *both* $\varphi$ *and* $\psi$ (thereby, in a monotonic setting, validating more entailments than either the premise $\varphi$ does or the premise $\psi$ does) but that are also the *weakest ones to do so* (thereby validating stronger entailments than stronger premises do), the natural idea of combining $\varphi$ and $\psi$ as conclusions is the one of having conclusions that are (typically, jointly) *weaker* in that they are yielded by *either* $\varphi$ *or* $\psi$ (thereby, in a monotonic setting, validating more entailments than either the conclusion $\varphi$ does or the conclusion $\psi$ does) but that are also the *strongest ones to be such* (thereby validating stronger entailments than weaker conclusions do). Thirdly, if, in checking for the requirement of *upwards falsity preservation* (which, as I'll explain in the next paragraph, for Weir is just as sound a requirement on logical consequence as *downwards truth preservation* is), the relevant status for the conclusions is that *all of them are false*, in checking for downwards truth preservation the relevant status for the conclusions should be the *negation* thereof—that is, assuming De Morgan quantificational principles that routinely go unchallenged by systems of naive truth, that *some of them are not false* (which, incidentally, given the strong kind of system of naive truth I'll advocate for in Sect. 4.4, is in turn equivalent with the status that *some of them are true*). I should stress that all this is compatible with a *formal* definition of logical consequence that employs nonconjunctive and nondisjunctive resources in representing premise or conclusion combination—the point is only that, however the formal definition might go, the resulting consequence relation should *informally be interpreted* so that premises are combined conjunctively and conclusions disjunctively. In fact, as we'll see in the second next paragraph, Weir himself favours one such formal definition, and, consequently with what I've just said, throughout I assume that the resulting consequence relation is informally interpreted so that premises are combined conjunctively and conclusions disjunctively. (Weir informs me that he would not interpret his consequence relation so that conclusions are combined disjunctively: while that might stave off some of my points (but notice that, of those, some arguably only rely on interpreting the consequence relation as delivering as conclusions a set of alternatives each of which one would like to deny, in which case the point will go through also on Weir's nonstandard interpretation), it would thrust his view straight against the obstacles raised by the considerations I've developed in this footnote.)

[10] Should you be tempted to interpret instead '$\varphi$ has value 1 in model $M$' as '$\varphi$ is true in $M$', consider that, in the kind of system in question, a Liar sentence saying of itself that it is not true (Sect. 4.4) cannot have value 1 in any model, and so would not be true in any model: since, presumably, if a sentence does not have (the representor of) a certain semantic status *in any model*, it does not have that semantic status *simpliciter*, you would then be forced to grant that *that Liar sentence is not true*, which, in the kind of system in question, quickly leads to disaster. Even once you've overcome that temptation, the idea of a best, an average and a worst value of course and in due course invites the formulation of *strengthened semantic paradoxes*. In the case of the semantic paradoxes, Weir's idea is that the classical metatheory we're going through, while *propaedeutically*

With this background in place, one usual definition of logical consequence out of the Łukasiewicz 3-valued valuation scheme is:

(B $\Rightarrow$ B) $\Gamma \vdash \Delta$ is valid iff, for every admissible interpretation,[11] if every member of $\Gamma$ has best value, some member of $\Delta$ has best value.

However, Weir thinks that (B $\Rightarrow$ B) is, among other things, *too strong* in that it invalidates e.g. the law of *exhaustion*:

(EXH) $\varnothing \vdash \varphi, \neg\varphi$,

which, as I'll explain in the sixth next paragraph, is a necessary element of operational classicism. Weir also thinks that (B $\Rightarrow$ B) is *too weak* in that, while it guarantees preservation of best value from the premises to the conclusions, it does not guarantee preservation of worst value from the conclusions to the premises.[12] Indeed, if one is starting from (B $\Rightarrow$ B), it is the addition of preservation of worst value from the conclusions to the premises that most straightforwardly validates e.g. the metaentailment of the *reduction theorem*:

(RT) If $\varphi \vdash \varnothing$ is valid, $\varnothing \vdash \neg\varphi$ is valid,

which, as I'll explain in footnote 21, is also a necessary element of operational classicism (I go to some depth into the philosophical justification for (RT) even in the context of the paradoxes of selfreference in Zardini 2015a, pp. 469–473).

*useful*, does not provide an *accurate description of semantic reality*, which is only provided by a nonclassical metatheory that has no truck with any truth values other than truth and falsity (cf footnote 16) (AYE). I'll follow Weir in sticking to the propaedeutic stance in this paper.

[11] On Weir's account (e.g. Weir 1998b, pp. 226–227), an *admissible* interpretation must satisfy a further condition in addition to the one of respecting the Łukasiewicz 3-valued valuation scheme, whose details are however immaterial for the purposes of this paper.

[12] As a lover of semantic classicism, Weir insists on the general *symmetry between truth and falsity* (AYE), and so on the specific idea that *upwards falsity preservation* should be just as necessary for logical consequence as *downwards truth preservation* is (AYE). However, maybe because of the temptation discussed in footnote 10, Weir typically treats this idea as interchangeable with the idea that upwards *worst-value* preservation should be just as necessary to logical consequence as downwards *best-value* preservation. That is actually a significant slide, given the significant differences in Weir's system between *truth* and *best value* and between *falsity* and *worst value* (footnote 10), and given that, with such differences made clear, the latter idea is much less compelling than the former (see footnote 16 for further details on truth preservation and falsity preservation). (In addition to the consideration advanced in footnote 10, here is another, more general consideration in favour of the existence of the significant differences just mentioned: given (EXH), either $\varphi$ is true or $\varphi$ is false even if $\varphi$ has average value and so neither best nor worst value.) Notice that those differences concern the gaps *from* something's being true *to* its having best value and *from* something's being false *to* its having worst value, whereas it can extremely plausibly be assumed that there are no gaps in the *converse* directions. The gaps that do exist are however sufficient to drive a wedge between the two kinds of ideas mentioned above: for example, downwards truth preservation does not require downwards best-value preservation, since a conclusion might be true without having best value.

One would expect that Weir's solution to the first problem is to lower the standard for validity to:

(B $\Rightarrow$ A) $\Gamma \vdash \Delta$ is valid iff, for every admissible interpretation, if every member of $\Gamma$ has best value, some member of $\Delta$ has at least average value.

But that is not quite Weir's solution, which deviates from (B $\Rightarrow$ A) in a specific range of cases:

(B $\Rightarrow$ B$^{\mathrm{W}}$) $\Gamma \vdash \Delta$ is valid iff, for every admissible interpretation, if every member of $\Gamma$ has best value, then [$\Delta \neq \varnothing$ and, for every $\varphi \in \Delta$, if every member of $\Delta \setminus \{\varphi\}$ has worst value, $\varphi$ has best value].

That is puzzling in that it is not the most natural lowering of the standard for validity, which is instead represented by (B $\Rightarrow$ A). To facilitate a comparison between (B $\Rightarrow$ A) and (B $\Rightarrow$ B$^{\mathrm{W}}$), notice that the former is equivalent with:

(B $\Rightarrow$ A$^{\mathrm{W}}$) $\Gamma \vdash \Delta$ is valid iff, for every admissible interpretation, if every member of $\Gamma$ has best value, then [$\Delta \neq \varnothing$ and, for every $\varphi \in \Delta$, if every member of $\Delta \setminus \{\varphi\}$ has worst value, $\varphi$ has at least average value].

It is then clear that (B $\Rightarrow$ A) represents not only a *more natural* lowering of the standard for validity from (B $\Rightarrow$ B) than (B $\Rightarrow$ B$^{\mathrm{W}}$) does (in that, *re* conclusions, it agrees with (B $\Rightarrow$ B) structure-wise, only changing the target value in order to validate things like (EXH)), but also a *less demanding* one (and therefore one that gets closer to the set of entailments that are classically valid, which in turn makes it more likely that it preserves operational classicism). So why on earth is Weir going for (B $\Rightarrow$ B$^{\mathrm{W}}$) instead?[13]

I can extract from Weir's writings three explicit reasons for his choice. Firstly, Weir claims that, *if one accepts all the premises of a valid entailment and denies all its conclusions but one, one must accept the remaining conclusion, and, if one denies all the conclusions of a valid entailment and accepts all its premises but one,*

[13] There would also be the lowering of the standard for validity represented by:

(A $\Rightarrow$ A) $\Gamma \vdash \Delta$ is valid iff, for every admissible interpretation, if every member of $\Gamma$ has at least average value, some member of $\Delta$ has at least average value

(well known e.g. from Priest 2006). But, again, (B $\Rightarrow$ A) represents not only a *more natural* lowering of the standard for validity from (B $\Rightarrow$ B) than (A $\Rightarrow$ A) does (in that, *re* premises, it agrees with (B $\Rightarrow$ B), no change being required there in order to validate things like (EXH)), but also a *less demanding* one (and therefore one that gets closer to the set of entailments that are classically valid, which in turn makes it more likely that it preserves operational classicism). Both aspects are exemplified in the fact that (B $\Rightarrow$ A) but not (A $\Rightarrow$ A) validates the law of *explosion*:

(EXPL) $\varphi, \neg\varphi \vdash \varnothing$,

which, as I'll explain in the fifth next paragraph, is also a necessary element of operational classicism.

*one must deny the remaining premise*.[14] That's a normative principle that will smell foul to most friends of nontransitivity, and I actually don't think that even Weir can comfortably endorse it *in full generality*, but, since the matter is tangential to the present dialectic (for I do think that Weir can comfortably endorse a *restricted* version of the principle that can play the same role in the present dialectic), I leave its discussion to a footnote (and refer to Zardini 2015d for a fuller treatment).[15] What is crucial is rather that this *normative* fact about acceptance and denial has no straightforward bearing on the *semantic* issue of what value the remaining conclusion should have. Presumably, Weir's background thought here is that at least one entailment declared valid by (B $\Rightarrow$ A) is such that it is possible that all its premises have best value and all its conclusions have worst value except for one having average value, in which case one would plausibly be allowed to accept all

[14] Throughout, such principles should be understood as *restricted to suitably idealised subjects* (cf Zardini 2015e, pp. 383–385; 2016b, pp. 465–470 for related discussions). Even so idealised, such principles are likely false for the *epistemological* reasons emerging with the *Preface paradox* (Makinson 1965, see Zardini 2021a for an extended discussion).

[15] The principle in question entails that one can (and indeed must) effectively use the metaentailment of *transitivity with side premises*:

(TRANS$^{SP}_{\vdash}$) If $\Gamma_0 \vdash \varphi$ and $\Gamma_1, \varphi \vdash \psi$ are valid, $\Gamma_1, \Gamma_0 \vdash \psi$ is valid

(which too is denied by Weir) in working out what further to *accept* from what one already *accepts*. This consequence should immediately ring some alarm bells for those who in general have qualms about transitivity. Moreover, such a consequence forecloses accepting theories that, according to Weir, are *consistent* but would be *inconsistent if (TRANS$^{SP}_{\vdash}$) were valid* (e.g., in addition to many instances of a contradiction $\varphi$ & $\neg\varphi$, the principle of tolerance (modulo positive and negative cases) and a paradoxical selfreferential sentence (modulo naive truth), see Sects. 4.3 and 4.4 respectively for the details): oddly enough, such *consistent* theories would at the same time be *logically unacceptable* (since, *by the sheer normative force of logic*, their acceptance would lead to the acceptance of an inconsistency). While *consistent but (for some reason or other) in-principle unacceptable* theories plausibly exist *in general* due to some sort or other of *epistemic clash* (witness *Moore's paradox*, Moore 1942, pp. 540–543), for such a relatively *simple* language as the one *in the present context* there would seem to be *no gap from consistency to in-principle acceptability*, and anyway the existence of consistent but *in-principle* unacceptable theories would still be a long way from the existence of consistent but *logically* unacceptable theories. Finally, such a principle about *categorical* acceptance and denial presumably stands or falls together with an analogous principle extended to cover also *hypothetical* acceptance and denial (attitudes that are at play when one e.g. accepts *only for the sake of argument* a certain *supposition* to work out its *consequences* and accept them *only under that supposition*). For the same reason, such an extended principle however entails that one can (and indeed must) effectively use (TRANS$^{SP}_{\vdash}$) in working out what further to *suppose* from what one already *supposes*—at this point, given the extent to which (TRANS$^{SP}_{\vdash}$) would thus govern one's reasoning, any claim to the effect that it is nevertheless "invalid" would sound rather hollow. (Fine, while (TRANS$^{SP}_{\vdash}$) would thus govern one's downwards reasoning with hypothetical *acceptance*, it would still not govern the premises and deny all the conclusions but the one having average value, and, if one proceeded to do so, by the relevant normative principle one would then be under pressure to accept a sentence that has average value. But whether that is a bad consequence depends crucially on *how the average value is understood*. Now, it's true that Weir has a propensity to gloss the average value as "*indeterminacy*", and that indeterminacy, at least if understood as absence of a fact of the matter (NAE), is typically a bar to acceptance, but more important than this *psychological* fact is the *philosophical* fact that (EXH) is supposed to be valid, and so that, for every $\varphi$, as a matter of logic, there are only two *alternatives* ($\varphi$ and $\neg\varphi$) which might both have average value: since one is presumably allowed to accept at least one of the only two alternatives left open by logic, *average value can be no bar to acceptance* (cf footnote 12). If so, it's hard to see what's so bad in the consequence envisaged above that, under certain circumstances, one would be under pressure to accept a sentence that has average value. Secondly, Weir observes that, in the case of *single-conclusion* entailments, *($B \Rightarrow B^W$) but not ($B \Rightarrow A$) entails that valid entailments*

one's upwards reasoning with hypothetical *denial*: however, if one can only suppose $\varphi$ on pain of one's supposition's being inconsistent—if $\varphi$ cannot consistently be supposed to hold—is it really helpful to adopt a notion of validity according to which [$\varphi \vdash \varnothing$ may nevertheless be "invalid", and so $\varphi$ may nevertheless be "consistent", simply because one still needn't deny $\varphi$]? The point of the notion of consistency would seem to have directly to do with *logical possibility of truth* rather than with *logical nonnecessity of falsity*. Having said that, I don't mean to suggest that it is really possible that $\varphi$ cannot consistently be supposed to hold while one still needn't deny $\varphi$: as a lover of the classical essence of negation, keeping fixed the first conjunct I'd say that the second conjunct is due to an insufficiently strong theory of negation.) Dual considerations hold essentially by substituting transitivity with side conclusions for transitivity with side premises, denial for acceptance and *vice versa*, contingency (understood as logical possibility of falsity) for consistency, an excluded middle $\varphi \vee \neg\varphi$ for a contradiction $\varphi \,\&\, \neg\varphi$, the negations of the principle of tolerance and of a paradoxical selfreferential sentence for their *neganda* respectively, falsity for truth and *vice versa*. I should add that we can restrict the principle in question in a *nontransitivity-friendly* way, to the effect that, if one accepts *for noninferential reasons* all the premises of a valid entailment and denies *for noninferential reasons* all its conclusions but one, one must accept the remaining conclusion, and, if one denies *for noninferential reasons* all the conclusions of a valid entailment and accepts *for noninferential reasons* all its premises but one, one must deny the remaining conclusion (where one accepts (denies) $\varphi$ for *noninferential reasons* iff one accepts (denies) $\varphi$ not because it is a conclusion (premise) of a valid entailment all of whose premises (conclusions) one accepts (denies) and all of whose other conclusions (premises) one denies (accepts)). Given that at least one entailment declared valid by ($B \Rightarrow A$) is such that it is possible that all its premises have best value and all its conclusions have worst value except for one having average value, in which case, in an appropriate setting, one would plausibly be allowed to accept *for noninferential reasons* all the premises and deny *for noninferential reasons* all the conclusions but the one having average value, and, if one proceeded to do so, by the restricted normative principle one would then be under pressure to accept a sentence that has average value, as will be clear from the second next sentence in the main text such a restricted principle can play the same role in the present dialectic as the unrestricted principle.

*preserve best value*.[16] But how can that possibly matter for one who is happy with the idea that, in the case of *multiple-conclusion* entailments, *valid entailments can lead from best value to alternatives each of which has average value*? If it's OK if sometimes there are only two (or three, or four, or five...) alternatives each of which has average value, how can it not be OK if sometimes there is only one alternative which has average value? Thirdly, Weir points out that, given that in the philosophical applications to the semantic paradoxes and to the paradoxes of vagueness there will be sentences that in all admissible interpretations have average value (footnote 11), letting $\alpha$ be some such sentence (B $\Rightarrow$ A) entails that both $\mathsf{t} \vdash \alpha$ and $\alpha \vdash \mathsf{f}$ are valid, and so, since $\mathsf{t} \vdash \mathsf{f}$ is invalid, *even* $(TRANS_{\vdash})$ *is invalid*. But, again, how can that possibly matter for one who is happy with the idea that (TRANS$^{\text{SPSC}}_{\vdash}$) is invalid? Indeed, the acute reader will recall the third point of the fifth last paragraph to the effect that (TRANS$^{\text{SPSC}}_{\vdash}$) is an *extremely natural* extension

[16] Developing further a theme already emerged in footnote 12, Weir has a tendency to treat the idea of *truth preservation* as interchangeable with the idea of *best-value preservation to a single conclusion*, which I regard as misconceived for two reasons. Firstly, it misguidedly entails that the notion of preservation only applies in cases where there is *one* conclusion, which is arguably not the case since that notion arguably applies also in cases where there is *more than one* conclusion, by requiring that *some* conclusion have the relevant value (dually to how the notion of preservation obviously applies not only in cases where there is *one* premise, but also in cases where there is *more than one* premise, by requiring that *every* premise have the relevant value, see the truth-theoretic variation of the first reason developed in footnote 9). (It is understandable that Weir should have such a tendency, for $(B \Rightarrow B^W)$ *does not secure anything recognisable as preservation in the case of multiple premises*—a conditional preservation is not a preservation!) Secondly, as already noted with more detail in footnotes 10, 12, it harmfully confuses *truth* with *best value*, thereby conferring spurious plausibility to the idea that logical consequence requires best-value preservation. On a plausible conception of truth (naive truth, see Sect. 4.4 for details on both content and notation), if $\varphi$ behaves nonclassically, *so does* '$\varphi$' *is true'*, so that we should not expect the notion of truth to come up *in the classical metatheory* of a nonclassical system—we should rather expect it to come up *in the nonclassical system itself*, and the requirement of truth preservation on logical consequence should consequently be formulated so that, if $\varphi_0, \varphi_1, \varphi_2 \ldots, \varphi_i \vdash \psi_0, \psi_1, \psi_2 \ldots, \psi_j$ is valid, $\varnothing \vdash (T\ulcorner\varphi_0\urcorner \,\&\, T\ulcorner\varphi_1\urcorner \,\&\, T\ulcorner\varphi_2\urcorner \ldots \,\&\, T\ulcorner\varphi_i\urcorner) \rightarrow (T\ulcorner\psi_0\urcorner \vee T\ulcorner\psi_1\urcorner \vee T\ulcorner\psi_2\urcorner \ldots \vee T\ulcorner\psi_j\urcorner)$ is valid (see Zardini 2018a, pp. 254–259 for more details on how to understand the requirement so formulated). Analogous observations hold for *falsity preservation*. Notice that, properly so formulated, the requirements of truth preservation and of falsity preservation are satisfied by the system mentioned in footnote 26, whose native metatheory works in terms of *connection* between different values rather than *preservation* of a single value. For good measure, they are also satisfied by the system mentioned in footnote 50, which constitutes a rather unique feature of that system *vis-à-vis* alternative systems of naive truth. [WARNING: THE NEXT SENTENCE CONTAINS A SPOILER] Unfortunately, those requirements are *not* satisfied by Weir's system, not because of a misbehaviour of *implication* (as is typically the case for systems of naive truth) and not even because of a misbehaviour of *truth* (as is typically the case for systems of nonnaive truth)—in other words, in Weir's system, it is both the case that, if $\varphi \vdash \psi$ is valid, $\varnothing \vdash \varphi \rightarrow \psi$ is valid, and that, if $\varphi_0, \varphi_1, \varphi_2 \ldots, \varphi_i \vdash \psi_0, \psi_1, \psi_2 \ldots, \psi_j$ is valid, $T\ulcorner\varphi_0\urcorner, T\ulcorner\varphi_1\urcorner, T\ulcorner\varphi_2\urcorner \ldots, T\ulcorner\varphi_i\urcorner \vdash T\ulcorner\psi_0\urcorner, T\ulcorner\psi_1\urcorner, T\ulcorner\psi_2\urcorner \ldots, T\ulcorner\psi_j\urcorner$ is valid—but because of a misbehaviour of *conjunction* and *disjunction*.

of ($\text{TRANS}_{\vdash}$), so that denial of ($\text{TRANS}_{\vdash}^{\text{SPSC}}$) should *extremely naturally* bring in its wake denial of ($\text{TRANS}_{\vdash}$).[17]

As a friendly suggestion, I would urge that the third reason was a near-miss. Given $\alpha$'s property, (B $\Rightarrow$ A) entails that $\mathsf{t} \vdash \neg\alpha$ is also valid, and so presumably one would have to accept both $\alpha$ and $\neg\alpha$ (footnote 45). (Dually, both $\alpha \vdash \mathsf{f}$ and $\neg\alpha \vdash \mathsf{f}$ are valid, and so presumably one would have to deny both $\alpha$ and $\neg\alpha$.) It's hard to imagine a more dramatic *violation of the classical essence of negation*. As a somewhat converse friendly suggestion, given the Łukasiewicz 3-valued valuation scheme, (B $\Rightarrow$ A) *forces in effect the validity of exactly the classically valid entailments*, and that is most definitely *not* what lovers of naive vagueness or of naive truth such as Weir and I were looking for: for the problem was exactly that, in classical logic, principles of naive vagueness and of naive truth are inconsistent (and, to add insult to injury, entail that Nessie exists). The idea was always that of preserving *the classical essence of the logical operations*—it was never that of preserving *all classically valid entailments*.

[17] And so it is utterly unsurprising that the scission operated by Weir between ($\text{TRANS}_{\vdash}$) and ($\text{TRANS}_{\vdash}^{\text{SPSC}}$) comes at the cost of denying some of the extremely natural assumptions under which ($\text{TRANS}_{\vdash}$) extends to ($\text{TRANS}_{\vdash}^{\text{SPSC}}$). In particular, going back to the notation of the fifth last paragraph, Weir is perplexingly committed to denying that the fact that the $\gamma \vdash \delta$-entailment is valid under the combination of truth of every member of $\Gamma_1 \cup (\Gamma_0 \setminus \{\gamma\})$ and falsity of every member of $\Delta_0 \cup (\Delta_1 \setminus \{\delta\})$ (which, model-theoretically, we can assume to consist, for Weir, in the fact that, for every model where every member of $\Gamma_1 \cup (\Gamma_0 \setminus \{\gamma\})$ has best value and every member of $\Delta_0 \cup (\Delta_1 \setminus \{\delta\})$ has worst value, if $\gamma$ has best value in the model, so does $\delta$ and, if $\delta$ has worst value in the model, so does $\gamma$) suffices for the fact that $\Gamma_1, \Gamma_0 \vdash \Delta_0, \Delta_1$ is valid. Even more in particular, it is instructive to observe that, under Weir's assumptions, the former fact fails to suffice for the latter fact basically because in certain cases $\Delta_0 = \{\delta\}$ while $\Delta_1 \setminus \{\delta\} = \varnothing$, so that, under the condition that every member of $\Gamma_1 \cup (\Gamma_0 \setminus \{\gamma\})$ has best value, while it is indeed the case that, under the condition that every member of $\Delta_0$ (i.e. $\delta$) has worst value, $\delta$ has best value and "*vice versa*", what really matters for logical consequence is that, under the condition that every member of $(\Delta_0 \cup \Delta_1) \setminus \{\delta\}$ has worst value, $\delta$ have best value: since $(\Delta_0 \cup \Delta_1) \setminus \{\delta\} = \varnothing$, the latter condition is vacuously satisfied, and so $\delta$ is in effect *categorically* required to have best value—a taller order than its having best value *hypothetically* on its having worst value (for an example, let $\delta$ be $\varphi \vee \neg\varphi$). This brings out *the absolutely crucial role* that Weir's assumption of a set-based format (footnote 5) plays in bringing it about that ($\text{TRANS}_{\vdash}^{\text{SPSC}}$) is invalid in spite of ($\text{TRANS}_{\vdash}$)'s validity: it is that assumption that unnaturally identifies the occurrence of a sentence as a side conclusion ($\delta$ in $\Delta_0$, a conclusion additional to $\varphi$ in the first entailment of ($\text{TRANS}_{\vdash}^{\text{SPSC}}$)) with its occurrence as a main conclusion ($\delta$ in $\Delta_1$, a conclusion of ($\Gamma_1$ and) $\varphi$ in the second entailment of ($\text{TRANS}_{\vdash}^{\text{SPSC}}$)), and so that unnaturally identifies the side conclusions without the main conclusion ($\Delta_0$) with the side conclusions without themselves ($\Delta_0 \setminus \{\delta\}$), and hence that unnaturally requires the main conclusion ($\delta$) to have best value instead of its having it conditionally on the side conclusions' ($\Delta_0$) having worst value. More trenchantly, under Weir's assumptions, *transitivity is invalid only because contraction* (to be properly introduced in the next sentence) *is valid* (and only because its validity *unnaturally obliterates the structural distinction between side conclusions and main conclusions*). If one instead insists on preserving that natural distinction, one will deny the structural metaentailment of *contraction*:

(CONTR) If $\Gamma, \varphi, \varphi \vdash \Delta$ is valid, $\Gamma, \varphi \vdash \Delta$ is valid; if $\Gamma \vdash \Delta, \varphi, \varphi$ is valid, $\Gamma \vdash \Delta, \varphi$ is valid

With (B $\Rightarrow$ B$^{W}$)'s template having being so motivated, Weir's solution to the second problem is to raise the standard for validity to:

(B $\Rightarrow$ B$^{W}$|W$^{B}$ $\Leftarrow$ W) $\Gamma \vdash \Delta$ is valid iff, for every admissible interpretation, [if every member of $\Gamma$ has best value, then [$\Delta \neq \varnothing$ and, for every $\varphi \in \Delta$, if every member of $\Delta \setminus \{\varphi\}$ has worst value, $\varphi$ has best value]], and, [if every member of $\Delta$ has worst value, then [$\Gamma \neq \varnothing$ and, for every $\varphi \in \Gamma$, if every member of $\Gamma \setminus \{\varphi\}$ has best value, $\varphi$ has worst value]],

which is obviously impeccable once his solution to the first problem has been granted. For the purposes of this paper, we can rest content with (B $\Rightarrow$ B$^{W}$|W$^{B}$ $\Leftarrow$

---

and consequently assume that premises and conclusions are each combined into, and operations on them understood in terms of, *multisets* rather than *sets* as well as reformulate Weir's definition of logical consequence in terms of *sentence occurrences* rather than *sentences* (throughout, at the relevant places, I presuppose this alternative format). There will then no longer be a bar to the fact that the $\gamma \vdash \delta$-entailment is valid under the combination of best value of every member of $\Gamma_1 \cup (\Gamma_0 \setminus \{\gamma\})$ and worst value of every member of $\Delta_0 \cup (\Delta_1 \setminus \{\delta\})$ sufficing for the fact that $\Gamma_1, \Gamma_0 \vdash \Delta_0, \Delta_1$ is valid, and so (TRANS$_{\vdash}$) will once more extend to (TRANS$_{\vdash}^{SPSC}$) (the resulting system—but not its fundamental motivation uncovered in this footnote—is briefly mentioned by Weir 2015, p. 104: while I've just argued it to be a more natural system than Weir's own favoured system, all the main problems I'll raise for his nontransitive system—save for those depending on intersubstitutability principles—arise for this noncontractive system just as well, which I'd therefore find inferior to the noncontractive system mentioned in footnote 50). Notice that, starting from "Even more in particular" in this footnote, dual considerations hold concerning side premises and main premises (with $\Gamma_1 = \{\gamma\}$ while $\Gamma_0 \setminus \{\gamma\} = \varnothing$), and that especially those considerations also explain the restriction in place on certain inference principles in Weir's *natural-deduction calculus* for his system. In more detail, for the two-premise rules (which Weir calls '*nonminimax*') where the premises can all have nonworst value while the conclusion has worst value (e.g. *modus ponens*), Weir's natural-deduction calculus places the restriction—which smells awfully a lot like a reservation about (CONTR)!—that *every assumption on which both premises depend not have average value* (in order not to horrify typesetters with this already XXXL footnote, that's a bit rough in a couple of respects, in a way however that does not affect the essence of the explanation). As Weir (2005), pp. 227–228 emphasises (and as Zardini 2015d, pp. 233–237 in effect explains in a more general setting), given the architecture of a natural-deduction calculus, that is not really a restriction *on the rule itself*, but a restriction *on extending a correct derivation by means of that rule*—that is, in fact, a restriction on transitivity (*qua* applied to that rule as its second step) (AYE). In Weir's writings, however, such a restriction typically comes rather out of the blue, at best supported by a single-case induction from the consideration of one such rule (chained together with another rule) seemingly chosen at random (e.g. Weir 1998b, pp. 225–226). We can now understand better why that is the right restriction: for, dually to what we've explicitly seen above, transitivity will be invalid for nonminimax rules exactly when, if it were valid, given the other assumptions on which the two premises depend, those premises could individually be replaced by, and jointly be contracted to, a single assumption on which they both depend and which, with all the other assumptions' having best value, satisfies the condition of having worst value hypothetically on its having best value (for, if the assumption in question has best value, since every other assumption on which one premise depends has best value, that premise has best value, and so, since the conclusion has worst value, the other premise has worst value, and hence, since every other assumption on which the latter premise depends has best value, the assumption in question has worst value) but does not satisfy the condition of categorically having worst value—that is, an assumption that has average value.

W) as codifying Weir's system (footnote 11): we'll henceforth refer to it as **W** and to its consequence relation as $\vdash_{\mathbf{W}}$.

It's easy to see that ($\text{TRANS}_{\vdash}$) is valid in **W**. To see that, on the contrary, ($\text{TRANS}_{\vdash}^{\text{SPSC}}$) is invalid in **W**, consider that the entailment of *addition*:

(ADD) $\varphi \vdash \varphi \vee \psi; \varphi \vdash \psi \vee \varphi$

is valid in **W**, and recall that (EXH) is also valid in **W**; by two application of ($\text{TRANS}_{\vdash}^{\text{SPSC}}$), it would then follow that the law of *excluded middle*:

(LEM) $\varnothing \vdash \varphi \vee \neg\varphi$

is valid in **W**, which however it's easy to see that it isn't. It's also easy to see that *the very style of definition* instantiated by ($\text{B} \Rightarrow \text{B}^{\text{W}}|\text{W}^{\text{B}} \Leftarrow \text{W}$) (with its standard for a single premise not being lower than its standard for a single conclusion, with its conditions on premises and conclusions variously automatically extending to supercollections thereof and with its set-based format) builds in the structural entailment of *reflexivity*:

(REFL) $\varphi \vdash \varphi$,

the structural metaentailment of *monotonicity*:

(MON) If $\Gamma_0 \vdash \Delta_0$ is valid, $\Gamma_1, \Gamma_0 \vdash \Delta_1, \Delta_0$ is valid

and (CONTR) (footnote 17). Therefore, **W**'s substructurality is, for better or worse, *extremely limited.*

Weir likes to label **W** 'neoclassical', so let's see if that label is earned—that is, in conformity with what I've announced in Sect. 4.1, let's see if **W** preserves the classical essence of the logical operations.[18] Starting with *negation*, this includes (EXH) (which encodes the feature of the classical essence of negation according

---

[18] I should add that, both in writing and conversation, Weir himself displays a more *pragmatic* approach than the one I'm presupposing in this paper (NAE), whose bottom line is that the *strength, simplicity, elegance* etc. of classical logic are ideals to which we should strive to *get as close as possible* while preserving naive vagueness and naive truth (whereas things like "the classical essence of the logical operations" can suck it). Properly pursuing this thread would of course lead to a vast range of *methodological* issues, which lies beyond the scope of the paper. However, let me make a couple of (brisk) observations. Firstly, from the point of view of the pragmatic approach, it makes a lot of sense to try to hold to ($\text{TRANS}_{\vdash}$) *even once* ($\textit{TRANS}_{\vdash}^{\textit{SPSC}}$) *has been jettisoned*; yet, I've argued that that combination of views is not very stable. This is just an instance of the general fact that, if one arbitrarily tries to get as close as possible to classical logic *with no independent idea of what is true*, one risks to end up embracing a *congeries of claims that do not cohere very well together*. (Indeed, if a theory manifestly doesn't work, why on earth think that the truth must lie as close as possible to it?!) Classical logic certainly contains a lot of true elements, but then the aim should be the *specific* one of *preserving those elements* rather than the *generic* one of *getting as close as possible to classical logic*. Secondly, it's actually not at all clear in what respects (if any) **W** "gets as close as possible to classical logic". It's true that, *by adding as extra premises all the relevant instances of a certain schema,* **W** *can mimic all the classically valid entailments*, but that's a property that obviously doesn't even guarantee that a system gets anywhere *close* to classical logic and that indeed **W** *shares with many other* traditional philosophically motivated revisions of classical logic (starting from intuitionistic logic!). Slightly more promising is the fact

to which *the negation of* $\varphi$ *covers every way in which* $\varphi$ *may fail to hold*) and (EXPL) (footnote 13) (which encodes the feature of the classical essence of negation according to which *the negation of* $\varphi$ *only covers ways in which* $\varphi$ *may fail to hold*), which are both valid in **W**. On the proviso that there are no further features of the classical essence of negation (to be rejected in footnote 21), I therefore conclude that **W**'s negation is fully classical.

Moving on to *implication*, its classical essence includes the metaentailment of the *single-premise single-conclusion deduction theorem*:

(DT) If $\varphi \vdash \psi$ is valid, $\varnothing \vdash \varphi \rightarrow \psi$ is valid

(which encodes the feature of the classical essence of implication according to which *the implication from* $\varphi$ *to* $\psi$ *is licenced by the entailment from* $\varphi$ *to* $\psi$) and the entailment of *modus ponens*:

(MP) $\varphi, \varphi \rightarrow \psi \vdash \psi$

(which encodes the feature of the classical essence of implication according to which *the implication from* $\varphi$ *to* $\psi$ *licences the entailment from* $\varphi$ *to* $\psi$), which are both valid in **W**. On the proviso that there are no further features of the classical essence of implication (also to be rejected in footnote 21), I therefore conclude that **W**'s implication is fully classical.

Matters get more glitteringly baroque when we move—of all things!—to *conjunction*.[19] It's true that the entailments of *adjunction*:

(ADJ) $\varphi, \psi \vdash \varphi \,\&\, \psi$

(which encodes the feature of the classical essence of conjunction according to which $\varphi$ *together with* $\psi$ *as facts are sufficient for the conjunction of* $\varphi$ *with* $\psi$) and *simplification*:

(SIMPL) $\varphi \,\&\, \psi \vdash \varphi$; $\varphi \,\&\, \psi \vdash \psi$

(which encodes the feature of the classical essence of conjunction according to which *the conjunction of* $\varphi$ *with* $\psi$ *is sufficient for* $\varphi$ *and is sufficient for* $\psi$) are valid in **W**. But that won't satisfy a real lover of operational classicism. For she knows that the perfect beauty of such a classicism requires that, just as it is also in the essence of implication *to correlate in the object-language with the metalanguage*

---

that **W** *preserves all the compositional principles of the logical operations* (about to be exposed in the main text). But the significance of that fact is also lessened not only by *its holding also for many other* traditional philosophically motivated revisions of classical logic (starting from certain versions of quantum logics!), but also by the additional fact that **W** *does not preserve structure-correlating, close relatives of those principles that arguably encode more fundamental features of the logical operations* (footnote 21): since *the latter principles are valid not only in classical logic, but also in a few other nonclassical logics* (see e.g. footnotes 26, 50), at least in that significant respect, then, **W** does not get as close as possible to classical logic.

[19] I could—and, in the next paragraph, will—make the same point with disjunction, but I prefer to make it first with conjunction as it can be made in the more usual and familiar single-conclusion setup.

*relation between premise and conclusion* in the way achieved by the metaentailment of the *deduction theorem with side premises and side conclusions*:

($DT^{SPSC}$) If $\Gamma, \varphi \vdash \Delta, \psi$ is valid, $\Gamma \vdash \Delta, \varphi \rightarrow \psi$ is valid

and the metaentailment of *second-step major-premise transitivity of modus ponens*:

($MP^{2TRANS^{MAJ}}$) If $\Gamma \vdash \Delta, \varphi \rightarrow \psi$ is valid, $\Gamma, \varphi \vdash \Delta, \psi$ is valid,[20]

it be also in the essence of conjunction *to correlate in the object-language with the metalanguage relation between premise and premise* in the way achieved by the metaentailment of *first-step transitivity of adjunction*:

($ADJ^{1TRANS}$) If $\Gamma, \varphi \,\&\, \psi \vdash \Delta$ is valid, $\Gamma, \varphi, \psi \vdash \Delta$ is valid

and the metaentailment of *dissociation*:

(DISS) If $\Gamma, \varphi, \psi \vdash \Delta$ is valid, $\Gamma, \varphi \,\&\, \psi \vdash \Delta$ is valid.

Alas, it is at this point that, without meaning to be dissing it, I think that the neoclassical building of **W** crumbles (sorry Alan), for (DISS) is systematically invalid in **W**, as can be verified in what is for a lover of operational classicism

[20] It's true that, given the classical essence of implication, the *sheer fact* that $\varphi \rightarrow \psi$ holds is (so much) *weaker* than the *sheer fact* that $\varphi$ entails $\psi$ (for example, 'If Scotland won the World Cup four times, Scotland won the World Cup more times than Italy' holds, but in no reasonable sense does 'Scotland won the World Cup four times' entail 'Scotland won the World Cup more times than Italy', see Zardini 2022, p. 1590, footnote 13 for some discussion). However, in ($DT^{SPSC}$) and ($MP^{2TRANS^{MAJ}}$), you must pay heed both to the fact that $\varphi \rightarrow \psi$ is a *conclusion* and to the role of *side premises* and *side conclusions*. Once you do that, you'll see that the fact that $\varphi \rightarrow \psi$ is *valid under the assumptions of the truth of all side premises* $\Gamma$ *and of the falsity of all side conclusions* $\Delta$ correlates with the fact that, *under the same assumptions*, $\varphi$ entails $\psi$ (so that, in the offending example, what correlates is the fact that 'If Scotland won the World Cup four times, Scotland won the World Cup more times than Italy' is valid under the (as a matter of fact, correct) assumption of the truth of 'Scotland did not win the World Cup four times' and what is correlated with is the fact that, under the same assumption, 'Scotland won the World Cup four times' does entail, by (EXPL), 'Scotland won the World Cup more times than Italy'). Therefore, in terms of entailment, the fact that $\varphi \rightarrow \psi$ holds is to be interpreted not in the sense that $\varphi \vdash \psi$ is valid, but in the sense that, *for some side premises* $\Gamma$ *that hold and side conclusions* $\Delta$ *that fail to hold*, $\Gamma, \varphi \vdash \Delta, \psi$ is valid. Implication is, in this sense, "*de facto* entailment". (Indeed, it's easy to see that *de facto* entailment is equivalent with the features of the classical essence of implication according to which *the implication from* $\varphi$ *to* $\psi$ *is necessary and sufficient for the absence of [*$\varphi$*'s holding together with* $\psi$*'s failing to hold as facts] (or, equivalently in a classical framework, the presence of [*$\varphi$*'s failing to hold together with* $\psi$*'s holding as alternatives]). Right-to-left.* Suppose that either $\varphi$ fails to hold or $\psi$ holds. If $\varphi$ fails to hold, by the classical essence of negation $\neg\varphi$ holds, and so, since, by (EXPL) and (MON), $\neg\varphi, \varphi \vdash \psi$ holds, $\varphi$ *de facto* entails $\psi$. If $\psi$ holds, by the classical essence of negation $\neg\psi$ fails to hold, and so, since, by (EXH) and (MON), $\varphi \vdash \neg\psi, \psi$ holds, $\varphi$ *de facto* entails $\psi$. *Left-to-right.* If, for some side premises $\Gamma$ that hold and side conclusions $\Delta$ that fail to hold, $\Gamma, \varphi \vdash \Delta, \psi$ were valid, but $\varphi$ held together with $\psi$'s failing to hold, there would be an entailment all of whose premises hold and all of whose conclusions fail to hold, which is impossible.)

an especially poignant way by observing that (EXPL) is valid in **W** whereas (*pace* what the title of Weir 2004 suggests to naive ears) the law of *noncontradiction*:

(LNC) $\varphi \,\&\, \neg\varphi \vdash \varnothing$

is not—what a barbarism. What's the point of implication's correlating with the turnstile (at least in the way afforded by the validity of (DT) and (MP)) if conjunction does not correlate with the comma on the left? I therefore conclude that **W**'s conjunction is not fully classical.

Given the duality between conjunction and disjunction in **W** (AYE), the same kind of situation obtains with respect to *disjunction*. It's true that (ADD) (which encodes the feature of the classical essence of disjunction according to which $\varphi$ *is sufficient for the disjunction of* $\varphi$ *with* $\psi$ *and is sufficient for the disjunction of* $\psi$ *with* $\varphi$) and the entailment of *abjunction*:

(ABJ) $\varphi \vee \psi \vdash \varphi, \psi$

(which encodes the feature of the classical essence of disjunction according to which *the disjunction of* $\varphi$ *with* $\psi$ *is sufficient for* $\varphi$ *together with* $\psi$ *as alternatives*) are valid in **W**. But that also won't satisfy a real lover of operational classicism. For she knows that the perfect beauty of such a classicism requires that, just as it is also in the essence of implication to correlate in the object-language with the metalanguage relation between premise and conclusion in the way achieved by ($\mathrm{DT}^{\mathrm{SPSC}}$) and ($\mathrm{MP}^{\mathrm{2TRANS}^{\mathrm{MAJ}}}$), and just as it is also in the essence of conjunction to correlate in the object-language with the metalanguage relation between premise and premise in the way achieved by ($\mathrm{ADJ}^{\mathrm{1TRANS}}$) and (DISS), it be also in the essence of disjunction *to correlate in the object-language with the metalanguage relation between conclusion and conclusion* in the way achieved by the metaentailment of *association*:

(ASS) If $\Gamma \vdash \Delta, \varphi, \psi$ is valid, $\Gamma \vdash \Delta, \varphi \vee \psi$ is valid

and the metaentailment of *second-step transitivity of abjunction*:

($\mathrm{ABJ}^{\mathrm{2TRANS}}$) If $\Gamma \vdash \Delta, \varphi \vee \psi$ is valid, $\Gamma \vdash \Delta, \varphi, \psi$ is valid.

Alas, it is at this point that, without making indecent puns on the label of the metaentailment, I think that the ruins of the neoclassical building of **W** further crumble, for (ASS) is systematically invalid in **W**, as can be verified in what is for a lover of operational classicism an especially poignant way by observing that (EXH) is valid in **W** whereas (LEM) is not—what a barbarism. What's the point of implication's correlating with the turnstile (at least in the way afforded by the validity of (DT) and (MP)) if disjunction does not correlate with the comma on the right? I therefore conclude that **W**'s disjunction is not fully classical.[21]

[21] Once such features of the classical essence of conjunction and disjunction are in view, we can see that also **W**'s negation and implication are not fully classical. Starting with negation, it is also in its essence *to correlate in the object-language with the metalanguage property of a premise's being inconsistent* in the way achieved by the metaentailment of *inclusion*:

## 4.3 The Sorites Paradox

Take 'bald' as our sample vague expression, let $Bi$ be short for 'A man with number $i$ hairs is bald' and $i'$ short for $i+1$. *Naive vagueness* is then the equivalence between $Bi$ and $Bi'$ coupled with the uncontroversial $B0$ and $\neg B100{,}000$. While, being for the purposes of this paper wholly unproblematic, the right-to-left direction of such an equivalence need not detain us further, its left-to-right direction is naturally understood, informally, in terms of *the holding of the former guaranteeing the holding of the latter*, and, formally, in terms of the implications $Bi \rightarrow Bi'$, with the proviso that $\rightarrow$ obeys (MP). (This is a *tolerance*-oriented understanding of naive vagueness (Wright 1975). A different, [*no-sharp-boundary*]-oriented understanding of naive vagueness (Wright 1987) is, informally, in terms of *the holding of the*

---

(INCL) If $\Gamma, \varphi \vdash \Delta$ is valid, $\Gamma \vdash \Delta, \neg\varphi$ is valid

and the metaentailment of *exclusion*:

(EXCL) If $\Gamma \vdash \Delta, \neg\varphi$ is valid, $\Gamma, \varphi \vdash \Delta$ is valid

(with clarifications analogous to those entered for implication in footnote 20). Alas, it is also at this point that I think that the ruins of the neoclassical building of **W** further crumble, for both (INCL) and (EXCL) are systematically invalid in **W**, as can be verified by noticing that $\varphi \rightarrow \neg\varphi, \varphi \vdash_{\mathbf{W}} \neg\varphi$ is valid but neither $\varphi \rightarrow \neg\varphi \vdash_{\mathbf{W}} \neg\varphi$ nor $\varphi \rightarrow \neg\varphi, \varphi \vdash_{\mathbf{W}} \varnothing$ are. Denying the proviso entered in the third last paragraph, I therefore conclude that, after all, not even **W**'s negation is fully classical. Moving on to implication, as I've already mentioned in the last paragraph, it is also in its essence *to correlate in the object-language with the metalanguage relation between premise and conclusion* in the way achieved by ($\mathrm{DT}^{\mathrm{SPSC}}$) and ($\mathrm{MP}^{\mathrm{2TRANS}^{\mathrm{MAJ}}}$). Alas, it is at this point that I think that whatever is left of the neoclassical building of **W** finally crumbles, for both ($\mathrm{DT}^{\mathrm{SPSC}}$) and ($\mathrm{MP}^{\mathrm{2TRANS}^{\mathrm{MAJ}}}$) are systematically invalid in **W**, as can be verified by noticing that $\varphi \rightarrow (\varphi \rightarrow \psi), \varphi \vdash_{\mathbf{W}} \varphi \rightarrow \psi, \psi$ is valid but neither $\varphi \rightarrow (\varphi \rightarrow \psi) \vdash_{\mathbf{W}} \varphi \rightarrow \psi$ nor $\varphi \rightarrow (\varphi \rightarrow \psi), \varphi \vdash_{\mathbf{W}} \psi$ are. Denying the proviso entered in the second last paragraph, I therefore conclude that, after all, not even **W**'s implication is fully classical. With such features of the classical essence of the logical operations in view, I'd like to make two remarks. Firstly, as the astute reader will have noticed, the two members of each pair of the structure-correlating metaentailments are one the *converse* of the other one (ideally, their names should reflect this, whereas the ones I've gone for in this paper (e.g. 'second-step major-premise transitivity of *modus ponens*') often connect instead with familiar terminology, thereby however suggesting a distorted, transitivity-based idea of what the best justification is for the metaentailment in question). What is less trivial to observe in this connection is that many other attempts at characterising logical operations in terms of a pair of a "*how strong*"-principle and a "*how weak*"-principle (arguably including the one by Weir 1986) have been much less *harmonious* than this. Secondly, for each logical operation, it's easy to see that, together with (REFL) and (MON), *the structure-correlating metaentailments entail (but are not entailed by) the compositional principles* I've also been mentioning as features of the classical essence of the logical operations (those concerning which **W** scores so well; notice that, among those, the ones for implication should be understood so that "entailment" is *de facto* entailment (footnote 20), and hence so that these principles too count as compositional). It is natural to take this fact as evidence for the idea that the structure-correlating metaentailments encode *more fundamental* features of the logical operations. I should perhaps add that I myself wouldn't take even the structure-correlating metaentailments presented in the paper to be rock bottom, but I leave for other occasions the exploration of *even more fundamental correlations between structural elements and logical operations* and the question of *which of these two relata is prior* (see Zardini 2022, pp. 1586–1593 and Zardini 2021b, pp. 507–516; 2023b respectively).

*former ruling out the failure to hold of the latter*, and, formally, in terms of the negations $\neg(Bi \,\&\, \neg Bi')$. For the purposes of the paper, I'll mostly focus on the tolerance-oriented understanding of naive vagueness.)

Among the paradoxes of vagueness, historically and theoretically prominent is the *Sorites paradox* (see Oms and Zardini 2019a for an introduction and Oms and Zardini 2019b for a guide). Here is a typical version of the paradox. By (MP), $B0, B0 \rightarrow B1 \vdash B1$ and $B1, B1 \rightarrow B2 \vdash B2$ are valid, and so, by ($\text{TRANS}_{\vdash}^{\text{SPSC}}$), $B1 \rightarrow B2, B0 \rightarrow B1, B0 \vdash B2$ is valid. But, by (MP), $B2, B2 \rightarrow B3 \vdash B3$ is also valid, and so, by ($\text{TRANS}_{\vdash}^{\text{SPSC}}$), $B2 \rightarrow B3, B1 \rightarrow B2, B0 \rightarrow B1, B0 \vdash B3$ is valid. Applying this routine another 99, 997 times, we reach the conclusion that $B99{,}999 \rightarrow B100{,}000, B99{,}998 \rightarrow B99{,}999, B99{,}997 \rightarrow B99{,}998 \ldots, B0 \rightarrow B1, B0 \vdash B100{,}000$ is valid. Since $B100{,}000$ is obviously false and $B0$ obviously true, by the kind of nontransitivity-friendly normative principle mentioned in footnote 15 one cannot accept in the intended way all instances of $Bi \rightarrow Bi'$, contrary to naive vagueness.

One might think that **W** is well positioned to vindicate naive vagueness in the presence of the Sorites paradox, since the typical version of the paradox rehearsed in the last paragraph makes blatant use of ($\text{TRANS}_{\vdash}^{\text{SPSC}}$)—exactly the strong transitivity principle **W** denies. However, on further reflection, things turn out to be less rosy. For one thing, it's easy to see that (as adumbrated in footnote 15) $B0, \neg B100, 000 \vdash_{\mathbf{W}} B0 \,\&\, \neg B1, B1 \,\&\, \neg B2, B2 \,\&\, \neg B3 \ldots, B99, 999 \,\&\, \neg B100, 000$ is valid. In **W**, keeping fixed positive and negative cases, there is only one kind of alternative: a sharp boundary! For another thing, it's easy to see that, in the Łukasiewicz 3-valued valuation scheme, there is no interpretation where $B0$, $\neg B100{,}000$ and the instances of $Bi \rightarrow Bi'$ all have best value. True, it's also easy to see that, in the Łukasiewicz 3-valued valuation scheme, there are interpretations where $B0$, $\neg B100{,}000$ have best value and no instance of $Bi \rightarrow Bi'$ has worst value, but that is not a great consolation for naive vagueness for at least two reasons. Firstly, while I've briefly argued in Sect. 4.2 that, contrary to common glosses in terms of indeterminacy as absence of a fact of the matter, we should not interpret the average value in **W** as a bar to acceptance, it remains the case that it is disappointing for a lover of naive vagueness that some instance of tolerance (which for her is clearly true) should have *the same value as a borderline case* (which for her is not clearly true)—worse, *the same value as its own negation* (which entails a sharp boundary—the "*aff-yer-heid*"-[kind-of-thing] they'd accept south of the border, e.g. Williamson 1994).[22] Secondly, it's easy to see that *the theses constitutive of naive vagueness are jointly inconsistent in* **W**: letting $\text{N}^{\rightarrow}$

[22] I think that, for the lover of naive vagueness, this simple point is as decisive as it can be, but let me also mention a further, more complicated one. In a nontransitive framework, it is natural to interpret the average value as a value that, while *good enough for acceptance*, is not *good enough for a premise of a valid entailment that warrants acceptance of its conclusion* (see Zardini 2015d, pp. 255–264 for an in-depth discussion of this difference). But that implies that those instances of tolerance can never be used to infer $Bi'$ from $Bi$, which is also disappointing for a lover of naive vagueness (to guarantee the possibility of such a use for every instance of tolerance was after all the

be $\{B0 \rightarrow B1, B1 \rightarrow B2, B2 \rightarrow B3 \ldots, B99{,}999 \rightarrow B100{,}000\}$, it is the case that $B0, \neg B100{,}000, \mathrm{N}^{\rightarrow} \vdash_{\mathbf{W}} \varnothing$ is valid (if, living in the paraconsistent new age, that still leaves you somewhat cool, consider that, by (MON), it entails that, for every $\varphi$, it is the case that $B0, \neg B100{,}000, \mathrm{N}^{\rightarrow} \vdash_{\mathbf{W}} \varphi$ is valid, so that naive vagueness entails e.g. that Rangers F.C. won the last Champions League). Naturally understood (as in the last paragraph), the Sorites paradox targets naive vagueness by allegedly showing that, keeping fixed uncontroversial positive and negative cases, *the instances of tolerance are, in spite of their immense plausibility, jointly untenable*: on this understanding, **W** surrenders naive vagueness to the Heaper just as the grimmest classical logician (no names!) does. True, it's also easy to see that *the conjunction* of the theses constitutive of naive vagueness is *not* inconsistent in **W** (Weir 2015, p. 103): letting $\nu^{\rightarrow}$ be $(B0 \rightarrow B1) \,\&\, (B1 \rightarrow B2) \,\&\, (B2 \rightarrow B3) \ldots \,\&\, (B99{,}999 \rightarrow B100{,}000)$, it is the case that $B0 \,\&\, \neg B100{,}000 \,\&\, \nu^{\rightarrow} \vdash_{\mathbf{W}} \varnothing$ is invalid. That fact is unsurprising given the invalidity of (DISS) in **W** observed in Sect. 4.2, and, expanding on what I've already urged in that discussion, it's better taken as bad news for conjunction in **W** rather than as good news for naive vagueness in **W**: in **W**, the conjunction of some sentences falls "logically short" of their totality.[23]

One might still hope that, in **W**, the conjunction of some sentences does not fall *a lot* "logically short" of their totality, and in particular that, in **W**, $B0 \,\&\, \neg B100{,}000 \,\&\, \nu^{\rightarrow}$ can serve reasonably well as an *Ersatz* for naive vagueness. However, it arguably serves very poorly as such an *Ersatz*, since it does not even entail $B1$! As far as naive vagueness is concerned, the general problem here is that $Bi \,\&\, (Bi \rightarrow Bi') \vdash_{\mathbf{W}} Bi'$ is invalid: within a conjunction, implication no longer obeys *modus ponens*, and so it is now without the force required to establish such a strong equivalence between $Bi$ and $Bi'$ as to ensure that the holding of the one guarantees the holding of the other (as *per* naive vagueness). And that problem

---

point of adding 'with the proviso that $\rightarrow$ obeys (MP)' in the characterisation of naive vagueness in the second last paragraph).

[23] Notice that the move to the Łukasiewicz *continuum*-valued valuation scheme (Łukasiewicz and Tarski 1930), on which every instance of $Bi \rightarrow Bi'$ can have a value that is at least extremely close to the best possible value in the valuation scheme, would only address the *first* but not the *second* reason presented in the main text. And, on reflection, even the first reason would arguably not be addressed very satisfactorily: for one thing, by the same token, also some (for virtually everyone) gruesome instance of the "*paratonicity*" principle $Bi' \rightarrow \neg Bi$ (where neither is it clear that $Bi'$ does not hold nor is it clear that $\neg Bi$ holds) has a value that is at least extremely close to the best possible value (should one be tempted to move to a partially ordered lattice as a value space, one ought to keep in mind that doing so in such a way as to get the right results in this kind of case generates a great risk of validating (LEM)); for another thing, the analogue of the first reason *with a [no-sharp-boundary]-oriented understanding of naive vagueness* remains unaddressed (should one try to address it by appealing to the "*strong*" versions of conjunction and disjunction available in the Łukasiewicz framework, together with the—only apparently minor—blemish that a lot of instances of $Bi \rightarrow Bi'$ cannot have the best possible value that would lead to the results, using again the "strong" versions of conjunction and disjunction, that one *or* the other of those instances does not hold and that one *or* the other of the instances of $Bi \,\&\, \neg Bi'$ holds).

arises because, very generally, while (MP) is valid in **W**, the entailment of *one-premise modus ponens*:

(MP$^{\mathrm{OP}}$) $\varphi \,\&\, (\varphi \rightarrow \psi) \vdash \psi$

is not (see Zardini 2013b for a defence—unsurprisingly, essentially based on (DISS)—of the claim that (MP$^{\mathrm{OP}}$) should stand together with (MP)), and, even more generally, *many entailments are no longer valid once their premises are conjoined* (thereby precipitating the invalidity of (DISS)).[24] What can consistently be accepted in **W** is therefore a far cry from anything recognisable as naive vagueness.[25]

While negative, I take it that the foregoing discussion is helpful in that it helps to make emerge some desiderata of naive vagueness: neither $B0, \neg B100{,}000, \mathrm{N}^{\rightarrow} \vdash \varnothing$ (or, for that matter, $B0, \mathrm{N}^{\rightarrow} \vdash B100{,}000$) nor $B0 \,\&\, \neg B100{,}000 \,\&\, \nu^{\rightarrow} \vdash \varnothing$ (or, for that matter, $B0 \,\&\, \nu^{\rightarrow} \vdash B100{,}000$) should be valid, whereas both $Bi, \mathrm{N}^{\rightarrow} \vdash Bi'$ and $Bi \,\&\, \nu^{\rightarrow} \vdash Bi'$ should be valid. Indeed, I think that the second, positive kind of desideratum can be strengthened to the requirement that both $Bi, \mathrm{N}^{\rightarrow} \vdash Bi' \,\&\, \nu^{\rightarrow}$ and $Bi \,\&\, \nu^{\rightarrow} \vdash Bi' \,\&\, \nu^{\rightarrow}$ be valid (for future reference, let the latter entailment be (NAIV)): given $Bi$, coherence demands that we should be able to infer $Bi'$ *together with the naive idea licencing that inference*—that is, we should be able to infer $Bi' \,\&\, \nu^{\rightarrow}$.

It is this set of desiderata of naive vagueness that opens the way to a *limitative result* showing that *the kind of substructurality characterising* **W** *is not radical enough for naive vagueness*. By the second, positive kind of desideratum (in particular, (NAIV)), $B0 \,\&\, \nu^{\rightarrow} \vdash B1 \,\&\, \nu^{\rightarrow}$ and $B1 \,\&\, \nu^{\rightarrow} \vdash B2 \,\&\, \nu^{\rightarrow}$ are valid, and so, by (TRANS$_{\vdash}$), $B0 \,\&\, \nu^{\rightarrow} \vdash B2 \,\&\, \nu^{\rightarrow}$ is valid. But, by (NAIV), $B2 \,\&\, \nu^{\rightarrow} \vdash B3 \,\&\, \nu^{\rightarrow}$ is also valid, and so, by (TRANS$_{\vdash}$), $B0 \,\&\, \nu^{\rightarrow} \vdash B3 \,\&\, \nu^{\rightarrow}$ is valid. Applying this routine another 99,997 times, we reach the conclusion that $B0 \,\&\, \nu^{\rightarrow} \vdash B100{,}000 \,\&\, \nu^{\rightarrow}$ is valid, which clashes with the first, negative kind of desideratum (assuming throughout that the metaentailment of *second-step transitivity of simplification*:

(SIMPL$^{\mathrm{2TRANS}}$) If $\Gamma \vdash \Delta, \varphi \,\&\, \psi$ is valid, $\Gamma \vdash \Delta, \varphi$ and $\Gamma \vdash \Delta, \psi$ are valid

---

[24] This circumstance raises the question of *how to interpret conjunction in* **W**, since it is so drastically weaker than its two conjuncts together. One might jokingly say that, in **W**, $\varphi \,\&\, \psi$ really means 'It'd be nice if both $\varphi$ and $\psi$ were true', but that would be an unfair comment since (SIMPL) is valid in **W**. Indeed, *validity of (SIMPL) accompanied by invalidity of (DISS) is arguably characteristic of free-choice disjunction* (see von Wright 1951, p. 7, footnote 1 for an early mention and Zardini 2015c, pp. 500–501; 2016a, p. 321, footnote 51 for a recent take which emphasises the role of (SIMPL) and (DISS)), so that would be a natural place to start in further research on this question (and is a natural place to stop in this paper).

[25] Even focusing on $\nu^{\rightarrow}$, its acceptance in **W** basically amounts to a licence to *use only one of its conjuncts, arbitrarily chosen, only once*, no more (cf footnote 24)! As a lover of naive vagueness, I certainly wouldn't deny that that is some significant progress with respect to a dire epistemicist outlook, but it still falls a lot short of *standing by each and every conjunct of* $\nu^{\rightarrow}$ *together*, which I argue in Zardini (2019b), pp. 174–176 to be a stance that is absolutely crucial for naive vagueness.

is valid; anyway, since $B100{,}000 \,\&\, \nu^{\rightarrow}$ is obviously false, by the kind of nontransitivity-friendly normative principle mentioned in footnote 15 one cannot accept $B0 \,\&\, \nu^{\rightarrow}$, contrary to naive vagueness). Therefore, sticking to naive vagueness, something in the argument must fail, but this only involves (NAIV) and ($\text{TRANS}_{\vdash}$); since, as I've briefly argued in the last paragraph, the former is also part of naive vagueness, sticking to naive vagueness the latter must fail. A nontransitive solution to the paradoxes of vagueness is only faithful to naive vagueness if it denies ($\text{TRANS}_{\vdash}$) over and above ($\text{TRANS}_{\vdash}^{\text{SPSC}}$). Indeed, since (NAIV) is part of naive vagueness, this shows that *any solution whatsoever* to the paradoxes of vagueness is only faithful to naive vagueness if it denies ($\text{TRANS}_{\vdash}$): *naive vagueness requires the denial of even the most basic transitivity principle* (Zardini 2019b, pp. 173–176 offers further arguments building up to this conclusion), and, in the sense I'm about briefly to explain in a footnote, *such a denial suffices to solve the Sorites paradox.*[26]

## 4.4 The Liar and Curry Paradoxes

Take 'true' as our sample semantic expression, let $Ts$ be short for 'Sentence $s$ is true' and $\ulcorner\varphi\urcorner$ short (or long) for a singular term denoting $\varphi$. *Naive truth* is then the equivalence between $\varphi$ and $T\ulcorner\varphi\urcorner$. At first sight, such an equivalence is also naturally understood, informally, in terms of *the holding of the one guaranteeing the holding of the other*, and, formally, in terms of the implications $\varphi \rightarrow T\ulcorner\varphi\urcorner$ and $T\ulcorner\varphi\urcorner \rightarrow \varphi$, with the proviso that $\rightarrow$ obeys (MP). (This is a *convergence*-oriented understanding

[26] For what it's worth, my own take on all this is that naive vagueness should be preserved and classical logic should be revised exactly along nontransitive lines (as cried out by the *little-by-little* nature of the Sorites paradox), but, because of the points made in this section, in such a way as to deny ($\text{TRANS}_{\vdash}$) over and above ($\text{TRANS}_{\vdash}^{\text{SPSC}}$). My own way of doing this is to follow Weir's groundbreaking move of letting the conclusions of a valid entailment have a worse value than its premises (Weir 1998b, p. 222 is making history when he writes: "My approach here will be to abandon the rather conservative (in its holding to a form of bivalence) designated/undesignated dichotomy"), but, as *per* my discussion in Sect. 4.2, to go even further in that direction by letting that be the case even in the case of *one* conclusion, so as to give up the rather conservative (in its holding to a form of monism) *best-value preservation* and look rather for a *from-best-value-to-average-value connection* (concerning Weir's insistence on the symmetry between downwards best-value preservation and upwards worst-value preservation, notice that downwards from-best-value-to-average-value connection automatically guarantees upwards from-worst-value-to-average-value connection). As *per* my friendly suggestion in that discussion, that cannot satisfactorily—neither for the lover of operational classicism nor for the lover of naive vagueness—be done while maintaining the simple 3-valued space and Łukasiewicz 3-valued valuation scheme (or, for that matter, the simple *continuum*-valued space and Łukasiewicz *continuum*-valued valuation scheme, see footnote 23), so that a more complex value space and valuation scheme are required. As I show in Zardini (2008a); 2008b, pp. 93–173; 2009; 2014a; 2015c; 2019b, a more complex semantics is indeed available that delivers a nontransitive logic where the relevant instances of ($\text{TRANS}_{\vdash}$) are invalid (and so where naive vagueness is consistent) and where all the features of the classical essence of the logical operations (i.e., as *per* the discussion in Sect. 4.2, (INCL), (EXCL), ($\text{DT}^{\text{SPSC}}$), ($\text{MP}^{\text{2TRANS}^{\text{MAJ}}}$), ($\text{ADJ}^{\text{1TRANS}}$), (DISS), (ASS) and ($\text{ABJ}^{\text{2TRANS}}$)) are preserved.

of naive truth (Aristotle, *Metaph.*, Γ, 1011b25). A different, *nondivergence*-oriented understanding of naive truth (Zardini 2013a) is, informally, in terms of *the holding of the one ruling out the failure to hold of the other*, and, formally, in terms of the negations $\neg(\varphi \,\&\, \neg T\ulcorner\varphi\urcorner)$ and $\neg(T\ulcorner\varphi\urcorner \,\&\, \neg\varphi)$. For the purposes of this paper, I'll mostly focus on the convergence-oriented understanding of naive truth.)

Among the semantic paradoxes, historically and theoretically prominent is the *Liar paradox* (see Zardini 2014c for an introduction and a guide). Here is a typical version of the paradox. Because of the usual battery of philosophical considerations,[27] it can be assumed that there is a sentence $\lambda$ identical with $\neg T\ulcorner\lambda\urcorner$. By (EXH), $\varnothing \vdash \lambda, \neg\lambda$ (i.e. $\varnothing \vdash \lambda, \neg\neg T\ulcorner\lambda\urcorner$) is valid and so, by full intersubstitutability of $\varphi$ with $\neg\neg\varphi$,[28] $\varnothing \vdash \lambda, T\ulcorner\lambda\urcorner$ is valid. But, by (MP), $T\ulcorner\lambda\urcorner, T\ulcorner\lambda\urcorner \to \lambda \vdash \lambda$ is also valid, and so, by ($\text{TRANS}^{\text{SPSC}}_{\vdash}$), $T\ulcorner\lambda\urcorner \to \lambda \vdash \lambda, \lambda$ is valid, and hence, by (CONTR), $T\ulcorner\lambda\urcorner \to \lambda \vdash \lambda$ is valid. Applying a similar routine to $\neg\lambda$, we reach the conclusion that $\lambda \to T\ulcorner\lambda\urcorner \vdash \neg\lambda$ is also valid, and so, by the metaentailment of *second-step transitivity of adjunction*:

($\text{ADJ}^{\text{2TRANS}}$) If $\Gamma_0 \vdash \Delta_0, \varphi$ and $\Gamma_1 \vdash \Delta_1, \psi$ are valid, $\Gamma_0, \Gamma_1 \vdash \Delta_0, \Delta_1, \varphi \,\&\, \psi$ is valid

$T\ulcorner\lambda\urcorner \to \lambda, \lambda \to T\ulcorner\lambda\urcorner \vdash \lambda \,\&\, \neg\lambda$ is valid. Since $\lambda \,\&\, \neg\lambda$ is obviously false, by the kind of nontransitivity-friendly normative principle mentioned in footnote 15 one cannot accept in the intended way both $T\ulcorner\lambda\urcorner \to \lambda$ and $\lambda \to T\ulcorner\lambda\urcorner$, contrary to naive truth.

One might think that **W** is well positioned to vindicate naive truth in the presence of the Liar paradox, since the typical version of the paradox rehearsed in the last paragraph makes blatant use of ($\text{TRANS}^{\text{SPSC}}_{\vdash}$)—exactly the strong transitivity principle **W** denies.[29] However, on further reflection, things turn out to be less rosy.

---

[27] These concern the *apparent general possibility of the existence of an expression containing (in any singular-term position within the expression) a singular term referring to that very expression*. I'll save you trite examples, and only emphasise that such *philosophical* considerations are available quite independently of the question whether the existence of such expressions is provable in the operative *formal* system of syntax.

[28] Throughout, talk of "full" intersubstitutability is really *restricted to nonattitudinal contexts* (for example, 'Intuitionism holds that it is not the case that it is not the case that either Goldbach's Conjecture is true or it is not' is true, but 'Intuitionism holds that either Goldbach's Conjecture is true or it is not' is false).

[29] Actually, as I'll explain in the sixth next paragraph, **W** also denies *in general* full intersubstitutability of $\varphi$ with $\neg\neg\varphi$. As will be clear from that explanation, the *particular* instance occurring explicitly in the last paragraph is valid in **W**; however, interestingly enough, contrariwise, the *particular* instance of ($\text{TRANS}^{\text{SPSC}}_{\vdash}$) occurring only implicitly in the last paragraph (from the validity of $\varnothing \vdash \neg\lambda, \lambda$ and $\lambda \to T\ulcorner\lambda\urcorner, \lambda \vdash T\ulcorner\lambda\urcorner$ to the validity of $\lambda \to T\ulcorner\lambda\urcorner \vdash \neg\lambda, T\ulcorner\lambda\urcorner$) is valid in **W**, whereas the *particular* instance of full intersubstitutability of $\varphi$ with $\neg\neg\varphi$ occurring only implicitly in the last paragraph (from the validity of $\lambda \to T\ulcorner\lambda\urcorner \vdash \neg\lambda, T\ulcorner\lambda\urcorner$ to the validity of $\lambda \to T\ulcorner\lambda\urcorner \vdash \neg\lambda, \neg\neg T\ulcorner\lambda\urcorner$) is invalid in **W**. That is not an especially good news for ($\text{TRANS}^{\text{SPSC}}_{\vdash}$), for the validity of $T\ulcorner\lambda\urcorner \to \lambda \vdash \lambda$ is bad enough (as will become clear in the eighth next paragraph), but it does show that the *real issue* for **W** is not *specifically* transitivity, but *any* principle that, given Weir's assumption of a set-based format (footnote 17), allows one to replace

Because of the usual battery of philosophical considerations, it can be assumed that there is a sentence $\kappa$ identical with $T\ulcorner\kappa\urcorner \to \varepsilon$ (where $\varepsilon$ means that Edward II defeated Robert the Bruce at Bannockburn in 1314). Now, at second sight, the naive-truth equivalence between $\varphi$ and $T\ulcorner\varphi\urcorner$ is naturally understood, informally, in terms of *whatever holds of the one holding of the other*, and, formally, in terms of the full intersubstitutability of $\varphi$ with $T\ulcorner\varphi\urcorner$ (so that, in particular, if $\Gamma_0 \vdash \Delta_0$ and $\Gamma_1 \vdash \Delta_1$ only differ in that $\varphi$ or $T\ulcorner\varphi\urcorner$ occur in certain positions in some member of $\Gamma_0$ or $\Delta_0$ whereas $T\ulcorner\varphi\urcorner$ or $\varphi$ respectively occur in those positions in $\Gamma_1$ and $\Delta_1$, then $\Gamma_0 \vdash \Delta_0$ is valid iff $\Gamma_1 \vdash \Delta_1$ is). Letting $\mathbf{W^T}$ be the system such that $\Gamma \vdash_{\mathbf{WT}} \Delta$ is valid iff $\mathrm{N}^{\to}, \Gamma \vdash_{\mathbf{W}} \Delta$ is (where $\mathrm{N}^{\to}$ is now $\{\varphi \to T\ulcorner\varphi\urcorner, T\ulcorner\varphi\urcorner \to \varphi : \varphi$ is in the language$\}$), since, by (MP), $T\ulcorner\kappa\urcorner, T\ulcorner\kappa\urcorner \to \varepsilon \vdash_{\mathbf{WT}} \varepsilon$ is valid, if full intersubstitutability[30] were valid in $\mathbf{W^T}$, it would follow that $T\ulcorner\kappa\urcorner, T\ulcorner\kappa\urcorner \vdash_{\mathbf{WT}} \varepsilon$—that is, $T\ulcorner\kappa\urcorner \vdash_{\mathbf{WT}} \varepsilon$—is valid.[31] That would be disastrous, for then, by (DT), $\varnothing \vdash_{\mathbf{WT}} T\ulcorner\kappa\urcorner \to \varepsilon$ would be valid, and so, by full intersubstitutability, $\varnothing \vdash_{\mathbf{WT}} T\ulcorner\kappa\urcorner$ would be valid, and hence, by the metaentailment of *modus ponens for logical truths*:

(MP$^{\mathrm{LT}}$) If $\varnothing \vdash \varphi$ is valid and $\varnothing \vdash \varphi \to \psi$ is valid, $\varnothing \vdash \psi$ is valid,

$\varnothing \vdash_{\mathbf{WT}} \varepsilon$ would be valid.[32]

Since (DT) and (MP$^{\mathrm{LT}}$) are valid in $\mathbf{W^T}$, full intersubstitutability (must be and indeed) is invalid in $\mathbf{W^T}$.[33] In one, possibly less worrying direction, this points to

multiple conclusions (premises) by a single conclusion (premise)—even if the principle ought to be as harmless as full intersubstitutability of $\varphi$ with $\neg\neg\varphi$ ought to be! (Of course, one might try to analyse an intersubstitutability principle in terms of transitivity and certain entailments—so that e.g. full intersubstitutability of $\varphi$ with $\neg\neg\varphi$ would be analysed as (TRANS$^{\mathrm{SPSC}}_{\vdash}$) applied to the relevant entailment and the entailments $\varphi \vdash \neg\neg\varphi$ and $\neg\neg\varphi \vdash \varphi$—but, granting for the sake of argument such an unlikely analysis, it would be the case that the *real issue* for $\mathbf{W}$ is not transitivity as applied to *specific* entailments (e.g. (EXH) and (MP)), but transitivity as applied to *any* entailments in such a way that, given Weir's assumption of a set-based format, it allows one to replace multiple conclusions (premises) by a single conclusion (premise)—even if the entailments ought to be as harmless as $\varphi \vdash \neg\neg\varphi$ and $\neg\neg\varphi \vdash \varphi$ ought to be!)

[30] Henceforth, occurrences of 'intersubstitutability' and its relatives without specification of the *substituentia*/*substituenda* are understood as concerning the intersubstitutability of $\varphi$ with $T\ulcorner\varphi\urcorner$.

[31] Alternatively, if full intersubstitutability were valid in $\mathbf{W^T}$, it would follow that $\kappa, \kappa \vdash_{\mathbf{WT}} \varepsilon$ —that is, $\kappa \vdash_{\mathbf{WT}} \varepsilon$—is valid, from which everything to follow would also follow *mutatis mutandis*.

[32] While valid in many systems of truth (including $\mathbf{W^T}$), the second full-intersubstitutability step is not as crucial as the first one: even without it, $\mathbf{W^T}$ would have to accept $T\ulcorner\kappa\urcorner \to \varepsilon$, and so at the same time accept $\kappa$—what worse offence to naive truth can one imagine than to assert that something is the case the truth of a sentence saying which is sufficient for Edward II's having defeated Robert the Bruce at Bannockburn in 1314?

[33] Never mind the merits of this solution to the version of Curry's paradox for *naive truth and implication* (which I'm about to discuss), it should be mentioned that it would not seem very *robust*. As is well known (Beall and Murzi 2013), there is a version of Curry's paradox for *naive validity*, which only employs the analogues of (MP) and (DT) for a validity predicate (*plus* (CONTR) and (TRANS$_{\vdash}$), see Zardini 2013c; 2014b for some critical discussion). In effect, *the naivety of naive validity is fully absorbed in the analogues of (MP) and (DT) for a validity predicate*—contrary to the case of naive truth with its implications $\varphi \to T\ulcorner\varphi\urcorner$ and $T\ulcorner\varphi\urcorner \to \varphi$, in the case of naive validity

a *technical weakness* of $\mathbf{W^T}$, since many systems of naive truth do validate full intersubstitutability. In another, possibly more worrying direction, full intersubstitutability is arguably required by *the concept of truth exceptionlessly fulfilling a well known expressive function*,[34] which is in turn required by a *quasideflationist* conception of truth according to which [at least one of the functions of the concept of truth is the expressive one and that concept exceptionlessly fulfils that function].[35] Therefore, $\mathbf{W^T}$ clashes with quasideflationism about truth. How badly does this clash reflect on $\mathbf{W^T}$? Well, if the reason in favour of naive truth in the first place is precisely quasideflationism about truth (as is the case e.g. for Field 2008 and would seem to be the case for Weir 2005, who actually would seem to assume steadfast deflationism about truth (NAE)), the clash is indeed arguably fatal. In my view (for whose defence I can here do no better than refer to Zardini 2015b), quasideflationism about truth is however so deeply problematic that it hardly makes appealing anything it requires.

---

there is no further principle at play in the paradox (other than CONTR) and ($\text{TRANS}_{\vdash}$) which may require applications of ($\text{TRANS}_{\vdash}^{\text{SPSC}}$). Therefore, keeping fixed (CONTR) and ($\text{TRANS}_{\vdash}$) (as $\mathbf{W^T}$ does), either of the analogues of (MP) and (DT) for a validity predicate must be invalid. But such a commitment of $\mathbf{W^T}$ is problematic in at least three (related) respects. Firstly, it would seem *at best rather inelegant* to give two quite different solutions to the version of Curry's paradox for naive truth and implication and to its version for naive validity. Secondly, if one has made peace with the idea that either of the analogues of (MP) and (DT) for a validity predicate is invalid, one should presumably be able also to make peace with the idea that either of (MP) and (DT) is invalid, contrary to the fact that one of the big selling points of $\mathbf{W^T}$ against other systems of naive truth (all fully structural systems with full intersubstitutability) was precisely *not* to make peace with the latter idea and validate (MP) and (DT) (AYE). Thirdly, just as (MP) and (DT) are indeed crucial for a prominent notion of *implication*, it is arguable that their analogues for a validity (or, better (Zardini 2021b, p. 518), sufficiency) predicate are crucial for a prominent notion of naive *validity* (or, better, *sufficiency*).

[34] The claim is not obvious and there has been a tendency in the literature to assume it without giving sufficient reasons in its favour (see Gupta and Standefer 2018 for an insightful questioning of the assumption). Here's a nifty argument for it (see Zardini 2022, p. 1600, footnote 38 for a clumsier one). Assume that a biimplication $\Leftrightarrow$ licences full intersubstitutability of its two sides (perhaps because $\varphi \Leftrightarrow \psi$ means something along the lines of 'Its being the case that $\varphi$ is the same thing as its being the case that $\psi$'). Presumably, the usual argument for thinking that intersubstitutability is valid inside a biimplication (i.e. we may want to express a sufficient and necessary condition consisting of infinitely many or unknown subconditions) is as good as ever under that assumption (i.e. we may want to express that something is the same thing as something consisting of infinitely many or unknown conditions). Since $\varphi \Leftrightarrow \varphi$ surely holds, that entails that $\varphi \Leftrightarrow T\ulcorner\varphi\urcorner$ holds, and so, by the full intersubstitutability of the two sides of $\Leftrightarrow$, we get the full intersubstitutability of $\varphi$ with $T\ulcorner\varphi\urcorner$.

[35] Notice that *steadfast deflationism about truth* differs from *quasideflationism about truth* by replacing in the formulation of the latter '*at least one of the* functions' with '*the* function', while *deflationism about truth* is the first component of steadfast deflationism about truth (after all, this is a mean old world and the concept of truth might have a function that it does not exceptionlessly fulfil). Notice also that, *at least under a minimal conception of a function of a concept*, the first component of quasideflationism about truth would be much less controversial as, under such a conception, it would boil down to the extremely plausible claim that, *in some cases, the concept of truth can be used to achieve certain expressive purposes (with no need of pragmatics)*.

If quasideflationism about truth fails, is there any good reason left for thinking that a system of naive truth should validate full intersubstitutability? I think there is, and it goes as follows. Full intersubstitutability guarantees the *existence of sentences that are tantamount to (qua fully intersubstitutable with) logical operations on them* (for example, the existence of a sentence fully intersubstitutable with its negation).[36] In turn, such double-faced sentences are not only fun to have—they're virtually forced by *(impredicative) higher-order quantification*. For every context $\Phi(\ldots)$, presumably there is an abstract text where, for some $\Psi(\ldots)$-role in the text, the unique $\Psi(\ldots)$-thing is something to the effect that $\Phi(\ldots)$ holds of the unique $\Psi(\ldots)$-thing's being the case; with $\iota P\mathrm{X}(P)$ meaning the same as 'The unique $\mathrm{X}(\ldots)$-thing is the case' (and with $\Leftrightarrow$ as in footnote 34), it is then apparent that $\Phi(\iota P\Psi(P)) \Leftrightarrow \iota P\Psi(P)$ holds (for example, letting $\Phi(\ldots)$ be 'It is great that...', presumably there is an abstract text $t$ such that, letting $\Psi(\ldots)$ be 'The first thing said by $t$ is that...', the first thing said by $t$ is that it is great that the first thing said by $t$ is the case; it is then apparent that its being great that the first thing said by $t$ is the case is the same thing as the first thing said by $t$ being the case[37]).[38] We can indeed use this feature of higher-order quantification to generate *nonsemantic paradoxes of selfreference*. For example, letting $\Psi(\ldots)$ be 'The most enigmatic thing said by the Declaration of Arbroath is that...', and suppose that the most enigmatic thing said by the Declaration of Arbroath is (a hitherto unnoticed marginal annotation to the effect that) $\iota P\Psi(P) \rightarrow \varepsilon$; it is then apparent that $(\iota P\Psi(P) \rightarrow \varepsilon) \Leftrightarrow \iota P\Psi(P)$ holds. Since, by (MP), $\iota P\Psi(P), \iota P\Psi(P) \rightarrow \varepsilon \vdash \varepsilon$ is valid, by full

[36] Standard selfreference (as *per* footnote 27) only guarantees the existence of sentences that are identical (and so, *a fortiori*, intersubstitutable) with logical operations on *a sentence to the effect that they are true* (for example, the existence of a sentence identical with the negation of a sentence to the effect that it is true). That yields the existence of the kind of sentences just mentioned in the main text if full intersubstitutability is valid (to wit, for every context $\Phi(\ldots)$, standard selfreference guarantees the existence of a sentence $\varphi$ that is identical (and so, *a fortiori*, intersubstitutable) with $\Phi(T\ulcorner\varphi\urcorner)$; by full intersubstitutability, that is intersubstitutable with $\Phi(\varphi)$).

[37] I'm understanding $t$ so that it is *definitional* for it that the first thing said by it is that it is great that the first thing said by $t$ is the case (that's the point of taking $t$ to be an *abstract* text rather than a *historical* one), which is enough to guarantee the intersubstitutability of 'It is great that the first thing said by $t$ is the case' with 'The first thing said by $t$ is the case' in *most* contexts—arguably even *more* than those where, to take the example of a historical text, 'Old Scotland wants no watery stuff that splashes around in small wooden bowls' (which says—in a more flaccid way—the same thing as what is said by the fourth and third last lines of Burns' *Address to a Haggis* '*Auld Scotland wants nae skinking ware that jaups in luggies*') is intersubstitutable with 'What is said by the fourth and third last lines of Burns' *Address to a Haggis* is the case'. It might not be enough to guarantee such intersubstitutability in *some recherché hyperintensional (but nonattitudinal, see footnote 28)* context, but these are not needed for paradox to strike, and so can safely be ignored for the purposes of this paper.

[38] Those who (like me) suffer from reifying tendencies might just as well take $\Leftrightarrow$ to be $=$. That would arguably spoil the rhetoric to the effect that higher-order quantification is not quantification over objects (since *identity arguably only makes sense between objects*), but that obscurantist rhetoric presupposes a hierarchy of orders in being which clashes with the illuminated naive idea that all beings are equal—we're all fellow objects.

intersubstitutability of the two sides of $\Leftrightarrow$ it is the case that $\iota P\Psi(P), \iota P\Psi(P) \vdash \varepsilon$—that is, by (CONTR), $\iota P\Psi(P) \vdash \varepsilon$—is valid. That is disastrous, for then, by (DT), $\varnothing \vdash \iota P\Psi(P) \rightarrow \varepsilon$ is valid, and so, by full intersubstitutability of the two sides of $\Leftrightarrow$, it is the case that $\varnothing \vdash \iota P\Psi(P)$ is valid, and hence, by ($\mathrm{MP}^{\mathrm{LT}}$), $\varnothing \vdash \varepsilon$ is valid. Wherever this paradox fails, it doesn't fail at the full-intersubstitutability steps—but all the other deductive steps are valid in $\mathbf{W}^{\mathbf{T}}$ (see Zardini 2008c, pp. 563–566; 2018b, pp. 189–190 for similar higher-order paradoxes).

Setting now aside the significance of higher-order quantification, it is important to note that the invalidity of full intersubstitutability emerging with $\kappa$ is of a relatively benign kind. It is true that it concerns the substitution of a *stand-alone* sentence (rather than of an *embedded* sentence), but it is a stand-alone sentence occurring *as a premise*. Now, one strand in naive thinking is that there is no big difference between $\varphi_0, \varphi_1, \varphi_2 \ldots, \varphi_i \vdash \psi_0, \psi_1, \psi_2 \ldots, \psi_j$ being valid and $(\varphi_0 \mathbin{\&} \varphi_1 \mathbin{\&} \varphi_2 \ldots \mathbin{\&} \varphi_i) \rightarrow (\psi_0 \vee \psi_1 \vee \psi_2 \ldots \vee \psi_j)$ holding (with &, $\rightarrow$ and $\vee$ appropriately understood, see Zardini 2014b, pp. 371–382; 2018a, pp. 251–259 for various ways of pushing this idea to extreme consequences; however, I hasten to add that the point to follow really only requires such a minimal connection between being *a premise* and being *(a conjunct of) an antecedent of an implication* as to be congenial also to classical thinking). But, if so, the invalidity of full intersubstitutability in question can in effect be seen as a failure *in the context of an antecedent of an implication*, and so, keeping in mind the connection between $\varphi \rightarrow \mathsf{f}$ and $\neg\varphi$, as a failure *in a negative context* (footnote 39 gives a general characterisation of positive and negative contexts), which, as evidenced by the difference between e.g. the very implausible 'Newtonian mass is not relativistic mass' and the very plausible ' 'Newtonian mass is relativistic mass' is not true' (see Field 1973 for philosophical considerations on 'Newtonian mass' and suchlike singular terms), is precisely one kind of context where full intersubstitutability is extremely problematic for completely independent reasons (Zardini 2015b, pp. 39–45). Given the more specific features of the case at hand, there is however only so much solace to be had from this observation. Firstly, what is extremely problematic for completely independent reasons is *substituting* $\varphi$ *for* $T\ulcorner\varphi\urcorner$ in a negative context (from ' 'Newtonian mass is relativistic mass' is not true' to 'Newtonian mass is not relativistic mass'), but the case at hand is (also) a case of failure of *substituting* $T\ulcorner\varphi\urcorner$ *for* $\varphi$ in a negative context (from $T\ulcorner\kappa\urcorner, T\ulcorner\kappa\urcorner \rightarrow \varepsilon \vdash_{\mathbf{W^T}} \varepsilon$ to $T\ulcorner\kappa\urcorner, T\ulcorner\kappa\urcorner \vdash_{\mathbf{W^T}} \varepsilon$). Secondly, in the case at hand, $\mathbf{W}^{\mathbf{T}}$ is effectively inviting us to think that *an implication* together with its antecedent entails its consequent, but that *the truth of the implication* together with its antecedent doesn't—one needn't be a fan of full intersubstitutability to smell a whiff of sophistry in such an invitation (see footnote 32 for a somewhat converse situation).

On yet further reflection, things turn out to be even less rosy. At third sight, the naive-truth equivalence between $\varphi$ and $T\ulcorner\varphi\urcorner$ is naturally understood, informally, in terms of *the holding of the one being tantamount to the holding of the other*, and, formally, in terms of the intersubstitutability of $\varphi$ with $T\ulcorner\varphi\urcorner$ *in positive contexts*.[39] Since, by (EXH), $\varnothing \vdash_{\mathbf{WT}} T\ulcorner\lambda\urcorner, \neg T\ulcorner\lambda\urcorner$ is valid, if positive intersubstitutability were valid in $\mathbf{W}^{\mathbf{T}}$, it would follow that $\varnothing \vdash_{\mathbf{WT}} T\ulcorner\lambda\urcorner, T\ulcorner\lambda\urcorner$—that is, $\varnothing \vdash_{\mathbf{WT}} T\ulcorner\lambda\urcorner$—is valid. That would be disastrous, for then, by positive intersubstitutability, $\varnothing \vdash_{\mathbf{WT}} \neg T\ulcorner\lambda\urcorner$ would be valid, and so, by the metaentailment of *explosion for logical truths*:

(EXPL$^{\mathrm{LT}}$) If $\varnothing \vdash \varphi$ and $\varnothing \vdash \neg\varphi$ are valid, $\varnothing \vdash \varnothing$ is valid

$\varnothing \vdash_{\mathbf{WT}} \varnothing$ would be valid.[40]

Since (EXPL$^{\mathrm{LT}}$) is valid in $\mathbf{W}^{\mathbf{T}}$, positive intersubstitutability (must be invalid and indeed) is invalid in $\mathbf{W}^{\mathbf{T}}$. In one, possibly less worrying direction, this points to a *technical weakness* of $\mathbf{W}^{\mathbf{T}}$, since many systems of naive truth do validate positive intersubstitutability. In another, possibly more worrying direction, the failure in question of positive intersubstitutability means that something of the form $\varnothing \vdash_{\mathbf{WT}} \varphi, \psi$ is valid but $\varnothing \vdash_{\mathbf{WT}} \varphi, T\ulcorner\psi\urcorner$ is not. That will already strike many as totally puzzling: how can one accept, in addition to the sentence $\varphi$, the sentence $\psi$ as the only alternative, but not accept $T\ulcorner\psi\urcorner$ as the only alternative?[41] But I think that even more people will be so struck by putting the point in a slightly different way. One usual gloss on one's accepting that $\varnothing \vdash \varphi, \psi$ is valid is that, *conditionally on* denying $\varphi$, one accepts $\psi$.[42] Therefore, conditionally on denying $T\ulcorner\lambda\urcorner$, $\mathbf{W}^{\mathbf{T}}$ accepts $\neg T\ulcorner\lambda\urcorner$, but does not accept $T\ulcorner\lambda\urcorner$. Notice that, while the former conjunct in the last sentence is, to say the least, utterly unsurprising (and applicable

[39] For the purposes of this paper, we can work with the following *doubly recursive* understanding of *positive* and *negative contexts*. *Base clause*: stand-alone sentences as conclusions (premises) are in a positive context (negative context). *Recursive clauses*: sentences in a positive context (negative context) further embedded under negation or as antecedents of an implication are in a negative context (positive context); sentences in a positive context (negative context) further embedded as consequents of an implication or as conjuncts of a conjunction or as disjuncts of a disjunction are in a positive context (negative context).

[40] While valid in many systems of truth (including $\mathbf{W}^{\mathbf{T}}$), the second positive-intersubstitutability step is not as crucial as the first one: even without it, $\mathbf{W}^{\mathbf{T}}$ would have to accept $T\ulcorner\lambda\urcorner$, and so at the same time accept the *negandum* of $\lambda$—what worse offence to naive truth can one imagine than to assert that a sentence saying something that is incompatible with what is the case is true?

[41] Throughout this discussion, I assume that facts about entailment correlate with facts about acceptance (under every circumstance, by a suitably idealised subject as *per* footnote 14, in the mould of the nontransitivity-friendly normative principle mentioned in footnote 15), and leave implicit those three parenthetical modifications.

[42] That gloss is arguably vindicated if we assume that the relevant subject endorses $\mathbf{W}$. (Indeed, as will be clear from the argument, the only condition on the system is that it validate both (EXPL) and (EXH).) *Left-to-right.* If one accepts that $\varnothing \vdash \varphi, \psi$ is valid, one accepts $\varphi$ and $\psi$ as the only alternatives. Since (EXPL) is valid in $\mathbf{W}$, conditionally on denying $\varphi$, one presumably rules out $\varphi$. But, in general, if one accepts only at most two alternatives, conditionally on ruling out one of them one accepts the other one. Therefore, conditionally on denying $\varphi$, one accepts $\psi$. *Right-to-left.* Since (EXH) is valid in $\mathbf{W}$, one accepts $\varphi$ and $\neg\varphi$ as the only alternatives. Therefore, since, conditionally on denying $\varphi$, one accepts $\psi$, one presumably accepts $\varphi$ and $\psi$ as the only alternatives, and so one accepts that $\varnothing \vdash \varphi, \psi$ is valid.

to virtually every system), it is the second conjunct that is totally puzzling (and inapplicable to virtually every system of naive truth): how, can one, conditionally on denying $\varphi$, accept $\psi$ but not accept $T\ulcorner\psi\urcorner$—that is exactly the kind of truth-theoretic incoherence with which systems of naive truth typically charge certain classical systems (e.g. Field 2008, pp. 115–228)! Maybe, in this case, such an incoherence would be less acute if the denial of $\varphi$ were inconsistent (assuming that inconsistencies somehow do not entail everything), for, maybe, the principles of naive truth break down under certain inconsistent suppositions. But, setting aside the fact that such defeasibility would discriminate the principles of naive truth from the usual principles of logic, the denial of $T\ulcorner\lambda\urcorner$ is perfectly consistent in $\mathbf{W^T}$, for $\neg T\ulcorner\lambda\urcorner \vdash_{\mathbf{W^T}} \varnothing$ is invalid (AYE).[43,44] In $\mathbf{W^T}$, it is possible that something is the case without a sentence saying it being true.[45]

[43] On almost all systems of naive truth, conditionally on denying $T\ulcorner\lambda\urcorner$, *both* $\neg T\ulcorner\lambda\urcorner$ *and* $T\ulcorner\lambda\urcorner$ are accepted. On some of these systems, contradictory sentences are jointly rejected and so $\neg T\ulcorner\lambda\urcorner$ is rejected; on some others of these systems, contradictory sentences are accepted and so $\neg T\ulcorner\lambda\urcorner$ need not be rejected. Just about the only system of naive truth I know of on which the connection between denying $T\ulcorner\lambda\urcorner$ and accepting both $\neg T\ulcorner\lambda\urcorner$ and $T\ulcorner\lambda\urcorner$ is severed is my own system (recalled in footnote 50), on which, conditionally on denying $T\ulcorner\lambda\urcorner$, *either* $\neg T\ulcorner\lambda\urcorner$ *or* $T\ulcorner\lambda\urcorner$ are accepted but *not both* (see especially Zardini 2023c).

[44] One might think that, while the denial of $T\ulcorner\lambda\urcorner$ is perfectly consistent in $\mathbf{W^T}$, nevertheless it—i.e. $\neg T\ulcorner\lambda\urcorner$, i.e. $\lambda$—cannot be accepted in $\mathbf{W^T}$. A natural argument in support of the thought would be that, firstly, $\lambda \vdash_{\mathbf{W^T}} T\ulcorner\lambda\urcorner$ is valid, so that, if one accepted $\lambda$, one would have to accept $T\ulcorner\lambda\urcorner$ and, secondly, by (EXPL) and (MON), $T\ulcorner\lambda\urcorner, \neg T\ulcorner\lambda\urcorner \vdash_{\mathbf{W^T}} \varepsilon$ is valid, so that, since one would be accepting $T\ulcorner\lambda\urcorner$ and $\neg T\ulcorner\lambda\urcorner$, one would have to accept $\varepsilon$. However, such an argument *patently presupposes the normative principle* that footnote 15 argues (partly by variously bolstering precisely the route from consistency to logical acceptability) to be extremely problematic in a nontransitive framework (and falls flat at the 'secondly'-step—which is crucial in explaining why contradictory sentences cannot be jointly accepted—*if reformulated with the restricted normative principle* that footnote 15 proposes as more friendly to a nontransitive framework, see Sect. 4.2 for discussion of related issues).

[45] It is natural at this point to wonder what the prospects are for a nontransitive approach that, contrary to Weir's, preserves positive and indeed full intersubstitutability. As I've flagged at the relevant places, the resulting system (a version of which is explored by Ripley 2012) will have to deny both ($\mathrm{MP^{LT}}$) and ($\mathrm{EXPL^{LT}}$). Such denials directly entail that *logical truths do not obey modus ponens and explosion (i.e. modus ponens and explosion fail in logic)*, and so, fanfares about preserving "classical logic" notwithstanding, as I've already mentioned in Sect. 4.2, such denials are rebarbative to lovers of operational classicism. They are even more rebarbative once we assume that, from the point of view of the system, logical truths hold—which we must do anyway if the system is to vindicate a minimal understanding of naive truth (the first one considered in this section, both in its convergence-oriented understanding and in its nondivergence-oriented understanding), for otherwise the fact that e.g. $\varphi \to T\ulcorner\varphi\urcorner$ is a logical truth would no longer entail that the system accepts that, so that the system could no longer be regarded as a serious system of naive truth. Under that assumption, the system's denials of ($\mathrm{MP^{LT}}$) and ($\mathrm{EXPL^{LT}}$) imply that *an implication ($\kappa$) and its antecedent ($T\ulcorner\kappa\urcorner$) hold but its consequent ($\varepsilon$) fails to do so* and that *a sentence ($\lambda$) and its negation ($\neg\lambda$) both hold*, which implies a conception of implication and negation that is deeply rebarbative to lovers of operational classicism (Zardini 2013b, pp. 578–580).

Notice that very similar instances of positive intersubstitutability are valid in $\mathbf{W^T}$: for example, by (EXH), $\varnothing \vdash_{\mathbf{WT}} \lambda, \neg\lambda$ is valid, given which an instance of positive intersubstitutability that is valid in $\mathbf{W^T}$ entails that $\varnothing \vdash_{\mathbf{WT}} T\ulcorner\lambda\urcorner, \neg\lambda$—that is, $\varnothing \vdash_{\mathbf{WT}} T\ulcorner\lambda\urcorner, \neg\neg T\ulcorner\lambda\urcorner$—is valid. In $\mathbf{W^T}$, there are only two alternatives: either $T\ulcorner\lambda\urcorner$ holds or it is not the case that it is not the case that it holds. For many, and especially for lovers of involutive negation, it's hard to see how that can fall short of conceding that $T\ulcorner\lambda\urcorner$ holds.[46] Further, by (EXH), $\varnothing \vdash_{\mathbf{WT}} \neg\lambda, \neg\neg\lambda$ is valid, given which an instance of full intersubstitutability that is valid in $\mathbf{W^T}$ entails that $\varnothing \vdash_{\mathbf{WT}} \neg T\ulcorner\lambda\urcorner, \neg\neg\lambda$—that is, $\varnothing \vdash_{\mathbf{WT}} \neg T\ulcorner\lambda\urcorner, \neg\neg\neg T\ulcorner\lambda\urcorner$—is valid. In $\mathbf{W^T}$, there are only two further alternatives: either $\neg T\ulcorner\lambda\urcorner$ holds or it is not the case that it is not the case that it holds. For many, and especially for lovers of involutive negation, it's hard to see how that can fall short of further conceding that $\neg T\ulcorner\lambda\urcorner$ holds. Indeed, many, and especially lovers of involutive negation, only see a *syntactic* distinction between a sentence and its double negation, and, of the rest, many only see some sort of *hyperintensional semantic* distinction that in any case has no bearing (in nonhyperintensional contexts like those under discussion) on truth and logic. Any such position requires full intersubstitutability (at least in nonhyperintensional contexts) of $\varphi$ with $\neg\neg\varphi$, which however, as already announced in footnote 29, is invalid in $\mathbf{W^T}$: for example, while $\varnothing \vdash_{\mathbf{WT}} T\ulcorner\lambda\urcorner, \neg\neg T\ulcorner\lambda\urcorner$ is valid, $\varnothing \vdash_{\mathbf{WT}} T\ulcorner\lambda\urcorner, T\ulcorner\lambda\urcorner$—that is, $\varnothing \vdash_{\mathbf{WT}} T\ulcorner\lambda\urcorner$—is not (indeed, that shows that even positive intersubstitutability of $\varphi$ with $\neg\neg\varphi$ is invalid in $\mathbf{W^T}$). It should actually not be surprising that $\mathbf{W^T}$ draws such crucial distinctions without a difference, for, as we've seen in footnote 17, **W**'s crucial failures of transitivity are determined precisely by identifying occurrences of premises and by identifying occurrences of conclusions that are occurrences of the *same sentence, so that any syntactic distinction will block failures of transitivity and will allow consequences* (such as

[46] To be clear, the root of the problem here has nothing to do with $T\ulcorner\lambda\urcorner$ and $\neg\neg T\ulcorner\lambda\urcorner$ in particular: *every* pair of (possibly only syntactically) distinct sentences $\varphi$ and $\psi$ each of which is such that, in $\mathbf{W^T}$, it cannot have worst value is such that $\varnothing \vdash_{\mathbf{WT}} \varphi, \psi$ is valid (dually, *every* pair of (possibly only syntactically) distinct sentences $\varphi$ and $\psi$ each of which is such that, in $\mathbf{W^T}$, it cannot have best value is such that $\varphi, \psi \vdash_{\mathbf{WT}} \varnothing$ is valid). And so $\varnothing \vdash_{\mathbf{WT}} \varphi \vee \neg\varphi, \neg\varphi \vee \varphi$ is valid, $\varnothing \vdash_{\mathbf{WT}} \forall x\lambda, \forall y\lambda$ is valid, $\varnothing \vdash_{\mathbf{WT}} \kappa, \kappa^*$ (where $\kappa^*$ is distinct from $\kappa$ in that it replaces 'defeated' with 'won against') is valid... Having noted this, in the rest of this section, in addition to belabouring the consequences it has for intersubstitutability principles (footnote 47), I focus on what I think are particularly embarrassing instances of those general facts.

e.g. $\varnothing \vdash T\ulcorner\lambda\urcorner, \neg\neg T\ulcorner\lambda\urcorner$) *that are essentially just as bad as those licenced by transitivity* (such as e.g. $\varnothing \vdash T\ulcorner\lambda\urcorner$).[47,48]

On a final reflection, things turn out to be decidedly bleak. Forget about any version of intersubstitutability (which is driven by a convergence-oriented understanding of naive truth) and consider the nondivergence-oriented understanding of naive truth: it's easy to see that $\neg(\lambda \,\&\, \neg T\ulcorner\lambda\urcorner)$ is equivalent in **W** with $\neg\lambda$ and that $\neg(T\ulcorner\lambda\urcorner \,\&\, \neg\lambda)$ is equivalent in **W** with $\lambda$ (Heck 2012 first points out and exploits the fact that these equivalences are valid in the vast majority of logics applied to truth; Zardini 2013a first points out and exploits the fact that these equivalences are invalid in the noncontractive logic mentioned in footnote 50). Since neither $\varnothing \vdash_{\mathbf{W^T}} \neg\lambda$ nor $\varnothing \vdash_{\mathbf{W^T}} \lambda$ are valid, it follows that neither $\varnothing \vdash_{\mathbf{W^T}} \neg(\lambda \,\&\, \neg T\ulcorner\lambda\urcorner)$ nor $\varnothing \vdash_{\mathbf{W^T}} \neg(T\ulcorner\lambda\urcorner \,\&\, \neg\lambda)$ are valid[49] (contrary to the fact that both $\varnothing \vdash_{\mathbf{W^T}} \lambda \to T\ulcorner\lambda\urcorner$ and $\varnothing \vdash_{\mathbf{W^T}} T\ulcorner\lambda\urcorner \to \lambda$ are valid). Therefore, while $\mathbf{W^T}$ vindicates at least a minimal convergence-oriented understanding of naive truth (the first one considered in this section), it does not vindicate the nondivergence-oriented understanding of naive truth. Indeed, not only it does *not vindicate* it; it *repudiates* it, as it accepts that

---

[47] I've exemplified the problem with the distinction between a sentence and its *double negation*, but, with its root now in view, it should be clear that the problem can be exemplified with several other distinctions (cf footnote 46)—for instance, by considering the distinction between a sentence and its *conjunction with the logical-truth constant* $\mathsf{t}$. Many, and especially lovers of the identity-style interaction of conjunction with the logical-truth constant, only see a *syntactic* distinction between a sentence and its conjunction with the logical-truth constant, and, of the rest, many only see some sort of *hyperintensional semantic* distinction that in any case has no bearing (in nonhyperintensional contexts like those under discussion) on truth and logic. Any such position requires full intersubstitutability (at least in nonhyperintensional contexts) of $\varphi$ with $\varphi \,\&\, \mathsf{t}$, which however is invalid in $\mathbf{W^T}$: for example, while $\kappa \,\&\, \mathsf{t}, T\ulcorner\kappa\urcorner \to \varepsilon \vdash_{\mathbf{W^T}} \varepsilon$ is valid, $\kappa, T\ulcorner\kappa\urcorner \to \varepsilon \vdash_{\mathbf{W^T}} \varepsilon$—that is, $\kappa \vdash_{\mathbf{W^T}} \varepsilon$—is not (ditto for the distinction between a sentence and its *disjunction with the logical-falsehood constant* $\mathsf{f}$).

[48] One might think that the problem is essentially due to the *multiple-conclusion* setup, but a dual problem can be posed simply in a *multiple-premise* one. For example, by (EXPL), $\lambda, \neg\lambda \vdash_{\mathbf{W^T}} \varnothing$ is valid, given which an instance of full intersubstitutability that is valid in $\mathbf{W^T}$ entails that $T\ulcorner\lambda\urcorner, \neg\lambda \vdash_{\mathbf{W^T}} \varnothing$—that is, $T\ulcorner\lambda\urcorner, \neg\neg T\ulcorner\lambda\urcorner \vdash_{\mathbf{W^T}} \varnothing$—is valid. Now, a plausible requirement on logical consequence (also according to Weir) is the one of upwards falsity preservation, so interpreted (but not by Weir) that, if $\Gamma \vdash \varnothing$ is valid, it requires that some members of $\Gamma$ be false (just as, if $\varnothing \vdash \Gamma$ is valid, downwards truth preservation requires that some members of $\Gamma$ be true, see footnote 9 for further related discussions). If so, in $\mathbf{W^T}$, there are only two alternatives: either $\neg T\ulcorner\lambda\urcorner$ holds or it is not the case that it is not the case that it holds. For many, and especially for lovers of involutive negation, it's hard to see how that can fall short of conceding that $\neg T\ulcorner\lambda\urcorner$ holds. Further, by (EXPL), $\neg\lambda, \neg\neg\lambda \vdash_{\mathbf{W^T}} \varnothing$ is valid, given which an instance of positive intersubstitutability that is valid in $\mathbf{W^T}$ entails that $\neg T\ulcorner\lambda\urcorner, \neg\neg\lambda \vdash_{\mathbf{W^T}} \varnothing$—that is, $\neg T\ulcorner\lambda\urcorner, \neg\neg\neg T\ulcorner\lambda\urcorner \vdash_{\mathbf{W^T}} \varnothing$—is valid. For the reason explained in the third last sentence, that implies that, in $\mathbf{W^T}$, there are only two further alternatives: either $\neg\neg T\ulcorner\lambda\urcorner$ holds or it is not the case that it is not the case that it holds. For many, and especially for lovers of involutive negation, it's hard to see how that can fall short of further conceding that $\neg\neg T\ulcorner\lambda\urcorner$ (and $T\ulcorner\lambda\urcorner$) holds.

[49] As we've had the opportunity to appreciate in the last paragraph (especially footnote 47), equivalence does not always licence intersubstitutability in $\mathbf{W^T}$; however, for the reason seen in that paragraph, it clearly does so in the present cases (the same point applies for the substitutions of equivalents to follow).

*the nondivergence-oriented understanding of naive truth is inconsistent* (for, since, by (EXPL), $\neg\lambda, \lambda \vdash_{\mathbf{W^T}} \varnothing$ is valid, $\neg(\lambda \,\&\, \neg T\ulcorner\lambda\urcorner), \neg(T\ulcorner\lambda\urcorner \,\&\, \neg\lambda) \vdash_{\mathbf{W^T}} \varnothing$ is valid). Notice that this not an *incompleteness*-style case where, as in the second last paragraph, $\mathbf{W^T}$ simply rejects a principle that, by the lights of naive truth, one would expect it to accept; it is an *unsoundness*-style case where $\mathbf{W^T}$ accepts a principle that, by the lights of naive truth, one would expect it to reject and indeed to deny.

Dually, but even more startlingly, it's easy to see that $\lambda \,\&\, \neg T\ulcorner\lambda\urcorner$ is equivalent in $\mathbf{W}$ with $\lambda$ and that $T\ulcorner\lambda\urcorner \,\&\, \neg\lambda$ is equivalent in $\mathbf{W}$ with $\neg\lambda$. Since, by (EXH), $\varnothing \vdash_{\mathbf{W^T}} \lambda, \neg\lambda$ is valid, it follows that $\varnothing \vdash_{\mathbf{W^T}} \lambda \,\&\, \neg T\ulcorner\lambda\urcorner, T\ulcorner\lambda\urcorner \,\&\, \neg\lambda$ is valid. In $\mathbf{W^T}$, there are only two alternatives: either something is the case but a sentence saying it is not true or a sentence is true but what it says is not the case. Therefore, not only, negatively, $\mathbf{W^T}$ accepts that the nondivergence of $\varphi$ from $T\ulcorner\varphi\urcorner$ is *not possible* (as *per* the conclusion of the last paragraph); positively, it accepts that *there is* divergence of $\varphi$ from $T\ulcorner\varphi\urcorner$.[50]

---

[50] For what it's worth, my own take on all this is that naive truth should be preserved and classical logic should be revised, but not along nontransitive lines: preserving the desideratum of full intersubstitutability (for the reason explained in sixth last paragraph) and the other structural principles, (MP) and (DT) already lead to $\varnothing \vdash \kappa$ and $\kappa \vdash \varepsilon$, and (EXH) and (EXPL) already lead to $\varnothing \vdash \lambda$ and $\lambda \vdash \varnothing$—by then, operational classicism, and indeed the much more minimal ideal of *nonselfrefutation* (of not declaring that one's view—$\kappa$—entails that Edward II defeated Robert the Bruce at Bannockburn in 1314, or of not declaring that one's view—$\lambda$—is inconsistent), has already been irreparably wrecked (see also footnote 45). I believe that classical logic should rather be revised along noncontractive lines (as cried out by the *double-face* nature of the Liar and Curry paradoxes, which exploit the fact that, under naive-truth conditions, certain sentences are tantamount to their own negation, or to the implication from them to another sentence). As I show in Zardini (2011); 2014b; 2016a; 2023a; 2023c, a suitable noncontractive logic is available where naive truth is consistent and where the classical essence of the logical operations (i.e., as *per* the discussion in Sect. 4.2, (INCL), (EXCL), ($\mathrm{DT^{SPSC}}$), ($\mathrm{MP^{2TRANS^{MAJ}}}$), ($\mathrm{ADJ^{1TRANS}}$), (DISS), (ASS) and ($\mathrm{ABJ^{2TRANS}}$)) is preserved. Weir does mention noncontractive approaches in a few places (see e.g. Weir 2015, p. 99), and powerfully emphasises (as does Zardini 2019a, pp. 156–157, footnote 3) how problematic the usual resource-theoretic justification is in a context where one is supposed to be dealing with *entertainment* (acceptance, supposition, rejection etc.) of *propositions* rather than *use* of *resources* (AYE). However, contrary to what Weir's scepticism suggests, *there are strong general reasons—independent of which logic is correct—for thinking that there is a difference between the rational role of entertaining a proposition once and the rational role of entertaining it twice*. (I won't insist on the *ad hominem* point that, for the reason seen in the second last paragraph (especially footnote 47), in $\mathbf{W^T}$ too $\varphi$ is not fully intersubstitutable with $\varphi \,\&\, \varphi$, given which it is hard not to draw a difference between the rational role of entertaining a proposition once and the rational role of entertaining it twice (cf Weir (2015), p. 104)!) For example, how can one argue that $\varphi$ entails $\varphi \,\&\, \varphi$? The general principle for inferring a conjunction would seem (ADJ); letting $\varphi = \psi$, (ADJ) is immediately triggered by supposing $\varphi$ *and by further supposing* $\varphi$ (i.e. by supposing $\varphi$ twice), but not by only supposing $\varphi$ (i.e. by supposing $\varphi$ once). You might be tempted to reply to this observation by saying that one can trigger (ADJ) by supposing $\varphi$ once and then *use the very same supposition on the one hand to infer that the first premise of (ADJ) holds* and *on the other hand to infer that the second premise of (ADJ) holds*. But such a reply gives the game away (twice). Firstly, since the way of triggering (ADJ) I initially described is apparently possible, the reply at best shows that, *by bringing in extra means*, by supposing $\varphi$ once one can get the same end results as one *immediately* (i.e. without bringing in extra means) gets by supposing $\varphi$ twice, but, by doing so, only confirms the difference between the rational role of entertaining a proposition once

## References


Beall, J. C., & Murzi, J. (2013). Two flavors of Curry's paradox. *The Journal of Philosophy, 110*, 143–165.

Birkhoff, G., & von Neumann, J. (1936). The logic of quantum mechanics. *The Annals of Mathematics, 37*, 823–843.

Brouwer, L. (1908). De onbetrouwbaarheid der logische principes. *Tijdschrift voor wijsbegeerte, 2*, 152–158.

Chihara, C. (1979). The semantic paradoxes: A diagnostic investigation. *The Philosophical Review, 88*, 590–618.

Field, H. (1973). Theory change and the indeterminacy of reference. *The Journal of Philosophy, 70*, 462–481.

Field, H. (2008). *Saving truth from paradox*. Oxford University Press.

Gentzen, G. (1934). Untersuchungen über das logische Schließen I. *Mathematische Zeitschrift, 39*, 176–210.

Gupta, A., & Standefer, S. (2018). Intersubstitutivity principles and the generalization function of truth. *Synthese, 195*, 1065–1075.

Heck, R. (2012). A liar paradox. *Thought, 1*, 36–40.

Łukasiewicz, J. (1920). O logice trójwartościowej. *Ruch Filozoficzny, 5*, 170–171.

Łukasiewicz, J., & Tarski, A. (1930). Untersuchungen über den Aussagenkalkül. *Comptes rendus des séances de la Société des Sciences et des Lettres de Varsovie, 23*, 30–50.

Makinson, D. (1965). The paradox of the preface. *Analysis, 25*, 205–207.

Moore, G. (1942). A reply to my critics. In P. Schilpp (Ed.), *The philosophy of G.E. Moore* (pp. 535–677). Northwestern University Press.


---

and the rational role of entertaining it twice. Secondly, since the first premise of (ADJ) is nothing but $\varphi$ and the second premise of (ADJ) is nothing but $\varphi$, the reply does nothing but shift the real action from *supposing* $\varphi$ *once* to *inferring* $\varphi$ *twice* (on the one hand inferring $\varphi$ to secure the first premise of (ADJ) and on the other hand inferring $\varphi$ to secure the second premise of (ADJ))—since, contrary to inferring $\varphi$ twice, inferring $\varphi$ once would not get one much further in triggering (ADJ) than where one is by supposing $\varphi$ once, this further confirms the difference between the rational role of entertaining a proposition once and the rational role of entertaining it twice. To be clear, this is not a reason for thinking that contraction is invalid; it is, as I've said, a reason for thinking that there is a difference between the rational role of entertaining a proposition once and the rational role of entertaining it twice—a difference that can be endorsed while maintaining that e.g. what follows from supposing a proposition once includes what follows from supposing it twice, and so while maintaining that contraction is valid. (To apply this point to an issue that has emerged in footnote 17, it is only such a difference that is required for distinguishing the occurrence of a sentence as a side conclusion (premise) from its occurrence as a main conclusion (premise). What then creates the unnatural short circuit out of which the invalidity of transitivity arises in **W** is to adopt an *idea of logical consequence* (embodied in ($\mathrm{B} \Rightarrow \mathrm{B}^{\mathrm{W}}|\mathrm{W}^{\mathrm{B}} \Leftarrow \mathrm{W}$)) for which the difference between entertaining a proposition once and entertaining it twice would naturally be crucial (ironically, given Weir's scepticism about any such difference) while at the same time adopting a *format for logical consequence* where such a difference *cannot have any import for entailments* (as is forced by the validity of contraction)—indeed, where it *is not even representable* (as is forced by a set-based format).) Having said this, once we draw a difference between the rational role of entertaining a proposition once and the rational role of entertaining it twice, there are also strong reasons for thinking that it does have import on logical consequence: I've mentioned in footnote 17 and in this footnote how the idea of logical consequence embodied in ($\mathrm{B} \Rightarrow \mathrm{B}^{\mathrm{W}}|\mathrm{W}^{\mathrm{B}} \Leftarrow \mathrm{W}$) cries out for that, and I've myself developed a different account of logical consequence (mainly in Zardini 2019a) where that is the result of a natural and attractive view of what kind of states of affairs are expressed by sentences that give rise to semantic paradoxes.

Moruzzi, S., & Zardini, E. (2007). Conseguenza logica. In A. Coliva (Ed.), *Filosofia analitica* (pp. 157–194). Carocci.
Oliver, A., & Smiley, T. (2016). *Plural logic* (2nd edn.). Oxford University Press.
Oms, S., & Zardini, E. (2019a). An introduction to the sorites paradox. In S. Oms & E. Zardini (Eds.), *The sorites paradox* (pp. 3–18). Cambridge University Press.
Oms, S., & Zardini, E. (Eds.). (2019b). *The sorites paradox*. Cambridge University Press.
Oms, S., & Zardini, E. (2021). Inclosure and intolerance. *Notre Dame Journal of Formal Logic, 62*, 201–220.
Orlov, I. (1928). Isčislenije sovmestnosti predloženij. *Matematičeskij sbornik, 35*, 263–286.
Priest, G. (2006). *In contradiction* (2nd edn.). Oxford University Press.
Priest, G. (2010). Inclosures, vagueness, and self-reference. *Notre Dame Journal of Formal Logic, 51*, 69–84.
Ripley, D. (2012). Conservatively extending classical logic with transparent truth. *The Review of Symbolic Logic, 5*, 354–378.
von Wright, G. (1951). Deontic logic. *Mind, 60*, 1–15.
Weir, A. (1986). Classical harmony. *Notre Dame Journal of Formal Logic, 27*, 459–482.
Weir, A. (1993). Putnam, Gödel and mathematical realism. *International Journal of Philosophical Studies, 1*, 255–285.
Weir, A. (1996). Ultramaximalist minimalism! *Analysis, 56*, 10–22.
Weir, A. (1998a). Naïve set theory is innocent! *Mind, 107*, 763–798.
Weir, A. (1998b). Naïve set theory, paraconsistency and indeterminacy: Part I. *Logique et Analyse, 161–163*, 219–266.
Weir, A. (1999a). Naïve set theory, paraconsistency and indeterminacy: Part II. *Logique et Analyse, 167–168*, 283–340.
Weir, A. (1999b). The force of reason: Why is logic compelling? In T. Childers (Ed.), *The Logica yearbook 1999* (pp. 37–52). FILOSOFIA-ΦΙΛΟΣΟΦΙΑ.
Weir, A. (2000). Token relativism and the liar. *Analysis, 60*, 156–170.
Weir, A. (2002). Rejoinder to Laurence Goldstein on the liar. *Analysis, 62*, 26–34.
Weir, A. (2004). There are no true contradictions. In G. Priest, J. C. Beall, & B. Armour-Garb (Eds.), *The law of non-contradiction* (pp. 385–417). Oxford University Press.
Weir, A. (2005). Naive truth and sophisticated logic. In B. Armour-Garb & J. C. Beall (Eds.), *Deflationism and paradox* (pp. 218–249). Oxford University Press.
Weir, A. (2015). A robust non-transitive logic. *Topoi, 34*, 99–107.
Williamson, T. (1994). *Vagueness*. Routledge.
Wright, C. (1975). On the coherence of vague predicates. *Synthese, 30*, 325–365.
Wright, C. (1987). Further reflections on the sorites paradox. *Philosophical Topics, 15*, 227–290.
Yablo, S. (1985). Truth and reflection. *Journal of Philosophical Logic, 14*, 297–348.
Zardini, E. (2008a). A model of tolerance. *Studia Logica, 90*, 337–368.
Zardini, E. (2008b). *Living on the slippery slope. The nature, sources and logic of vagueness*. Ph.D. Thesis, University of St Andrews.
Zardini, E. (2008c). Truth and what is said. *Philosophical Perspectives, 22*, 545–574.
Zardini, E. (2009). Towards first-order tolerant logics. In O. Prozorov (Ed.), *Philosophy, mathematics, linguistics: Aspects of interaction* (pp. 35–38). Russian Academy of Sciences Press.
Zardini, E. (2011). Truth without contra(di)ction. *The Review of Symbolic Logic, 4*, 498–535.
Zardini, E. (2013a). It is not the case that [*P* and 'It is not the case that *P*' is true] nor is it the case that [*P* and '*P*' is not true]. *Thought, 1*, 309–319.
Zardini, E. (2013b). Naive *modus ponens*. *Journal of Philosophical Logic, 42*, 575–593.
Zardini, E. (2013c). Naive logical properties and structural properties. *The Journal of Philosophy, 110*, 633–644.
Zardini, E. (2014a). Evans tolerated. In K. Akiba & A. Abasnezhad (Eds.), *Vague objects and vague identity* (pp. 327–352). Springer.
Zardini, E. (2014b). Naive truth and naive logical properties. *The Review of Symbolic Logic, 7*, 351–384.

Zardini, E. (2014c). És la veritat una mentida? Perspectives sobre les paradoxes semàntiques. *Anuari de la Societat Catalana de Filosofia, 25*, 181–202.
Zardini, E. (2015a). Getting one for two, or the contractors' bad deal. Towards a unified solution to the semantic paradoxes. In T. Achourioti, K. Fujimoto, H. Galinon, & J. Martínez (Eds.), *Unifying the philosophy of truth* (pp. 461–493). Springer.
Zardini, E. (2015b). The opacity of truth. *Topoi, 34*, 37–54, .
Zardini, E. (2015c). ∀ and $\omega$. In A. Torza (Ed.), *Quantifiers, quantifiers, and quantifiers: Themes in logic, metaphysics, and language* (pp. 489–526). Springer.
Zardini, E. (2015d). Breaking the chains. Following-from and transitivity. In C. Caret & O. Hjortland (Eds.), *Foundations of logical consequence*, pp. 221–275. Oxford University Press.
Zardini, E. (2015e). Truth, demonstration and knowledge: A classical solution to the paradox of knowability. *Theoria: An International Journal for Theory, History and Foundations of Science, 30*, 365–392.
Zardini, E. (2016a). Restriction by noncontraction. *Notre Dame Journal of Formal Logic, 57*, 287–327.
Zardini, E. (2016b). No state *a priori* known to be factive is mental. *Philosophical Issues, 26*, 462–492.
Zardini, E. (2018a). Generalised Tarski's thesis hits substructure. In N. Kellen, N. Pedersen, & J. Wyatt (Eds.), *Pluralisms in truth and logic* (pp. 237–276). Palgrave MacMillan.
Zardini, E. (2018b). The role of utterances in Bradwardine's theory of truth. *Recherches de théologie et philosophie médiévales, 85*, 155–193.
Zardini, E. (2019a). Instability and contraction. *Méditations hégéliennes* I. *Journal of Philosophical Logic, 48*, 155–188.
Zardini, E. (2019b). Non-transitivism and the sorites paradox. In S. Oms & E. Zardini (Eds.), *The sorites paradox* (pp. 168–186). Cambridge University Press.
Zardini, E. (2021a). Closed without boundaries. *Synthese, 199*, 641–679.
Zardini, E. (2021b). Substructural approaches to paradox: An introduction to the special issue. *Synthese, 199*, 493–525.
Zardini, E. (2022). The final cut. *Journal of Philosophical Logic, 51*, 1583–1611.
Zardini, E. (2023a). One, and only one. ms.
Zardini, E. (2023b). Paradox and substructurality. ms.
Zardini, E. (2023c). Unstable knowledge. ms.

# Chapter 5
# The Umpire's Dilemma and the Ashes of Realism

**Neil Tennant**

**Abstract** Radford (Analysis 45:109–111, 1985) poses a prima-facie problem for the anti-realist or intuitionist who holds that all truths are knowable yet refuses to assert (or even denies) that all declarative sentences have determinate truth-values—values that might be independent of our means for determining what they are. This study sets out the Umpire's Dilemma and explores the prospect for an anti-realist solution of the problem that it poses.

**Keywords** Cricket · Umpire's dilemma · Anti-realism · Knowability · Doubt

## 5.1 On a Personal Note

Alan was so kind as to write as follows, in his Preface to *Proof through Truth*:

> To start fairly near the beginning, my earliest inspiration in turning to philosophy of mathematics was Neil Tennant, who, as a young lecturer, turned my head away from Bradley towards Quine (though later reflection discerns more in common between the two than I initially would have suspected); towards logic (implanting a firm bias in favour of the Gentzen/Prawitz approach); and towards philosophy of mathematics. I hope he does not count his work with me as a failure because I remained, then and now, immune to the attractions of intuitionism and relevance. Following a widespread tradition, I bite the hand which fed me in Chap. 6, where I criticize the kind of appeal to idealization which Neil makes in accounting for our mathematical knowledge. (Weir, 2010), p. vi

It gives me unadulterated joy, as a happy victim of philosophical patricide, to show (or at least try to, in due course) that the old codger might have had some philosophical kevlar under his vest, which Alan's viewfinder might not have revealed when he had me in his sights. The reader might be expecting from me a stern critical study of Chap. 6 in his book designed to persuade Alan to abandon

N. Tennant (✉)
Department of Philosophy, The Ohio State University, Columbus, OH, USA
e-mail: tennant.9@osu.edu



A. Rieger, S. Leuenberger (eds.), *Themes from Weir: A Celebration of the Philosophy of Alan Weir*, Synthese Library 484,
https://doi.org/10.1007/978-3-031-54557-3_5

his puzzling combination of a Byzantine mereology of physical tokens, a refusal to countenance abstract types, and a happy embrace of completed infinities, *including* infinite *proofs*. But that would strike an altogether discordant note in a *Festschrift*. We are gathered here between these covers, as it were, to celebrate Alan's wonderful example of how to philosophize, regardless of the doctrinal outcome.

Let me begin with some reminiscences of my own about those earlier days with Alan. We all know him to be a brave and kind man of fantastic good humour. Nowhere was his character on better display than back in the early summer of 1976, when Alan was one of four Edinburgh students whom I was accompanying to Dartmouth on our annual exchange between the two Philosophy Departments. We flew into Newark late, and the group opted for a night in a local motel—except for Alan. He was so anxious to get to Dartmouth that he struck out on his own, taking a taxi to the downtown Greyhound station in NYC. There his taxi driver relieved him, at knifepoint, of his wallet and traveller's cheques. Alan's subsequent conversation with a ticket clerk inside the station was overheard by a fellow Scot. 'Dinnae worry, lad!' the stranger said. He paid for Alan's ride to Hanover, and gave Alan his address in New Orleans for repayment. The first thing I had to do for Alan on arrival in Hanover was help him send the money to this kind compatriot. It was his overriding concern; he did not succumb to any post-traumatic stress at having his life threatened while being mugged. During our visit at Dartmouth, he was also the group's uproarious entertainer. It was Alan who chivvied us to gather to watch Startrek each week; it was Alan who regaled us with his discoveries in the vast aisles of the meat and poultry department of the local supermarket. To this day I remember the Renfrewshire brogue of his exclamation 'Breasts! Thighs!' in feigned disbelief at the upfront coarseness of American advertising.

## 5.2 Introducing an Underappreciated Problem for the Anti-Realist

Cricket, even if Wittgenstein might not have appreciated the fact, is not only the most important game, but also the most important form of life. Dummettian anti-realism is inspired by Wittgensteinian doctrines about meaning as use, manifestationism and rule-following. So Dummettian anti-realism should be able to resolve the dilemma to be set out below. Not to be able to do so would, philosophically speaking, just not be cricket.

### 5.2.1 A Little Lesson on Some Rules of Cricket (for American, and Perhaps Even Certain Scottish, Readers)

Two ways in which a batsman (i.e., striker) can be dismissed (or *Out*) are to be *Caught* or to be *Leg Before Wicket (LBW)*.

The batsman is out *Caught* if the ball, upon being properly bowled, makes above-ground contact with his bat or gloves, and is thereafter caught by a fielder without the ball making prior or simultaneous contact with the ground. A good example of this would be a case where the ball, only ever so slightly deflected after nicking the edge of the bat, is caught by the wicketkeeper, who stands behind the wicket. Afficionados call this being *Caught behind.*[1]

The batsman is out *LBW* if two conditions are met: (i) the ball strikes any part of his[2] body (*without* making prior or simultaneous contact with his bat) on a line between the two wickets,[3] or on the off-side;[4] and (ii) if the batsman had not been in the way, and the ball had been allowed to complete its trajectory, then it would have struck the batsman's wicket.[5]

Usually the epistemic situation with *LBW* cases is very unclear for the batsman himself, both with respect to condition (i) and with respect to condition (ii). He is not well-positioned to judge, even in certain quite clear cases, that he is indeed out *LBW*. The bowling-and-fielding team does not rely on any presumed confluence of honour and relevant omniscience on the part of the batsman. Instead, they appeal to the umpire to deliver the verdict of *LBW*.[6]

Umpires always act under the dictate 'If in doubt, give *Not Out*.'

---

[1] If the batsman does not immediately 'walk' (as he would be honour-bound to do if he himself knew that he had indeed been *Caught behind*), the bowling-and-fielding team has to appeal to the umpire if they wish to have a verdict of *Caught behind*. This they do by screaming 'Huzzat?', which the Quinean radical interpreter would be within a thick edge of rendering as 'How's that?'. It is an indeterminacy not yet fully resolved whether the apostrophe-'s' is short for 'is' or 'was' or the timeless 'is'. Even the totality of all possible behavioural evidence on the part of cricketers and umpires might underdetermine the metaphysics of time within that community. This might explain why cricket can appear to be so tedious, to the uninitiated radical interpreter.

[2] The use of the masculine pronoun is justified by appeal to the ease of use of 'batsman' as opposed to 'bats*person*'.

[3] Each wicket consists of three parallel and upright *stumps*, topped off with two *bails*.

[4] This is the side to which the batsman's toes point when he assumes his stance to face the bowler.

[5] Batsmen hardly ever 'give themselves' *Out* in the case of a possible *LBW* dismissal. A rare case where the batsman might be honour-bound to walk would be where he had stepped right back to the wicket, and had taken a full-toss delivery on his pads right in front of the wicket.

[6] This is done with a theatrical and histrionic yell of *'Huzzat!'* (radically interpretable as *'How's that?'*), usually from high in mid-air and with arms flung upwards. The umpire, if he agrees, raises a single finger at the batsman, who must then walk. This raised finger is *not* radically interpretable as any kind of insult. There is absolutely nothing American about the act.

### 5.2.2 The Umpire's Dilemma

The ground has now been prepared for the statement of our dilemma. Here (in our own words) is the deeply problematic situation that Radford asked his reader to consider.

> The umpire is certain that the ball strikes the batsman's leg pad in line with the stumps, and would not hesitate to give him out *LBW* if that were all. He is also certain that the ball went from the pad straight into the wicket-keeper's gloves. What he is not certain about is whether the ball nicked the bat before hitting the pad. Since he is not certain that it did, he cannot give the batsman out *Caught*. But since he is not certain that it did not, he cannot give the batsman out *LBW*. So although he knows that the batsman has broken one of those two laws, he cannot give him out as the laws stand (there is no 'disjunctive' way of being out, *Caught*-or-*LBW*). Video replays offer no help to the 3rd umpire—they do not reveal whether or not the ball nicked the bat, no matter how closely he looks.
> Problem for the anti-realist or intuitionist: this is a case of the umpire knowing that the batsman has broken either the *LBW*-law or the *Caught*-law $[K(p \vee q)]$, without either knowing $p$ or knowing $q$ $[\neg Kp, \ \neg Kq]$.

## 5.3 First Anti-Realist Response

The anti-realist might respond as follows.

> The question as to whether or not
>
> $N$: the ball nicked the bat
>
> has a discoverable answer. And that gives the anti-realist a way out. The umpire already knows
>
> $\Delta$: the ball went straight into the wicket-keeper's gloves
>
> and knows that
>
> $\Delta, N \vdash$ *Caught (behind)*; whence, obviously,
>
> $\Delta, N \vdash$ *LBW* $\vee$ *Caught (behind)*.
>
> The umpire also knows that
>
> $\Gamma$: the ball struck the batsman's pads in line with the stumps, and would otherwise have hit the wicket
>
> and knows that
>
> $\Gamma, \neg N \vdash$ *LBW*; whence, obviously,
>
> $\Gamma, \neg N \vdash$ *LBW* $\vee$ *Caught (behind)*.
>
> So by Dilemma on the decidable proposition $N$, the anti-realist concludes that
>
> $$\Delta, \Gamma \ \vdash \ LBW \vee Caught\ (behind).$$

**Realist Rejoinder** That is not satisfying. The answer (*LBW*, or *Caught*, as the case may be) is not discoverable by the umpire. Yet he does know that *LBW* $\vee$ *Caught*.

**Anti-Realist's Rebuttal** It is not important, in so far as this is to be taken as a *philosophical* problem rather than a mere problem for umpiring procedures, that it be the *umpire himself* who can get himself into a position to know which of *LBW*, or *Caught*, is true. The umpire's decision methods, given his responsibility to make timely decisions, need to be highly feasible. They cannot be expected to match the power or range of methods that would guarantee a result *in principle*, with no limits on the time or energy that might have to be expended in order to arrive at a decision. To repeat: we are seeking a resolution of the *philosophical* problem. We are not seeking a resolution of the problem, for the umpire, of how best to arrive at a timely judgement. So all that is required, philosophically, is that it be possible, in principle, for someone suitably placed and with the totality of evidence that would be available to her from being so placed, to discover, *eventually*, exactly which one of the propositions *LBW*, or *Caught*, is true. *In this sense*, there is a way one could discover which of *LBW*, *Caught*, is true. And that assurance is all that the umpire would need in order to be able, with a clear conscience, and on the evidence currently available to him, to give the batsman *Out*. The umpire's reasoning would be as follows:

$$\dfrac{\dfrac{\dfrac{\Delta \quad \overline{N}^{\,(1)}}{\mathit{Caught}}}{\mathit{Out}} \qquad \dfrac{\dfrac{\Gamma \quad \overline{\neg N}^{\,(1)}}{\mathit{LBW}}}{\mathit{Out}}}{\mathit{Out}}\;(1)$$

The step of constructive dilemma (labelled (1)) is licit because $N$ ('the ball nicked the bat') is a claim that is decidable in principle. Given that the umpire knows both $\Delta$ and $\Gamma$, his verdict of *Out* (courtesy of the reasoning just regimented) is subject to no doubt. He cannot be 'in doubt'; so he does not need to give 'Not Out'.

**Further Realist Objection** This proposal entails a new (and impermissible) category of dismissal, which would have to be recorded in the scorebook. The new category is *LBW or Caught*. But the rules of cricket do not allow for such disjunctive categories. The umpire is required to have arrived at a non-disjunctive determination—an *atomic* judgement, if you will—of *the exact way in which the batsman has been dismissed.*

## 5.4 Second Anti-Realist Response

So be it. You are saying that the rules implicitly require dismissals to be recorded not just as

*Out, LBW*

or

*Out, Caught*

but rather as

The umpire *knows that* the batsman is *Out, LBW*

or

The umpire *knows that* the batsman is *Out, Caught.*

As alternatives to 'knows that' here, one might consider also 'has determined that', or 'is morally certain that', or 'believes beyond any reasonable doubt that'. What is important is only that the attitudinal phrase be one that does *not* apply to either of the two disjuncts in the circumstances described in the Umpire's Dilemma.

Let us use $K$ for the kind of operator under consideration here. Let $u$ be the umpire. The suggestion is that the rules implicitly require, for licit dismissal, that the dismissal of the batsman as described in the Dilemma be recordable either in the form $uK(LBW)$ or in the form $uK(Caught)$. And, given the circumstances described, this fails to be the case even though $uK(LBW \vee Caught)$ (so the realist would have it, as well as the anti-realist who is prepared to treat as bivalent statements that are decidable in-principle).

**Renewed Realist Objection** So now the anti-realist faces a problem that takes the following general form.

One wishes to claim the consistency of the set

$$\{\neg xKp, \neg xKq, xK(p \vee q)\}.$$

(The umpire does *not* know that the batsman is *LBW*; does *not* know that the batsman is *Caught*; but *does* know that the batsman is *either LBW or Caught.*) But, on the anti-realist construal of knowledge of a disjunction, this is impossible. Anyone who knows that $p \vee q$ should know *which* of $p$, $q$ is true. That is, the following is a principle of (anti-realist) epistemic logic:

$$\frac{xK(p \vee q)}{xKp \vee xKq}.$$

One will then have a proof of the *in*consistency of the aforementioned set:

$$\dfrac{\dfrac{xK(p \vee q)}{xKp \vee xKq} \quad \dfrac{\neg xKp \quad \overline{xKp}^{(1)}}{\bot} \quad \dfrac{\neg xKq \quad \overline{xKq}^{(1)}}{\bot}}{\bot}{\scriptstyle(1)}$$

**Anti-Realist Rejoinder** There is an equivocation here on the force of the epistemic operator $K$. The sense of $K$ in which the umpire does not know that the batsman is out *LBW*, and does not know that the batsman is out *Caught*, is the *occurrent* sense of 'knows, right now, with conclusive justification, and with no further investigations of matters of fact, or drawings of inferences, being required'. Let us designate this as **K**, and reserve the italic $K$ for the 'laxer' kind of knowledge. This is the kind of knowledge (of a conclusion) that can be attained by taking steps of constructive dilemma on decidable propositions on which no decision has yet been reached. Note that, accordingly, it is much *harder* for a claim of the form $\neg xKp$ to be true. For this would require (on the anti-realist construal of negation) that one be able to demonstrate the absurdity, *modulo* what one already knows, of the hypothesis that $xKp$. The umpire surely cannot be represented as not knowing that the batsman is *LBW* (or as not knowing that the batsman is *Caught*) in *this*, weaker, sense of knowledge. For someone *might*, say, come forward with hugely magnified, slow-motion videography of the event, showing conclusively, one way or the other, whether the bat nicked the ball, and hence which exact category of dismissal applies. And to assert both $\neg xKp$ and $\neg xKq$ is (mistakenly) to rule out such a possibility.

The anti-realist would therefore reply to the realist's renewed objection above by pointing out that consistency is being claimed only for the set

$$\{\neg x\mathbf{K}p, \neg x\mathbf{K}q, xK(p \vee q)\}.$$

(The umpire does not *occurrently* know that the batsman is *LBW*; does not *occurrently* know that the batsman is *Caught*; but *does* know, in the weaker sense just described, that the batsman is *either LBW or Caught*.) Furthermore the anti-realist refuses to endorse the epistemic principle of inference

$$\frac{xK(p \vee q)}{xKp \vee xKq}.$$

What he *will* concede is only the weaker inferential principles

$$\frac{x\mathbf{K}(p \vee q)}{x\mathbf{K}p \vee x\mathbf{K}q}$$

and

$$\frac{xK(p \vee q)}{\Diamond\exists y\, yKp \vee \Diamond\exists y\, yKq}.$$

Of a disjunction known by $x$ (in the laxer sense $K$) one can only say: either its first disjunct is *knowable*, or its second disjunct is. And what is meant here by a propositon's being *knowable* is just that it be possible for *someone or other* (not necessarily $x$ himself) to know the proposition in question.

## 5.5 Different Construals of Doubt

'When in doubt, give *Not Out*.' Cricket is the nursery for criminal law. But what kind of doubt is at issue here? Presumably, if one thinks of credences or subjective probabilities, a proposition $p$ is in doubt (for the thinker in question) if $\Pr(p)$ is not high enough. What is 'high enough'? Presumably a threshold such as .7 or .9 or .99, or … . Whatever it is, one can assume that a value of .5 is not high enough. This is the value that a rational agent would assign to any proposition on which he was no more confident that it was true rather than that it was false, or *vice versa*.

Let us assume that the umpire is in this state of mind with regard to the proposition $N$ (that the bat nicked the ball). He *heard* no tell-tale 'click' of willow on leather; he *saw* no apparent deflection of the flight of the ball. *Yet* he could see that the ball was awfully close to the bat, and indeed could see no daylight between bat and ball. For the umpire, it could simply go either way.

This ensures that he is indeed in a state of doubt with regard to the claim *LBW*, and with regard to the alternative claim *Caught*. Yet he is in no state of doubt (see the proof above) with regard to the disjunctive claim *either LBW or Caught*.

Given all this, what should be his state of mind regarding the claim *Out* itself? In the proof above, we completed each case-proof by inferring the broad claim *Out* from the respective category-of-dismissal claims *LBW* and *Caught*. But perhaps one cannot infer *Out* so effortlessly in this way. Perhaps the relevant material (or Sellarsian) inference rules should not be

$$\frac{LBW}{Out} \quad \text{and} \quad \frac{Caught}{Out}$$

but rather the following, which require chapter-and-verse for any verdict of *Out*:

$$\frac{LBW \qquad not\ Caught}{Out} \quad \text{and} \quad \frac{Caught \qquad not\ LBW}{Out}$$

If *these* were the rules against the background of which doubt about *Out* were to be assessed, the umpire would be in a state of justifiable doubt about *Out*. This is because *not Caught* is *not* something he is in a position to affirm; nor is *not LBW*.

So the maxim should perhaps be

Give *not Out* if in doubt as to which exact category-of-dismissal applies.

Thus (so this line of thought continues) the umpire is enjoined to give *Not Out* even when possessed of the little proof above. The problem with that proof was that it invoked the wrong material rules of inference.

But the proof can be repaired. Consider the following:

$$\dfrac{\dfrac{\dfrac{\Delta \quad \overline{N}^{(1)}}{Caught} \qquad \dfrac{\Delta \quad \overline{N}^{(1)}}{\neg LBW}}{Out} \qquad \dfrac{\dfrac{\Gamma \quad \overline{\neg N}^{(1)}}{LBW} \qquad \dfrac{\Gamma \quad \overline{\neg N}^{(1)}}{\neg Caught}}{Out}}{Out}\,(1)$$

The modified, 'category-specific' rules are employed here. The rub is that any nick of ball on bat rules out *LBW*; whereas the absence of contact between bat and ball rules out *Caught*! So the modified rules afford no progress with the original problem. That problem, as we now appreciate more clearly, is generated by a constructive dilemma whose two horns are statements of matters of fact, not statements about umpiring states of mind. The modified rules would have to be modified even further:

$$\frac{LBW \quad \text{Umpire knows: } \textit{not Caught}}{Out}; \quad \text{and}$$

$$\frac{Caught \quad \text{Umpire knows: } \textit{not LBW}}{Out}$$

Now, from $\Delta$ and $N$ one cannot infer that the umpire *knows that* $\neg LBW$. And from $\Gamma$ and $\neg N$ one cannot infer that the umpire *knows that* $\neg Caught$.

But can umpiring states of mind enter constitutively into the determination of category-of-dismissal (the making-it-the-case-that the batsman was out this way, rather than that)?

This is really a matter for philosophizing cricketers and cricketing philosophers to decide.[7] We create our categories by means of our rules. There is no mind-independent category of dismissal 'out there' to be determined by the umpire-as-discoverer. Rather, the category of dismissal depends, at least in part, on norm-governed states of mind of the thinker whose job it is to apply the categories in question. The umpire does not make the rules (even though the umpire is *always right*).[8] Nor, however, does the umpire 'merely' apply them to a realm of mind-independent and fully determinable facts-of-the-matter. Rather, it would appear that one could make a strong case for the view that the rules require that umpiring states of mind enter constitutively into the determination of the category-of-dismissal to which a cricketing event is to be assigned.

This will be a verdict accepted only under protest by metaphysical realists, among whom are all Australian philosophers, as Devitt (1996), at p. x seems to imply:

> I have always been a realist about the external world. Such realism is common in Australia. Some say that Australian philosophers are born realists. I prefer to attribute our realism to nurture rather than nature. David Armstrong has suggested (lightly) that the strong sunlight and harsh brown landscape of Australia force reality upon us. In contrast, the mists and gentle green landscape of Europe weaken the grip on reality.

This is hard to reconcile with professional cricketers from downunder spending the antipodean winter in the gentler, greener landscapes of England and claiming to be Kanty cricketers. Perhaps *they* are reconciled to the anti-realist conclusion reached above, and Australian philosophers like Devitt need to bring their philosophizing about $X$ in line with the actual practice of $X$-ers. The resolution of the Umpire's Dilemma, if the rules do indeed demand determinate categories of dismissal to be assigned, is to give *not Out*.[9]

---

[7] With Paul Grice and Jack Smart having passed on, one wonders who might step up to follow that opening partnership in pursuit of such investigations.

[8] Another deep and dignified lesson for the bewildered American reader, who lives in the land of instant replays, John McEnroe tantrums, CNN moms and dads, referee-bashers, and constant readiness for other forms of revolutionary dissent.

[9] Thanks are owed to Simon Beck for raising my awareness of the Umpire's Dilemma in Radford (Analysis 45:109–111, 1985).

## References

Devitt, M. (1996). *Realism and truth* (2nd edn.). Princeton University Press.
Radford, C. (1985). The umpire's dilemma. *Analysis, 45*, 109–111.
Weir, A. (2010). *Truth through proof: A formalist foundation for mathematics*. Clarendon Press.

# Chapter 6
# Abstraction and Modest Reflection

**Roy T. Cook**

**Abstract** In "Neo-fregeanism: An Embarrassment of Riches" (Notre Dame J Formal Logic 44(1):13–48, 2003) Alan Weir introduces a number of formal constraints on abstraction principles that have become central to subsequent discussions of the Bad Company Problem, including stability and (Field and Caesar-neutral) conservativeness. In this essay we return to another notion introduced in that paper that has received much less attention—*modest reflection*—and we demonstrate that a natural version of this criterion is (i) a plausible constraint that acceptable abstraction principles must meet, (ii) distinct from all other such constraints proposed in the subsequent literature, yet (iii) intimately connected to a recent puzzle raised by Heck (and intimately connected to Heck's own preferred solution to this puzzle).

**Keywords** Abstraction · Distractions · Modest reflection · Logicism · Neo-logicism · Monotonicity · Conservativeness

## 6.1 Neo-Logicism, Convervativeness, and Modesty

According to neo-logicism, we can provide definitions of mathematical concepts, and provide epistemological access to the abstract objects falling under those mathematical concepts, via abstraction principles.[1] An abstraction principle is a formula of the form:

$$\mathsf{A}_E : (\forall\alpha)(\forall\beta)[@_E(\alpha) = @_E(\beta) \leftrightarrow E(\alpha, \beta)]$$

[1] For more details and defenses of neo-logicism, see Hale and Wright (2001), the essays in Cook (2007), Cook (2009), Fine (2002), and Linnebo (2018).

R. T. Cook (✉)
University of Minnesota, Minneapolis, MN, USA
e-mail: Cookx432@umn.edu



A. Rieger, S. Leuenberger (eds.), *Themes from Weir: A Celebration of the Philosophy of Alan Weir*, Synthese Library 484,
https://doi.org/10.1007/978-3-031-54557-3_6

where $@_E$ denotes a non-logical function mapping arguments of the type ranged over by $\alpha$ ( = the type ranged over by $\beta$) to objects, and where $E(\alpha, \beta)$ is an equivalence relation on entities of that type.

Here we shall restrict our attention to the case that has received the most attention in the literature on abstraction principles—conceptual abstraction principles—where $\alpha$ and $\beta$ range over unary concepts. Thus, we can rewrite our schema for abstraction principles as:

$$\mathsf{A}_E : (\forall X)(\forall Y)[@_E(X) = @_E(Y) \leftrightarrow E(X, Y)]$$

Abstraction principles serve to implicitly define the relevant mathematical concepts via providing identity conditions for the abstract objects falling under those concepts (i.e., objects in the range of $@_E$) in terms of the equivalence relation $E(X, Y)$.[2]

The two most well-known and well-studied abstraction principles are Hume's Principle and Basic Law V. Hume's Principle:

$$\mathsf{HP} : (\forall X)(\forall Y)[\#(X) = \#(Y) \leftrightarrow X \approx Y]$$

(where $X \approx Y$ abbreviates the purely logical second-order claim that there is a one-one onto mapping from $X$ to $Y$) provides a neo-logicist definition of the concept CARDINAL NUMBER, and is the paradigm instance of an acceptable abstraction principle. Basic Law V:

$$\mathsf{BLV} : (\forall X)(\forall Y)[\S(X) = \S(Y) \leftrightarrow (\forall x)(X(z) \leftrightarrow Y(z))]$$

on the other hand—which purports to similarly define the concept SET or EXTENSION—is inconsistent (in second-order logics with sufficiently strong comprehension principles), hence unacceptable.[3] A comparison between the status of Hume's Principle and of Basic Law V highlights a significant problem that the neo-logicist must solve:

> **The Bad Company Problem:** The Neo-logicist must provide a philosophically principled account of the line that distinguishes the acceptable abstraction principles (e.g., HP) from the unacceptable principles (e.g., BLV).

---

[2] To simplify matters, we will also assume that the equivalence relations $E(X, Y)$ occurring on the right-hand side of abstraction principles are purely logical (albeit often higher-order).

[3] We will assume so-called standard semantics for second-order logic throughout this essay. Hence, the full comprehension schema:

> Comp: For any formula $\Phi(y)$ not containing $X$ free:
>
> $$(\exists X)(\forall y)[X(y) \leftrightarrow \Phi(y)]$$

is valid, see Shapiro (1992) for details.

The Bad Company Problem is, of course, not the only challenge that the neo-logicist faces, but solving it is critical to the project, since we cannot know *which* mathematical theories can be reconstructed along neo-logicist lines (and whether *all* mathematical theories can be so reconstructed), nor can we know how such constructions are to proceed, until we know which abstraction principles are, in fact, acceptable.

The majority of work on the Bad Company Problem has focused on characterizing abstraction principles in terms of the collection of cardinals at which they are satisfiable. Alan Weir contributed to this debate in "Neo-Fregeanism: An Embarassment of Riches" (Weir, 2003), introducing (among other things) two *conservativeness* conditions. First, we have *Caesar-neutral conservativeness* (Weir, 2003, 23):[4]

**Definition 6.1.1** $\mathsf{A}_E$ is *Caesar-neutral conservative* if, for any theory $\mathcal{T}$ and formula $\Phi$ not containing @, and any primitive predicate $\Psi(x)$ not occurring in $\mathcal{T}$ or $\Phi$, if:

$$\mathsf{A}_E, \mathcal{T}^{\Psi(x)} \models \Phi^{\Psi(x)}$$

then:

$$\mathcal{T} \models \Phi$$

C-CON = the class of Caesar-conservative abstraction principles.

Second, we have *Field-conservativeness* (Weir, 2003, 22–23):[5]

**Definition 6.1.2** $\mathsf{A}_E$ is *Field-conservative* if and only if, for any theory $\mathcal{T}$ and formula $\Phi$ not containing @, if:

$$\mathsf{A}_E, \mathcal{T}^{\neg(\exists Y)(x=@_E(Y))} \models \Phi^{\neg(\exists Y)(x=@_E(Y))}$$

then:

$$\mathcal{T} \models \Phi$$

F-CON = the class of Field-conservative abstraction principles.

The idea that abstraction principles must be conservative in some sense has been at the center of the debate surrounding the Bad Company Problem ever since, based

[4] Cook and Linnebo (2018) call this condition *weak conservativeness*. Given any formula $\Phi$ and unary predicate $\Psi(x)$ (where $\Psi(x)$ might be a second-order variable), $\Phi^{\Psi(x)}$ is the *relativization* of (the quantifiers of) $\Phi$ to $\Psi(x)$.

[5] Cook and Linnebo (2018) call this *strong conservativeness*.

in part on Crispin Wright's endorsement of (an informal version of) the idea in "On the Philosophical Significance of Frege's Theorem":

> A legitimate abstraction, in short, ought to do no more than introduce a concept by fixing truth conditions for statements concerning instances of that concept ... How many sometime, someplace zebras there are is a matter between that concept and the world. No principle which merely assigns truth conditions to statements concerning objects of a quite unrelated, abstract kind—and no legitimate second-order abstraction can do any more than that—can possibly have any bearing on the matter. What is at stake ... is, in effect, conservativeness in (something close to) the sense of that notion deployed in Hartry Field's exposition of his nominalism. (Wright, 1997, 296)

Due to a minor (but rather "infectious") mistake in Weir (2003), however, it was not noted until recently (Cook & Linnebo, 2018) that Weir's two conservativeness conditions are distinct.[6] As a consequence, it was only recently noted that determining whether one of these conservativeness conditions (Field-conservativeness) holds of an abstraction principle involves considering not only the cardinality of domains at which an abstraction principle is satisfiable, but also the number of abstracts whose existence is entailed by that abstraction principle on such domains—again, see Cook and Linnebo (2018).

Weir's paper also introduces another condition that abstraction principles might meet: *modest reflection* (Weir, 2003, 30):

**Definition 6.1.3** $\mathsf{A}_E$ *reflects modestly* if and only if, for every formula $\Phi$ not containing $@_E$, if:

$$\mathsf{A}_E \models \Phi$$

then:

$$\mathsf{A}_E \models \Phi^{(\exists Y)(x=@_E(Y))}$$

In other words, an abstraction principle reflects modestly if and only if, for any formula that it entails, it also entails that formula restricted to the abstracts (i.e., restricted to the image of the abstraction operator $@_E$).

The underlying intuition behind modest reflection (which we will investigate in more detail below) is that anything that an abstraction principle entails about the universe as a whole ought to be guaranteed in virtue of properties of the subdomain of abstracts. In what follows we will see that (a restricted version of) modest reflection (what we will call *modest logical reflection*, or *modest $\mathcal{L}$-reflection*) is, indeed, a plausible constraint on acceptable abstraction principles (despite Weir's

[6] The false claim that Field-conservativeness and Caesar-neutral conservativeness are equivalent was either repeated, or implicitly assumed, in many, many articles and book chapters between 2003 and 2018, including (*mea culpa*) Cook (2009, 2012, 2016). Weir does not explicitly claim that the two are equivalent, but many readers (including myself) inferred the equivalence of the two conditions from Weir's incorrect **Theorem 4.2**, (Weir, 2003, 24) which states that any unbounded abstraction principle is conservative in both senses.

rejection of the condition). Unlike Weir's two conservativeness conditions (and likely due to Weir's unsympathetic treatment), modest reflection has received little attention in the literature on neo-logicism, and hence the ultimate purpose of this paper is investigating this condition and its relation to those conditions that have been proposed as constraints on acceptable abstraction.[7]

The paper will proceed as follows: In Sect. 6.2 we will rehearse some well-known facts about the various constraints that can be imposed on an abstraction principle via considering only the size of domains on which the principle is satisfiable. Then, in Sect. 6.3, we will review the state of the art with regard to a second kind of condition on abstraction principles: those conditions that require attending, not only to the size of the domains on which an abstraction principle is satisfiable, but also to the number of abstracts that must exist in each such domain. The majority of the material in these two sections will consist of a review of extant results, but Sect. 6.3 will conclude with the introduction of a novel device—a means for representing all of the information relevant to determining whether an abstraction principle falls into any of these categories—that we will call the *cardinality profile* of an abstraction principle.

Then, in Sect. 6.4, we will proceed to our main topic, showing that a natural (and plausible) version of Weir's modest reflection condition—the aforementioned requirement that abstraction principles be modestly $\mathcal{L}$-reflecting—is equivalent to a particular (and particularly simple) constraint on the cardinality profile of an abstraction principle, and we will use this fact to demonstrate that modest reflection is distinct from any of the other constraints on acceptable abstraction principles that have been proposed in the literature elsewhere. Then, in Sect. 6.5, we will examine how Weir's modest reflection fares with regard to a puzzle raised in the 2012 postscript to Heck (1992) (spoiler alert: it does quite well), and briefly investigate the connections between modest reflection and (one way of formulating) Heck's proposed solution to this same puzzle. Finally, in Sect. 6.6 we will tie up some loose ends.

## 6.2 Abstraction and the Number of Objects

As noted in the introduction, the majority of work on the Bad Company Problem has focused on various conditions that can be imposed on an abstraction principle via consideration of the class of cardinals at which that abstraction principle is satisfied.

[7] As far as I can tell, the modest reflection constraint formulated in Weir (2003), which is motivated by a comment in Wright (1999), makes only two subsequent appearances—MacBride (2003) and Ebert and Shapiro (2009)—and neither paper does more than mention the condition in passing.

Thus, we begin with a careful definition of what it means to say that an abstraction principle is satisfiable at a cardinal:

**Definition 6.2.1** An abstraction principle $\mathsf{A}_E$ is *$\kappa$-satisfiable* if and only if there is a model $\mathcal{M} = \langle \Delta, \mathcal{I} \rangle$ such that $|\Delta| = \kappa$ and:

$$\mathcal{M} \vDash \mathsf{A}_E$$

A rather trivial-seeming fact is worth emphasizing here, since it plays a surprisingly prominent role in some of the proofs given in later sections: No abstraction principle can be 0-satisfiable, even if our formal semantics allowed for models with empty first-order domains, since on an empty domain there must be an abstract corresponding to the empty property.

The literature on acceptable abstraction has focused for the most part on the following conditions (or conditions equivalent to them), which can be formulated solely in terms of the cardinals $\kappa$ for which a particular abstraction principle is $\kappa$-satisfiable:[8]

**Definition 6.2.2** An abstraction principle $\mathsf{A}_E$ is *satisfiable* if and only if there is a cardinal $\kappa$ such that $\mathsf{A}_E$ is $\kappa$-satisfiable. $\mathsf{SAT}$ = the class of satisfiable abstraction principles.

**Definition 6.2.3** An abstraction principle $\mathsf{A}_E$ is *unbounded* if and only if, for any cardinal $\gamma$, there is a $\kappa \geq \gamma$ such that $\mathsf{A}_E$ is $\kappa$-satisfiable. $\mathsf{UNB}$ = the class of unbounded abstraction principles.

**Definition 6.2.4** An abstraction principle $\mathsf{A}_E$ is *stable* if and only if there is some cardinal $\gamma$ such that, for all $\kappa \geq \gamma$, $\mathsf{A}_E$ is $\kappa$-satisfiable. $\mathsf{STB}$ = the class of stable abstraction principles.

**Definition 6.2.5** An abstraction principle $\mathsf{A}_E$ is *strongly stable* if and only if there is some cardinal $\gamma$ such that, for any $\kappa$, $\mathsf{A}_E$ is $\kappa$-satisfiable if and only if $\kappa \geq \gamma$. $\mathsf{S\text{-}STB}$ = the class of strongly stable abstraction principles.

[8] The reason for calling the target of **Definition 6.2.6** *conservativeness* is simple: this condition is easily seen to be equivalent to the following *unrelativized* (semantic) notion of conservativeness: $\mathsf{A}_E$ is *inferentially conservative* if and only if, for any theory $\mathcal{T}$ and formula $\Phi$ not containing $@_E$, if:

$$\mathsf{A}_E, \mathcal{T} \vDash \Phi$$

then:

$$\mathcal{T} \vDash \Phi$$

As we shall see, inferential conservativeness entails Caesar-neutral conservativeness (but not Field-conservativeness), but it is too strong to be a plausible constraint on acceptable abstraction, since not even Hume's Principle is inferentially conservative.

$$\mathsf{CON} \longrightarrow \mathsf{S\text{-}STB} \longrightarrow \mathsf{STB} \longrightarrow \mathsf{UNB} \longrightarrow \mathsf{SAT}$$

**Fig. 6.1** Satisfiability implications

**Definition 6.2.6** An abstraction principle $\mathsf{A}_E$ is *conservative* if and only if for every cardinal $\kappa > 0$ , $\mathsf{A}_E$ is $\kappa$-satisfiable. CON = the class of conservative abstraction principles.

**Definition 6.2.7** If $\mathsf{A}_E \in \mathsf{STB}$, then the *stabilization point* of $\mathsf{A}_E$ ($\mathsf{stab}(\mathsf{A}_E)$) is the least cardinal $\kappa$ such that $\mathsf{A}_E$ is $\gamma$-satisfiable for all $\gamma \geq \kappa$.

Figure 6.1 sums up the relationships between these five conditions on abstraction principles, where we can read the arrows either as entailment between the various conditions (e.g. conservativeness entails strong stability, but not vice versa) or as subsethood between the classes (e.g., that CON is a proper subset of S-STB).[9]

Finally, as promised, one of Weir's conservativeness conditions turns out to be equivalent to unboundedness, at least if we make the following additional assumption:[10]

*The Maximal Theories Assumption*: Given any cardinal $\kappa$, there is a theory $\mathcal{T}$ such that:

1. There is a model $\mathcal{M} = \langle \Delta, \mathcal{I} \rangle$ such that $\mathcal{M} \vDash \mathcal{T}$.
2. For any model $\mathcal{M} = \langle \Delta, \mathcal{I} \rangle$ such that $\mathcal{M} \vDash \mathcal{T}$, $|\Delta| \geq \kappa$.

---

[9] The reader is encouraged to consult (Cook & Linnebo, 2018) for a good survey of these results, and references to the earlier works that proved some of them.

[10] Weir also introduces a stronger notion defined in terms of Caesar-neutral conservativeness:

> $\mathsf{A}_{E_1}$ is *Caesar-neutral irenic* if and only if $\mathsf{A}_{E_1}$ is Caesar-neutral conservative and, for any $\mathsf{A}_{E_2}$ such that $\mathsf{A}_{E_2}$ is Caesar-neutral conservative, there is a model $\mathcal{M} = \langle \Delta, \mathcal{I} \rangle$ such that:
>
> $$\mathcal{M} \vDash \mathsf{A}_{E_1}$$
>
> and:
>
> $$\mathcal{M} \vDash \mathsf{A}_{E_2}$$
>
> C-IRN is the class of Caesar-neutral irenic abstraction principles.

Caesar-neutral irenicity (again, assuming the Maximal Theories Assumption) turns out to be equivalent to one of the five cardinality-based conditions considered above:

$$\mathsf{C\text{-}IRN} = \mathsf{STB}$$

This is less interesting than the identity between C-CON and UNB, however, since there seems to be no independent reason for thinking that an abstraction principle should be Caesar-neutral irenic beyond the thought that it should be stable, plus the equivalence between C-IRN. and STB.

The Maximal Theories Assumption is most easily met by assuming that, for any cardinal $\kappa$, we can extend our language with $\kappa$-many distinct constants $c_\alpha$ where $\alpha$ is an ordinal less than $\kappa$. Then the theory $\mathcal{T}$ in question is just:

$$\mathcal{T} = \{\text{“}c_\alpha \neq c_\beta\text{”} : \alpha \neq \beta\}$$

Since conservativeness (of either variety) is intended to guarantee that acceptable abstraction principles not conflict with the truths entailed by *any* theory, in *any* possible language, this assumption seems not only unobjectionable but philosophically mandated.

With the Maximal Theories Assumption in hand, we obtain:

**Theorem 6.2.8** C-CON = UNB.

***Proof*** See Cook and Linnebo (2018, 65–66). □

As we shall see, however, the second conservativeness requirement—Field-conservativeness—is not equivalent to any of the conditions listed above, and cannot be reformulated merely in terms of the cardinals $\kappa$ at which an abstraction principle is satisfiable. Thus, our analysis of abstraction principles needs to marshal additional resources.

## 6.3 Abstraction and the Number of Abstracts

As noted in the introduction, what is required in order to understand what, exactly, Weir's second sort of conservativeness—Field-conservativeness—amounts to is to consider, not only the cardinals at which an abstraction principle is satisfiable, but also the number of abstracts that are guaranteed to exist on domains of those cardinalities. One straightforward way to begin this examination is to ask whether models of abstraction principles are *full* at a particular cardinal:

**Definition 6.3.1** $\mathsf{A}_E$ is $\kappa$-*full* if and only if there is a model $\mathcal{M} = \langle \Delta, \mathcal{I} \rangle$ such that $|\Delta| = \kappa$ and, for any such model $\mathcal{M}$, we have:

$$|\{x \in \Delta : (\exists Y \subseteq \Delta)(\mathcal{I}(@_E(Y)) = x\}| = \kappa$$

More simply put, an abstraction principle $\mathsf{A}_E$ is $\kappa$-full just in case it is $\kappa$-satisfiable, and any model of $\mathsf{A}_E$ whose domain is $\kappa$-sized contains $\kappa$-many abstracts of the type introduced by $\mathsf{A}_E$.

Just as we formulated various conditions based on $\kappa$-satisfiability in the previous section, we can formulate similar conditions on abstraction principles in terms of this notion of $\kappa$-fullness. First, and most obvious: we can follow the pattern of the previous section, focusing on whether the class of cardinals at which

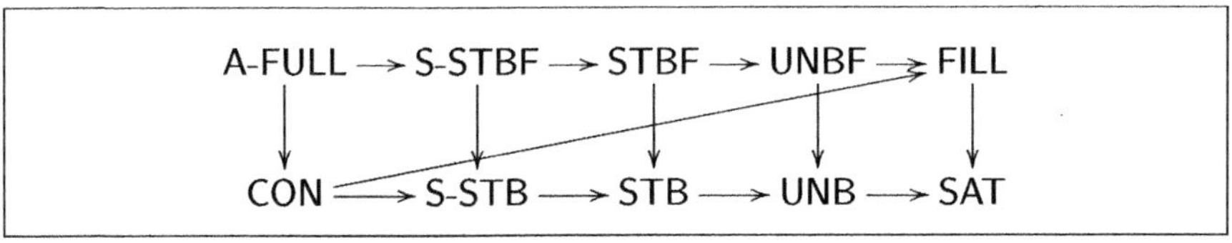


**Fig. 6.2** Satisfiability and simple fullness implications

an abstraction principle is full is non-empty, or unbounded, etc., obtaining the following conditions:

**Definition 6.3.2** An abstraction principle $\mathsf{A}_E$ is *fillable* if and only if there is a cardinal $\kappa$ such that $\mathsf{A}_E$ is $\kappa$-full. FILL = the class of fillable abstraction principles.

**Definition 6.3.3** An abstraction principle $\mathsf{A}_E$ is *unboundedly full* if and only if, for any cardinal $\gamma$, there is a $\kappa \geq \gamma$ such that $\mathsf{A}_E$ is $\kappa$-full. UNBF = the class of unboundedly full abstraction principles.

**Definition 6.3.4** An abstraction principle $\mathsf{A}_E$ is *stably full* if and only if there is some cardinal $\gamma$ such that, for all $\kappa \geq \gamma$, $\mathsf{A}_E$ is $\kappa$-full. STBF = the class of stably full abstraction principles.

**Definition 6.3.5** An abstraction principle $\mathsf{A}_E$ is *strongly stably full* if and only if there is some cardinal $\gamma$ such that, for any $\kappa$, $\mathsf{A}_E$ is $\kappa$-full if and only if $\kappa \geq \gamma$. S-STBF = the class of strongly stably full abstraction principles.

**Definition 6.3.6** $\mathsf{A}_E$ is *absolutely full* if and only if $\mathsf{A}_E$ is $\kappa$-full for any $\kappa > 0$. A-FULL = the class of absolutely full abstraction principles.

Figure 6.2 sums up the implications between these new notions and the criteria from Sect. 6.2.[11]

These criteria—despite their obvious structural relationship to the satisfiability conditions discussed in Sect. 6.2—have not turned out to be of particular *philosophical* interest within studies of the Bad Company Problem.[12] In order to obtain fullness-related conditions on abstraction principles that are of philosophical

[11] The inferences in Fig. 6.2 are all obvious consequences of definitions except, perhaps, the arrow from CON to FILL. This entailment holds because any abstraction principle in CON is 1-satisfiable, and any 1-satisfiable abstraction principle generates exactly one abstract on singleton domains, hence any abstraction principle in CON is 1-full, hence in FILL.

[12] The investigator who is interested merely in *mathematically* interesting phenomena arising with regard to abstraction principles might well find study of these heretofore ignored conditions fruitful, however. Hence their inclusion here.

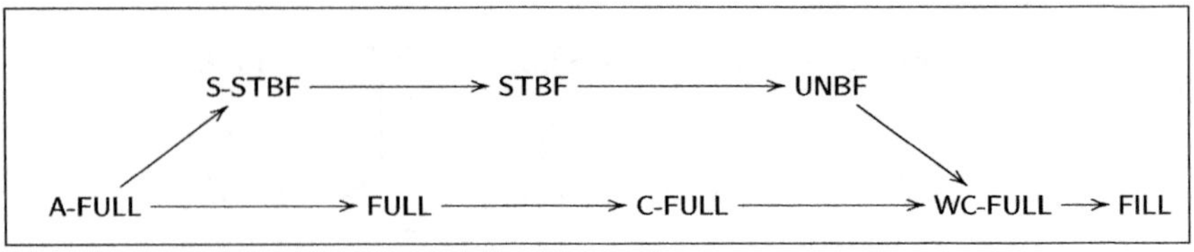


**Fig. 6.3** More fullness implications

interest, the notion of a *critical point* (which is a generalization of *stabilization point*, introduced in the previous section) will be useful:[13]

**Definition 6.3.7** $\kappa$ is a *critical point* of $\mathsf{A}_E$ if and only if $\mathsf{A}_E$ is $\kappa$-satisfiable and there is a $\gamma < \kappa$ such that, for all $\lambda$ where $\gamma \leq \lambda < \kappa$, $\mathsf{A}_E$ is not $\lambda$-satisfiable. $\mathsf{crit}(\mathsf{A}_E)$ is the set of critical points of $\mathsf{A}_E$.

The following criteria formulated in terms of $\kappa$-fullness and critical points will be useful in what follows:

**Definition 6.3.8** $\mathsf{A}_E$ is *weakly critically full* if and only if for any $\kappa \in \mathsf{crit}(\mathsf{A}_E)$, there is a $\gamma \geq \kappa$ such that $\mathsf{A}_E$ is $\gamma$-full. WC-FULL = the class of weakly critically full abstraction principles.

**Definition 6.3.9** $\mathsf{A}_E$ is *critically full* if and only if $\mathsf{A}_E$ is $\kappa$-full for any $\kappa \in \mathsf{crit}(\mathsf{A}_E)$. C-FULL = the class of critically full abstraction principles.

**Definition 6.3.10** $\mathsf{A}_E$ is *full* if and only if $\mathsf{A}_E$ is $\kappa$-full for any $\kappa$ such that $\mathsf{A}_E$ is $\kappa$-satisfiable. FULL = the class of full abstraction principles.

Figure 6.3 summarizes the implications between these various notions of fullness, and Fig. 6.4 codifies the relationships between the various combinations of satisfiability conditions from the previous section and the philosophically interesting fullness conditions presented in the present section.[14]

Weir's second notion of conservativeness, Field-conservativeness, turns out to be equivalent to unboundedness (= Caesar-neutral conservativeness) plus weak critical fullness (the Maximal Theories Assumption is again needed):

[13] Note that, since no abstraction principle is 0-satisfiable, if an abstraction principle is conservative (i.e., satisfiable at every cardinal $\kappa > 0$), then 1 is its only critical point. So *every* satisfiable abstraction principle has at least one critical point (the least cardinal $\kappa$ such that it is $\kappa$-satisfiable).

[14] The equivalences between CON, WC-FULL $\cap$ CON, and C-FULL $\cap$ CON are due to the fact that, if an abstraction principle is in CON, then its only critical point is 1, and any model of an abstraction principle with a unit domain is full. The equivalence between A-FULL and FULL $\cap$ CON is obvious.

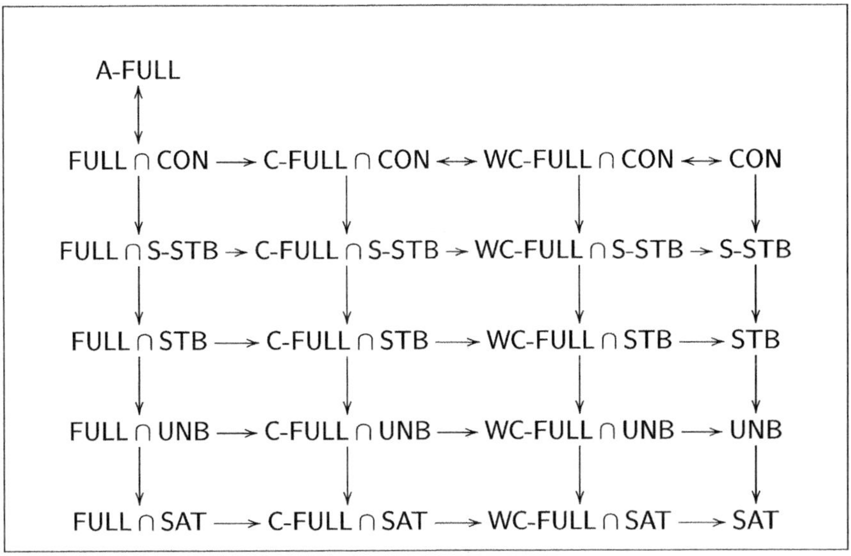


**Fig. 6.4** Fullness and satisfiabilty implications

**Theorem 6.3.11** $\mathsf{F\text{-}CON} = \mathsf{UNB} \cap \mathsf{WC\text{-}FULL}$.[15]

Thus ends (for the most part) our summary of extant results—now for something new. Each of the criteria considered in this section and the previous one can be understood in terms of two conditions: the size of the domain of models of an

[15] Weir also introduces a stronger notion defined in terms of Field-conservativeness:

> $\mathsf{A}_{E_1}$ is *Field-irenic* if and only if $\mathsf{A}_{E_1}$ is Field-conservative and, for any $\mathsf{A}_{E_2}$ such that $\mathsf{A}_{E_2}$ is Field-conservative, there is a model $\mathcal{M} = \langle \Delta, \mathcal{I} \rangle$ such that:
>
> $$\mathcal{M} \models \mathsf{A}_{E_1}$$
>
> and:
>
> $$\mathcal{M} \models \mathsf{A}_{E_2}$$
>
> $\mathsf{F\text{-}IRN}$ is the class of Field-irenic abstraction principles.

Field irenicity (assuming the Maximal Theories Assumption) turns out to be equivalent to the intersection of one of our satisfiability-based conditions and one of our fullness-based conditions:

$$\mathsf{F\text{-}IRN} = \mathsf{STB} \cap \mathsf{WC\text{-}FULL}$$

Like the corresponding result about Caesar-neutral irenicity, there seems to be little independent reason for thinking that an abstraction principle should be Field irenic apart from this condition's equivalence to stability plus weak critical fullness (and any independent justification these latter two criteria might have).

abstraction principle $\mathsf{A}_E$, and the size of the collection of abstracts introduced by that abstraction principle on domains of that size. In short, for each model $\mathcal{M} = \langle \Delta, \mathcal{I} \rangle$ of $\mathsf{A}_E$, if we know $|\Delta|$ and we know:

$$|\{x \in \Delta : (\exists Y \subseteq \Delta)(\mathcal{I}(@_E(Y)) = x\}|$$

then we know everything we need to know in order to determine whether an abstraction principle is satisfiable, unbounded, stable, strongly stable, conservative, fillable, unboundedly full, stably full, strongly stably full, absolutely full, weakly critically full, critically full, or full. We can codify this information as follows. First, a preliminary definition:

**Definition 6.3.12** Given a domain $\Delta$, $X \subseteq \Delta$, and (purely logical) equivalence relation $E(X, Y)$:

$$[X]_E^\Delta =_{df} \{Y : Y \subseteq \Delta \,\&\, E(X, Y)\}$$

We now define the cardinality profile of an abstraction principle as follows:

**Definition 6.3.13** Given an abstraction principle $\mathsf{A}_E$, the *cardinality profile* of $\mathsf{A}_E$ is the function:[16]

$$\mathsf{prof}_{\mathsf{A}_E} : \mathsf{Card} \rightarrow \mathsf{Card}$$

where (for any domain $\Delta$ of cardinality $\kappa$):

$$\mathsf{prof}_{\mathsf{A}_E}(\kappa) = |\{[X]_E^\Delta : X \subseteq \Delta\}|$$

More informally, $\mathsf{prof}_{\mathsf{A}_E}$ is the function that maps each cardinal number $\kappa$ to the number of distinct equivalence classes generated by $E(X, Y)$ on domains of size $\kappa$.[17] We can now give equivalent formulations of each of the criteria we have considered so far in terms of cardinality profiles:[18]

[16] Here $\mathsf{Card}$ is the proper class of cardinal numbers.

[17] The point made earlier in this essay regarding the fact that no abstraction principle can be 0-satisfiable can now be reformulated nicely in terms of cardinality profiles: For any abstraction principle $\mathsf{A}_E$ whatsoever:

$$\mathsf{prof}_{\mathsf{A}_E}(0) = 1 > 0$$

[18] An important note: In order to know that there is an abstraction principle in one or another of the classes defined above, it is not enough merely to demonstrate that there is a function $f : \mathsf{Card} \rightarrow \mathsf{Card}$ with the corresponding property. In addition, we need to know that there *is* some abstraction principle with that function as its cardinality profile, since not every function $f : \mathsf{Card} \rightarrow \mathsf{Card}$ corresponds to an abstraction principle expressible in a standard countable language for higher-order logic (as a simple cardinality argument will demonstrate).

- $\mathsf{A}_E$ is $\kappa$-satisfiable if and only if $\mathsf{prof}_{\mathsf{A}_E}(\kappa) \leq \kappa$.
- $\mathsf{A}_E \in \mathsf{SAT}$ if and only if there is a $\kappa$ such that $\mathsf{prof}_{\mathsf{A}_E}(\kappa) \leq \kappa$.
- $\mathsf{A}_E \in \mathsf{UNB}$ if and only if, for every $\kappa$, there is a $\gamma \geq \kappa$ such that $\mathsf{prof}_{\mathsf{A}_E}(\gamma) \leq \gamma$.
- $\mathsf{A}_E \in \mathsf{STB}$ if and only if there is a $\kappa$ such that, for every $\gamma \geq \kappa$, $\mathsf{prof}_{\mathsf{A}_E}(\gamma) \leq \gamma$.
- $\mathsf{A}_E \in \mathsf{S\text{-}STB}$ if and only if there is a $\kappa$ such that, for every $\gamma$, $\gamma \geq \kappa$ if and only if $\mathsf{prof}_{\mathsf{A}_E}(\gamma) \leq \gamma$.
- $\mathsf{A}_E \in \mathsf{CON}$ if and only, for every $\kappa > 0$, $\mathsf{prof}_{\mathsf{A}_E}(\kappa) \leq \kappa$.
- $\mathsf{A}_E$ is $\kappa$-full if and only if $\mathsf{prof}_{\mathsf{A}_E}(\kappa) = \kappa$.
- $\mathsf{A}_E \in \mathsf{FILL}$ if and only if there is a $\kappa$ such that $\mathsf{prof}_{\mathsf{A}_E}(\kappa) = \kappa$.
- $\mathsf{A}_E \in \mathsf{UNBF}$ if and only if, for every $\kappa$, there is a $\gamma \geq \kappa$ such that $\mathsf{prof}_{\mathsf{A}_E}(\gamma) = \gamma$.
- $\mathsf{A}_E \in \mathsf{STBF}$ if and only if there is a $\kappa$ such that, for every $\gamma \geq \kappa$, $\mathsf{prof}_{\mathsf{A}_E}(\gamma) = \gamma$.
- $\mathsf{A}_E \in \mathsf{S\text{-}STBF}$ if and only if there is a $\kappa$ such that, for every $\gamma$ $\gamma \geq \kappa$ if and only if $\mathsf{prof}_{\mathsf{A}_E}(\gamma) = \gamma$.
- $\mathsf{A}_E \in \mathsf{A\text{-}FULL}$ if and only if, for every $\kappa > 0$, $\mathsf{prof}_{\mathsf{A}_E}(\kappa) = \kappa$.
- $\mathsf{A}_E \in \mathsf{WC\text{-}FULL}$ if and only if, for every $\kappa$ such that $\kappa \in \mathsf{crit}(\mathsf{A}_E)$, there is a $\gamma \geq \kappa$ such that $\mathsf{prof}_{\mathsf{A}_E}(\gamma) = \gamma$.
- $\mathsf{A}_E \in \mathsf{C\text{-}FULL}$ if and only if, for every $\kappa$ such that $\kappa \in \mathsf{crit}(\mathsf{A}_E)$, $\mathsf{prof}_{\mathsf{A}_E}(\kappa) = \kappa$.
- $\mathsf{A}_E \in \mathsf{FULL}$ if and only if, for every $\kappa$ such that $\mathsf{prof}_{\mathsf{A}_E}(\kappa) \leq \kappa$, $\mathsf{prof}_{\mathsf{A}_E}(\kappa) = \kappa$.

The use of cardinality profiles to codify each of these conditions is mathematically neat. But the payoff comes, not in codifying these conditions (which are already relatively well-understood), but in providing a simple formulation of a condition on abstraction principles which has, unlike the conditions listed above, not received much in the way of formal analysis: Weir's notion of modest reflection.[19] Thus, we now turn to our examination of this relatively unstudied notion.

## 6.4 Modest Reflection

Weir's examination of modest reflection begins with consideration of the following passage, this time from Wright's "Is Hume's Principle Analytic?":

> An abstraction is Modest if its addition to any theory with which it is consistent results in no consequences—whether proof- or model-theoretically established—for the ontology of the combined theory which cannot be justified by reference to the consequences for its own abstracts. And again, *justification* is the crucial point: an abstraction may fail this constraint even though every consequence it has for the ontology of the combined theory may be seen to *follow from* things it entails about its proper abstracts; in particular, it will not count if, as in the case of the Limit-accessible Distraction, a consequence for the combined ontology is needed as a lemma in the proof that the abstracts have a property from which that very consequence follows. (Wright, 1999, 330)

[19] There is an additional advantage to cardinality profiles: their use allows us to prove significant theorems via carrying out cardinality profile computations. We adopt this strategy in many proofs given below.

As noted in the introduction, Weir (temporarily ignoring the second portion of this passage, and its somewhat mysterious distinction between what an abstraction principle *justifies* and what it *entails*) proposes the following as one way to understand Wright's suggestion that abstraction principles be *modest* (Weir, 2003, 30):

**Definition 6.4.1** $\mathsf{A}_E$ *reflects modestly* if and only if, for every formula $\Phi$ not containing $@_E$, if:

$$\mathsf{A}_E \models \Phi$$

then:

$$\mathsf{A}_E \models \Phi^{(\exists Y)(x=@_E(Y))}.$$

Weir's paper, as its title suggests, focuses on *distractions*—that is, on abstraction principles of the form:

$$(\forall X)(\forall Y)[\S(X) = \S(Y) \leftrightarrow ((\mathsf{Bad}(X) \wedge \mathsf{Bad}(Y)) \vee (\forall x)(X(z) \leftrightarrow Y(z)))]$$

and he proves that any distraction where all singleton concepts are $\mathsf{Good}$ ($= \neg\mathsf{Bad}$) reflects modestly (we will look at a generalization of this result below). Since there are clearly unacceptable distractions of this sort, he rejects modest reflection as a viable necessary and sufficient condition for abstraction principle acceptability.[20] After briefly considering various ways to reformulate the principle in a manner that somehow incorporates Wright's distinction between justification and entailment, Weir abandons modest reflection and moves on to consider other candidate criteria for acceptability, and, other than being mentioned in passing a handful of times (e.g., MacBride (2003) and Ebert and Shapiro (2009)), the notion of modest reflection disappears from the literature.

This, however, is too quick. The fact that many clearly unacceptable abstraction principles reflect modestly is merely evidence that modest reflection is not a sufficient condition for acceptability—it is no way implies that modest reflection is not a necessary condition. And this is important, for two reasons.

[20] The distraction principle where:

$$\mathsf{Bad}(X) = (\exists y)(\exists z)(y \neq z \wedge X(y) \wedge X(z) \wedge \mathsf{Succ})$$

(where $\mathsf{Succ}$ abbreviates the purely logical (second-order) claim that that the cardinality of the universe is a successor cardinal) is an example of the sort of distraction that Weir has in mind here. Singleton concepts (and the empty concept) are $\mathsf{Good}$ on any domain at which such a principle is satisfiable. Furthermore, there can only be $\mathsf{Bad}$ concepts if the cardinality of the domain is a successor cardinal. Thus, such a principle will reflect modestly yet be satisfiable only at successor cardinals.

First, modest reflection (and the restricted version of this notion that we will concentrate on below: modest logical reflection) turns out to be distinct from all of the rather more well-studied potential criteria for abstraction principle acceptability surveyed in the two previous sections. Thus, even if modest reflection is not sufficient for acceptability, it seems worth careful study since it might capture an important or interesting, but otherwise neglected, aspect of the mathematical behavior of abstraction principles.

Second, the beginning of Wright's comments quoted above seem compelling (from here on out we will charitably ignore the somewhat less clear latter part of the passage). Any novel consequence that an acceptable abstraction principle has for the entire domain should, in *some* sense, be justified in terms of consequences that that principle entails regarding the subdomain of abstracts. A particularly compelling instance of this thought is as follows: Letting $\Phi_{\geq \kappa}$ be the purely logical claim that there are at least $\kappa$-many objects (for any $\kappa$ definable in second- or higher-order logic), it seems plausible that, if an abstraction principle $\mathsf{A}_E$ entails $\Phi_{\geq \kappa}$—that is, it entails that there are at least $\kappa$-many objects—then it ought to secure this consequence via guaranteeing that there are at least $\kappa$-many abstracts—that is, $\mathsf{A}_E$ ought to entail $\Phi_{\geq \kappa}^{(\exists Y)(x=@_E(Y))}$. Generalizing this thought slightly: for any purely logical claim, if it is entailed by an abstraction principle, then its relativization to the abstracts ought to also be entailed by that abstraction principle. Thus, we should require that abstraction principles satisfy *modest logical reflection*:

**Definition 6.4.2** $\mathsf{A}_E$ is *modestly $\mathcal{L}$-reflecting* if and only if, for every purely logical formula $\Phi$, if:

$$\mathsf{A}_E \models \Phi$$

then:

$$\mathsf{A}_E \models \Phi^{(\exists Y)(x=@_E(Y))}$$

MLR = the class of modestly $\mathcal{L}$-reflecting abstraction principles.

It turns out that modest reflection is equivalent to the following condition, formulated in terms of cardinality profiles:[21]

**Definition 6.4.3** $\mathsf{A}_E$ is *idempotent* if and only if, for any $\kappa$, if:

$$\mathsf{prof}_{\mathsf{A}_E}(\kappa) \leq \kappa$$

[21] In mathematics, an *idempotent* function is a function $f : \Delta \rightarrow \Delta$ such that, for any $x \in \Delta$, $f(f(x)) = f(x)$.

then:

$$\mathsf{prof}_{\mathsf{A}_E}(\mathsf{prof}_{\mathsf{A}_E}(\kappa)) \leq \mathsf{prof}_{\mathsf{A}_E}(\kappa)$$

$\mathsf{IDEM}$ = the class of idempotent abstraction principles.

Intuitively, an abstraction principle $\mathsf{A}_E$ is idempotent if and only if, for any cardinal $\kappa$, if $\mathsf{A}_E$ is $\kappa$-satisfiable, then $\mathsf{A}_E$ is $\gamma$-satisfiable where $\gamma$ is the cardinality of the collection of abstracts on any model whose domain is $\kappa$-sized.

The proof that these two definitions are equivalent—that is, the idempotent abstraction principles are exactly the abstraction principles that modestly $\mathcal{L}$-reflect—is straightforward. First, a definition:

**Definition 6.4.4** Given a (closed) formula $\Phi$, the *Ramsification* of $\Phi$:

$$\mathsf{R}(\Phi)$$

is the result uniformly replacing each primitive non-logical expression with a variable of the same type, and then binding all resulting free variables with prenex existential quantifiers.

The particular case of interest in the present context is the Ramsification of an abstraction principle. Since we are assuming that the equivalence relations on the right-hand side of our abstraction principles are purely logical, the only non-logical vocabulary in a conceptual abstraction principle:

$$\mathsf{A}_E : (\forall X)(\forall Y)[@_E(X) = @_E(Y) \leftrightarrow E(X, Y)]$$

is the abstraction operator $@_E$, thus, $\mathsf{R}(\mathsf{A}_E)$ is just:

$$(\exists \mathcal{F})(\forall X)(\forall Y)[\mathcal{F}(X) = \mathcal{F}(Y) \leftrightarrow E(X, Y)]$$

where $\mathcal{F}$ is a *third*-order variable ranging over unary functions mapping concepts to objects. Now the theorem:[22]

**Theorem 6.4.5** $\mathsf{MLR} = \mathsf{IDEM}$

***Proof*** ($\rightarrow$) Assume, for arbitrary abstraction principle $\mathsf{A}_E$, that $\mathsf{A}_E \in \mathsf{MLR}$. Clearly:

$$\mathsf{A}_E \vDash \mathsf{R}(\mathsf{A}_E)$$

[22] Note the point at which the purely logical character of $\Phi$ is used. Whether Weir's original definition of modest reflection is equivalent to idempotence remains an open question.

But then:

$$\mathsf{A}_E \models \mathsf{R}(\mathsf{A}_E)^{(\exists Y)(x=@_E(Y))}$$

Assume that $\kappa$ is any cardinal such that:

$$\mathsf{prof}_{\mathsf{A}_E}(\kappa) \leq \kappa$$

Then $\mathsf{A}_E$ is $\kappa$-satisfiable, so there is a model $\mathcal{M} = \langle \Delta, \mathcal{I} \rangle$ where $|\Delta| = \kappa$ and:

$$\mathcal{M} \models \mathsf{A}_E$$

Then:

$$\mathcal{M} \models \mathsf{R}(\mathsf{A}_E)^{(\exists Y)(x=@_E(Y))}$$

This entails that there is a model $\mathcal{M}^* = \langle \Delta^*, \mathcal{I}^* \rangle$ where:

$$\Delta^* = \{x \in \Delta : \exists Y \subset \Delta \;\&\; x = \mathcal{I}(@_E(Y))\}$$

such that:

$$\mathcal{M}^* \models \mathsf{A}_E$$

Clearly, $|\Delta^*| = \mathsf{prof}_{\mathsf{A}_E}(\kappa)$, and, since $\mathsf{A}_E$ is $|\Delta^*|$-satisfiable (i.e., $\mathsf{prof}_{\mathsf{A}_E}(\kappa)$-satisfiable), it follows that:

$$\mathsf{prof}_{\mathsf{A}_E}(\mathsf{prof}_{\mathsf{A}_E}(\kappa)) \leq \mathsf{prof}_{\mathsf{A}_E}(\kappa)$$

So $A_E \in \mathsf{IDEM}$. Hence $\mathsf{MLR} \subseteq \mathsf{IDEM}$.

($\leftarrow$) Assume $\mathsf{A}_E \in \mathsf{IDEM}$. So, for any cardinal $\kappa$, if:

$$\mathsf{prof}_{\mathsf{A}_E}(\kappa) \leq \kappa$$

then:

$$\mathsf{prof}_{\mathsf{A}_E}(\mathsf{prof}_{\mathsf{A}_E}(\kappa)) \leq \mathsf{prof}_{\mathsf{A}_E}(\kappa)$$

Let $\Phi$ be a purely logical formula such that:

$$\mathsf{A}_E \models \Phi$$

Let $\mathcal{M} = \langle \Delta, \mathcal{I} \rangle$ be any model such that:

$$\mathcal{M} \models \mathsf{A}_E$$

Let $\gamma = |\Delta|$, and:

$$\Delta^* = \{x \in \Delta : \exists Y \subset \Delta \,\&\, x = \mathcal{I}(@_E(Y))\}$$

Then, since:

$$\mathsf{prof}_{\mathsf{A}_E}(\gamma) \leq \gamma$$

(i.e., $\mathsf{A}_E$ is $\gamma$-satisfiable), we have:

$$\mathsf{prof}_{\mathsf{A}_E}(\mathsf{prof}_{\mathsf{A}_E}(\gamma)) \leq \mathsf{prof}_{\mathsf{A}_E}(\gamma)$$

(i.e., $\mathsf{A}_E$ is $\mathsf{prof}_{\mathsf{A}_E}(\gamma)$-satisfiable). But $\mathsf{prof}_{\mathsf{A}_E}(\gamma) = |\Delta^*|$, so there is a model $\mathcal{M}^* = \langle \Delta^*, \mathcal{I}^* \rangle$ such that:

$$\mathcal{M}^* \models \mathsf{A}_E$$

Hence:

$$\mathcal{M}^* \models \Phi$$

But, since $\Phi$ purely logical, we have:

$$\mathcal{M}^* \models \Phi \text{ iff } \mathcal{M} \models \Phi^{(\exists Y)(x = @_E(Y))}$$

Hence:

$$\mathcal{M} \models \Phi^{(\exists Y)(x = @_E(Y))}$$

Since $\mathcal{M}$ was arbitrary:

$$\mathsf{A}_E \models \Phi^{(\exists Y)(x = @_E(Y))}$$

So $\mathsf{A}_E \in \mathsf{MLR}$. Since $\mathsf{A}_E$ was arbitrary, this provides $\mathsf{IDEM} \subseteq \mathsf{MLR}$ □

Both Hume's Principle and (perhaps somewhat more surprisingly) Basic Law V modestly $\mathcal{L}$-reflect.[23] The inclusion of Basic Law V in $\mathsf{MLR}$ is a consequence of the following somewhat uninteresting result:

**Theorem 6.4.6** *For any abstraction principle* $\mathsf{A}_E \notin \mathsf{SAT}$, $\mathsf{A}_E \in \mathsf{MLR}$.

[23] The fact that all unsatisfiable principles are in $\mathsf{MLR}$ further emphasizes the fact that modest $\mathcal{L}$-reflection is, at best, a necessary condition on acceptability.

***Proof*** For any $\mathsf{A}_E \notin \mathsf{SAT}$, we have that, for all cardinals $\kappa$:

$$\mathsf{prof}_{\mathsf{A}_E}(\kappa) > \kappa$$

Membership in $\mathsf{IDEM}$ (hence in $\mathsf{MLR}$) follows trivially. □

The following is immediate:

**Corollary 6.4.7** $\mathsf{BLV} \in \mathsf{MLR}$

Thus, many of our results, if they are to be interesting, will need to be formulated in terms of $\mathsf{MLR} \cap \mathsf{SAT}$ rather than merely in terms of $\mathsf{MLR}$. Here is a simple proof that Hume's Principle modestly $\mathcal{L}$-reflects:

**Theorem 6.4.8** $\mathsf{HP} \in \mathsf{MLR}$

***Proof*** We have:

$$\mathsf{prof}_{\mathsf{HP}}(\kappa) \leq \kappa \text{ if and only if } \kappa \geq \aleph_0$$

and (via Frege's Theorem):

$$\aleph_0 \leq \mathsf{prof}_{\mathsf{HP}}(\kappa) \text{ if and only if } \kappa \geq \aleph_0$$

Assume, for arbitrary $\gamma$, that:

$$\mathsf{prof}_{\mathsf{HP}}(\gamma) \leq \gamma$$

Then $\gamma \geq \aleph_0$, so (by applying the first fact again):

$$\aleph_0 \leq \mathsf{prof}_{\mathsf{HP}}(\gamma)$$

Hence:

$$\mathsf{prof}_{\mathsf{HP}}(\mathsf{prof}_{\mathsf{HP}}(\gamma)) \leq \mathsf{prof}_{\mathsf{HP}}(\gamma)$$

So $\mathsf{HP} \in \mathsf{IDEM} = \mathsf{MLR}$. Clearly $\mathsf{HP} \in \mathsf{SAT}$. Hence $\mathsf{HP} \in \mathsf{MLR} \cap \mathsf{SAT}$. □

Our next result is a generalization of Weir's main theorem regarding modest reflection.[24] The proof given here is somewhat simpler than Weir's proof of the same result, due to our use of cardinality profiles:

[24] Although, as already noted, Weir actually proves that any distraction where all singleton concepts are $\mathsf{Good}$ is in $\mathsf{MLR}$, the key idea underlying his proof is that such distractions are full (Weir, 2003, 31).

**Theorem 6.4.9** $\mathsf{FULL} \subsetneq \mathsf{MLR} \cap \mathsf{SAT}$

***Proof*** Assume that $\mathsf{A}_E \in \mathsf{FULL}$. So, for any cardinal $\kappa$, if:

$$\mathsf{prof}_{\mathsf{A}_E}(\kappa) \leq \kappa$$

then:

$$\mathsf{prof}_{\mathsf{A}_E}(\kappa) = \kappa.$$

Assume, for some arbitrary $\gamma$, that:

$$\mathsf{prof}_{\mathsf{A}_E}(\gamma) \leq \gamma$$

Then, since $\mathsf{A}_E \in \mathsf{FULL}$:

$$\mathsf{prof}_{\mathsf{A}_E}(\gamma) = \gamma$$

Hence:

$$\mathsf{prof}_{\mathsf{A}_E}(\mathsf{prof}_{\mathsf{A}_E}(\gamma)) = \mathsf{prof}_{\mathsf{A}_E}(\gamma)$$

So:

$$\mathsf{prof}_{\mathsf{A}_E}(\mathsf{prof}_{\mathsf{A}_E}(\gamma)) \leq \mathsf{prof}_{\mathsf{A}_E}(\gamma)$$

So $\mathsf{A}_E \in \mathsf{IDEM} = \mathsf{MLR}$. Since $\mathsf{A}_E$ was arbitrary, we have $\mathsf{FULL} \subseteq \mathsf{MLR}$. Clearly $\mathsf{FULL} \subsetneq \mathsf{SAT}$. So $\mathsf{FULL} \subseteq \mathsf{MLR} \cap \mathsf{SAT}$. $\mathsf{HP} \in \mathsf{MLR} \cap \mathsf{SAT}$, but $\mathsf{HP} \notin \mathsf{FULL}$. So $\mathsf{FULL} \neq \mathsf{MLR} \cap \mathsf{SAT}$. □

Modest $\mathcal{L}$-reflection is also entailed by conservativeness:

**Theorem 6.4.10** $\mathsf{CON} \subsetneq \mathsf{MLR} \cap \mathsf{SAT}$

***Proof*** Assume $\mathsf{A}_E \in \mathsf{CON}$. So, for every $\kappa > 0$:

$$\mathsf{prof}_{\mathsf{A}_E}(\kappa) \leq \kappa$$

Assume, for some arbitrary $\gamma$, that:

$$\mathsf{prof}_{\mathsf{A}_E}(\gamma) \leq \gamma$$

Both $\mathsf{prof}_{\mathsf{A}_E}(\gamma)$ and $\gamma$ are greater than 0, hence:

$$\mathsf{prof}_{\mathsf{A}_E}(\mathsf{prof}_{\mathsf{A}_E}(\gamma)) \leq \mathsf{prof}_{\mathsf{A}_E}(\gamma)$$

So $\mathsf{A}_E \in \mathsf{IDEM} = \mathsf{MLR}$. Since $\mathsf{A}_E$ was arbitrary, we have $\mathsf{CON} \subseteq \mathsf{MLR}$. Clearly, $\mathsf{CON} \subseteq \mathsf{SAT}$. So $\mathsf{CON} \subseteq \mathsf{MLR} \cap \mathsf{SAT}$. $\mathsf{HP} \in \mathsf{MLR} \cap \mathsf{SAT}$, but $\mathsf{HP} \notin \mathsf{CON}$. So $\mathsf{CON} \neq \mathsf{MLR} \cap \mathsf{SAT}$. □

Thus, every conservative abstraction principle modestly $\mathcal{L}$-reflects, as does every full abstraction principle. But the combination of the next strongest satisfiability condition with the next strongest fullness condition is insufficient to guarantee modest $\mathcal{L}$-reflection:

**Theorem 6.4.11** $\mathsf{C\text{-}FULL} \cap \mathsf{S\text{-}STB} \not\subseteq \mathsf{MLR} \cap \mathsf{SAT}$

***Proof*** Consider:

$$\mathsf{A}_{\mathsf{E}_1} : (\forall X)(\forall Y)[@_{\mathsf{E}_1}(X) = @_{\mathsf{E}_1}(Y) \leftrightarrow (\Phi_{>\aleph_0} \vee X \approx Y)]$$

where $\Phi_{>\aleph_0}$ abbreviates the purely logical claim that the cardinality of the universe is greater than $\aleph_0$. Then:

$$\mathsf{prof}_{\mathsf{A}_{\mathsf{E}_1}}(\kappa) \leq \kappa \text{ if and only if } \kappa \geq \aleph_0$$

Hence $\mathsf{A}_{E_1} \in \mathsf{S\text{-}STB}$, with stabilization point (and only critical point) $\aleph_0$. Additionally:

$$\mathsf{prof}_{\mathsf{A}_{\mathsf{E}_1}}(\aleph_0) = \aleph_0$$

hence $\mathsf{A}_{\mathsf{E}_1} \in \mathsf{C\text{-}FULL}$. But:

$$\mathsf{prof}_{\mathsf{A}_{\mathsf{E}_1}}(\kappa) = 1 \text{ if and only if } \kappa \geq \aleph_1$$

Hence:

$$\mathsf{prof}_{\mathsf{A}_{\mathsf{E}_1}}(\aleph_1) = 1 \leq \aleph_1$$

but:

$$\mathsf{prof}_{\mathsf{A}_{\mathsf{E}_1}}(\mathsf{prof}_{\mathsf{A}_{\mathsf{E}_1}}(\aleph_1)) = \mathsf{prof}_{\mathsf{A}_{\mathsf{E}_1}}(1) > 1 = \mathsf{prof}_{\mathsf{A}_{\mathsf{E}_1}}(\aleph_1)$$

So $\mathsf{A}_{\mathsf{E}_1} \notin \mathsf{IDEM} = \mathsf{MLR}$. □

These results suffice to guarantee the novelty of modest $\mathcal{L}$-reflection (relative to the satisfiability- and fullness-based conditions that have been of central concern in discussions of the Bad Company Problem), insofar as modest $\mathcal{L}$-reflection fails to be identical to any combination of the conditions detailed in Fig. 6.4 above. We conclude this comparison with the following two results, which further emphasize the novelty of modest $\mathcal{L}$-reflection:

**Theorem 6.4.12** $\mathsf{MLR} \cap \mathsf{SAT} \not\subset \mathsf{UNB}$

***Proof*** Consider:

$$\mathsf{A}_{\mathsf{E}_2} : (\forall X)(\forall Y)[@_{\mathsf{E}_2}(X) = @_{\mathsf{E}_2}(Y) \leftrightarrow ((\Phi_{\aleph_0} \wedge X \approx Y) \\ \vee (\forall z)(X(z) \leftrightarrow Y(z)))]$$

where $\Phi_{\aleph_0}$ abbreviates the purely logical claim that the universe is countably infinite. We have:

$$\mathsf{prof}_{\mathsf{A}_{\mathsf{E}_2}}(\aleph_0) = \aleph_0$$

and:

$$\mathsf{prof}_{\mathsf{A}_{\mathsf{E}_2}}(\kappa) > \kappa \text{ if } \kappa \neq \aleph_0$$

Hence, if:

$$\mathsf{prof}_{\mathsf{A}_{\mathsf{E}_2}}(\gamma) \leq \gamma$$

then $\gamma = \aleph_0$, so:

$$\mathsf{prof}_{\mathsf{A}_{\mathsf{E}_2}}(\mathsf{prof}_{\mathsf{A}_{\mathsf{E}_2}}(\gamma)) = \mathsf{prof}_{\mathsf{A}_{\mathsf{E}_2}}(\aleph_0) = \mathsf{prof}_{\mathsf{A}_{\mathsf{E}_2}}(\gamma)$$

Hence:

$$\mathsf{prof}_{\mathsf{A}_{\mathsf{E}_2}}(\mathsf{prof}_{\mathsf{A}_{\mathsf{E}_2}}(\gamma)) \leq \mathsf{prof}_{\mathsf{A}_{\mathsf{E}_2}}(\gamma)$$

So $\mathsf{A}_{\mathsf{E}_2} \in \mathsf{IDEM} = \mathsf{MLR}$. Clearly $\mathsf{A}_{\mathsf{E}_2} \in \mathsf{SAT}$ So $\mathsf{A}_{\mathsf{E}_2} \in \mathsf{MLR} \cap \mathsf{SAT}$. But $\mathsf{A}_{\mathsf{E}_2} \notin \mathsf{UNB}$. □

**Theorem 6.4.13** $\mathsf{MLR} \cap \mathsf{SAT} \not\subset \mathsf{WC\text{-}FULL} \cap \mathsf{SAT}$

***Proof*** Consider:

$$\mathsf{A}_{\mathsf{E}_3} : (\forall X)(\forall Y)[@_{\mathsf{E}_3}(X) = @_{\mathsf{E}_3}(Y) \leftrightarrow ((\Phi_{\aleph_0} \wedge X \approx Y) \\ \vee (\Phi_{\aleph_2} \wedge X \approx Y) \\ \vee (\forall z)(X(z) \leftrightarrow Y(z)))]$$

where $\Phi_{\aleph_0}$ abbreviates the purely logical claim that the cardinality of the universe is $\aleph_0$ and $\Phi_{\aleph_2}$ abbreviates the purely logical claim that the cardinality of the universe is $\aleph_2$.

$$\mathsf{prof}_{\mathsf{A}_{\mathsf{E}_3}}(\aleph_0) = \aleph_0$$

$$\mathsf{prof}_{\mathsf{A}_{\mathsf{E}_3}}(\aleph_2) = \aleph_0$$

and:

$$\mathsf{prof}_{\mathsf{A}_{\mathsf{E}_3}}(\kappa) > \kappa \text{ if } \kappa \neq \aleph_0 \text{ and } \kappa \neq \aleph_2$$

Hence, if:

$$\mathsf{prof}_{\mathsf{A}_{\mathsf{E}_3}}(\gamma) \leq \gamma$$

then $\gamma = \aleph_0$ or $\gamma = \aleph_2$. Either way:

$$\mathsf{prof}_{\mathsf{A}_{\mathsf{E}_3}}(\mathsf{prof}_{\mathsf{A}_{\mathsf{E}_3}}(\gamma)) = \mathsf{prof}_{\mathsf{A}_{\mathsf{E}_3}}(\aleph_0) = \aleph_0 = \mathsf{prof}_{\mathsf{A}_{\mathsf{E}_3}}(\gamma)$$

Hence:

$$\mathsf{prof}_{\mathsf{A}_{\mathsf{E}_3}}(\mathsf{prof}_{\mathsf{A}_{\mathsf{E}_3}}(\gamma)) \leq \mathsf{prof}_{\mathsf{A}_{\mathsf{E}_3}}(\gamma)$$

So $\mathsf{A}_{\mathsf{E}_3} \in \mathsf{IDEM} = \mathsf{MLR}$. Clearly $\mathsf{A}_{\mathsf{E}_3} \in \mathsf{SAT}$. So $\mathsf{A}_{\mathsf{E}_3} \in \mathsf{MLR} \cap \mathsf{SAT}$. But $\aleph_2 \in \mathsf{crit}(\mathsf{A}_{\mathsf{E}_3})$, but there is no $\kappa \geq \aleph_2$ such that:

$$\mathsf{prof}_{\mathsf{A}_{\mathsf{E}_3}}(\kappa) = \kappa$$

Hence $\mathsf{A}_{\mathsf{E}_3} \notin \mathsf{WC\text{-}FULL}$, hence $\mathsf{A}_{\mathsf{E}_3} \notin \mathsf{WC\text{-}FULL} \cap \mathsf{SAT}$. □

Thus, modest $\mathcal{L}$-reflection is not only distinct from any of the conditions surveyed in Sects. 6.2 and 6.3 above (and distinct from any combination of these conditions), it is in a certain sense *very* different from them. One way to see this is to note that, if we were to incorporate the class of satisfiable modestly $\mathcal{L}$-reflecting abstraction principles into our chart of previously studied satisfiability- and fullness-based conditions (i.e., Fig. 6.4), the result would look something like Fig. 6.5.

It turns out, however, that there is an additional type of restriction that we can impose on abstraction principles—a kind that has received even less attention than fullness-related conditions—that will help us to get a better grasp on modest reflection: monotonicity. It is to this topic that we now turn.

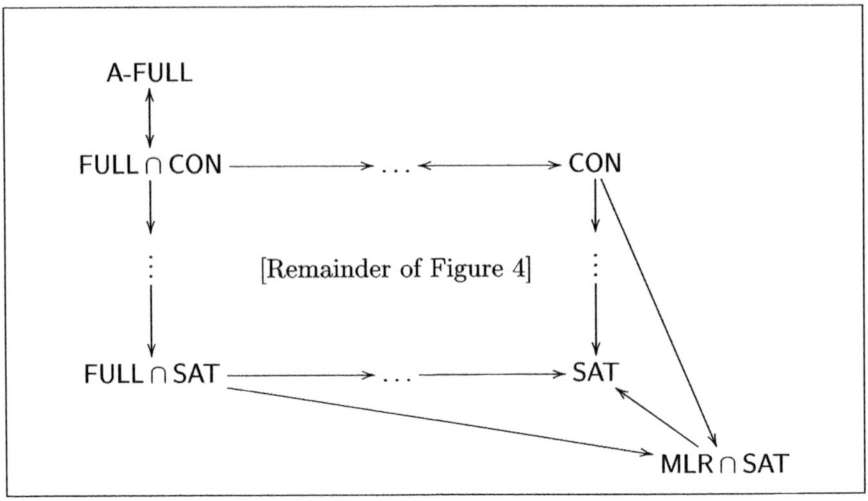


**Fig. 6.5** Adding modest $\mathcal{L}$-reflection to the picture

## 6.5 Modest $\mathcal{L}$-Reflection and Monotonicity

In "On the Consistency of Second-order Contextual Definitions" (Heck, 1992) Richard Kimberly Heck considers a particular, and particularly worrying, special case of the Bad Company Problem—a case that is closely related to the worries motivating Weir's modest $\mathcal{L}$-reflection. Heck asks us to consider "Fishy" (Heck, 1992, 232) abstraction principles of the form:

$$\mathsf{Fish}_\Phi : (\forall X)(\forall Y)[@(X) = @(Y) \leftrightarrow (\Phi \vee (\forall z)(X(z) \leftrightarrow Y(z)))]$$

Given a particular formula $\Phi$, we will call the abstraction principle obtained via substituting $\Phi$ into the above schema the *$\Phi$-Fishy abstraction principle* (or $\mathsf{Fish}_\Phi$).

Via a judicious choice of $\Phi$, we can use this template to construct abstraction principles that are satisfiable, unbounded, stable or strongly stable, or even conservative, but all such principles will have one thing in common: Given any Fishy abstraction principle $\mathsf{Fish}_\Phi$ and any $\kappa$ such that $\mathsf{Fish}_\Phi$ is $\kappa$-satisfiable, $\mathsf{Fish}_\Phi$ will entail the existence of exactly one abstract on a domain of size $\kappa$.

Since we are interested more in the semantic behavior of abstraction principles than in their syntactic form, however, it is useful to provide an alternative definition of the kind of abstraction principle Heck has in mind. We will formulate this definition directly in terms of cardinality profiles, thus:[25]

[25] Heck's discussion suggests a more general version of the problem that arises if we consider abstraction principles that always generate *fewer* abstracts than there are objects in the universe on

**Definition 6.5.1** $\mathsf{A}_E$ is *Fishy* if and only if, for any $\kappa$, if:

$$\mathsf{prof}_{\mathsf{A}_E}(\kappa) \leq \kappa$$

then:

$$\mathsf{prof}_{\mathsf{A}_E}(\kappa) = 1$$

FISH is the class of Fishy abstraction principles.

Note that not every abstraction principle that is in FISH is necessarily a $\Phi$-Fishy abstraction principle, but for any $\Phi$, the $\Phi$-Fishy abstraction principle is in FISH.

The first thing to notice is that not every Fishy abstraction principle is necessarily unacceptable. For example, if $\Phi$ is a logical truth, then there is every reason to think that the $\Phi$-Fishy abstraction principle is acceptable:

**Theorem 6.5.2** *Given any logical truth $\Phi$:*

$$\mathsf{Fish}_\Phi \in \mathsf{C\text{-}FULL} \cap \mathsf{CON} \cap \mathsf{MLR}$$

***Proof*** If $\Phi$ is a logical truth, then:

$$\mathsf{prof}_{\mathsf{Fish}_\Phi}(\kappa) = 1 \leq \kappa \text{ for all } \kappa > 0$$

Hence $\mathsf{Fish}_\Phi \in \mathsf{CON}$. 1 is the only critical point of $\mathsf{Fish}_\Phi$, and:

$$\mathsf{prof}_{\mathsf{Fish}_\Phi}(1) = 1$$

Hence $\mathsf{Fish}_\Phi \in \mathsf{C\text{-}FULL}$. Finally, assume:

$$\mathsf{prof}_{\mathsf{Fish}_\Phi}(\gamma) \leq \gamma$$

---

any domain on which they are satisfiable (other than, *per impossibile*, on the singleton domain), rather than always generating a *single* abstract. We can capture this idea as follows:

$\mathsf{A}_E$ is *Generalized Fishy* if and only if, for any $\kappa > 1$ such that:

$$\mathsf{prof}_{\mathsf{A}_E}(\kappa) \leq \kappa$$

we have:

$$\mathsf{prof}_{\mathsf{A}_E}(\kappa) < \kappa$$

G-FISH is the class of generalized Fishy abstraction principles.

The reader is encouraged to verify that the main points of our discussion of Fishy abstraction principles can be transferred to Generalized Fishy principles with little modification.

Then $\gamma > 0$, so:

$$\mathsf{prof}_{\mathsf{Fish}_\Phi}(\gamma) = 1$$

hence:

$$\mathsf{prof}_{\mathsf{Fish}_\Phi}(\mathsf{prof}_{\mathsf{Fish}_\Phi}(\gamma)) = \mathsf{prof}_{\mathsf{Fish}_\Phi}(1) = 1 = \mathsf{prof}_{\mathsf{Fish}_\Phi}(\gamma)$$

Thus:

$$\mathsf{prof}_{\mathsf{Fish}_\Phi}(\mathsf{prof}_{\mathsf{Fish}_\Phi}(\gamma)) \leq \mathsf{prof}_{\mathsf{Fish}_\Phi}(\gamma)$$

So $\mathsf{Fish}_\Phi \in \mathsf{IDEM} = \mathsf{MLR}$. Thus, $\mathsf{Fish}_\Phi \in \mathsf{C\text{-}FULL} \cap \mathsf{CON} \cap \mathsf{MLR}$. □

$\mathsf{C\text{-}FULL} \cap \mathsf{CON} \cap \mathsf{MLR}$ is much narrower than *any* of the extant proposals for acceptability in the literature. For example, Cook (2012) argues (in effect—see footnote 6!) that abstraction principles should be strongly stable and Field-conservative.

The situation is of course more complicated if $\Phi$ is not a logical truth. In order to help us sort out the situation, we introduce two stronger Fishiness notions:

**Definition 6.5.3** $\mathsf{A}_E$ is *Moderately Fishy* if and only if there is a $\kappa$ such that:

$$\mathsf{prof}_{\mathsf{A}_E}(\kappa) > \kappa$$

and, for any $\kappa$ such that :

$$\mathsf{prof}_{\mathsf{A}_E}(\kappa) \leq \kappa$$

we have:

$$\mathsf{prof}_{\mathsf{A}_E}(\kappa) = 1$$

$\mathsf{M\text{-}FISH}$ is the class of Moderately Fishy abstraction principles.

**Definition 6.5.4** $\mathsf{A}_E$ is *Strongly Fishy* if and only if:

$$\mathsf{prof}_{\mathsf{A}_E}(1) > 1$$

and, for any $\kappa$ such that:

$$\mathsf{prof}_{\mathsf{A}_E}(\kappa) \leq \kappa$$

we have:

$$\mathsf{prof}_{\mathsf{A}_E}(\kappa) = 1$$

$\mathsf{S\text{-}FISH}$ is the class of Strongly Fishy abstraction principles.

Simply put, an abstraction principle is Moderately Fishy if and only if it only provides a single abstract on domains on which it is satisfiable, and it is not $\kappa$-satisfiable for some $\kappa$, and an abstraction principle is Strongly Fishy if and only if it only provides a single abstract on domains on which it is satisfiable, and it entails the existence of at least two objects (hence is not 1-satisfiable). The following is straightforward:

**Theorem 6.5.5** $\mathsf{S\text{-}FISH} \subsetneq \mathsf{M\text{-}FISH} \subsetneq \mathsf{FISH}$

***Proof*** The inclusions are immediate consequences of definitions. The following facts secure the non-identities:

- $\mathsf{Fish}_{(\forall x)(x=x)} \in \mathsf{FISH}$, $\mathsf{Fish}_{(\forall x)(x=x)} \notin \mathsf{M\text{-}FISH}$
- $\mathsf{Fish}_{(\forall x)(\forall y)(x=y)} \in \mathsf{M\text{-}FISH}$, $\mathsf{Fish}_{(\forall x)(\forall y)(x=y)} \notin \mathsf{S\text{-}FISH}$

□

Modest $\mathcal{L}$-reflection rules out satisfiable Strongly Fishy abstraction principles:[26]

**Theorem 6.5.6** *If* $\mathsf{A}_E \in \mathsf{S\text{-}FISH}$, *then* $\mathsf{A}_E \notin \mathsf{MLR} \cap \mathsf{SAT}$.

***Proof*** Assume that $\mathsf{A}_E \in \mathsf{S\text{-}FISH}$. If $\mathsf{A}_E \notin \mathsf{SAT}$, then we are done. So assume that $\mathsf{A}_E \in \mathsf{SAT}$. Thus, there is a $\kappa$ such that:

$$\mathsf{prof}_{\mathsf{A}_E}(\kappa) \leq \kappa$$

So:

$$\mathsf{prof}_{\mathsf{A}_E}(\kappa) = 1$$

Since:

$$\mathsf{prof}_{\mathsf{A}_E}(1) > 1$$

it follows that $\kappa > 1$. But then:

$$\mathsf{prof}_{\mathsf{A}_E}(\mathsf{prof}_{\mathsf{A}_E}(\kappa)) = \mathsf{prof}_{\mathsf{A}_E}(1) > 1 = \mathsf{prof}_{\mathsf{A}_E}(\kappa)$$

Hence $\mathsf{A}_E \notin \mathsf{IDEM} = \mathsf{MLR}$. So $\mathsf{A}_E \notin \mathsf{MLR} \cap \mathsf{SAT}$ □

But all Fishy abstraction principles that are not Strongly Fishy are modestly $\mathcal{L}$-reflecting:

**Theorem 6.5.7** *If* $\mathsf{A}_E \in \mathsf{FISH}$ *and* $\mathsf{A}_E \notin \mathsf{S\text{-}FISH}$, *then* $\mathsf{A}_E \in \mathsf{MLR} \cap \mathsf{SAT}$.

[26] Note that the following is not true:

If $\mathsf{A}_E \in \mathsf{S\text{-}FISH}$, then $\mathsf{A}_E \notin \mathsf{MLR}$.

since, for any contradiction $\Phi$, $\mathsf{Fish}_\Phi \in \mathsf{S\text{-}FISH} \cap \mathsf{MLR}$.

***Proof*** Assume that $\mathsf{A}_E \in \mathsf{FISH}$ and $\mathsf{A}_E \notin \mathsf{S\text{-}FISH}$. So, for every $\kappa$, if:

$$\mathsf{prof}_{\mathsf{A}_E}(\kappa) \leq \kappa$$

then:

$$\mathsf{prof}_{\mathsf{A}_E}(\kappa) = 1$$

In addition:

$$\mathsf{prof}_{\mathsf{A}_E}(1) = 1$$

So $\mathsf{A}_E \in \mathsf{SAT}$. Additionally, for any cardinal $\gamma > 0$, if:

$$\mathsf{prof}_{\mathsf{A}_E}(\gamma) \leq \gamma$$

then:

$$\mathsf{prof}_{\mathsf{A}_E}(\gamma) = 1$$

Hence:

$$\mathsf{prof}_{\mathsf{A}_E}(\mathsf{prof}_{\mathsf{A}_E}(\gamma)) = \mathsf{prof}_{\mathsf{A}_E}(1) = 1 \leq \gamma$$

Hence $\mathsf{A}_E \in \mathsf{IDEM} = \mathsf{MLR}$. □

The following corollary is worth stating explicitly:

**Corollary 6.5.8** *If* $\mathsf{A}_E \in \mathsf{M\text{-}FISH}$ *but* $\mathsf{A}_E \notin \mathsf{S\text{-}FISH}$, *then* $\mathsf{A}_E \in \mathsf{MLR}$.

Thus, modest $\mathcal{L}$-reflection rules out the Strongly Fishy abstraction principles, but does not on its own rule out the merely Moderately Fishy ones. This outcome does not seem implausible: if our worry is that we should not judge abstraction principles to be acceptable if they entail that the universe contains a certain number of objects without providing a corresponding number of abstracts, then it seems like it is the Strongly Fishy principles that are the real problem, since it is the Strongly Fishy principles that entail that the universe contains more than one object. Thus, modest $\mathcal{L}$-reflection filters out the most obviously objectionable Fishy abstraction principles, demonstrating that this constraint is not merely intuitively plausible. In addition, requiring that acceptable abstraction principles satisfy this criterion constitutes genuine progress towards a solution to the Bad Company Problem.

Nevertheless, Heck is right that there is something "fishy" about even the Moderately Fishy abstraction principles. But the fact that modest $\mathcal{L}$-reflection on its own does not rule these out need not worry us overly much. After all, we have already noted that modest $\mathcal{L}$-reflection cannot be a sufficient condition for acceptability (since unsatisfiable principles modestly $\mathcal{L}$-reflect), and thus a full solution to the Bad Company Problem will need to supplement modest $\mathcal{L}$-reflection with other criteria.

One such criteria that is immensely plausible—Field conservativeness—will, when combined with modest $\mathcal{L}$-reflection, rule out many of the Moderately Fishy abstraction principles. But another option is provided by Heck's own much stronger proposal for ruling out the problematic Fishy abstraction principles (actually, they suggest a number of distinct but closely related solutions, see Heck (1992, 233)). First, we need the following definition:[27]

**Definition 6.5.9** $\mathsf{A}_E$ is *monotonic* if and only if, for any $\kappa$ and $\gamma$ such that $\kappa \leq \gamma$, if:

$$\mathsf{prof}_{\mathsf{A}_E}(\kappa) \leq \kappa$$

$$\mathsf{prof}_{\mathsf{A}_E}(\gamma) \leq \gamma$$

then:

$$\mathsf{prof}_{\mathsf{A}_E}(\kappa) \leq \mathsf{prof}_{\mathsf{A}_E}(\gamma)$$

Heck motivates the idea that acceptable abstraction principles should be monotonic by asking:

> How could the fact that there are more objects of other sorts affect how many abstracts there are? (Heck, 1992, 233)

[27] In addition, there is a more substantial understanding of monotonicity, which requires not only that the number of abstracts not decrease when "moving" from a domain to a larger domain, but that in addition no particular abstracts disappear. We can formulate this stronger notion as follows:

$\mathsf{A}_E$ is *intrinsically monotonic* if and only if:

$$\mathsf{A}_E \models (\forall X)(\forall Y)(\forall Z)((X \subseteq Z \wedge Y \subseteq Z) \rightarrow (E(X, Y) \leftrightarrow E(X, Y)^Z))$$

In other words, an abstraction principle is intrinsically monotonic if and only if two concepts are related to each other by the equivalence relation in question on a domain if and only if they are related to each other on any domain that contains them as sub-concepts. Constructing abstraction principles that are monotonic in the sense discussed above, but fail to be intrinsically monotonic, is not difficult—for example:

$$(\forall X)(\forall Y)[@(X) = @(Y) \leftrightarrow (\mathsf{Succ} \wedge X \approx Y$$
$$\vee \mathsf{Limit} \wedge (\neg X) \approx (\neg Y))]$$

Intrinsic monotonicity is definitely worthy of considerable attention, but a proper analysis would go well beyond the cardinality-profile-centric methods mobilized in this essay, and hence must wait for another time, but see Cook and Linnebo (2018) for a very brief discussion.

They then suggest that the acceptable abstraction principles are those principles that are both monotonic and critically full.[28] This criterion, like modest $\mathcal{L}$-reflection, rules out the Strongly Fishy abstraction principles (in fact, critical fullness alone does the job):

**Theorem 6.5.10** *If* $\mathsf{A}_E \in \text{S-FISH} \cap \text{SAT}$, *then* $\mathsf{A}_E \notin \text{C-FULL}$.

***Proof*** Assume $\mathsf{A}_E \in \text{S-FISH} \cap \text{SAT}$. Then:

$$\text{prof}_{\mathsf{A}_E}(1) > 1$$

and, for any $\kappa$ such that :

$$\text{prof}_{\mathsf{A}_E}(\kappa) \leq \kappa$$

we have:

$$\text{prof}_{\mathsf{A}_E}(\kappa) = 1$$

[28] More carefully, Heck introduces *stabilization points*:

> Say that $\mu$ is a stabilization point if $P$ is satisfiable at $\mu$ and either (i) $\mu$ is a successor cardinal, $\lambda^+$, where $P$ is not satisfiable at $\lambda$, or (ii) $\mu$ is a limit and, for all $\lambda < \mu$ there is [a] $\nu$ between $\lambda$ and $\mu$ such that $P$ is not satisfiable at $\nu$. Then $P$ is acceptable if it is stable and, at every stabilization point $\mu$ in every model of cardinality $\mu$, there are $\mu$ abstracts of the sort characterized by $P$. (Heck, 1992, 233, fn. 9)

We make this more precise as follows (and we call these *Heck points* so as to avoid confusion with stabilisation points introduced in Sect. 6.2 above):

> $\kappa$ is a *Heck point* of $\mathsf{A}_E$ if and only if $\mathsf{A}_E$ is $\kappa$-satisfiable and for every $\gamma < \kappa$, there is a $\lambda$ such that $\gamma \leq \lambda < \kappa$ and $\mathsf{A}_E$ is not $\lambda$-satisfiable. $\text{heck}(\mathsf{A}_E)$ is the set of Heck points of $\mathsf{A}_E$.

Heck suggests that acceptable abstraction principles must be both monotonic (in the sense defined above) and full at all Heck points—that is, they must be monotonic and *Heck full*:

> $\mathsf{A}_E$ is *Heck full* if and only if, for any $\kappa \in \text{heck}(\mathsf{A}_E)$ :
>
> $$\text{prof}_E(\kappa) = \kappa$$
>
> H-FULL = the class of Heck full abstraction principles.

Interestingly, although Heck Fullness is a stronger condition than critical fullness:

$$\text{H-FULL} \subsetneq \text{C-FULL}$$

it turns out that, when we restrict our attention to monotonic abstraction principles, the two notions collapse:

$$\text{H-FULL} \cap \text{MONO} = \text{C-FULL} \cap \text{MONO}$$

(Proofs of these results are left to the reader). Hence the simplified treatment above.

Also, since $\mathsf{A}_E \in \mathsf{SAT}$, there is a $\gamma$ such that:

$$\mathsf{prof}_{\mathsf{A}_E}(\gamma) \leq \gamma$$

Let $\delta$ be the least cardinal such that:

$$\mathsf{prof}_{\mathsf{A}_E}(\delta) \leq \delta$$

But then $\delta \in \mathsf{crit}(\mathsf{A}_E)$, $\delta > 1$, and:

$$\mathsf{prof}_{\mathsf{A}_E}(\delta) = 1$$

Hence:

$$\mathsf{prof}_{\mathsf{A}_E}(\delta) < \delta$$

So $\mathsf{A}_E \notin \mathsf{C\text{-}FULL}$. □

In order to deal with Moderately Fishy abstraction principles, the following definition will be useful:[29]

**Definition 6.5.11** An abstraction principle $\mathsf{A}_E$ is *strongly anti-stable* if and only if there is a cardinal $\kappa$ such that, for any $\gamma$:

$$\mathsf{prof}_{\mathsf{A}_E}(\gamma) \leq \gamma \text{ if and only if } 0 < \gamma < \kappa.$$

$\mathsf{S\text{-}ASTB}$ = the class of strongly anti-stable abstraction principles.

Intuitively, an abstraction principle $\mathsf{A}_E$ is strongly anti-stable if and only if it is "strongly downwards closed"—that is, there is a cardinal $\kappa$ such that $\mathsf{A}_E$ is $\gamma$-satisfiable at all and only those non-zero cardinals $\gamma$ less than $\kappa$. The following facts are worth noting explicitly:

**Theorem 6.5.12** $\mathsf{A}_E \notin \mathsf{S\text{-}ASTB}$ *if and only if, for any cardinal* $\kappa$*, there is a cardinal* $\gamma$ *such that either* $0 < \gamma < \kappa$ *and:*

$$\mathsf{prof}_{\mathsf{A}_E}(\gamma) > \gamma$$

*or* $\gamma \geq \kappa$ *and:*

$$\mathsf{prof}_{\mathsf{A}_E}(\gamma) \leq \gamma$$

***Proof*** Straightforward consequence of the definition of strong anti-stability. □

[29] Note that unsatisfiable abstraction principles are strongly anti-stable, hence abstraction principles that fail to be strongly anti-stable are satisfiable.

**Theorem 6.5.13** *If* $\mathsf{A}_E \in \mathsf{S\text{-}ASTB} \cap \mathsf{SAT}$, *then* $\mathsf{A}_E \notin \mathsf{S\text{-}FISH}$.

***Proof*** If $\mathsf{A}_E \in \mathsf{S\text{-}ASTB} \cap \mathsf{SAT}$, then:

$$\mathsf{prof}_{\mathsf{A}_E}(1) = 1$$

So $\mathsf{A}_E \notin \mathsf{S\text{-}FISH}$. □

We can now prove that every strongly anti-stable Moderately Fishy abstraction principle is critically full and monotonic. The following lemma is worth noting separately, since it applies to all satisfiable, strongly anti-stable abstraction principles, monotonic or not:

**Theorem 6.5.14** $\mathsf{S\text{-}ASTB} \cap \mathsf{SAT} \subsetneq \mathsf{C\text{-}FULL}$.

***Proof*** Assume $\mathsf{A}_E \in \mathsf{S\text{-}ASTB} \cap \mathsf{SAT}$. So there is a $\kappa$ such that, for all $\gamma$:

$$\mathsf{prof}_{\mathsf{A}_E}(\gamma) \leq \gamma \text{ if and only if } 0 < \gamma < \kappa$$

So $\mathsf{crit}(\mathsf{A}_E) = \{1\}$. Also:

$$\mathsf{prof}_{\mathsf{A}_E}(1) = 1$$

Thus $\mathsf{A}_E \in \mathsf{C\text{-}FULL}$ . Thus $\mathsf{S\text{-}ASTB} \cap \mathsf{SAT} \subseteq \mathsf{C\text{-}FULL}$. But, for any logical truth $\Phi$, $\mathsf{Fish}_\Phi \in \mathsf{C\text{-}FULL}$ but $\mathsf{Fish}_\Phi \notin \mathsf{S\text{-}ASTB}$. □

Hence:

**Corollary 6.5.15** $\mathsf{M\text{-}FISH} \cap \mathsf{S\text{-}ASTB} \cap \mathsf{SAT} \subsetneq \mathsf{C\text{-}FULL} \cap \mathsf{MONO}$.

***Proof*** $\mathsf{M\text{-}FISH} \subsetneq \mathsf{MONO}$, hence $\mathsf{M\text{-}FISH} \cap \mathsf{S\text{-}ASTB} \cap \mathsf{SAT} \subseteq \mathsf{C\text{-}FULL} \cap \mathsf{MONO}$. For non-identity, again consider $\mathsf{Fish}_\Phi$ where $\Phi$ is any logical truth. □

Every Moderately Fishy abstraction principle that is not anti-stable fails to be critically full, however:

**Theorem 6.5.16** *If* $\mathsf{A}_E \in \mathsf{M\text{-}FISH}$ *but* $\mathsf{A}_E \notin \mathsf{S\text{-}ASTB}$, *then* $\mathsf{A}_E \notin \mathsf{C\text{-}FULL}$.

***Proof*** Assume that $\mathsf{A}_E \in \mathsf{M\text{-}FISH}$ and $\mathsf{A}_E \notin \mathsf{S\text{-}ASTB}$. Then there is a $\kappa$ such that:

$$\mathsf{prof}_{\mathsf{A}_E}(\kappa) > \kappa$$

and, for any $\kappa$ such that:

$$\mathsf{prof}_{\mathsf{A}_E}(\kappa) \leq \kappa$$

we have:

$$\mathsf{prof}_{\mathsf{A}_E}(\kappa) = 1$$

Let $\gamma$ be the least cardinal such that:

$$\mathsf{prof}_{\mathsf{A}_E}(\gamma) > \gamma$$

Then there is a cardinal $\delta$ such that either $0 < \delta < \gamma$ and:

$$\mathsf{prof}_{\mathsf{A}_E}(\delta) > \delta$$

or $\delta \geq \gamma$ and:

$$\mathsf{prof}_{\mathsf{A}_E}(\delta) \leq \delta$$

The first of these contradicts our choice of $\gamma$, thus there is a $\delta$ such that $\delta \geq \gamma$ and:

$$\mathsf{prof}_{\mathsf{A}_E}(\delta) \leq \delta$$

Let $\mu$ be the least cardinal such that $\mu \geq \gamma$ and:

$$\mathsf{prof}_{\mathsf{A}_E}(\mu) \leq \mu$$

$\mu > \gamma$, so $\mu > 1$ and $\mu \in \mathsf{crit}(\mathsf{A}_E)$. But:

$$\mathsf{prof}_{\mathsf{A}_E}(\mu) = 1$$

Hence, $\mathsf{A}_E \notin \mathsf{C\text{-}FULL}$. □

Thus, requiring acceptable abstraction principles to be monotonic and critically full rules out many principles that do not get ruled out by modest $\mathcal{L}$-reflection—specifically, those that are Moderately Fishy but not strongly anti-stable.[30] Thus, Heck's proposal rules out more of the "Fishy" principles than modest $\mathcal{L}$-reflection.

On careful inspection, however, it is apparent that it is critical fullness, and not monotonicity, that is incompatible with the Moderately Fishy abstraction principles that are not strongly anti-stable. So what work is monotonicity doing in Heck's proposal? The answer is simple: Monotonicity, in conjunction with critical fullness, guarantees modest $\mathcal{L}$-reflection, although neither monotonicity nor critical fullness alone implies modest $\mathcal{L}$-reflection:

**Theorem 6.5.17** $\mathsf{MONO} \not\subset \mathsf{MLR}$

***Proof*** $\mathsf{Fish}_{(\exists x)(\exists y)(x \neq y)} \in \mathsf{MONO}$, but $\mathsf{Fish}_{(\exists x)(\exists y)(x \neq y)} \notin \mathsf{MLR}$ □

**Theorem 6.5.18** $\mathsf{C\text{-}FULL} \not\subset \mathsf{MLR}$

[30] A simple strategy for constructing such principles: Consider the $\mathsf{Fish}_{\Phi}$ principle for any purely logical $\Phi$ where (i) $\Phi$ is true on the one-element model, and (ii) there are cardinals $\gamma, \kappa$ such that $\gamma < \kappa$, $\Phi$ is false on domains of cardinality $\gamma$, and $\Phi$ true on domains of size $\kappa$.

***Proof*** Consider:

$$\mathsf{A}_{\mathsf{E}_4} : (\forall X)(\forall Y)[@_{\mathsf{E}_4}(X) = @_{\mathsf{E}_4}(Y) \leftrightarrow ((\Phi_{\aleph_0} \wedge X \approx Y) \\ \vee (\Phi_{\aleph_1}) \\ \vee (\forall z)(X(z) \leftrightarrow Y(z)))]$$

where $\Phi_{\aleph_0}$ abbreviates the purely logical claim that the cardinality of the universe is $\aleph_0$ and $\Phi_{\aleph_2}$ abbreviates the purely logical claim that the cardinality of the universe is $\aleph_2$. Then $\mathsf{crit}(\mathsf{A}_{\mathsf{E}_4}) = \{\aleph_0\}$ and:

$$\mathsf{prof}_{\mathsf{A}_{\mathsf{E}_4}}(\aleph_0) = \aleph_0$$

So $\mathsf{A}_{\mathsf{E}_4} \in \mathsf{C\text{-}FULL}$. But:

$$\mathsf{prof}_{\mathsf{A}_{\mathsf{E}_4}}(\aleph_1) = 1 < \aleph_1$$

Yet:

$$\mathsf{prof}_{\mathsf{A}_{\mathsf{E}_4}}(1) > 1$$

So $\mathsf{A}_{\mathsf{E}_4} \notin \mathsf{IDEM} = \mathsf{MLR}$. □

**Theorem 6.5.19** $\mathsf{MONO} \cap \mathsf{C\text{-}FULL} \subsetneq \mathsf{MLR}$

***Proof*** Assume $\mathsf{A}_E \in \mathsf{MONO} \cap \mathsf{C\text{-}FULL}$, and assume that:

$$\mathsf{prof}_{\mathsf{A}_E}(\kappa) \leq \kappa$$

Assume, for *reductio*, that:

$$\mathsf{prof}_{\mathsf{A}_E}(\mathsf{prof}_{\mathsf{A}_E}(\kappa)) > \mathsf{prof}_{\mathsf{A}_E}(\kappa)$$

Let $\gamma$ be the least cardinal such that $\gamma > \mathsf{prof}_{\mathsf{A}_E}(\kappa)$ and:

$$\mathsf{prof}_{\mathsf{A}_E}(\gamma) \leq \gamma$$

Then $\gamma \in \mathsf{crit}(\mathsf{A}_E)$. So:

$$\mathsf{prof}_{\mathsf{A}_E}(\gamma) = \gamma$$

Also:

$$\gamma \leq \kappa$$

So by monotonicity:

$$\mathsf{prof}_{\mathsf{A}_E}(\gamma) \leq \mathsf{prof}_{\mathsf{A}_E}(\kappa)$$

But this contradicts:

$$\mathsf{prof}_{\mathsf{A}_E}(\gamma) = \gamma > \mathsf{prof}_{\mathsf{A}_E}(\kappa)$$

Thus:

$$\mathsf{prof}_{\mathsf{A}_E}(\mathsf{prof}_{\mathsf{A}_E}(\kappa)) \leq \mathsf{prof}_{\mathsf{A}_E}(\kappa)$$

Hence $\mathsf{A}_E \in \mathsf{IDEM} = \mathsf{MLR}$. So $\mathsf{MONO} \cap \mathsf{C\text{-}FULL} \subseteq \mathsf{MLR}$. For non-identity, consider:

$$\begin{aligned}\mathsf{A}_{\mathsf{E}_5} : (\forall X)(\forall Y)[@_{\mathsf{E}_5}(X) = @_{\mathsf{E}_5}(Y) \leftrightarrow ((\Phi_{<(\aleph_{\aleph_1}+1)} \wedge X \approx Y) \\ \vee\ \Phi_{(\aleph_{\aleph_1}+1)} \wedge (\forall z)(X(z) \leftrightarrow Y(z)) \\ \vee\ \Phi_{>(\aleph_{\aleph_1}+1)} \wedge (\mathsf{Inf}(X) \wedge \mathsf{Inf}(Y)) \vee X \approx Y)\end{aligned}$$

where $\Phi_{<(\aleph_{\aleph_1}+1)}$ abbreviates the purely logical claim that the cardinality of the universe is less than $\aleph_{\aleph_0+1}$ (and similarly for $\Phi_{(\aleph_{\aleph_1}+1)}$ and $\Phi_{>(\aleph_{\aleph_1}+1)}$), and $\mathsf{Inf}(X)$ abbreviates the purely logical claim that $X$ holds of infinitely many members of the domain. Then $\mathsf{crit}(\mathsf{A}_{\mathsf{E}_5}) = \{\aleph_0, \aleph_{\aleph_1+2}\}$, but:

$$\mathsf{prof}_{\mathsf{A}_{\mathsf{E}_5}}(\aleph_{\aleph_1+2}) = \aleph_0$$

So $\mathsf{A}_\mathsf{E} \notin \mathsf{C\text{-}FULL}$. Also:

$$\mathsf{prof}_{\mathsf{A}_{\mathsf{E}_5}}(\aleph_{\aleph_1}) = \aleph_1 > \aleph_0 = \mathsf{prof}_{\mathsf{A}_{\mathsf{E}_5}}(\aleph_{\aleph_1+2})$$

Thus $\mathsf{A}_{\mathsf{E}_5} \notin \mathsf{MONO}$. But, for any $\kappa$ such that:

$$\mathsf{prof}_{\mathsf{A}_{\mathsf{E}_5}}(\kappa) \leq \kappa$$

we have:

$$\mathsf{prof}_{\mathsf{A}_{\mathsf{E}_5}}(\kappa) = \aleph_0$$

(if $\kappa < \aleph_{\aleph_1}$ or $\kappa > \aleph_{\aleph_1+1}$) or:

$$\mathsf{prof}_{\mathsf{A}_{\mathsf{E}_5}}(\kappa) = \aleph_1$$

**Fig. 6.6** Fullness, monotonicity, and modest $\mathcal{L}$-reflection

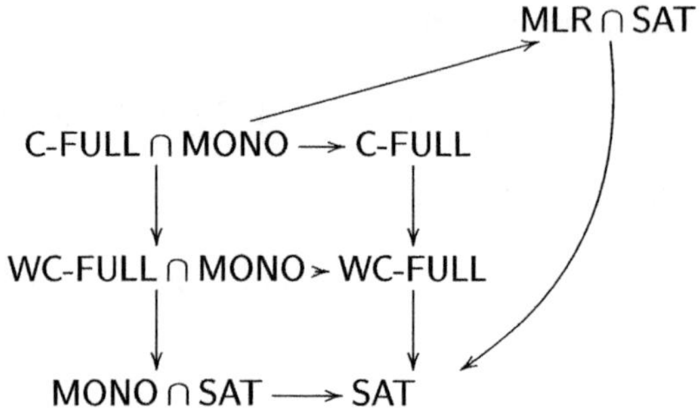


(if $\kappa = \aleph_{\aleph_1}$). Since:

$$\text{prof}_{\mathsf{A}_{\mathsf{E}_5}}(\aleph_1) = \text{prof}_{\mathsf{A}_{\mathsf{E}_5}}(\aleph_0) = \aleph_0$$

we have:

$$\text{prof}_{\mathsf{A}_{\mathsf{E}_5}}(\text{prof}_{\mathsf{A}_{\mathsf{E}_5}}(\kappa)) = \aleph_0 \leq \text{prof}_{\mathsf{A}_{\mathsf{E}_5}}(\kappa)$$

and hence $\mathsf{A}_{\mathsf{E}_5} \in \mathsf{IDEM} = \mathsf{MLR}$ □

Thus, Heck's suggested criterion entails, but is not entailed by, modest $\mathcal{L}$-reflection. In addition, we note that the fact that modest $\mathcal{L}$-reflection does not entail monotonicity is an immediate corollary:

**Corollary 6.5.20** $\mathsf{MLR} \not\subset \mathsf{MONO}$

***Proof*** Consider again $\mathsf{A}_{E_5}$ from the proof of **Theorem 6.5.19**. $\mathsf{A}_{E_5} \in \mathsf{MLR}$ but $\mathsf{A}_{E_5} \notin \mathsf{MONO}$. □

Thus, although modest $\mathcal{L}$-reflection is distinct from any of the conditions that we can formulate in terms of satisfiability, fullness, and monotonicity, consideration of monotonicity allows us to narrow down much more tightly the location of the modest $\mathcal{L}$-reflecting abstraction principles in logical space, as indicated by Fig. 6.6.

In short, there turn out to be deep connections between Weir's modest $\mathcal{L}$-reflection constraint and Heck's suggestion that acceptable abstraction principles be monotonic and critically full.

## 6.6 Loose Ends Tied

As we have now seen, Weir's notion of modest reflection (on our slightly amended, modest $\mathcal{L}$-reflection reading of the idea) is plausible, distinct from competing proposals, and helps address a novel problem faced by the proponent of neo-logicist foundations of mathematics—Heck's Fishy abstraction principles. Thus, it is unfortunate that this idea has received so little attention in the literature until now.

The unfortunate aspects of this oversight are not limited simply to the fact that this particular restriction could have, but was not, included in various examinations of criteria that acceptable abstraction principles must meet, or of the Bad Company problem more generally. In addition, Weir's proposal, understood in full, amounts not only to a new criterion for abstraction principle acceptability, but also to a *new way of thinking about* such criteria.

All of the extant proposals that have been made with respect to the Bad Company problem since Weir's 2003 paper—that is, all of the various criteria surveyed in Sects. 6.2 and 6.3 above—fit a very specific pattern: Very loosely put, given an abstraction principle $\mathsf{A}_E$, satisfaction each of these criteria by $\mathsf{A}_E$ involves a relationship that must hold between a cardinal number $\kappa$ and the value of the cardinality profile of $\mathsf{A}_E$ applied to $\kappa$—that is, $\mathsf{prof}_{\mathsf{A}_E}(\kappa)$). Thus, the satisfiability criteria in Sect. 6.2 all involved determining at which cardinals the former number is no greater than the latter number, and the fullness criteria of Sect. 6.3 involved determining at which cardinals the former number is identical to the latter number. In short, whether or not any of these criteria hold of an abstraction principle $\mathsf{A}_E$ depends solely on the relationship between $\kappa$ and $\mathsf{prof}_{\mathsf{A}_E}(\kappa)$.

Once we see that modest $\mathcal{L}$-reflection is equivalent to idempotence (i.e., $\mathsf{IDEM} = \mathsf{MLR}$), then it becomes clear that modest $\mathcal{L}$-reflection involves a different and more complex pattern. In order to know that an abstraction principle is modestly $\mathcal{L}$-reflecting, it is not enough merely to know how $\kappa$ and $\mathsf{prof}_{\mathsf{A}_E}(\kappa)$ are related to each other, for each cardinal $\kappa$. In addition, we need to know how $\mathsf{prof}_{\mathsf{A}_E}(\mathsf{prof}_{\mathsf{A}_E}(\kappa))$ is related to both $\mathsf{prof}_{\mathsf{A}_E}(\kappa)$ and to $\kappa$. Bluntly put, modest $\mathcal{L}$-reflection, unlike all of the satisfiability and fullness criteria of Sects. 6.2 and 6.3 above, is not a condition that holds merely in virtue of the number of abstracts that are generated on each size of domain (i.e., in virtue of the relationship between $\kappa$ and $\mathsf{prof}_{\mathsf{A}_E}(\kappa)$ for each cardinal $\kappa$). In addition, modest $\mathcal{L}$-reflection involves the relationship between the number of abstracts generated on domains of one size and the number of abstracts generated on domains *of another cardinality* ((i.e., the relationship between $\mathsf{prof}_{\mathsf{A}_E}(\mathsf{prof}_{\mathsf{A}_E}(\kappa))$ and $\mathsf{prof}_{\mathsf{A}_E}(\kappa)$ for each cardinal $\kappa$). In this sense, it is a deeper, more complex condition than any of the satisfiability or fullness criteria covered in Sects. 6.2 and 6.3, and it is not surprising that it has taken a good bit of work to get a decent grasp on what, exactly, modest $\mathcal{L}$-reflection amounts to.

Of course, monotonicity is another condition on abstraction principles that involves direct comparison between $\mathsf{prof}(\kappa_1)$ and $\mathsf{prof}(\kappa_2)$ for distinct domain sizes $\kappa_1$ and $\kappa_2$. And monotonicity has begun to get a bit of attention in the literature on the Bad Company Problem (again, see, e.g., the postscript to Heck (1992) and Cook and Linnebo (2018)). But the attention "lavished" on monotonicity is recent—a full decade or more after the publication of Weir (2003). Thus, the lack of attention that Weir's modest reflection proposal has received represents a very real missed opportunity in terms of the direction that research on abstraction has taken over the intervening fifteen years. Had more effort been expended on exploring this idea, it might have led us much more quickly to recognize and consider the general class of conditions on abstraction principles to which it belongs—a class including both modest $\mathcal{L}$-reflection and monotonicity, and a class whose importance is only now becoming apparent. While there is no way that we can roll back the clock and correct this mistake, it is hoped that this essay at least helps to get us back to where we should have been all along: taking modest $\mathcal{L}$-reflection seriously as a constraint on acceptable abstraction principles.[31]

## References


Bueno, O., & Linnebo, Ø. (Eds.). (2009). *New waves in philosophy of mathematics*. Basingstoke: Palgrave.

Cook, R. (Ed.). (2007). *The Arché papers on the mathematics of abstraction*. Dordrecht: Springer.

Cook, R. (2009), New waves on an old beach: Neo-Fregean philosophy of mathematics today. In Bueno and Linnebo (2009) (pp. 13–34).

Cook, R. (2012). Conservativeness, stability, and abstraction. *The British Journal for the Philosophy of Science, 63*(3), 673–696.

Cook, R. (2016). Conservativeness, cardinality, and bad company. In Ebert and Rossberg (2016) (pp. 223–246).

Cook, R., & Linnebo, Ø. (2018). Cardinality and acceptable abstraction. *Notre Dame Journal of Formal Logic, 59*(1), 61–74.

Ebert, P., & Rossberg, M. (2016). *Abstractionism: Essays in philosophy of mathematics*. Oxford: Oxford University Press.

Ebert, P., & Shapiro, S. (2009). The good, the bad, and the ugly. *Synthese, 170*, 415–441.

Fine, K. (2002). *The limits of abstraction*. Oxford: Oxford University Press.

Hale, B., & Wright, C. (2001). *The Reason's proper study*. Oxford: Oxford University Press.

Heck, R. (1992). On the consistency of second-order contextual definitions. *Noûs, 26*, 491–494. Reprinted with a Postscript in Heck (2012) (pp. 227–236).

Heck, R. (Ed.). (1997). *Language, thought and logic: Essays in honour of Michael Dummett*. Oxford: Oxford University Press.

Heck, R. (2012). *Frege's theorem*. Oxford: Oxford University Press.

Linnebo, Ø. (2018). *Thin objects: An abstractionist account*. Oxford: Oxford University Press.

MacBride, F. (2003). Speaking with shadows: A study of neo-logicism. *The British Journal for Philosophy of Science, 54*(1), 103–163.

Shapiro, S. (1992). *Foundations with foundationalism: The case for second-order logic*. Oxford: Oxford University Press.

Weir, A. (2003). Neo-Fregeanism: An embarrassment of riches? *Notre Dame Journal of Formal Logic, 44*(1), 13–48.

Wright, C. (1997). On the philosophical significance of Frege's theorem. In Heck (1997) (pp. 201–244).

Wright, C. (1999). Is Hume's principle analytic? *Notre Dame Journal of Formal Logic, 40*, 6–30.


[31] Thanks are due to Øystein Linnebo and Richard Kimberly Heck for conversations related to the material that eventually became this essay. Thanks are also owed to Alan as well, for providing a detailed and excellent set of comments on an early draft of this essay, but, more importantly, for being a wonderful colleague and an even better friend.

# Chapter 7
# Mass Nouns and Plural Logic

**Eric Snyder and Stewart Shapiro**

**Abstract** According to singularism, 'the students' refers to a single collective entity, e.g. a sum or set. In contrast, according to pluralism, 'the students' plurally refers to multiple students at once, through the primitive relation of plural reference. Although it was originally designed exclusively for plural nouns, this paper addresses whether plural reference can be extended so as to provide an empirically adequate semantics for mass nouns, such as 'the furniture', as well, as certain pluralists have maintained. Our contention is that extant pluralist analyses of mass nouns suffer from two major setbacks. First, while they may or may not explain the semantic commonalities between plural and mass nouns, they do not explain their semantic differences. Secondly, pluralism is inconsistent with the possibility that mass nouns are gunky, i.e. fail to have atomic parts. However, we provide three kinds of semantic arguments showing that the gunkiness of at least some mass nouns is not only possible, but plausible.

**Keywords** Singularism · Pluralism · Mass nouns · Mereology · Plural logic

## 7.1 Introduction

There are two prominent views regarding definite plurals like 'the students' in (1):

(1) The students gathered around the teacher.

According to *singularism*, 'the students' refers to a single collective entity, typically a set or a mereological sum, and (1) is true if that entity has the property of gathering

E. Snyder (✉)
Ashoka University, Sonipat, India
e-mail: Eric.snyder@ashoka.edu.in

S. Shapiro
Ohio State University, Columbus, OH, USA
e-mail: Shapiro.4@osu.edu

 
A. Rieger, S. Leuenberger (eds.), *Themes from Weir: A Celebration of the Philosophy of Alan Weir*, Synthese Library 484,
https://doi.org/10.1007/978-3-031-54557-3_7

around the teacher. There is one reference relation—*singular reference*—which both definite plurals and singular definites like 'the student' exploit. Accordingly, there are just individual variables—$x, y, z, \ldots$—ranging over both singular individuals like John or that student and pluralities of individuals like John and Mary or the students, viewed as single, collective entities.

In contrast, according to *pluralism*, 'the students' refers to multiple students simultaneously, and (1) is true if those students together have that same collective property of gathering around the teacher. There are at least two reference relations—singular reference and *plural reference*—employed by different kinds of nouns—singular nouns and plural nouns. In *plural logic*, those are regimented using different variables. Typically, there are singular variables $x, y, z, \ldots$ ranging over singular individuals, and plural variables $xx, yy, zz, \ldots$ ranging over individuals in a primitively plural way. That is, plural variables range over multiple individuals at once, not sums or sets, or any other singular entities.[1]

Singularism is the predominant view within linguistic semantics,[2] but pluralism is the received view among philosophers and logicians who invoke plurals.[3] This is due largely to the influence of Boolos (1984, 1985), who developed the notion of plural reference. His main goal was to make sense of set theory, without invoking paradox. Consider (2a,b):

(2) a. If there is a set that does not contain itself, then there are some sets among which is every set that does not contain itself.

b. If there is a set that does not contain itself, then there is a set of all sets that do not contain themselves.

As Boolos observes, (2a) at least appears to express a trivial truth, whereas (2b) leads directly to Russell's Paradox. Accordingly, Boolos contends that if 'some sets' in (2a) is interpreted as a set of sets, as some forms of singularism might suggest, then (2a,b) would be semantically indistinguishable, thus leading to incoherency in virtue of the truth of (2a).

Boolos (1984, 1985) did not intend to contribute directly to the semantics of natural language. However, more recent pluralists have been concerned with natural language as such. The program is to extend Boolos' notion plural reference to analyze ordinary plural talk, giving truth-conditions to natural language utterances, in context, containing plurals.

Here we contrast pluralism with a version of singularism that takes mereological sums or fusions to be the "singular entities" that plural definites denote. We dub the program *mereological singularism*; details will follow. For present purposes, we use term *plurality* for the semantic value of a plural noun, intending it to be neutral

[1] See Linnebo (2017). Alternatively, it is possible to formulate pluralism with only one reference relation, namely plural reference, where singular reference is a limiting case. For details, see Florio and Linnebo (2021).

[2] There are exceptions, e.g. Higginbotham and Schein (1989).

[3] See Carrara et al. (2016) for a nice summary of the singularism/pluralism debate.

between singularism and pluralism. So according to mereological singularism, pluralities are mereological sums.

To be sure, "plurality" is a count noun in ordinary English. Since pluralists take plural reference to be primitive, they will not understand our talk of "pluralities" literally, as referring to entities in their own right, to things called "pluralities". Rather, it will be viewed as only a manner of speaking, ultimately to be paraphrased into the plural idiom.

One attractive feature of mereological singularism is that it affords a uniform analysis of nouns. Specifically, count nouns ('student','students') and mass nouns ('furniture', 'mud') can both be analyzed via mereological notions like sum and parthood. Indeed, most contemporary analyses of the count/mass distinction within linguistic semantics do just that, and so are versions of mereological singularism.

This raises an interesting question: Although plural reference was originally designed as an analysis of count nouns, could this same primitive reference relation be used to also analyze mass nouns? There are at least three reasons why a pluralist analysis of mass nouns may be independently desirable, from the pluralist perspective.

First, there are many well-known semantic commonalities between plural nouns and mass nouns.[4] For example, unlike singular count nouns, plural and mass nouns both appear to be *cumulative*, licensing the following characteristic entailments.

(3) a. $x$ is an apple and $y$ is an apple $\nvDash$ $x$ together with $y$ is an apple
b. $x, \ldots z$ are apples and $x', \ldots z'$ are apples $\vDash$ $x, \ldots, z$ and $x', \ldots, z'$ together are apples
c. $x$ is water and $y$ is water $\vDash$ $x$ together with $y$ is water

Also, plural and mass nouns can both occur bare, i.e. without overt accompanying determiners, in argument positions. Depending on the predicate, this can lead to kind-interpretations of the noun, as with (4a,b).

(4) a. Water is a liquid.
b. {??Raccoon is/Raccoons are} mammalian.

Third, plural and mass nouns pattern alike with respect to certain quantificational expressions they are acceptable with, as witnessed in (5a,b).

(5) a. All {??coin is/change is/coins are} exchangeable.
b. Every {coin is/??change is/??coins are} exchangeable.

Ideally, a semantics for nouns would explain these commonalities. This suggests that insofar as plural reference is appropriate for plural nouns, it also ought to be appropriate for mass nouns.

Secondly, some of the original theoretical motivations for adopting pluralism extend straightforwardly to mass nouns. For example, Rayo (2002) originally argued

[4] See e.g. Chierchia (1998a) and Lasersohn (2011).

against mereological singularism on the basis that it wrongly analyzes (6a,b) as truth-conditionally equivalent.

(6) a. The piles of grains are scattered.
b. The grains are scattered.

Specifically, if 'the piles of grains' and 'the grains' denote the same mereological sum, then (6a) is true just in case (6b), contrary to fact. Similarly, Nicolas (2008) argues that an analogous problem afflicts singularist analyses of mass nouns, as evidenced by (7a,b).

(7) a. The piles of furniture are scattered.
b. The furniture is scattered.

Again, if 'piles of furniture' and 'the furniture' denote the same sum, then mereological singularism would wrongly predict (7a,b) to be equivalent. Thus, if plural reference could similarly be used to analyze 'the piles of furniture' and 'the furniture' as non-coreferential terms, this would appear to secure an important theoretical advantage for pluralism.

Third, all else being equal, we ought to prefer semantic theories with broader empirical coverage. Thus, if mereological singularism but not pluralism is capable of providing an analysis of all nouns, that would be a good, albeit defeasible, reason for preferring the former.

Thus, the question pursued here is whether a pluralist analysis of mass nouns is viable. In our view, there are two major kinds of challenges facing such an analysis. The first is theoretical in nature, and concerns the viable possibility that at least some mass nouns, such as 'time' and 'space', are *gunky*. In current philosophical jargon, a substance is *gunky* if every part of it has a proper part. Similarly, a mass predicate is *gunky* if anything satisfying that predicate has a proper part satisfying the predicate.

In the language of mereology, something is an *atom* if it has no proper parts. This is defined in (8), where '$\sqsubseteq$' designates parthood.

(8) $\forall x.\, \mathtt{ATOM}(x) \leftrightarrow \forall y.\ y \sqsubseteq x \rightarrow x = y$

Classical extensional atomistic mereology is the theory consisting of the usual axioms of extensional mereology together with *the axiom of atomicity*.

(9) $\forall x.\exists y.\ y \sqsubseteq x \wedge \mathtt{ATOM}(y)$

This guarantees that all elements in a mereological structure ordered by parthood "bottom out" in atoms. Dropping (9) from classical extensional mereology thus allows for the possibility of gunky structures.

Since mereological singularism is formulated within classical extensional mereology, it is consistent with the possibility of gunky mass structures. In fact, certain prominent singularist analyses, such as Link (1998) and Landman (2020), explicitly argue that some mass nouns are gunky. Minimally, this suggests that analyses ought to be neutral with respect this possibility. After all, whether some mass nouns are gunky is ultimately an *empirical* matter. Thus, it would be a significant theoretical disadvantage of a theoretical framework if it precluded that possibility, in principle.

However, it is far from clear whether pluralism can plausibly tolerate gunkiness. After all, it *begins* with individuals, or the referents of singular count terms—this student, that piece of furniture, etc.—and construes the referents of plural terms via them. Pre-theoretically, at least, people and chairs are not gunky substances. This is why, presumably, we can regularly identify and distinguish them, count them, etc.

Nevertheless, whether plural logic can in fact tolerate gunkiness is partly a formal matter. The analog of (9) within plural logic is *the axiom of non-emptiness*, given in (10), where '$\prec$' is the primitive relation of *inclusion*.

(10) $\forall xx.\exists y.\ y \prec xx$

As the label suggests, this guarantees that "pluralities" are non-empty: any *things* include at least one *thing*. As we show elsewhere, it is possible to drop (10) from plural logic. Doing so, however, requires replacing the standard comprehension schema with one transcribing the standard comprehension schema within classical extensional mereology. Put differently, accommodating gunkiness within the pluralist setting renders the two theories *notational variants*, or different ways of saying the same thing. We will return to this briefly in Sect. 7.5.

Our primary concern here is rather with the *empirical* challenges facing pluralist analyses of mass nouns, such as Nicolas (2008) and Cocchiarella (2009). Their claim is that mass nouns, despite being *syntactically* singular, are *semantically* plural. That is, like plural nouns, mass nouns denote primitive pluralities, thus exploiting the semantic similarities between plural and mass nouns noted above. Consequently, perhaps the most obvious empirical challenge to any such analysis is explaining the apparent semantic *differences* between plural and mass nouns, especially those like (11).

(11) a. Mary bought those {pieces of furniture/??furniture(s)}.
 b. Mary bought three {pieces of furniture/??furniture(s)}.
 c. How many {pieces of furniture/??furniture(s)} did Mary buy?
 d. Mary bought both {pieces of furniture/??furniture(s)}.

Traditionally, (11a–d) have been taken to have semantic import. Specifically, whereas plural nouns like 'pieces of furniture' denote pluralities of atoms, mass nouns like 'furniture' have a different denotation.[5] Yet if 'the furniture' plurally refers to some things, then since those things are pieces of furniture, intuitively, it is not at all clear how (11a–d) could reflect a *semantic* difference between the nouns involved.

Thus, the purpose of this paper is to develop what we take to be the most serious empirical challenges to a pluralist semantics for mass nouns, taking Nicolas's

[5] See e.g. Link (1983, 1998), Chierchia (1998a,b), Rothstein (2017), and Landman (2020).

(2008) analysis as our guide. We argue that the proposed explanation of apparent semantic differences like those in (11) does not readily extend to other, similar differences between mass nouns, thus suggesting that a unified semantic explanation is required. However, providing one threatens to undermine Boolos' motivation for plural reference, namely making sense of our plural talk without understanding "pluralities" as things.

The rest of the paper is organized as follows. In the next section, we sketch basic versions of mereological singularism and pluralism, showing how they account for some (but not all) ordinary plural talk. In Sect. 7.3, we outline Nicolas's (2008) pluralist analysis of mass nouns. In Sect. 7.4, we pose three significant challenges for that analysis. We conclude in Sect. 7.5, where we summarize our conclusions and briefly return to gunkiness.

## 7.2 Basic Singularism and Pluralism

Consider the examples in (12a–c).

(12) a. The students gathered around the teacher.
b. The students smiled.
c. The students lifted the box.

Clearly, (12a) can only be understood as having a *collective interpretation*: the students gathered around the teacher *together*. In contrast, (12b) can only be understood as having a *distributive interpretation*: *each* of the students smiled. Finally, (12c) is ambiguous between a collective and distributive interpretation: either the students lifted the box together, or else each lifted the box individually. In what follows, we illustrate basic versions of mereological singularism and pluralism, by showing how they capture these different interpretations.

### *7.2.1 Basic Mereological Singularism*

We begin with the semantics of Link (1983). Count nouns are interpreted in a lattice obeying the principles of classical extensional atomistic mereology.[6]

(13) a. $[\![\text{student}]\!] = \lambda x.\, \texttt{student}(x)$
b. $[\![\text{students}]\!] = \lambda x.\, {}^{*}\texttt{student}(x)$
c. $[\![\text{the students}]\!] = \sigma x[{}^{*}\texttt{student}(x)]$
d. $[\![\text{Andy, Betty, and Charlie}]\!] = a \sqcup b \sqcup c$

[6] Link also postulates a second mereological lattice, which need not be atomic, for interpreting for mass nouns. Our focus here is just count nouns, however.

Singular count nouns like 'student', denote a set of individuated, or *atomic*, students. Plural nouns such as 'students' denote the closure of the atomic students under sum-formation, represented in (13b) by the *-operation. Following Sharvy (1980), 'the' is analyzed as a maximality-operation, represented in (13c) by '$\sigma$'. Accordingly, 'the students' refers to a salient maximal sum of atomic students (if there is one). Finally, conjunctive terms like 'Andy, Betty, and Charlie' reference the sum ($\sqcup$) of atoms named.

Suppose we have three atomic students: $a$, $b$, and $c$. We can sum these together to form sums of atomic students. Ordering these by parthood results is an *atomic join semilattice* structure like the following, where arrows represent *individual parthood*:

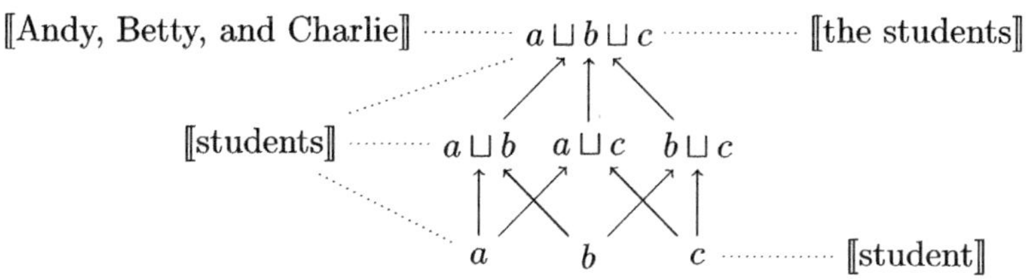


(7.1)

The bottom elements of the structure are atoms, while the remaining elements are proper sums. In Link's semantics, these enjoy the same semantic type and ontological status. They are all individuals, of type $e$, though different *sorts* of individuals: *atomic individuals* and *plural individuals*, respectively.

As such, sums may serve as arguments of various predicates, as illustrated in (14b).

(14) a. Andy and Betty gathered around the teacher.
b. $\texttt{gather}(a \sqcup b)$

While collective predicates apply to only sums, distributive predicates apply to just atoms. Hence, (15a) will be true if each atomic student slept.

(15) a. The students smiled.
b. $\texttt{smile}(\sigma x[{}^*\texttt{student}(x)])$

Finally, mixed predicates such as 'carry a piano' will be ambiguous between collective and distributive interpretations.

(16) a. The students carried a piano.
b. $\texttt{carry}(\sigma x[{}^*\texttt{student}(x)])$

Specifically, (16a) is true on the collective interpretation if the maximal sum of students has the property in question, and is true on the distributive interpretation if each atomic student has that property.

### 7.2.2 Basic Pluralism

We now turn to basic pluralism, as developed by Boolos (1984), Rayo (2002), and Oliver and Smiley (2016), among others.[7] While notations vary considerably, the following captures a core common to most treatments. In *plural logic*, there usually are two sorts of variables, corresponding to two sorts of reference: singular and plural reference. Singular variables $x$, $y$, ... range over individuals like John, and roughly play the role of singular pronouns such as 'this' and 'that' in natural language. There are also have plural variables $xx$, $yy$, ... that range plurally over individuals such as John and Mary or those people. These roughly play the role of plural pronouns such as 'these' and 'those'. There are also individual constants $a, b$, ... and plural constants $aa, bb$, ... The latter translate plural names such as 'The Hebrides'.

The key theme is that, unlike with singularism, pluralities are *not* to be viewed as things in their own right—sums, sets, aggregates, etc. Only individuals are recognized, and singular and plural reference are different, primitive ways of referring to individuals.

In addition to singular and plural terms, there is a primitive relation called *inclusion*, represented as '$\prec$', and pronounced as "is one of" or "is among". For example, if '$a$' refers to Skye and '$bb$' to the Hebrides, then '$a \prec bb$' is true since Skye is one of, or among, the Hebrides. There is also a defined relation, $\preccurlyeq$, between plural terms, also called "inclusion", but pronounced "are among":

(17) $aa \preccurlyeq bb \leftrightarrow_{df} \forall x.\, x \prec aa \rightarrow x \prec bb$

If '$aa$' refer to the Inner Hebrides and '$bb$' to the Hebrides, then $aa \preccurlyeq bb$ is true since everything among the Inner Hebrides is also among the Hebrides.

Additional logical operations can be introduced to translate more plural discourse, including plural descriptions. We follow Rayo (2002) in introducing '$\sigma$', defined in (18), to render sentences of the form 'the $\phi$s $\psi$'.

(18) $\psi(\sigma xx[\phi(xx)]) \leftrightarrow_{df} \exists xx.\forall y.\, y \prec xx \leftrightarrow \phi(y) \wedge \psi(xx)$

Thus, (19a) is translated as (19b), which is true if there are some things, who are all of the students, and those things together gathered around the teacher.

(19) a. The students gathered around the teacher.
b. $\texttt{gather}(\sigma xx[\texttt{student}(xx)])$

We can also analyze conjunctive terms like 'Andy and Betty' using '$\sigma$', following Oliver and Smiley (2016).

[7] See Linnebo (2017) for a more detailed overview.

(20) a. Andy and Betty gathered around the teacher.
b. $\mathtt{gather}(\sigma xx[a \prec xx \vee b \prec xx])$

This description picks out the things which Andy or Betty are exclusively among, i.e. Andy *and* Betty. Also, 'are among' translates as inclusion:

(21) a. Andy and Betty are among the students.
b. $\sigma xx[a \prec xx \vee b \prec xx] \preccurlyeq \sigma xx[\mathtt{student}(xx)]$

Here (21a) is true if every individual which Andy or Betty are exclusively among, namely Andy and Betty, is among the students.

Distributive predicates apply to each individual among a plurality. For example, (22a) is true if each student among the students smiled.

(22) a. The students smiled.
b. $\mathtt{smile}(\sigma xx[\mathtt{student}(xx)])$

Collective predicates are then defined as non-distributive predicates (see McKay (2006) and Oliver and Smiley (2016).) For example, (19a) is true only if at least one student among the students didn't gather around the teacher. Finally, mixed predicates are ambiguous between collective and distributive interpretations.

(23) a. Andy and Betty carried a piano.
b. $\mathtt{carry}(\sigma xx[a \prec x \vee b \prec xx])$

On the distributive interpretation, the property attributed holds of both Andy and Betty, while on the collective interpretation, it fails to hold of at least one of them.

## 7.3 Mass Nouns

We have seen how basic mereological singularism and basic pluralism capture at least some plural discourse. The question here is how these basic analyses might be extended to so as to provide empirically respectable denotations for mass nouns. After all, as mentioned, there are many semantic commonalities between plural and mass nouns, and so one might reasonably expect that both sorts of analyses could be extended to capture them.

Although there are numerous singularist analyses available, we'll illustrate using Chierchia (1998a,b) here, since the latter is similar in spirit to Nicolas's (2008) pluralist analysis presented below. In particular, both analyze mass nouns as inherently plural. However, Chierchia's singularist analysis distinguishes count and mass nouns structurally. Specifically, singular count nouns denote atoms, plural nouns denote proper sums, and mass nouns denote *both* atoms and sums. The resulting picture looks as follows:

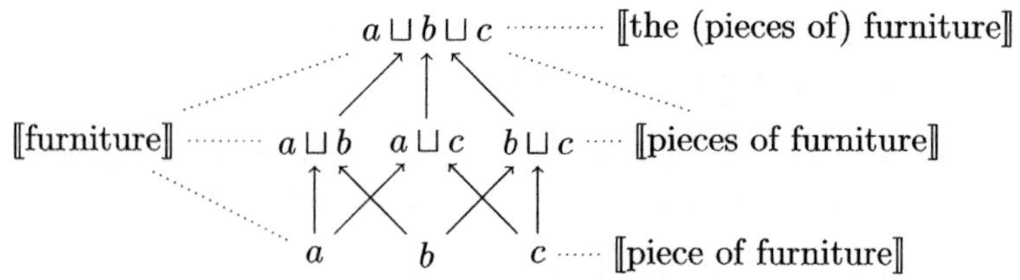


Thus, mass nouns *neutralize* the singular/plural contrast: they are neither strictly singular (atomic), nor strictly plural (non-atomic). Nevertheless, the key observation is that the same sums of atoms included in the extension of the plural noun 'pieces of furniture' are included in the extension of the mass noun 'furniture'. Hence, 'the pieces of furniture' and 'the furniture' are coreferential singular terms on Chierchia's analysis.

Chierchia's semantics for mass nouns is motivated primarily by the semantic parallels between so-called *object mass nouns*, such as 'furniture' and 'pottery', and corresponding plurals like 'pieces of furniture' and 'items of pottery', whose extensions contain countable atoms in the form of chairs, tea pots, etc. Nevertheless, the same analysis extends to so-called *substance mass nouns*, such as 'water' and 'information', whose extensions are less obviously atomic. To quote Chierchia (1998b, p. 347): "The extension of nouns like *water* is analogous to the one of nouns like *furniture*, the only difference being that what counts as a minimal portion of water is somewhat vague and may vary from context to context."

Similarly, on Nicolas's (2008) pluralist analysis, mass nouns are inherently plural, denoting primitive pluralities of individuals. Thus, just as 'the students' plurally refers to multiple students at once, 'the furniture' plurally refers to multiple pieces of furniture at once. Similarly, substance mass nouns denote multiple things at once, though it may be difficult to say what those things are, as Nicolas (2008, p. 19) intimates.

> We take it that, given the meaning of the noun, given how human cognition functions, and given the way the world is, there is an objective fact of the matter concerning what counts as wine. The greatest portion of wine in a bottle counts as wine, and so do many parts of this portion of wine. Perhaps there are minimal parts of wine that human beings can refer to. Or perhaps not. But in any case, what counts as wine is objectively fixed: it does not vary systematically with the way speakers may conceptualize what is in front of them.

Regardless, so long as some things are there, plural reference can be used to uniformly analyze mass nouns. The theoretical trade-off is that because the resulting analysis is not committed to "pluralities" as entities, it does not succumb to the problems attributed to mereological singularism in Sect. 7.1.

Consider a count noun like 'piece of furniture'. Intuitively, if some things are furniture, then each thing among them is furniture. Thus, count nouns are plausibly subject to Rayo's (2002) principle in (24).

(24) $\forall xx.\ P(xx) \leftrightarrow \forall y.\ y \prec xx \leftrightarrow P(y)$

Similarly, if some things are furniture, then each of one them is furniture, intuitively. Thus, on Nicolas' analysis, mass nouns are also subject to (24), which he calls *distributed reference*. Accordingly, (25a) is analyzed as (25b), where '$aa$' plurally refers to some deictically indicated things.

(25) a. This is furniture.
b. $\forall y.\ y \prec aa \rightarrow \texttt{furniture}(y)$

Hence, (25a) is true if each of the deictically indicated things is furniture. Incidentally, this explains the cumulativity of mass nouns: if $x$ and $y$ are both furniture, (24) guarantees that $x$ and $y$ together are furniture.

Now consider (26a), which is potentially ambiguous between a collective interpretation in (26b) and a distributive interpretation in (26c).

(26) a. The furniture costs \$500.
b. Altogether, the furniture costs \$500.
c. Each piece of furniture costs \$500.

In this respect, (26a) resembles (27a), which is similarly ambiguous.

(27) a. The chairs cost \$500.
b. Altogether, the chairs cost \$500.
c. Each chair costs \$500.

In order to capture both interpretations, Nicolas appeals to Gillon's (1992) notion of a *cover*, where a cover of a set $S$ is a contextually determined set of sets whose union is $S$. Predicates like 'cost \$500' apply to each cell of a cover. For example, if we have three chairs, say $a$, $b$, and $c$, the two covers relevant for (26a)/(27a) will be $S_1 = \{\{a, b, c\}\}$, thus resulting in (26b)/(27b), and $S_2 = \{\{a\}, \{b\}, \{c\}\}$, thus resulting in (26c)/(27c).

Nicolas's strategy is to define a pluralist analog of a cover to achieve the same result. Technically, this requires *superplural reference*. The relevant analogy is this: Just as individuals can stand in the primitive relation of plural reference to what we have loosely called "pluralities", "pluralities" can stand in a different, higher-order primitive relation of superplural reference to what we may loosely call "pluralities of pluralities".[8]

Formally, this is achieved by introducing (primitive) *levels* of terms, and relativizing inclusion to levels. Specifically, following Rayo (2006), along with first-level terms ($x, y, \ldots, a, b, \ldots$) and second-level terms ($xx, yy, \ldots aa, bb, \ldots$), we introduce third-level terms ($xxx, yyy, \ldots aaa, bbb, \ldots$), and then index inclusion by superscripts indicating the levels of the terms flanking $\preccurlyeq$. For example, $a \preccurlyeq^{1,2} bb$ indicates that $a$ is included in $bb$, while $bb \preccurlyeq^{2,3} ccc$ indicates that $bb$ are included in $ccc$.[9]

[8] See Linnebo and Nicolas (2008).

[9] In the framework we have been using, "$a \preccurlyeq^{1,2} bb$" would be just "$a \prec bb$".

With this in place, Nicolas defines an $M$-covering as follows, where $a \preccurlyeq^{1,3} ccc \leftrightarrow \exists yy.\, a \preccurlyeq^{1,2} yy \wedge yy \preccurlyeq^{2,3} ccc$, and '$\circ$' designates overlap:

(28) $ccc$ is an $M$-covering of $bb$ just in case:

(i) $\forall x.\, x \preccurlyeq^{1,3} ccc \rightarrow M(x)$
(ii) $\forall xx.\, xx \preccurlyeq^{2,3} ccc \rightarrow \neg\exists u, v.\, u \preccurlyeq^{1,2} bb \wedge v \preccurlyeq^{1,2} bb \wedge u \neq v \wedge u \circ v$
(iii) $\forall x.\exists y.\, y \preccurlyeq^{1,3} ccc \wedge x \circ y \leftrightarrow \exists w.\, w \preccurlyeq^{1,2} bb \wedge x \circ w$

Nicolas uses an example from Link (1998) to illustrate. Let $dd$ be the wine in the lower half of a given bottle; let $ee$ be the wine in the upper half; and let $ff$ be all of the wine in the bottle. Clause (i) requires that everything in $ccc$ satisfies $M$. Thus, a wine-covering including all of $dd$, $ee$, and $ff$ would meet this condition, since all three are wine. Clause (ii) requires that no "plurality" among $ccc$ overlaps. Thus, whereas a wine-covering including $dd$ and $ee$ meets condition (ii), a wine-covering including $dd$ and $ff$, or $ee$ and $ff$, would not. Finally, clause (iii) requires that the "pluralities" "constituting" $ccc$ do so exhaustively. Thus, whereas a wine-covering including $dd$ and $ee$ would meet (iii), as would one including just $ff$, one including just $dd$ or just $ee$ would not. In sum, then, $dd$ and $ee$ together would be a wine-covering for the wine in this bottle, as would $ff$ alone.

This notion of an $M$-cover can then be used to provide truth-conditions for examples like (26a). Specifically, it will be true just in case (29) is, where '$ccc$' names a contextually-determined furniture-cover.

(29) $\forall xx.\, xx \preccurlyeq^{2,3} ccc \rightarrow \texttt{cost-\$500}(xx)$

Suppose we have three pieces of furniture, $d$, $e$, and $f$. These will form eight potential "pluralities", including $d$, $e$, and $f$ themselves, along with five combinations. One such combination will be $d$, $e$, and $f$ together. Call it $aa$. In one case, the only plurality among $ccc$ is $aa$ itself. In that case, (29) is true if the furniture together costs \$500, i.e. the collective interpretation in (26b). In a second case, there are three things among $ccc$, namely $d$, $e$, and $f$. Given (29), this leads to the distributive interpretation in (26c).

To summarize, mass nouns are inherently plural, and as such, exploit the primitive relation of plural reference. Ambiguous predicates like 'cost \$500' are analyzed for mass nouns just as they would for plural nouns, in terms of covers. Meanings appropriate for determiners like 'some', 'all', and 'a lot of' can then be defined in terms of notions already introduced. We direct the reader to Nicolas (2008) for details.

Like Chierchia's singularist analysis, Nicolas' pluralist analysis of mass exploits the semantic commonalities between certain mass nouns and their plural counterparts. But what explains their apparent semantic *differences*, such as those repeated in (11a–d)?

(11a) Mary bought those {??furniture(s)/pieces of furniture}.

(11b) Mary bought three {??furniture(s)/pieces of furniture}.

(11c) How many {??furniture(s)/pieces of furniture} did Mary buy?

(11d) Mary bought both {??furniture(s)/pieces of furniture}.

According to Nicolas (2008, p. 3), (11a–d) apparently reflect a *morphosyntactic* distinction, and as such, have "no semantic import":

> A defining characteristic of mass nouns ... is that they are *invariable* in grammatical number, while count nouns ... can be used in the singular and in the plural. Depending on the language, this basic morphosyntactic difference between the two types of nouns is often supplemented by differences as to the determiners they can combine with ... The invariability of mass nouns suggests that *grammatical number has no semantic import with these nouns.*

In other words, differences like (11a–c) reflect a purely grammatical difference unique to English and other number-marking languages, and nothing about the meanings of the expressions involved.

This makes sense from the pluralist perspective. For one thing, there is quite a lot of cross-linguistic variability in how plurality is realized in natural language. For example, in certain languages arguably all nouns are lexically mass and become count only through use of a classifier, while in others all nouns are lexically count, even those which translate substance nouns in English.[10] Furthermore, and perhaps more importantly, if 'the furniture' plurally refers to pieces of furniture, then contrasts like (11a–d) seemingly *cannot* reflect a semantic difference in the nouns involved, as 'furniture' and 'pieces of furniture' would denote the same things.

## 7.4 Challenges for Pluralism

We have seen that it is possible to provide a pluralist analysis of mass nouns. By exploiting the semantic similarities between them and plural nouns, one may extend plural reference so as to deliver a pluralist semantics for all nouns. In this section, we develop what we take to be the primary empirical challenge to such an analysis: explaining the apparent semantic *differences* between various nouns. We will raise three issues illustrating this challenge. And while they are formulated specifically with respect to Nicolas's (2008) pluralist analysis, we believe they are suitably general, so that any pluralist analysis should reasonably address them.

### *7.4.1 Distributed Reference*

As mentioned, like plural nouns, mass nouns are subject to distributed reference on Nicolas' analysis, conforming to (24).

(24) $\forall xx.\ P(xx) \leftrightarrow \forall y.\ y \prec xx \leftrightarrow P(y)$

[10] See e.g. Chierchia (1998a,b), Rothstein (2017), and Landman (2020).

This explains the cumulativity of plural and mass nouns. Specifically, if some things *aa* are among some tables, and if some other things *bb* are among some other tables, then *aa* and *bb* are together among the tables. Similarly, if some things *aa* are among some furniture, and if some other things *bb* are among some other furniture, then *aa* and *bb* are together among the furniture.

However, mass nouns appear to provide counterexamples to distributed reference. Consider the lentil soup, which consists of lentils, water, onions, tomatoes, and spices. Since these are the ingredients, they exhaust the things which are the soup. For instance, the lentils are among those things. So, by (24), the lentils are soup. Yet lentils are not, by themselves, soup. Likewise for the other ingredients: water, onions, tomatoes, and spices are not, by themselves, soup. Thus, it would appear that *none* of the things included in the lentil soup are themselves lentil soup, contra (24).

Now consider the sugar water, which consists of just two ingredients: sugar and water. Apparently, then, the sugar is included in the sugar water, and so is the water. Yet neither the sugar itself nor the individual grains included within it are sugar water, nor is the water itself or the smallest bits of water included within it. Thus, again, none of the things apparently among the sugar water are sugar water, contra (24).

Of course, it may objected that the problem with these examples is not with (24), but rather with the assumption that the individual ingredients constituting the soup or the sugar water are actually included in them. After all, inclusion is a primitive relation, and so it needn't play the role of material constitution, as parthood does in singularist theories. Moreover, it is intuitively clear that the "individuals" among the soup and the sugar water are not individual lentils, onions, etc., or individual grains of sugar or bits of water. Rather, it is only *mixtures* of those things which are among the soup or sugar water, and mixtures are (primitive) *pluralities*.

However, there are two major challenges to this response. First, inclusion is *transitive*: if $x$ is among some things $yy$, and $yy$ are themselves among some things $zz$, then $x$ is among $zz$. So, if mixtures of lentils, onions, etc. are among the lentil soup, then since there are individual lentils, onions, etc. among those mixtures, the latter will also be among the lentil soup.

Secondly, it is easy to construct examples where a similar strategy will not succeed. Consider the following scenario, modeled on one suggested originally by Landman (2020).

Landman's Shop: Landman's shop sells both pottery and silverware, but in different departments. Among the pottery, you can buy a tea set, which consists of a tea pot, tea cup, and tea spoon. You can also buy the tea pot and tea cup together (for a better price) or separately (for a worse price). But if you want to buy the tea spoon separately, you'll need to go to the silverware department.

In the scenario described, the tea set is among the pottery, and the tea spoon is among the tea set. So by the transitivity of inclusion, the tea spoon is among the pottery. Yet the tea spoon is among the silverware, not the pottery. What's more, the other individuals among the tea set—the tea pot and the tea cup—are pottery. Thus, it will not do to claim that only pluralities are among the pottery.

It is instructive to consider how Landman's own singularist analysis deals with this and similar examples. The major theoretical innovation of Landman; Landman's (2011; 2020) semantics is to distinguish between what he calls the *base* of a noun, and the *body* formed from that base. Ultimately, this affords a useful four-way distinction relevant to the semantics of nouns: a noun is *count* if its base consists of non-overlapping elements, and is *mass* otherwise; a noun is (loosely) *neat* if its base consists of just non-overlapping atoms or sums of atoms, and is *mess* otherwise. The latter affords a linguistically important distinction between different kinds of mass nouns: *neat mass nouns*, such as 'furniture' and 'poultry' (otherwise known as "object mass nouns"), consist of a base with overlapping elements formed from non-overlapping atoms, while *mess mass nouns*, such as 'soup' and 'time' (otherwise known as "substance nouns"), consist of a base with overlapping elements formed from overlapping minimal parts or no minimal parts at all. The latter are *gunky*, in the sense indicated in Sect. 7.1.

In the previous example, we have a neat mass noun, namely 'pottery', which contrasts with the singular count noun 'piece of pottery'. Since the latter is count, it's base will consist entirely of (non-overlapping) atoms, i.e. entities which count as one piece of pottery. And in the scenario depicted, there are only two of these—the tea pot ($a$) and the tea cup ($b$)—since neither the tea spoon nor the tea set are pieces of pottery. Thus, we have:

(i) base of 'piece of pottery' = $\{a, b\}$
(ii) body of 'piece of pottery' = $\{a, b, a \sqcup b\}$
(iii) base atoms of 'piece of pottery' = $\{a, b\}$

The body of 'piece of pottery' is just the closure of the elements in the base under sum-formation, and the base atoms just are the base, as is characteristic of singular count nouns. Since the latter do not overlap, 'piece of pottery' is both count and neat. The base of plural 'pieces of pottery' will then be the body indicated in (ii), and so will consist of the same base atoms.

The semantically significant difference between 'piece of pottery' and 'pottery' will thus be that only the former contains overlapping elements in its base, thus making it mass. Specifically, its base will consist of the following four elements, including the tea set ($d$) but *not* the tea spoon ($c$):

(i) base of 'pottery' = $\{a, b, a \sqcup b, d\}$
(ii) body of 'pottery' = $\{a, b, d, a \sqcup b, \dots\}$
(iii) base atoms of 'pottery' = $\{a, b, d\}$

Here, the base represents the items you can buy in the pottery department. Because the tea pot and the tea cup ($a \sqcup b$) overlap with both the tea pot and the tea cup, 'pottery' is mass. As before, the body of 'pottery' is the closure of the elements in the base under sum-formation. The crucial observation is that the tea set is a base *atom*, rather than a sum of base atoms. That's because, unlike with mess mass nouns, Landman requires neat mass nouns to be *atomistic*: the elements in the body are either base atoms or else sums of base atoms. Thus, if the tea set were the sum $a \sqcup b \sqcup c$ included in the body, then 'pottery' would not be atomistic unless the tea spoon were also a base atom. In that case, it too would be pottery, contrary to fact.

In effect, the tea set is being analyzed as a *group*, in the sense of Landman (1989). Generally speaking, groups are pluralities construed as individuated entities in their own right, i.e. as *atoms*. Like all atoms, they count as one thing with respect to the predicate in question, in this case 'pottery'. And as atoms, they have no proper parts. Specifically, $c$ is not a proper part of $d$, thus ensuring that 'pottery' is atomistic.

Landman's analysis highlights two important points relevant to distributed reference. First, the base atoms of 'piece of pottery' needn't be identical to those of 'pottery'. That is, what counts as one with respect to a piece of pottery needn't be what counts as one with respect to pottery. In contrast, (24) requires that the same singular individuals among the pieces of pottery are among the pottery. To quote Nicolas (2008, p. 26):

> This semantics is faithful to the intuition that, if there are eight pieces of silverware on a table, the speaker refers to eight things at once when he says
>
> (105) The silverware that is on the table comes from Italy.
>
> and that in this particular case, he might just as well have said
>
> (106) The pieces of silverware that are on the table come from Italy.

Indeed, since the things among the silverware just are the pieces of silverware, 'the silverware' and 'the pieces of silverware' are coreferential plural terms. Likewise for 'the pottery' and 'the pieces of pottery', of course. But then each thing among the former is exactly each thing among the latter.

Secondly, Landman's analysis suggests that in order to prevent the tea spoon being among the pottery via the transitivity of inclusion, the tea set must be understood as an *individual*, i.e. something which is among the pottery but which itself includes no other things. However, this is difficult to accept from the pluralist perspective. After all, the tea set appears to be precisely the sort of thing that we can *plurally* reference: Rather than saying *that* tea set, we could instead say *those* tea set items. Thus, recognizing the tea set as an individual would be tantamount to endorsing *singularism*: a "plurality" would be a thing, ontologically on a par with the tea cup.

### 7.4.2 Non-Equivalence

As mentioned, Rayo (2002) originally argued against mereological singularism based on its apparent commitment to examples like (6a,b) being truth-conditionally equivalent, contrary to fact.

(6a) <u>The piles of grains</u> are scattered.

(6b) <u>The grains</u> are scattered.

Similarly, Nicolas (2008) motivates his pluralist analysis of mass nouns based on the non-equivalence of examples like (7a,b).

(7a) The piles of furniture are scattered.

(7b) The furniture is scattered.

Both examples appeal to the same purported problem for mereological singularism: if the pairs of underlined expressions refer to the same mereological sum, then (6a) is true just in case (6b) is, and likewise for (7a,b). Plausibly, however, (6a) can be true while (6b) isn't—the grains constituting the piles are not scattered. Similarly, the piles of furniture can be scattered even if the furniture constituting those piles is not.

On the other hand, pluralism needn't make such a prediction. There would appear to be two kinds of pluralist analyses available. On the first, 'the pile of grains' is what Oliver and Smiley (2016, p. 306) call a *pseudo-singular term*: though syntactically singular, it is semantically plural, coreferential with 'the grains'. To quote Oliver and Smiley:

> The other line of response, one we would put our money on, is to say that the pair really is the men, and the suit really is the cards, but 'the pair' and 'the suit' are not what they seem... 'The pair' and 'the suit' are not genuine singular terms but *pseudo-singular* terms, syntactically singular but semantically plural, denoting many things, not one.

Since the pile just is the grains of the pile, presumably the piles just are the grains of the piles. Thus, on this suggestion, 'the grains' and 'the piles of grains' are coreferential plural terms. Apart from making numerous false semantic predictions,[11] the obvious difficulty with this suggestion is that it succumbs to same problem attributed to mereological singularism: (6a,b) are seemingly wrongly predicted to be equivalent.

The second option would be to take the plural morpheme at face value. Intuitively, and speaking loosely, piles of grains are "pluralities of pluralities" of grains. For example, one pile may consist of thousands of red grains (the plural referent of 'those grains'), while another may consist of thousands of blue grains (the plural referent of 'these grains'). This naturally suggests understanding 'piles of grains' as a *superplural* term, one which primitive "pluralities" of grains are superplurally among. Thus, under this suggestion, (6a,b) would be respectively analyzed as (30a,b).

(30) a. $\texttt{scattered}(\sigma xxx[\texttt{pile-of-grain}(xxx)])$

b. $\texttt{scattered}(\sigma xx[\texttt{grain}(xx)])$

Clearly, the things superplurally referenced in (30a) may be distinct from those plurally referenced in (30b). Hence, (7a,b) are not equivalent. An exactly similar analysis is available for (7a,b), of course.

[11] See Snyder and Shapiro (2021).

(31) a. $\texttt{scattered}(\sigma xxx[\texttt{pile-of-furniture}(xxx)])$

b. $\texttt{scattered}(\sigma xx[\texttt{furniture}(xx)])$

Again, since the primitive "pluralities" superplurally referenced in (31a) may be distinct from the pieces of furniture plurally referenced in (31b), (7a,b) are not equivalent. If so, then pluralism would appear to have an obvious advantage over mereological singularism: it, unlike the latter, can distinguish non-equivalent pairs like (6a,b) and (7a,b).

Two points are worth highlighting about this argument. First, mereological singularism does not predict that (6a,b) or (7a,b) are equivalent. The problematic assumption is that underlined expressions refer to the same sums. This ignores a crucial fact about atoms: they are essentially *relative*. Indeed, Krifka (1989) defines a $P$-atom as follows.[12]

(32) $\forall x \forall P.\, \texttt{ATOM}(x, P) \leftrightarrow P(x) \wedge \forall y.\, y \sqsubseteq x \wedge P(y) \rightarrow y = x$

According to (32), the notion of a semantic atom is relative to at least a predicate. For example, a grain-atom is anything which is a grain and has no proper parts which are themselves grains. Similarly, a pile-of-grains-atom is anything which is a pile of grains and has no proper parts which are themselves piles of grains. Unlike with classical mereology, then, nothing is an atom *outright*, but only relative to at least a predicate.

However, relativizing atomicity to just predicates will not suffice. In fact, Rothstein (2017) forcefully argues that it must also be relativized to a context,[13] in order to account for so-called *divisive nouns*, such as 'fence'. Since 'fence' is a singular count noun, it denotes a set of atoms. However, which atoms those are depends on context. Suppose four neighbors build a backyard fence, as depicted in the following diagram.

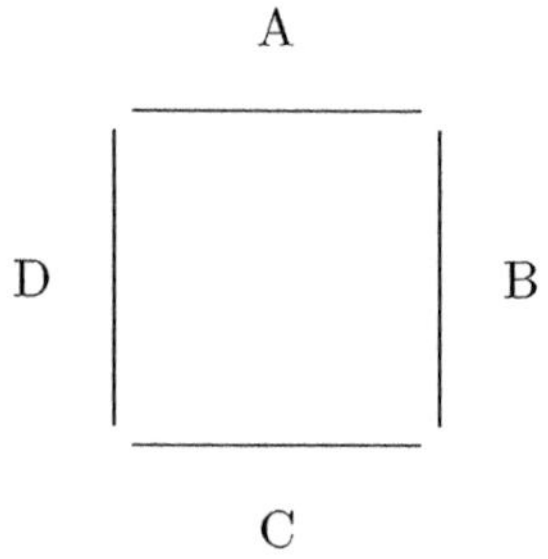


In this scenario, there are at least two plausible answers to the question 'How many fences were built?': one or four, but not both. Which of these is correct depends

[12] Actually, Krifka defines atomicity relative to both predicates and sorts. We ignore this here for simplicity.

[13] See also Snyder and Shapiro (2020).

on whether one views the connected fences as a whole fence or as separate. These correspond to a decision as to what counts as one fence, or a fence-*atom*, in context.

The problem with the argument above is that it ignores relativized atomicity. Specifically, though 'grain' and 'pile of grains' both plausibly denote sets of atoms, those sets are generally distinct. That's because 'pile of grains' is most plausibly a *group noun*, i.e. a count noun whose extension consists of groups.[14] Thus, on Landman's (2020) analysis, for instance, the respective bases will consist of different base atoms: grains and piles of grains, respectively. A fortiori, 'the grains' and 'the piles of grains' are not coreferential. Hence, (6a,b) are not predicted to be equivalent.

Something similar can be said about (7a,b). On Landman's analysis, both 'pile of furniture' and 'furniture' have bases containing atoms. However, those atoms are clearly different: the latter are piles (in the form of groups), while the former are typically pieces or groups (though not necessarily piles) of furniture. Clearly, then, the maximal sums of these base atoms are different. A fortiori, 'the piles of furniture' and 'the furniture' are not coreferential, and so (7a,b) are not equivalent.

Secondly, it is possible to construct a revenge form of the argument for pluralism. Contrast (33a,b) in the context provided.

<u>The Furniture Store</u>: The furniture store is attempting to clear out last year's merchandise, and the fastest way of doing that is to sell furniture in sets. Accordingly, prices of furniture sets have been reduced drastically, and furniture can only be bought in sets.

(33) a. <u>The pieces of furniture</u> are cheap.
b. <u>The furniture</u> is cheap.

Here, (33b), unlike (33a), can be interpreted as true. Consequently, (33a,b) are plausibly not truth-conditionally equivalent. Yet as we saw above, distributed reference requires that the same individuals among the pieces of furniture are also among the furniture. Thus, it appears that (33a,b) should be equivalent, after all.

Since 'the pieces of furniture' and 'the furniture' are both plural terms, the strategy suggested above for (6a,b) and (7a,b) will not extend to (33a,b). Rather, the obvious response appeals to covers. Recall that (26a) is similarly ambiguous between a distributive and collective interpretations, and these are to be distinguished via covers.

(26a) The furniture costs \$500.

Specifically, the collective interpretation of (26a) is true under the following cover: the pieces of furniture together—call them *aa*—are the only things superplurally among the things costing \$500. Similarly, (33b) could be true under the following cover: *aa* are the only things superplurally among the cheap things, assuming only

[14] See e.g. Rothstein (2017, Ch. 8).

one furniture set is left. But since these would *also* be the plural referent of 'the pieces of furniture', (33a) would also be true under this cover, apparently. Thus, it is difficult to see how covers could explain the non-equivalence of (33a,b).

Again, it is instructive to see how a singularist analysis might explain this. As we saw above, Landman's analysis requires that while the base of neat mass nouns contain overlapping elements, the atoms of certain bases may be groups. For example, just as the base of 'fence' may contain just a single fence-atom in the context above, ignoring that it could be divided into four distinct fences, the base of 'furniture' may contain just furniture-set-atoms, as *groups*, in this context, ignoring that these could be divided into individual furniture items. Moreover, these groups are distinct from the atoms in the base of 'piece of furniture'. Little wonder, then, that (33a,b) are not equivalent. As before, however, a similar explanation is difficult to accept from the pluralist perspective, since it would seemingly require understanding "pluralities" of furniture as individuals in their own right, thus threatening a collapse into singularism.

### 7.4.3 *Neat vs. Mess Mass Nouns*

Perhaps the most obvious challenge to a pluralist account of mass nouns is to explain their apparent semantic differences between count and mass nouns. As mentioned, these include contrasts like (11a–d), which Nicolas suggests are a reflection of morphosynactic properties of English, rather than differences in meaning.

(11a) Mary bought those {??furniture/pieces of furniture}.

(11b) Mary bought three {??furniture/pieces of furniture}.

(11c) How many {??furniture/pieces of furniture} did Mary buy?

(11d) Mary bought both {??furniture/pieces of furniture}.

Although we are skeptical that (11a–d) have no semantic import, our focus here is on a different but related issue: semantic differences between different mass nouns, specifically between neat and mess mass nouns. Since these presumably belong to the same lexical class, such differences seemingly must not reflect morphosyntax, but rather semantics.

Neat mass nouns resemble count nouns in a number of interesting respects. Perhaps the most well known concerns *measure selection*.[15] Suppose that Mary and John are moving house, and they want to determine how expensive the move will be. Mary has five large pieces of furniture, which fill two trucks, while John has ten smaller pieces of furniture, which fill only one truck. Now consider (34) in the following two contexts.

[15] There is a large literature on this topic. But see Rothstein (2017) and Landman (2020) for overviews and references.

Context 1: The moving company charges by the total number of items moved: the more items moved, the more expensive the move will be, regardless of the size or weight of those items.

Context 2: The moving company charges by the truckload: the more trucks needed to move furniture, the more expensive the move will be, regardless of how many items are moved.

(34) Mary has more furniture than John (... so it will be more expensive to move her furniture).

Clearly, (34) is true in the second context but not in the first. That's because 'more' selects for different measures in these contexts. In the first, cardinality is relevant, and so (34) is interpreted as (35a); in the second context, size is relevant, and so (34) is interpreted as (35b).

(35) a. The total number of Mary's furniture items exceeds John's.
b. The collective size of Mary's furniture items exceeds John's.

Notice that (34) is disambiguated by 'piece of furniture', which only allows for cardinality comparisons. Thus, (36) is unambiguously false in the scenario described.

(36) Mary has more pieces of furniture than John.

Likewise for count noun more generally: 'more N' typically compares cardinalities when N is count. In contrast, mess mass nouns typically select for other kinds of measures. Thus, (37) is unambiguously true in an analogous situation in which Mary has one large barrel containing 10 gallons of water, while John has five small barrels, each containing 1 gallon of water.

(37) Mary has more water than John.

Indeed, (37) can only be interpreted as claiming that total volume of Mary's water exceeds that of John's.

A second, related way in which neat mass nouns resemble count nouns but not mass nouns concerns a phenomenon known as *grinding*. Familiar examples are usually pretty gruesome, as with (38).

(38) After the accident, there was cat splattered all over the road.

Although 'cat' is normally count, it is coerced into having a mess mass interpretation in (38). Thus, suppose the accident involved a large cat which escaped a cage originally containing it and a much smaller cat. Then (39) is naturally interpreted as true, despite the relevant cardinalities involved.

(39) After the accident, there was more cat on the road than in the cage.

Indeed, as with 'more water' in (37), (39) is naturally interpreted as comparing the respective volumes of the two cats.

Since grinding involves coercing a non-mess mass noun into having a mess mass interpretation, it is generally impossible to grind mess mass nouns. Interestingly, however, Landman (2020) observes that it is possible to grind *neat* mass nouns. Consider (40), similar to an example from Landman.

(40) After the explosion, there was furniture all over the backyard. It must have been a powerful explosion, since there was more furniture outside the house than inside.

As Landman notes, the relevant measure for 'more furniture' here is volume: we are comparing the volume of the furniture debris to the volume of the (possibly intact) furniture remaining inside. Thus, (40) would remain true even in the following scenario: the exploded furniture outside originally consisted of nine, mostly dense pieces of furniture, while the furniture remaining intact inside consists of ten slightly larger but mostly hollow pieces of furniture. This shows that (40) is a genuine instance of grinding.

A third way in which neat mass nouns resemble count nouns but not mess mass nouns concerns distributivity. It is a characteristic feature of count nouns that they but not mass nouns are acceptable with distributive determiners such as 'each'.

(41) Each {piece of jewelry/??jewelry/??gold} was hidden.

Yet some distributive expressions are acceptable with certain mass nouns, namely distributive adjectives like 'big'. Schwarzschild (2011) calls these *stubbornly distributive*, since they do not allow for collective interpretations. So whereas 'expensive' in (42) is genuinely ambiguous, 'big' is not.

(42) The bracelets are {expensive/big}.

As Rothstein (2017) highlights, whereas neat mass nouns are generally acceptable with distributive adjectives, mess mass nouns are not.

(43) We hid the big {piece(s) of jewelry/jewelry/??gold}.

Thus, even if the only big pieces of jewelry are gold, we cannot acceptably use 'big gold' to describe a situation in which only those were hidden.

In sum, neat mass nouns resemble count nouns in a variety of ways that mess mass nouns do not, and this is something that a semantics for nouns should presumably explain. Moreover, since some of the differences between neat and mess nouns noted here closely resemble those between count and mass nouns in (11), e.g. both concern cardinality and distributivity, this suggests that a semantic explanation is in order for both cases. Again, however, it is hard to see what that could be if plural nouns like 'pieces of furniture' and mass nouns like 'furniture' denote the same things.

The data above also highlight a final difficulty for a pluralist analysis of mass nouns, namely specifying exactly what things are among the *M*s. Suppose a jeweler forms two gold rings, each weighing 10 grams, by melting a bracelet and necklace. According to Landman, since the nouns in (44) occur within a measure phrase, they have the same mess mass interpretation.

(44) 12 grams of the {rings/jewelry/gold} were from the bracelet.

This not only explains Rothstein's (2017) generalization that measure phrases are mass, but also why (45) could be true as a follow up to (44).

(45) So, more of the jewelry was from the bracelet.

Clearly, 'more of the jewelry' isn't comparing cardinalities here. Rather, it is comparing *parts* of the rings, measured in terms of volume, namely those from the bracelet and those from the necklace. If these are also the *things* among the gold, then apparently there are just as many things among the rings/jewelry. And recalling Nicolas' quote from above, "there is an objective fact" about what these are; they do not "vary systematically with the way speakers may conceptualize what is in front of them". Yet there are only two pieces of jewelry in the scenario described.

It is tempting to draw three related conclusions here. First, the extensions of mass nouns may vary across contexts, and so which things are among the *M*s may vary across contexts. Secondly, and in light of this, the things among the *Ms* may sometimes violate distributed reference. Specifically, in cases like (40), (44), and (45), the things among the furniture or jewelry are not furniture or jewelry. Third, inclusion is beginning to sound a lot like *parthood*. In fact, there is a strong suspicion that an empirically adequate pluralist semantics for mass nouns will be a notational variant of singularism, replacing parthood with inclusion, and sums with primitive pluralities. If so, then the case for adopting pluralism with respect to mass nouns will presumably depend largely on the strength of anti-singularist arguments like those mentioned in Sect. 7.4.2. But to be fair to Nicolas, he appears to no longer accept these arguments.[16]

## 7.5 Conclusion

We have argued that there are two significant challenges facing a pluralist analysis of mass nouns: allowing for the possibility of gunk, and making semantic sense of the similarities and differences between plural nouns and mass nouns. We have

[16] See Florio and Nicolas (2020).

focused exclusively on the latter here, using Nicolas's (2008) analysis to illustrate the challenges. We would like to close by briefly considering some challenges posed by gunkiness.

As mentioned in Sect. 7.1, classical extensional mereology is consistent with the possibility of gunky mass nouns: one simply drops the axiom of atomicity. Within plural logic, denying atomicity corresponds to denying non-emptiness, repeated in (10).

(10) $\forall xx.\exists y.\ y \prec xx$

Doing so, however, threatens to trivialize the standard plural comprehension schema, given in (46).

(46) $\exists u.\ \phi(u) \rightarrow \exists xx.\forall y.\ y \prec xx \leftrightarrow \phi(xx)$

Specifically, in gunky contexts, the antecedent of (46) would be false, not matter what $\phi$ is, and so (46) would fail to deliver any pluralities. As mentioned, it is possible to get around this by replacing (46) with a different comprehension schema, one which is a notational variant of the standard comprehension schema of classical extensional mereology.

One might reasonably wonder, however, whether denying (10) is even tolerable from the pluralist perspective. How could there be *some things* among which there is *no thing*? Again, the original pluralist idea was that we can make sense of plural talk without understanding "pluralities" as singular entities, or individuals. In reality, reference is always to individuals, but this is achieved in different ways—singularly and plurally. In the presence of gunkiness, however, there are no individuals to reference. It thus becomes rather mysterious how plural reference to *them* is achievable.

It is worth noting here that non-emptiness is usually defended on broadly linguistic grounds. For instance, speaking of a plural comprehension schema which allows for empty pluralities, Linnebo (2018, p. 67) says:

> Here we deviate slightly from plural locutions in natural language, where a plurality must consist of at least one and possibly two objects. This minor deviation can be justified pragmatically. Allowing an empty plurality is very convenient for the purposes of formalizing set theory. And even if plural locutions in natural language are in fact not used in this way, there might have been natural languages where they were.

Indeed, denying non-emptiness brings plural logic closer to standard set theory, which accepts an empty set. However, this is taken to be in conflict with how plurality is expressed in natural language. Thus, one potential takeaway for pluralism here is that, somewhat ironically perhaps, mass nouns may provide the linguistic evidence required to motivate a closer affinity between plural logic and standard set theory.

## References

Boolos, G. (1984). To be is to be a value of a variable (or to be some values of some variables). *The Journal of Philosophy, 81*(8), 430–449.

Boolos, G. (1985). Nominalist platonism. *The Philosophical Review, 94*(3), 327–344.

Carrara, M., Arapinis, A., & Moltmann, F. (2016). *Unity and plurality: Logic, philosophy, and linguistics*. Oxford University Press.

Chierchia, G. (1998a). Plurality of mass nouns and the notion of "semantic parameter". In S. Rothstein (Ed.), *Events and grammar*. Springer.

Chierchia, G. (1998b). Reference to kinds across languages. *Natural Language Semantics, 6*, 339–405.

Cocchiarella, N. B. (2009). Mass nouns in a logic of classes as many. *Journal of Philosophical Logic, 38*(3), 343–361.

Florio, S., & Linnebo, Ø. (2021). *The many and the one: A philosophical study of plural logic*. Oxford University Press.

Florio, S., & Nicolas, D. (2020). Plurals and mereology. *Journal of Philosophical Logic, 50*, 1–31.

Gillon, B. S. (1992). Towards a common semantics for english count and mass nouns. *Linguistics and Philosophy, 15*(6), 597–639.

Higginbotham, J., & Schein, B. (1989). Plurals. In *Proceedings of the north eastern linguistics society* (Vol. 19, pp. 161–75). University of Massachusetts at Amherst.

Krifka, M. (1989). Nominal reference, temporal constitution, and quantification in event semantics. In J. von Bentham, R. Bartsch, & P. von Emde Boas (Eds.), *Semantics and contextual expressions*. Foris.

Landman, F. (1989). Groups, i. *linguistics and Philosophy, 12*(5), 559–605.

Landman, F. (2011). Count nouns – mass nouns – neat nouns – mess nouns. *The Baltic Yearbook of Cognition, Logic, and Communication, 6*, 1–67.

Landman, F. (2020). *Iceberg semantics for mass nouns and count nouns: A new new framework for boolean semantics*. Springer.

Lasersohn, P. (2011). Mass nouns and plurals. In P. Portner, K. von Heusinger, & C. Mainborn (Eds.), *Semantics: noun phrases and verb phrases* (pp. 204–228). de Gruyter.

Link, G. (1983). The logical analysis of plurals and mass terms: A lattice-theoretic approach. In R. Bäuerle, C. Schwarze, & A. von Stechow (Eds.), *Meaning, use, and interpretation of language* (pp. 303–323). de Gruyter.

Link, G. (1998). *Algebraic semantics in language and philosophy*. SCLI Publications.

Linnebo, Ø. (2017). Plural quantification. In E. N. Zalta (Ed.), *The Standford encyclopedia of philosophy* (Summer 2017 Edition). https://plato.stanford.edu/archives/sum2017/entries/plural-quant.

Linnebo, Ø. (2018). *Thin objects: An abstractionist account*. Oxford University Press.

Linnebo, Ø., & Nicolas, D. (2008). Superplurals in English. *Analysis, 68*(3), 186–197.

McKay, T. (2006). *Plural predication*. Oxford University Press on Demand.

Nicolas, D. (2008). Mass nouns and plural logic. *Linguistics and Philosophy, 31*(2), 211–244.

Oliver, A., & Smiley, T. (2016). *Plural logic: Revised and enlarged*. OUP Oxford.

Rayo, A. (2002). Word and objects. *Noûs, 36*(3), 436–464.

Rayo, A. (2006). Beyond plurals. *Absolute generality* (pp. 220–254). Oxford University Press.

Rothstein, S. (2017). *Semantics for counting and measuring*. Cambridge University Press.

Schwarzschild, R. (2011). Stubborn distributivity, multiparticipant nouns and the count/mass distinction. In *Proceedings of NELS* (Vol. 39, pp. 661–678). Graduate Linguistics Students Association, University of Massachusetts.

Sharvy, R. (1980). A more general theory of definite descriptions. *The Philosophical Review, 89*(4), 607–624.

Snyder, E., & Shapiro, S. (2020). Mereological singularism and paradox. *Erkenntnis, 88*(1), 1–20.

Snyder, E., & Shapiro, S. (2021). Group nouns and pseudo-singularity. *Thought, 10*(1), 66–77.

# Chapter 8
# Informal Proof, Formal Proof, Formalism, and Fictionalism

**Mary Leng**

**Abstract** The fictionalist aims to avoid commitment to mathematical objects by replacing mathematical truth with fictional correctness: truth-in-the-story (of standard mathematics). For an axiomatically stated mathematical theory T, a sentence S is true-in-the-T-story if and only if it follows logically from the axioms of T. The formalist, on the other hand, seeks to avoid commitment to mathematical objects by replacing mathematical truth with formal derivability: a sentence S is true in a mathematical theory T if there is a derivation of S from the axioms of T. For first-order theories, and if the existence of a derivation is understood to mean the *in principle possibility* of deriving S from T, formalism and fictionalism would seem to coincide. However, in practice the two views typically diverge, with the divergence being due to different attitudes to modality. For fictionalists such as myself who embrace primitive modality, a modal notion of logical consequence allows for sentences in second order arithmetic to have objective 'truth-in-the-story' values even if they are not derivable. On the other hand as the most prominent contemporary advocate of formalism, Alan Weir wishes to develop an austere form of formalism without modal commitments, so that mathematical truth is not grounded in the *in principle possibility* of a derivation, but only in the existence of actual derivations. In this chapter I continue a long-standing debate between myself and Weir over the nominalistic acceptability of primitive modality, arguing that Weir's account of informal mathematics involves him in modal commitments whether he likes it or not.

**Keywords** Formalism · Fictionalism · Informal proof · Modality

I am a fictionalist about mathematics; Alan Weir is a formalist. We share a common enemy in the mathematical Platonist. But I think it was Alan who observed that our

M. Leng (✉)
Department of Philosophy, University of York, Heslington, UK
e-mail: Mary.Leng@york.ac.uk



A. Rieger, S. Leuenberger (eds.), *Themes from Weir: A Celebration of the Philosophy of Alan Weir*, Synthese Library 484,
https://doi.org/10.1007/978-3-031-54557-3_8

philosophical relationship to each other is much like that between the People's Front of Judea and the Judean People's Front. While from the outside our differences might look minimal, perhaps our most bitter disputes are with each other. In both cases, we take the view that the correctness of a mathematical result is a matter of its being a logical consequence of our mathematical axioms. But for the formalists (splitters!), the relevant notion of consequence is syntactic, so that mathematical correctness is a matter of derivability. On the other hand, for the (right thinking) fictionalist, a semantic notion of consequence is intended, grounded in the primitive modal notion of logical possibility (P follows from the sentences in the set Γ if it's not logically possible for the sentences in Γ to be true and P false). For the formalist Judean People's Front, this reliance on primitive modality makes the fictionalist as bad as – or maybe even worse than – their Platonist foes, who at least are happy to embrace their reliance on metaphysically dubious notions. (Weir makes this complaint in his (2014) review of my *Mathematics and Reality*.) On the other hand, the fictionalist People's Front of Judea accuses the formalist of suffering a form of false consciousness: they think they are getting away without any reliance on primitive modality, but for their account of mathematical truth to be plausible at all, for formalists truth shouldn't just correspond with proof simpliciter, but with proof in a *sound* derivation system – that is, a derivation system that respects modal facts about logical consequence. (This is the crux of the complaint I make in my (2015) review of Weir's *Truth without Proof*.) What follows is another round in this skirmish, prompted this time by consideration of Weir's (2016) account of informal proof in mathematics.

## 8.1 The Challenge to Hilbert's Thesis

Weir's, 2016 paper is a response to some sceptical worries about the prospects of finding formal correlates for informal mathematical proofs raised by Yehuda Rav (1999). Rav's paper offers a challenge to what he calls 'Hilbert's thesis', the proposition that "every conceptual proof can be converted into a formal derivation in a suitable formal system" (Rav, 1999, p. 11) This is clarified and expanded by Weir to "Hilbert's Thesis II", the claim that

> In any cogent mathematical practice there is a systematic process of transformation (not necessarily known to the practitioners) which turns any correct proof into a (suitably related) finite derivation in a formal system S. The system S in question is determined by the informal practice and its transformation process; in particular, the formal rules of S are rules which are implicit in the mathematical practice. (Weir, 2016, p. 30)

Rav's challenge to Hilbert's thesis (and with it the expanded Hilbert's Thesis II) is to offer reasons for thinking that there are informal proofs of mathematical results for which there is no formal correlate.

Why might we think that there are informal proofs that cannot be converted to a formal correlate? Rav's rich and wide-ranging discussion suggests several reasons in support of this claim, as follows:

1. [*Absence – or Irrelevance of Axioms*] Informal mathematics, Rav points out, often happens in a setting where there are no explicit axioms for the concepts we are working with. This can be the case in the development of theories prior to axiomatization, but also even in cases where axioms have been found (such as in the case of number theory), in practice mathematicians continue working with mathematical concepts without much attention to their axiomatic presentation (and are happy to make use of resources from other branches of mathematics as and when they find it helpful to do so): "As a matter of fact, the Queen of Mathematics—as Gauss called number theory—is rather promiscuous, opening her arms to algebraic, analytic, topological, geometrical, proof-theoretical, numerical, model-theoretical, combinatorial, and come as they may types of proofs and methodologies." (Rav, 1999, p. 16)
2. [*Infinite potential for expansion*] Rav asks us to suppose we read in a mathematical paper or monograph the claim that B follows from A (Rav proposes we symbolise this as '$A \rightarrow B$'). If we're not convinced of this we may stop and fill out the argument, coming up with a series of smaller steps: $A \rightarrow A_1, A_1 \rightarrow A_2, \ldots A_n \rightarrow B$, by means of which we are then convinced. "Explaining the structure of the argument to a student or non-specialist," Rav notes, "the other may still fail to see why, for instance, $A_1$ ought to follow from A. So again we interpolate $A \rightarrow A'$, $A' \rightarrow A_1$. *But the process of interpolations for a given claim has no theoretical upper bound.*" (Rav, 1999, p. 14, my emphasis). Since derivations are finitary objects, Hilbert's Thesis would require that this process of asking for clarification will terminate after finitely many steps, ending in a list of 'atomic' inferences that everyone will accept. But if Rav is right that there is no upper bound on the process of interpolations, then there is always potential to expand an informal proof into a longer proof that still does not meet the criteria of being a formal derivation.
3. [*Second-order logic*] In some branches of mathematics, Rav suggests, it seems clear that the underlying logic is second-order rather than first-order: "The standard theorems of group theory that one finds in books and papers on group theory are not even expressible in first-order predicate calculus. One just has to think about such fundamental concepts as normal subgroup, torsion group, finite group, composition series, or such famous theorems such as the Sylow theorems about p-groups, the Jordan-Holder theorem and the like, to realise that the implicit underlying logic of mainstream group theory is second-order logic" (Rav, 1999, p. 17). But if informal proof is tracking second-order logical consequence, then as there is no complete derivation system for second-order logic, we might expect there to be informal proofs that cannot be converted to formal derivations.
4. [*Gödelian Considerations*] Gödelian considerations provide a concrete example where we might think informal proof can convince us that something follows from the (2nd order) Peano axioms for arithmetic but is nevertheless not derivable from those axioms. For a given theory F, no informal proof of F's Gödel sentence G can be translated into a formal derivation (in F) of G, and yet reflection on G convinces us that it must be true in the standard model characterised by the second-order axioms. In this regard, Rav (1999, p. 14) approvingly quotes Kreisel (1970, footnote 22, p. 511) who "notes (with respect to the insight that proofs are of infinite character) that 'properly interpreted, Gödel's theorems can be used to support this insight, just as they are used to refute Hilbert's assumption that finite formal derivations reflect faithfully the structure of mathematical reasoning' (Kreisel's emphasis)."

To these challenges to Hilbert's thesis, Weir's discussion adds two more

5. [*Length of Derivations*] If, by 'derivation', we mean 'humanly producible concrete derivation', then many informal proofs will lack a corresponding formal derivation simply because a full derivation would be too long to write down. "If $S^k$ is the highest order system we ever use, there will be $S^k$ truths, expressed in short, graspable sentences, whose proofs contain more steps than the estimated number of neutrinos in the universe, truths for which there can be no concrete proof in other words." (Weir, 2016, p. 34)
6. [*Informal Proofs of Derivability*] Often our confidence in a mathematical result is due to metamathematical reasoning: for a given formal system F and sentence S, we present a metatheoretical argument for the existence of a derivation-in-F of S. Such an argument can be offered without pointing us to an actual F-derivation: it certainly need not take the form of an informal sketch of a formal derivation in F.

By showing how his formalist can neutralise the challenge from issues [5] and [6], Weir hopes to show how worries [2–4] can also be sidestepped. As for [1], for Weir this only gets its bite if we assume that, for a formalist who thinks that mathematical truth resides in formal proof, proof here has to involve derivation in an explicitly axiomatized formal system. "But", Weir suggests, "the point is not fatal to formalistic approaches so long as it is plausible that there is, implicit in mathematical practice, rules (and perhaps axioms never implicitly brought to the surface) which, when formalised, generate derivations of every truth of matrix theory, measure theory, etc." (Weir, 2016, p. 32)

What should a mathematical fictionalist, for whom mathematical correctness is a matter of following logically from mathematical axioms, think of worries [1–6]? Concern [1] looks as initially worrying for the fictionalist—who typically focusses on axioms as generative of mathematical fictions. In a (2010) paper on pre-axiomatic mathematical reasoning, I offer reasons for thinking that even prior to axiomatization, informal mathematical practice provides sufficient implicit constraints to pin down—to a reasonable degree at least—what is to count as 'true in the story'. So on this I am in agreement with Weir's response to Rav. [2–6], on the other hand, given the fictionalist's emphasis on logical consequence rather than formal derivation as determining mathematical correctness, look like they should be grist to the fictionalist's mill. For while derivation in a sound derivation system is one route to determining what follows from our mathematical assumptions, it is certainly not the only route, and we should expect that there may be plenty of means by which we can become convinced that a conclusion follows logically from our mathematical premises that do not proceed by sketching a formal derivation of that conclusion.

But Weir thinks that [2–6] are not decisive against his formalist claim that truth in mathematics consists in the existence of a formal derivation. His formalist fight back starts with a response to point [5], the claim that formal derivations that correspond to our informal mathematical arguments will typically be too long for us to write down. Weir's challenge to this point starts with a challenge to the assumption that what it is to produce a token of a formal derivation is to write down each step of the argument with no abbreviations. Not so, Weir claims. Starting with tokens of formulas in a system, Weir (2016, p. 35) suggests that if we introduce clear conventions on abbreviations, so that e.g. '$S^n0$' is used to abbreviate the string of symbols consisting of a token of '0' preceded by n tokens of 'S'), then '$S^{10}0$' and 'SSSSSSSSSS0' should be considered to be mere notational variants, two tokens of the same formal string of symbols. Allowing abbreviated formulas to count as tokens of formulas in a formal system already allows for huge reduction in length of strings of symbols in an individual line in a proof. Moving from here, Weir goes on to suggest that the same move can be made with derivations, so that abbreviated derivations can similarly count as tokens of formal derivations. Derivations that include very large numbers of repeated iterations of essentially the same step can then be tokened efficiently with very short abbreviated strings, vastly increasing the number of derivations for which a short finite token can be provided.

Strictly speaking, one might argue, we have not one formal system F but two, F and the abbreviated system F′. A derivation in an abbreviated formal system F′ is not an F-derivation, so how can a derivation in an abbreviated system convince us of the F-derivability of a claim? The solution to this worry relates to Weir's solution to concern [6] regarding metamathematics, to which we will return in a moment. For now let us see how the acceptance of abbreviated derivations might help with the outstanding worries [4], [3], and [2].

In relation to worries [4] and [3], Weir argues, if a short finite derivation token can be recognised to abbreviate an unfeasibly long derivation by means of indicating how each step of that derivation should go, then there is no principled reason to object to the existence of abbreviated tokens of infinite derivations. But doing this opens up resources beyond first order logic. In particular, Weir argues, although we can't write down all of its steps, we can easily abbreviate – in a way that shows exactly what each step should look like – a proof of the Gödel sentence using the ω-rule. Opening up the possibility of using abbreviated systems to convince ourselves of the existence of proofs with infinitely many steps opens up much stronger logics, which we may argue do a better job at approximating the second-order logical consequence relation.

In relation to worry [2], concerning the infinite potential for expanding informal proofs by elaborating on their steps, Weir thinks that allowing abbreviations of infinite proofs to count as derivations will also neutralise this objection. Here I am less convinced that Weir's solution will help. Weir notes that abbreviated derivations "must preserve the key feature, 'recursiveness', of our informal practice ... in the sense of grasp of iterable rules whose output can in turn be fed in as input to the self-same rules" (p. 40). It is this feature that allows Weir to suggest that abbreviated infinite derivations are no more problematic than abbreviated long finite ones. But the issue of indefinite interpolation that Rav points to is not a recursive procedure: we cannot simply look at a derivation without interpolations and know in virtue of its form how to fill out any intermediate steps. So simply allowing the existence of some infinite proofs (by virtue of allowing abbreviations of infinite proofs) does not tell us that infinite proofs of the kind Rav has in mind are acceptable for the formalist. A better response to this worry of Rav's, I think, might be to question whether it is plausible that the process of interpolation is genuinely infinitely extendible. Rav suggests that, for any derivation step '$A \rightarrow B$', it will be reasonable to ask for intermediate steps, and this is likely to be true for pretty much any proof in a maths journal. But should we be so quick to think that it will be true indefinitely, for proofs that have gone through many steps of painstaking clarification? If we get to a point in a proof where the step is a simple application of a rule of natural deduction (say, 'A & B $\rightarrow$ A'), is it still reasonable at that point to ask for further interpolating steps? My suspicion then is that Rav's worry [2] can be set aside, as it has not been established that the process of clarification will never terminate.

If, then, Weir is right that the formalist can appeal to abbreviated derivations to establish the truth of some derivability claims, then the remaining worries about informal proofs can be answered. What, though, of the worry mentioned above about the relation between abbreviated systems and the formal systems they abbreviate? Specifically, if for a formal system F the F-truths consist of those claims for which there exist an F-derivation, the existence of a derivation in another system, $F'$, that in some sense 'abbreviates' system F, tells us nothing of the actual existence of an F-derivation. So how can we use these abbreviations to tell us anything about F-truth?

This worry is a variation on worry [6] about metamathematics. That is, we are often convinced of the derivability of a claim in formal system F not through the production of an actual F-derivation, but through a metamathematical argument to the effect that an F-derivation exists. But a derivation in a metamathematical system F* of the existence of an F-derivation is not itself an F-derivation. For clarity, Weir uses the SMALL CAPS convention to indicate claims whose truth is to be taken at face value as grounded in concrete, mind-independent facts. The difficulty is that a metamathematical theory (or abbreviated proof theory) F* might prove that there exists an F-derivation of S even when it is not the case that THERE EXISTS AN F-DERIVATION OF S. If truth in a formal theory F consists in the EXISTENCE of a concrete F-derivation, as Weir's formalist requires, then a derivation in F* of the existence of an F-derivation of S is not a derivation of the truth (in F) of S.

Weir's solution to this worry is to liberalise somewhat his notion of truth in a formal theory. Rather than holding that the truth (in F) of S consists simply in the EXISTENCE of an F-derivation of S, instead he argues that the truth (in F) of S consists in the (lower-caps) existence of an F-derivation (where the truth of the claim that there exists an F-derivation of S may be established either by establishing the EXISTENCE of an F-derivation (by producing one), or by deriving, in an appropriately chosen metatheory F*, the existence of an F-derivation.

In order for this account to work, the theory we need to use to prove the existence of F-derivations needs to be appropriately chosen. Clearly not all choices of F* will do: an inconsistent F* will prove the existence of plenty of F-derivations, but would result in making truth in mathematics trivial. But consistency isn't, of course, the only requirement on our metamathematical theory. We need to make sure that it is closely enough connected to the formal theory F that it makes sense to think of it as 'talking about' derivability in F. Weir suggests the following adequacy criterion:

> If the literal content of sentences provable from F* is to tell us something about abstract proofs in F, we must be able at the least to *define* in F*, formal-proof-in-F (or at any rate

axiomatically characterize it) in a way which captures our intuitions about proof-in-F. (Weir, pp. 39–40)

What, though, does this adequacy criterion really amount to? What intuitions about 'proof-in-F' are meant to be captured?

For those willing to help themselves to a bit of modality, it is straightforward: if F* derives the existence of an F-proof of S, then it should be the case that S is *derivable* in F. We might wish to cash this out in terms of the 'possibility in principle' of producing an F-proof of S. If F* is a metamathematical system and F a first-order theory then in principle possibility might be thought of as just somewhat idealised human possibility (idealising to allow for enough time and materials to write the proof down). But if F* is an abbreviated system that allows for abbreviations of infinite proofs, the 'in principle possibility' here is not a matter of human possibility at all—if we're talking about a completable procedure of tokening the proof, we might try to tie this to superhuman abilities to complete supertasks—but it is controversial to say the least that such a derivation could be concretely completed.

Weir is no fan of the 'in principle possible', as this nice passage from his (2010) book makes clear:

> Now in trying to understand the IPP claim that it is 'in principle possible' for me to find out the truth-value of concrete indeterminables, I find myself interpreting it in a non-homophonic fashion, along the lines of:
>
> It is not in the least possible at all but I really, really regret this and wish it was possible. For that infernal impossibility refutes my philosophy of mathematics.
>
> Similarly, to say *s* is provable because a non-existent being can prove it—is this not just saying that it is not provable at all but in a way which hides this unpleasant fact from reflective self-consciousness (Weir, 2010, p. 182)

I certainly agree that resting on the 'in principle possibility' of producing a derivation does not soften the modal commitments of the claim. The derivability of S in F is not a matter of what's concretely possible, but rather a matter of what is permitted given the axioms and rules of F. But I see no other way of specifying an adequacy condition on F* that adequately captures, as Weir puts it, our intuitions about proof-in-F. A conservativeness requirement (according to which F* proves the existence of an F-proof of S only if THERE IS AN F-PROOF OF S) would defeat the point of Weir's introduction of metamathematics (and systems of abbreviated proofs) in order to preserve classical logic. I contend that the requirement that we characterize 'proof-in-F' in our metatheory in a way that captures our intuitions about proof-in-F is a requirement to embed the rules of F-proof in our formal system in such a way that it is derivable in F* that there is an F-proof of S only if S is F-provable. The modality is inescapable.

Weir, then, has shown us a way for a formalist to deal with the informal methods of standard mathematics in establishing mathematical truth. To the extent that informal proofs can be formalised as abbreviated proofs, we can consider these as concrete derivations of the existence (though not the EXISTENCE) of F-derivations of their theorems, just as metamathematical proofs can be utilized to show the existence (though not the EXISTENCE) of F-derivations. But, I contend, the cogency of this picture requires that we introduce some modality into the story, in the form of an adequacy criterion for our formal systems of abbreviated or metamathematical proofs: that is, they should only derive the existence of an F-derivation of S in cases where S is F-*derivable*. But if this is right, then the grounding of mathematical truth in the EXISTENCE of concrete derivations does not escape the kinds of problematic modal commitments Weir's formalist accuses the fictionalist of requiring. Splitter!

## References

Leng, M. (2010). *Mathematics and reality*. Oxford University Press.

Leng, M. (2015). Review of Alan Weir, *Truth through proof: A formalist foundation for mathematics*. *The Philosophical Quarterly, 65*(259), 288–292. https://doi.org/10.1093/pq/pqu071

Rav, Y. (1999). Why do we prove theorems? *Philosophia Mathematica, 7*, 5–41.

Weir, A. (2010). *Truth through proof: A formalist foundation for mathematics*. Clarendon Press.

Weir, A. (2014). Review of Mary Leng, *Mathematics and reality*. *British Journal for the Philosophy of Science, 65*(3), 657–664.

Weir, A. (2016). Informal proof, formal proof, formalism. *Review of Symbolic Logic, 9*(1), 23–43.

# Chapter 9
# Mathematical Naturalism and Revisionism About Mathematics

**Marianna Antonutti Marfori**

**Abstract** Anti-revisionism, according to which any revision to a practice should come autonomously from within the practice itself, is a key component of all naturalistic approaches to science and mathematics, yet it is rarely discussed in detail. In this paper, I focus on naturalistic anti-revisionism about mathematics. After clarifying the reasons why naturalists should be anti-revisionists, I examine two different ways of spelling out a criterion for mathematical anti-revisionism. The first criterion maintains that we should not revise mathematical practice on external (i.e. extra-mathematical) grounds. On the basis of case studies, I argue that this criterion is not supported by a sufficiently sharp distinction between revisions prescribed on internal vs. external grounds. The second criterion maintains that we should not countenance philosophical accounts of mathematics which involve abandoning accepted theorems. I argue that while in line with naturalistic tenets, this criterion does not fully capture what is important to the mathematical naturalist. This leads to the formulation of a third criterion, which adds to the second criterion the clause that we should not abandon established mathematical methodology.

**Keywords** Philosophy of mathematics · Mathematical naturalism · Mathematical revisionism · Anti-revisionism about mathematics

The writing of this paper was supported by the DFG within the ANR/DFG scheme, Project No. 390218268 (Gefördert durch die Deutsche Forschungsgemeinschaft (DFG)–Projektnummer 390218268), whose support is gratefully acknowledged.

M. Antonutti Marfori (✉)
University of Paris 1 Panthéon-Sorbonne, Paris, France
e-mail: Marianna.Antonutti-Marfori@univ-paris1.fr



A. Rieger, S. Leuenberger (eds.), *Themes from Weir: A Celebration of the Philosophy of Alan Weir*, Synthese Library 484,
https://doi.org/10.1007/978-3-031-54557-3_9

## 9.1 Introduction

In a much-quoted passage, Quine writes that "it is within science itself, and not in some prior philosophy, that reality is to be identified and described" (Quine 1981, p. 21). At least in a minimal sense, this type of naturalist approach is shared by many—perhaps even most—philosophers, to the point that we may reasonably wonder whether declaring oneself to be a naturalist still amounts to taking a substantial philosophical stance. Maddy (2007, p. 1) complains that "as more and more philosophers count themselves as naturalists, the term has come to mark little more than a vague science-friendliness".

However, even a cursory look at contemporary debates on science can reassure us that naturalism still counts for something. Consider, for example, the debates on whether homeopathy should be funded by public healthcare, or whether certain types of vaccinations should be mandatory. These debates (crucially, albeit not exclusively) rest respectively on considerations pertaining to whether homeopathy deserves to be included in our best scientific practices, and considerations pertaining to the significance of certain kinds of evidence. These debates involve an appeal to a certain normative conception of science, a conception that recognises disciplines such as homeopathy as pseudoscientific enterprises.[1]

It is important to note that the naturalist approach does not adopt a preconceived definition of what constitutes science: since, according to the naturalist, scientific methods are the only methods that can confer justification, any criticism of a pseudoscientific or anti-scientific enterprise can only be carried out by applications of scientific methods themselves. Such methods include observation, experimentation, theory formation, theory testing, and revision of theory on the basis of evidence—in brief, the methods whose validity is sanctioned by our current best science. Since these methods develop and improve over time, and with them our conception of what counts as a legitimate scientific method also develops, the naturalist does not commit to a definition of science that is given once and for all.[2]

In this sense, naturalism involves a certain degree of deference to experts and a recognition of the autonomy of science. This attitude supports the adoption of an anti-revisionist approach to scientific practice: revisions to science, where they prove necessary, should come autonomously from within the practice (see Sect. 9.2 below). More recent types of naturalism, which depart from Quinean naturalism in a number of places, argue that since mathematics is part of our best science, mathematics should enjoy the same trust placed in other kinds of scientific enterprises, such as physics or biology (Maddy, among others, argued extensively

[1] See e.g. Jerkert (2013) and Smith (2012).

[2] See e.g. Part I of Maddy (2007). For a discussion of the idealisations embedded in the Quinean notion of "best science", see also Fogelin (1997), Resnik (2005), and Antonutti Marfori (2020). It should be noted, however, that Quine's stance on the demarcation between science and pseudoscience, which is the inspiration for Maddy's, is not immune from serious issues; see e.g. Dieterle (1999), Weir (2014, §4), and Antonutti Marfori (2012).

for this kind of naturalism). Naturalists should therefore adopt an anti-revisionist approach also to mathematical practice.

This chapter seeks to better understand the anti-revisionism about mathematics to which the naturalist is committed, by trying to identify and clarify different anti-revisionist criteria. Section 9.3 begins by examining the reasons why anti-revisionism is central to mathematical naturalism, what precisely is meant by "anti-revisionism", and the widely-held view that naturalism constrains us from revising mathematics on grounds external to the practice. While this is an appealing characterisation, I argue that a closer examination of historical case studies suggests that the line between internal mathematical reasons and external philosophical reasons for revising mathematics is often more blurred than is usually considered. This criterion for anti-revisionism does not therefore provide adequately strong support for the naturalist's anti-revisionist attitude towards mathematics. In Sect. 9.4, I then consider a different criterion, not exclusively associated with naturalistic approaches but held widely in philosophy, that we should not countenance philosophical accounts of a scientific practice which involve abandoning principles which are deeply and widely held by scientists (so, in this case, that we should not countenance philosophical accounts of mathematics that involve revising the set of theorems accepted by the mathematical community). I argue that although this criterion should be accepted by the mathematical naturalist, it is too weak, and fails to capture what naturalists take to be important about anti-revisionism. I therefore propose a stronger criterion (Sect. 9.5), namely that the naturalist should also not seek to revise accepted mathematical methodology. In Sect. 9.6, I examine whether this criterion is too weak, and thus accomplishes little or nothing more than the minimal criterion does, or too strong, and thus rules out revisions that may be argued to be naturalistically acceptable, and I argue that it does neither of these things. In closing, I briefly assess what kinds of philosophical accounts of mathematics are compatible with naturalist views that accept anti-revisionism about mathematical methodology, and conclude that the space of possible views is quite broad, at least when it comes to the semantics and ontology of mathematics.

## 9.2 Naturalistic Anti-Revisionism

The deferential attitude embedded in naturalistic approaches is motivated by the beliefs that our current scientific methods are the best methods we know to make sense of our experience of the external world, and that our current scientific theories are the considered result of our best efforts to understand and systematise this evidence, together with the belief that there are no extra-scientific methods of discovering the nature of reality.

If a scientific practice is judged successful by a majority of practitioners, and this judgement is retained over time, it is reasonable to suppose that practitioners are doing something right, and are in fact advancing in their knowledge of the external world. This can be observed at the level of theories proposed to explain a set of results, at the level of methods to obtain such results, and at the level of whole scientific practices involving both a set of methods to collect data and a set of statements explaining and systematising that data. (Consider for example the case of alchemy, which, as a scientific practice, enjoyed a degree of scientific favour for some time, but was eventually abandoned as unscientific, while Perrin's atomic theory, albeit transformed and amended over time, has been retained by the scientific community). A central tenet of different types of naturalist approaches to science is thus that there is no higher authority than scientific method itself on which a critique of science could be based.

An important consequence of naturalism's reliance on the authority of our best science is the principle of abstaining from criticising a successful scientific practice on grounds that are external and allegedly superior to science, for example by prescribing or imposing on the practice goals and methodologies that are incompatible with those employed by practitioners.[3] Prototypical examples include political or ideological directions imposed on science which constrain the scientific process by limiting, or even positively changing, the theories and methods devised and employed by practitioners to make progress in our understanding of the world.

Naturalists are of course aware that revision in the light of evidence is the key element of the scientific method: there could be no progress in science without a constant revision of hypotheses and methods to obtain and test them. The naturalist is not supposed to discount a priori any hypothesis or methodology as non-scientific: she is supposed to test each hypothesis and methodology by using the best scientific tools that are available to her, and by applying scientific methods such as observation, experimentation, theory formation, theory testing, and revision of theory on the basis of evidence, and eventually criticise or understand that practice on the basis of the evidence so gathered.[4] So, a naturalist would argue that we should not recommend changes to evolutionary biology based on contemporary creationist theories of the history of life on the grounds that the latter is significantly less empirically supported than the former. Similar arguments can be made with respect to changes to astronomy predicated on astrological grounds, and changes to 'mainstream' medicine predicated on homeopathic grounds.[5]

[3] Throughout the paper, I will speak of "prescriptions" and "recommendations" for revisions to mathematical practice interchangeably. Even though a prescription has more normative force than a recommendation, from the naturalist's point of view, to make a distinction in the degree of force with which a revision is recommended is beside the point: anti-revisionism is deeply entrenched in the naturalist standpoint.

[4] The Quinean naturalist, for example, does not privilege any part of her inherited worldview, but continuously strives to understand, clarify, and improve the system from within (Quine 1981, p. 72).

[5] These are examples of extra-scientific criticism to science where the extra-scientific discipline is considered a pseudo-scientific, or an anti-scientific discipline. For an extensive discussion of what notions of science and pseudoscience are available to the naturalist, see Part I of Maddy (2007). For an overview of the debate on pseudoscience and anti-science, see Hansson (2021).

The presupposition that scientists are right until proven wrong leads the naturalists to contend that any revision to successful scientific practice demands scientific motivation. A highly revisionary restructuring of the theoretical tenets and methodology of a scientific practice would be very costly, and may unnecessarily hinder the practice's development while artificially curtailing its fruitful activities. It would likely also severely impact interdisciplinary work that centrally appeals to the methodology and theoretical tenets of the scientific practice undergoing revision. Such revisions would then be reasonable only if some specific problems in the foundation of the practice demanded such a radical restructuring in order for the practice to maintain its internal coherence and plausibility. A similar point applies also to smaller-scale revisions: even though the costs of carrying out localised changes to a practice are lower than those for radical revisions, enforcing changes that are not motivated by needs that arise in the practice itself would only impede the successful development of the practice.[6]

So, according to the naturalist, trust in science and its methodological autonomy supports a strong anti-revisionist attitude according to which only the revisions that are internal to the practice are naturalistically acceptable. Any external interference in a successful practice's tenets and methodology is unjustified and ultimately illegitimate. Since trust in science and its methodological autonomy warrants trust in mathematics and the autonomy of its methods, mathematical naturalism involves an anti-revisionary attitude towards mathematics in the same way in which scientific naturalism involves an anti-revisionary attitude towards successful scientific practices. But what exactly do we mean when we say that views such as intuitionism are revisionist about mathematical practice? There are different ways to be revisionist about mathematical practice: one can be revisionist about the logic, the ontology, the epistemology, the standards of proof, the interpretation of mathematical discourse, the theorems proved in the framework in question with respect to those proved in classical mathematics. Are all these types of revisionism equally bad from the naturalist's point of view? Since for most naturalists illegitimate revisions are those prescribed on external grounds, the answer to these questions rests on what precisely is a revision to the practice that is prescribed on external grounds, and why it goes against a naturalist conception of mathematics. The next section will address this issue in more detail.

## 9.3 Revising on External Grounds

Current science would be inconceivable without mathematics: this is surely evidence of the success of mathematics as a scientific enterprise. Mathematical methodology, like scientific methodology, has been revised over time according

[6] For a convincing argument to this effect, see Maddy (2007, 184; IV.2.iii). On the centrality of the anti-revisionist attitude for naturalism, see also Roland (2016, p. 294).

to the needs arising in its own development, and has been found adequate for the development of science over thousands of years. In fact, mathematics plays a central role in our understanding of the natural world, and as such it is deeply integrated in our scientific worldview. Many contemporary naturalists hold that since mathematics is a legitimate part of our best science, it should enjoy the same trust placed in other kinds of scientific enterprises.[7] This sentiment motivates the anti-revisionist attitude that pervades all kinds of naturalism, even though revision in the light of new evidence (obtained by employing the methods sanctioned by practitioners) is one of the core methodological tenets of all kinds of naturalism. So what sets revisions prescribed on external grounds apart from the naturalist's point of view?

An indiscriminate, large-scale revision of mathematical practice would look completely implausible to the naturalist, given its high level of success and the central role that it occupies in our pursuit of knowledge of the world. The cost of such a radical revision would be excessively high for mathematical practice, but also for scientific practice, as this revision would have pervasive consequences in all branches of science, seriously undermining our ability to do science at all. For the naturalist, the improbably high cost of a radical revision of mathematics would constitute evidence that such a revision is unwarranted, rather than evidence that our trust in current mathematics is unwarranted. This is also due to a fundamental difference in the progress of empirical sciences and mathematics: while the value of a revision to current scientific practice is assessed on the basis of the new theory's virtues such as enhanced predictive and explanatory power, mathematical knowledge progresses in a cumulative way. When a new mathematical theory starts developing, a key factor for its retention resides in the possibility of integration with previously established mathematical theories. Since the naturalistically acceptable revisions to scientific practices are those triggered by problems that arise within the development of that practice, many contemporary naturalists maintain that mathematics should not be criticised on extra-mathematical grounds, where extra-mathematical grounds include both scientific and philosophical grounds.

Quine famously argued that mathematics, like any other scientific enterprise, is subject to the authority and standards of scientific method (see e.g. Quine 1981, p. 182). In Quine's view, mathematical objects are unobservable entities on a par with electrons and black holes. Since confirmation is holistic, the existence of unobservables is confirmed or disconfirmed by means of our best scientific methods. Crucially, scientific method is conceived here as the method of the empirical sciences. This picture of science and the view of mathematics that ensues have been extensively criticised (for an overview of the debate and an influential critique, see e.g. the discussion in Maddy 2007).

The type of naturalism developed especially by Penelope Maddy in a number of works under the heading of *mathematical naturalism* (Maddy 1997), or *second philosophy of mathematics* (Maddy 2005, 2007), explicitly breaks with Quinean

[7] See Maddy (1997, 2007, 2011), Shapiro (1997), etc.

scientific naturalism in granting mathematics the same autonomy that scientific naturalism grants to science.[8] When we examine mathematical practice and its history, it is apparent that revisions to the set of accepted principles are motivated by reasons internal to the practice, and that mathematical statements are not regarded as confirmed or disconfirmed on the basis of their empirical applications (or lack thereof), neither within the mathematical nor within the scientific community. The history of mathematics also provides evidence to the effect that the development of mathematics is best served by not curtailing its freedom: practitioners' decisions are guided by what best serves the aims of the practice (see again Maddy 2007, 184; IV.2.iii). A consequence of this approach is that mathematics in general is not taken to be revisable in the light of empirical evidence, but only revisable on mathematical grounds.

Rather than expanding on the criticisms to Quinean naturalism, on which much has already been written, I will focus below on revisions prescribed on philosophical grounds, which mathematical naturalists are especially inclined to rule out. Most philosophical and foundational accounts of mathematics recommend, in some way or other, revisions to mathematical practice that are prescribed, at least partly, on the basis of philosophical reasons. To cite a well known example, Hermann Weyl famously declared in *Das Kontinuum* (Weyl 1918) that analysis was built on sand, and even though it could be rebuilt on "pillars of enduring strength" (Weyl 1918, 1994, p. 1), not all of mathematics would survive this reconstruction.

Mathematical revisionism prescribed on extra-mathematical grounds badly accords with the naturalist predicament that mathematics, as a successful practice, does not stand in need of philosophical defence or correction; it does not require philosophically loaded scepticism to set the standard for scrutiny for its theoretical posits or methods, nor does it require philosophical guidance to set research directions and priorities. So in the mathematical naturalist's investigation of mathematics there is no place for the rejection of any parts of established mathematical practice: the principles and methodologies that constitute accepted mathematical practices are precisely the things which make mathematics a successful enterprise, and underwrite the central role that mathematics occupies within science.

The legitimacy of revisions to mathematics prescribed on philosophical grounds has also been the object of debate,[9] but it will be argued in what follows that something has been missed in the adequacy of the distinction between philosophically-motivated vs. mathematically-motivated revisions to the practice, and the role that this distinction plays for mathematical naturalism. In order to do so, it is useful to consider some examples of philosophically-motivated revisions to mathematical practice in more detail.

Intuitionism is often cited as an example of a mathematical theory that prescribes changes to mathematics as it is practiced on the basis of a specific philosophical

[8] See also chapter 4 of Leng (2010), Roland (2016), chapter 2 of Shapiro (1997), and Weir (2005).

[9] Besides Maddy's work, see e.g. Burgess and Rosen (1997) and Daly and Liggins (2011).

view. For example, intuitionistic mathematics rejects the general validity of the principle of excluded middle, forbids the use of impredicative definitions, and holds that certain classically equivalent definitions correspond to distinct intuitionistic notions. The consequences for mathematical practice, which is largely classical, are significant. They involve the rejection of classical logic in favour of intuitionistic logic, and the rejection of the validity of many classical theorems whose truth is widely accepted in the mathematical community. Besides the radical consequences that an intuitionistic revision of mathematics would have for mathematical practice, equally significant consequences would impact scientific practice, which heavily relies on the use of classical mathematical methods and the appeal to classically valid theorems. It seems therefore that mathematical naturalists are correct in denying intuitionistic mathematics the naturalist stamp of approval. But what characterises intuitionistic mathematics as a revision to mathematics which is prescribed on philosophical, rather than mathematical grounds?

Intuitionistic mathematics historically arose, at least to some extent, in response to the set theoretic paradoxes, and thus out of a mathematical need to better understand the principles of correct reasoning concerning completed infinite totalities. Nonetheless, intuitionism—at least in the version proposed by Brouwer—is sustained by a specific philosophical account of the notions of intuition and construction from which it is inseparable. According to intuitionism, mathematics is the result of the free activity of the mind; this activity originates in the pure intuition of time. There is universal agreement in the literature on the fact that this philosophical view is a central component of the intuitionistic conception of mathematics, and that intuitionism prescribes radical revisions of mathematical practice based on this philosophical view.[10] One of the consequences of this view for mathematical practice is that a mathematical statement can be affirmed only if a corresponding construction for it can be given. This in turn implies that a mathematical statement becomes true only when an appropriate construction that proves its truth is experienced by the creating subject. Principles such as the excluded middle are therefore not generally valid in intuitionistic mathematics: a disjunction such as $(P \vee \neg P)$ is only valid if a construction proving $P$, or one proving $\neg P$, is given. Impredicative definitions, which are used liberally in mathematical practice, are rejected on the grounds that they are viciously circular. The rejection of the principle of excluded middle and of the use of impredicative definitions are two typical examples of mathematical consequences implied by the intuitionisitic ontological and epistemological approach to mathematics. More generally, the consequences of this approach are that intuitionistic mathematics contradicts classical mathematics in a fundamental way.

While these considerations help us to better understand why intuitionism is taken to prescribe philosophical revisions to mathematics, Brouwer was certainly not the only mathematician who expressed strong philosophical views on mathematics. The history of mathematics contains numerous examples of mathematicians who

[10] See e.g. Iemhoff (2020), Posy (2005, pp. 318–319), and van Atten (2017).

were also philosophers, and who used their own philosophical views to guide their mathematical practice. In the relatively recent past, a particularly famous instance of fruitful interplay between mathematical and philosophical work is provided by Gödel's claim that his mathematical realism lead him to the discovery of both the completeness theorem for first-order logic and the incompleteness theorems (see e.g. the discussion in Shapiro 1997, p. 31). Gödel, who likened the certainty of the existence of mathematical objects to that of physical objects (an analogy strongly rejected by Brouwer), used his realism also to justify the adoption of the law of excluded middle and impredicative definitions.

Brouwer and Gödel clearly held opposite views about many aspects of mathematics. Consider for example their respective positions with respect to the law of excluded middle and impredicative definitions. Brouwer recommended a revision to their widespread use in mathematical practice, while Gödel countenanced their acceptance and use. It appears, however, that both Brouwer's rejection of the law of excluded middle and of the use of impredicative definitions, and Gödel's acceptance of both, are similarly philosophically motivated on the grounds of their respective antirealist and realist attitudes towards mathematics. We may thus wonder what sets the two cases apart. Even though Gödel's defence of these principles on the basis of his mathematical realism cannot be characterised as genuinely revisionist, accepting the law of excluded middle and the use of impredicative definitions was not a given at the time: intuitionistic mathematics had gained considerable support since its introduction in Brouwer's PhD thesis in 1907, and with Brouwer's and Heyting's subsequent developments. The use of predicative definitions as a solution to the paradoxes had also gained credibility, especially thanks to the advocacy of Poincaré and Russell. The main difference, it seems, is that Brouwer's intuitionism requires a revision of established mathematical practice, while Gödel's realism does not.

From a naturalist point of view, the revisions to mathematical practice prescribed by intuitionistic mathematics are not sanctioned by mathematical standards. As is apparent from historical and contemporary mathematical practice, the foundational crisis did not dissuade most mathematicians from the use of the principle of the excluded middle and impredicative definitions, who evidently considered them to be secure and rigorous methods of proof. No further paradoxes and contradictions have arisen from their use, thereby prompting a revision of mathematical methodology based on purely mathematical considerations. A naturalist would thus argue that the success of current mathematics, validated by its central role in our best science, sanctions classical mathematics over intuitionistic mathematics.[11] Even if classical mathematics does not enjoy the type of justification enjoyed by intuitionistic mathematics, the 'limited success' of intuitionistic mathematics (relative to how widely it is practiced, rather than its internal coherence and justification) does not

[11] As Maddy pointed out (e.g. in Maddy 2007), this does not imply that mathematical naturalism sanctions Gödelian realism or any other positive philosophical view of mathematics.

support the deep and wide-ranging revisions to current mathematical practice which are prescribed by it.

What would the mathematical naturalist say about a philosophically motivated adoption of the law of excluded middle and the use of impredicative definitions, such as it was advocated by Gödel? The mathematical naturalist may likely argue that even though Gödel's own acceptance of the law of excluded middle and of the use of impredicative definitions was rooted in a realist view of mathematics, these were widely accepted in the practice in virtue of certain mathematical reasons, for example their fruitfulness in allowing the simpler, or more natural derivation of many previously accepted theorems. The philosophical reasons for one specific mathematician's adoption of principles also embraced by other practitioners at the time does not by itself support the assumption that a realist attitude was underlying the adoption of these principles by most of the practitioners in question.

It can be pointed out in reply, however, that a similar argument can be made on behalf of the intuitionist. The rejection of the law of the excluded middle and of impredicative definitions, advocated by Brouwer on philosophical grounds, can also be motivated exclusively on mathematical grounds. The fact that Brouwer's motivations for intuitionism were chiefly philosophical does not undermine the mathematical merits of intuitionistic mathematics, nor are Brouwer's philosophical motivations necessarily at the heart of every intuitionist mathematician's rejection of the law of excluded middle and of impredicative definitions. The intrinsic mathematical value of intuitionistic proofs can be appreciated regardless of any philosophical reasons that may have lead to the acceptance of intuitionistic mathematics. In this sense, it would be quite unreasonable for naturalists to completely disregard the importance of mathematicians' philosophical views even when they fruitfully guide their practice. The many examples of fruitful interaction between mathematicians' philosophical views and their practice in the history of mathematics should be sufficient reason for the naturalist to be liberal about whatever philosophical view a practitioner may have. A detailed discussion of such arguments goes unfortunately beyond the scope of this paper. Nonetheless, these arguments show that the line between philosophical and mathematical revisions to the practice is sufficiently blurred to question the validity of the naturalist criterion for anti-revisionism.

A more successful strategy for the naturalist may be to reply that even if classical proofs do not possess the same mathematical virtues as intuitionistic proofs, the limited portion of mathematical practice represented by intuitionistic mathematics does not support the deep and wide-ranging revisions to mathematical practice as a whole which are prescribed by intuitionisitic mathematics. So we can conclude that quite regardless of their philosophical motivation, what sets Brouwer's and Gödel's prescription to mathematical practice apart is the fact that Gödel's prescription is ultimately in line with mathematical practice, while Brouwer's is not. On the other hand, as already observed, Gödel's view may not be considered revisionist, or not sufficiently revisionist, to be put on a par with Brouwer's.

To consider a more revisionary proposal than Gödel's, we can go slightly earlier in time to the last decade of the nineteenth century and Hilbert's work in algebra. In 1890, Hilbert published a proof of what came to be known as *Hilbert's basis theorem* (Hilbert 1890). The highly innovative character of the proof resided in its purely existential nature: it is a proof by contradiction which establishes that every infinite system of homogeneous polynomials is generated by some finite basis, but does not provide an algorithm to construct this basis. The proof sparked an animated discussion amongst mathematicians of the time (Noether 1914; Klein 1926; McLarty 2012). Hilbert went on to employ non-constructive proofs to attack a range of problems in algebraic number theory, and also to advocate strongly for their use, even claiming that some such proofs "uncover the inner reason for the validity of the assertions", unlike more cumbersome constructive proofs of the same theorems (Hilbert 1928).[12] It may be argued that the reasons which led Hilbert to consider such a proof as unproblematic rested, at least to a good extent, on his philosophical views concerning the legitimacy of the use of principles such as the excluded middle in mathematical practice—or, more precisely, the use of infinitary principles of what Hilbert later called *ideal mathematics*—which presupposes the legitimacy of quantifying over completed infinite domains. In later years, Hilbert went on to develop his new foundational approach based on the distinction between real and ideal mathematics, and the justification of the latter by means of finitary methods with the collaboration of Bernays. This clearly shows that to Hilbert, the rigorous development of the philosophical foundations of his approach to mathematics was just as important as his mathematical work. It is crucial to stress in this context that Hilbert was not a revisionist about mathematical practice: his foundational programme was motivated precisely by the desire to justify infinitary reasoning as used in mathematical practice, but from the finitary point of view, which was guaranteed by its nature to be free from paradoxes and inconsistency.[13]

To evaluate the mathematical naturalist's criterion that a prescription to revise mathematical practice is only acceptable if motivated by reasons internal to the practice, it is useful to contrast Brouwer's, Gödel's, and Hilbert's attitudes towards of the law of excluded middle and the use of impredicative definitions. In all three cases, the proposed revisions are similarly philosophically motivated on the grounds of different philosophical views.[14]

---

[12] For a more extended discussion of this point, see Dreben and Kanamori (1997). As Dreben and Kanamori remark, Hilbert was not the first historically to use a merely existential proof. However, Hilbert was a forceful advocate of the use of such proofs, and the first to use them in a wide range of mathematical settings.

[13] See Zach (2019) for a survey of the primary and secondary literature on Hilbert's programme.

[14] It may be objected that the introduction of merely existential proofs is an *ad hoc* choice for a case study of mathematical revision insofar as it prescribes a methodological revision that does not involve the abandonment of previously accepted mathematics. The case study, however, is relevant for two reasons. Firstly, it concerns a proposed change to mathematical practice—and more precisely, to proving methods—that is recommended on illegitimate (extra-mathematical) grounds. Secondly, it is evident that the use of merely existential proofs clashed with the way in which mathematics was done until then, and there does not seem to be a principled distinction between revisions which limit previously established mathematical principles and methods, and revisions which extend them. The eventual inclusion and retention of merely existential proofs among accepted proving practices promoted a profound change that reshaped nineteenth century

We saw that from the naturalist's point of view, what is ultimately important are not the different reasons which are historically offered in support of the adoption of specific mathematical principles by specific mathematicians (which may also include extra-mathematical reasons, such as philosophical ones), but the mathematical reasons underlying the adoption and the retention of said principles in the practice. We are now in a position to better appreciate the rationale for the naturalist criterion that a revision to mathematical practice can only be legitimately recommended on mathematical grounds: of the many revisions of mathematical practice that are proposed, the ones that are legitimate (and thus have a chance of being retained by the mathematical community) are those proposed for reasons that arise in the ordinary practice of mathematicians, not those prescribed by philosophers on the basis of some ontological or methodological worry. The case studies considered, however, show that the criterion captures the notion of revision in question in a somewhat *ad hoc* manner: it does not appear to be the case that all the proposed revisions that are prescribed to some extent also on philosophical grounds are illegitimate. The case of merely existential proofs illustrates how some revisions proposed also on philosophical grounds are accepted and retained by the community. Furthermore, once a particular revision has been adopted for some time and it has become deeply integrated in mathematical practice, the importance of the philosophical motivations that contributed to its adoption are more easily forgotten. In addition to the case studies above, consider also the debate between Zermelo and Borel, Baire, and Lebesgue on the adoption of the axiom of choice (see e.g. Moore 1982). Now that the use of the axiom of choice has become unproblematic in contemporary set theory, only those set theorists with historical or philosophical leanings may be aware of the reasons for or against it which were debated at the time of its introduction. If revisions of mathematical practice proposed on both philosophical and mathematical grounds are accepted and retained by practitioners, it seems that the naturalist should recognise them a fortiori as legitimate.

A criterion that discerns acceptable revisions on the basis of the motivations that lead to introduce it—mathematical or philosophical, or a combination of both—ultimately fails to single out the revisions that are accepted and retained by the mathematical community. The analysis of the case studies above shows that the combination of philosophical and mathematical reasons advocated by Brouwer for the rejection of the principle of excluded middle and impredicative definitions do not substantially differ in nature from the reasons advocated by Hilbert for the use of merely existential proofs, even though the first proposed revision to the practice was not accepted, while the second was.

The naturalist could insist that if a certain revision of the practice is adopted by mathematicians, a fortiori there must have been some mathematical reasons for it, and these reasons must have been what was significant in the adoption of the new principles or new methods of proof in question. Not all mathematicians share

mathematics. See Sect. 9.6 below for a case study of a proposed revision to mathematical methodology which does not imply a change of the accepted theorems.

the same philosophical views about mathematics, or even have philosophical views about mathematics at all. On the other hand, most mathematicians share the same principles, methods, and standards of proof, and the acceptance of new principles or methods by the community, together with their integration with previously accepted ones, is what ultimately determines what is retained by the community as part of established mathematics. A similar strategy is advocated e.g. by Maddy, who argues that "[T]raditionally philosophical questions of ontology and epistemology are irrelevant to the methodology of mathematics" (Maddy 2007, 361), and that even "though metaphysical theories on the nature of mathematical truth and existence undeniably do turn up in such debates, they are not in fact decisive, they are in fact distractions from the underlying purely mathematical considerations at work." (Maddy 2007, 349). This, on Maddy's view, highlights a contrast with scientific practice, where theories about the nature of physical objects play a central role (*ibidem*, footnote 10). As a case study, Maddy cites Whitehead and Russell's reconstruction of mathematics without the use of impredicative definitions (Maddy 2007, 348). Regardless of the reasons that lead Whitehead and Russell to reject the use of impredicative definitions, their use became part of the practice because without them, we cannot reconstruct much of ordinary mathematics. Here Maddy cites Gödel's example that classical mathematics implies the existence of real numbers definable only by reference to the totality of real numbers (Gödel 1944, 127).

If we push this line further, however, the argument appears question begging: arguing for the necessity of reference to *all* real numbers presupposes that we have already assumed that the correct definition of the reals is the classical, non-constructive one. This assumption could be motivated by the importance of classical analysis to applied mathematics. This road is of course open to the mathematical naturalist, who recognises the paramount importance of mathematics to science. However, granting mathematics the same autonomy as science, the mathematical naturalist maintains that mathematical practice is developed and assessed on the basis of mathematical purposes and standards alone, not for its applicability in science; so the naturalist cannot motivate the assumption of the correctness of the impredicative definition of the reals exclusively on the basis of its usefulness in applications (see also Antonutti Marfori 2012). Rather, she needs to explain it by appealing to purely mathematical considerations, such as the fruitfulness of the classical definitions of the reals in the development of analysis, which is by all standards a very successful branch of mathematics. The success of this reply seems therefore to rest on the mathematical naturalist's ability to avoid any appeal to philosophical considerations in the course of doing so. In the case of predicative definitions, relevant philosophical considerations to avoid would include issues concerning ideals of rigour (in what sense is mathematics that uses impredicative definitions rigorous?), elegance or simplicity of proving methods (why are simpler proof engendered by the use of impredicative definitions preferable to more complex proofs which dispense from their use?), existence and consistency (if mathematical existence amounts to consistency, then the acceptance of impredicative definitions is unproblematic), and the relevance of theorems that are not predicatively justifiable for applicable mathematics, among others. The suggestion is that it would be hardly possible to disentangle locally the philosophical and mathematical arguments that feature in discussions of impredicative definitions found the literature, in a way that isolates purely mathematical reasons for adopting them that do not bring into play purely philosophical conceptions of mathematical definitions, methods, and so on. Similar arguments could be made about the use of the principle of excluded middle and the use of merely existential proofs. Unfortunately, it is not possible to discuss here each significant episode of introduction of new principles or methods of proof in mathematical practice individually in order to investigate the rationale for their adoption or rejection. In order to show that the objection above cannot easily be dismissed (that is, that the naturalist cannot completely shy away of any philosophical consideration in explaining the mathematical success of the impredicative definition of the reals), it is perhaps sufficient to point out that even the accounts which appeal to the mathematical fruitfulness of the adoption of a certain principle or method of proof are not easy to support without some appeal to philosophical considerations, ontological or methodological: after all, if the aim of mathematical practice was merely to maximise the derivation of results, inconsistent theories would be preferred to consistent ones. Even if we restrict our focus to consistent theories, it would still be difficult to motivate the superior fruitfulness of classical, impredicative definitions of the reals compared to predicative ones, and in general explain why their use was retained by the mathematical community, without engaging in any philosophical discussion at all, and keeping at all times a distinction between the naturalist's philosophical interpretation of the practice, and the purely mathematical reasons for the revision in question. The naturalist may still appeal to a distinction between the naturalistic philosopher's interpretation and explanation of the workings of mathematical practice, and the mathematical mechanisms governing it, but this would equally require that in each debate internal to the mathematical community, philosophical reasons are sufficiently sharply disentangled from mathematical ones and shown not to be decisive.

The question of whether philosophical inquiry is, or should be, part of the methods of inquiry accepted by the naturalist is a topic of discussion among the naturalists, which hinges on more complicated issues pertaining to whether philosophical and scientific methods are continuous with one another (see e.g. Clark 2016 for a sample of the recent debate on this matter). For reasons of space, I will not engage with this larger issue here. It suffices for the present argument to consider that there are two main relevant alternatives. If the methods of science and mathematics are discontinuous with the methods of philosophy on a global level, they are nonetheless locally entangled in the case studies analysed. As Weir (2005, 468) points out, the mathematical naturalist *à la* Maddy does not need locally sharp boundaries between philosophical, mathematical, and scientific methods, but nonetheless needs to concede that a higher level these are sufficiently discontinuous to allow talk of *philosophical* vs. *mathematical* vs. *scientific* methods. If the methods of science and mathematics and philosophy are continuous, this is all the more reason to think that the anti-revisionist criterion in question cannot do a sufficiently good job for the mathematical naturalist in distinguishing which revisions to the practice are legitimate and which ones are not.

A route out of this impasse could be for the naturalist to grant that some philosophical considerations can enter the justification of the adoption of certain principles or methods—when these are adopted by practitioners rather than imposed from the outside—while maintaining that philosophical considerations are not alone decisive in shaping mathematical practice. This would amount to adopting the more liberal view that mathematics is guided by a combination of mathematical and (often not explicitly formulated) philosophical considerations (for simplicity, I will henceforth refer to this view as the *liberal view*). It may be objected that adopting the liberal view simply moves the problem of drawing a distinction between the methods of different disciplines slightly further. What makes certain philosophical considerations, but perhaps not others, legitimate in assessing proposed revisions of mathematical practice? What distinguishes acceptable philosophical considerations from, for example, political or theological considerations which the naturalist thinks should not bear a weight on mathematical practice? The naturalist may ultimately accept that no global criterion can be given once and for all for what counts as a legitimate revision. Each case should be evaluated locally, in line with the naturalist contention that science and scientific methods (including here pure and applied mathematics and their methods) cannot be defined by necessary and sufficient conditions, as their conception develops over time with the development of science itself. At times of deep changes and radical revisions, like in the mathematics of the late nineteenth and early twentieth century with the foundational crisis and the emergence of abstract methods, it is possible that some philosophical considerations may be invoked in the justification of proposed revisions, even though it is never those philosophical considerations alone which drive and guide the need for the revision of the practice, and it is not possible to sharply distinguish those considerations from illegitimate ones at a global level.

Does the adoption of the liberal view succeed in fixing the criterion for anti-revisionism according to which we should reject revisions prescribed on external grounds? The liberal view was motivated by the consideration that the revisions accepted and retained by practitioners do not necessarily line up with the revisions prescribed on purely mathematical grounds: sometimes proposed revisions supported by a combination of philosophical and mathematical reasons are accepted and retained by the practice. The liberal view fixes this problem. However, it may now be objected that the amended criterion overgenerates to the point that it includes all proposed cases of mathematical revisions: presumably, each case of proposed revision of the practice—including both the revisions that are retained, and those that are eventually rejected—could be motivated by a suitable combination of mathematical and philosophical considerations. In order to identify which revisions to the practice are legitimately recommended, the criterion still crucially depends on the analysis of which proposed revision is eventually accepted or rejected by the majority of practitioners. Mathematical practice is not unitary: at any point in time, many voices inhabit the community, presumably including some which prescribe revisions of the practice on the basis of political or theological considerations that the naturalist would deem illegitimate (consider for example how the study of infinity in mathematics was at various points in history associated with the devil or with different mystical doctrines). This brings us back to the considerations made on the case studies of Brouwer's and Gödel's respective rejection and acceptance of the law of excluded middle and impredicative definitions: the most relevant feature of legitimate revisions seems to be whether they are eventually retained or rejected by practitioners. These kind of evaluations can only be made after some time has passed and they can be assessed against the course that subsequent mathematics has taken. In some cases, it is only possible to see that certain notions or methods are retained by the practice after several decades, or even several centuries they have been proposed (consider for example the cases of non-Euclidean geometries and the development of non-standard analysis).

After a long detour, the mathematical naturalist is left with no adequate criterion to discern which proposed revisions to the practice are naturalistically acceptable beyond ascertaining which ones were in fact accepted. In the next section, I will discuss a different criterion for anti-revisionism, and I will evaluate its fit for the mathematical naturalist's purposes.

## 9.4 Retaining Accepted Theorems

One criterion for mathematical revisionism characterises revisionist approaches as those which prescribe changes to the extent of ordinary mathematics.[15] Acceptance of this criterion is not especially tied to naturalistic approaches, and it is widely appealed to in philosophy. For example, the requirement that a framework for mathematics has the ability to "recover" ordinary mathematics provides a common criterion for assessing the success of a given foundational programme. When a proposed framework is formalised, it becomes possible to prove precisely the extent to which said foundation captures the theorems of classical mathematics. This view is explicitly advocated for example in Whitehead and Russell's *Principia Mathematica*:

> [T]he chief reason in favour of any theory on the principles of mathematics [...] must lie in the fact that the theory in question enables us to deduce ordinary mathematics. (Whitehead and Russell 1910, p. v)

The requirement to recover ordinary mathematics has a clear intuitive meaning, and it is often not defined explicitly. However, it can be interpreted in several ways. For example, it can refer to a formal theory's ability to capture the structure of proofs in informal mathematical reasoning in one or more subfields of mathematical practice, or it can refer to a formal theory's ability to extensionally capture the consequences of the axiomatic principles accepted in the subfield in question.

[15] Since ordinary mathematics is largely classical, I will be using "classical mathematics" and "ordinary mathematics" interchangeably.

Without further qualifications, "recovering classical mathematics" seems most often used in the second sense, as the minimal requirement that a formal theory for an area of mathematics proves all the theorems of that area that are accepted by the current mathematical community.

When not understood as a constraint on foundations for mathematics, but more generally as a criterion which philosophical accounts of mathematics should satisfy, this requirement appears to be in line with the naturalist's reasons for adopting anti-revisionism. The apparent success of mathematics warrants the acceptance of the principles and methods that are widely use in ordinary mathematics and retained over time. Testing this criterion against the case studies considered earlier can help us to assess whether it also captures the cases of mathematical revisionism which the naturalist aims to capture. The criterion correctly identifies intuitionism as a type of mathematical revisionism, since intuitionism is not merely quietist about classical theorems for which we have no intuitionistic proofs, but rejects classical theorems which contradict intuitionistic axioms such as Brouwer's continuity principle.[16] Predicativism is also identified as a revisionist account, since predicativists usually consider only the natural numbers as a given completed totality, and only accept theorems which can be proved starting from predicative definitions. The extent of predicative mathematics provably excludes classical theorems such as Kruskal's theorem, which cannot be proved without the use of impredicative definitions.[17] Hilbert's recommendation to revise the methods of proof used in nineteenth-century mathematical practice to include merely existential proofs is correctly not identified as a revisionist proposal, because it extended, rather than restricted, the extent of ordinary mathematics at that time, and more importantly, it did not prescribe the abandonment of any theorems which were widely accepted by the mathematical community at the time.

There are, of course, many aspects of mathematical practice which one can be revisionist about, not just the set of currently-accepted theorems, and a revision to that set of theorems can result from different kinds of revisionist approaches. For example, one can be revisionist about the ontology of current mathematical practice, rather than its epistemology or methodology. One can advocate a revision to the semantics of mathematics rather than its syntax (i.e. ordinary mathematical discourse), or one can be revisionist about classical logic. Much of the time, a recommendation to revise one of these aspects of the practice goes hand in hand with a recommendation to revise other aspects of the practice as well: a revision to the ontology of mathematics that excludes certain objects referred to in mathematical discourse, e.g. large cardinals, would typically come together with a revision to the semantics of mathematical discourse, telling us how to interpret

[16] See van Atten (2017) for the notion of (strong) Brouwerian counterexamples and Iemhoff (2020) for an introduction to Brouwer's continuity principle.

[17] The extent of predicative mathematics is broadly agreed to have been determined by the proof-theoretic analysis of Feferman and Schütte (Feferman 1964; Schütte 1964; Schütte 1965). The unprovability of Kruskal's theorem in predicative mathematics is a result due to Harvey Friedman; for details see Simpson (1985) and Rathjen and Weiermann (1993).

the discourse in set-theoretic practice that seemingly refers to those objects. On the other hand, the adoption of a certain model-theoretic semantics for mathematical discourse would likely go together with a revision of the ontology: consider e.g. the case of a finitistic semantics for mathematics. Once again, intuitionism seems to be the paradigm of mathematical revisionism: it prescribes a reinterpretation of mathematical discourse, and an according change in the logic, which go together with fundamental changes in the epistemology and methodology of mathematics (only intuitionistic proofs are a source of mathematical knowledge), and a change in the ontology (only those objects that can be constructed by intuitionistically sanctioned methods can be considered to exist).

The combinations of the aspects of the practice that are the supposed object of revision determine different kinds of mathematical revisionism. Many of the ensuing types of revisionism also produce a change in the set of accepted theorems. However, the requirement of "recovering mathematics", understood as the criterion that a non-revisionist account of mathematics should not exclude any theorem accepted as such by contemporary mathematical practice, is not sensitive to these differences. Should we conclude that mathematical naturalism does not have anything distinctive to add to the conventional criterion for revisionism? If so, this would undermine the central role that anti-revisionism plays in mathematical naturalism. In the last part of the paper, I will argue that the mathematical naturalist should not consider all types of revisionism as equivalent, and that in fact naturalism can provide a distinctive perpective on this issue.

## 9.5 A Revised Naturalist Criterion

In Sect. 9.2, I discussed how the naturalistic commitment to anti-revisionism stems from a deep trust in our best science, and in scientific method as our best method to pursue knowledge of the world. We may therefore be led to think that there is more to the mathematical naturalist's strong commitment to anti-revisionism that the mere preservation of the set of theorems that are accepted by a given mathematical community. In the naturalist's view, mathematics does not amount to a list of theorems, but is the product of a human activity, a collective effort to generate more results starting from existing ones, to find new connections among them, and now and again, to extend this knowledge by adding new principles and new methods of proof. The first criterion for naturalistic anti-revisionism, according to which revisions prescribed on extra-mathematical grounds are illegitimate, was itself inspired by the goal of getting to the root of what the naturalist considers problematic about revisionism. Namely, anti-revisionism as the mere preservation of accepted theorems does not recognise that what makes mathematics a successful enterprise is the fact that there is a reasonably definite set of principles and methods of proof that constitute "mathematical practice", and that mathematicians typically agree on what counts as a correct application of a specific method. This thesis often goes under the name of *methodological naturalism* (see e.g. Weir 2005, p. 460).

Methodological naturalism does not involve commitment to a specific worldview, but to a set of methods as a reliable way to obtain knowledge, and to what can be known according to those methods. In this sense, methodological naturalism is neutral about ontology, albeit within the constraint that a naturalistic ontology should not include entities whose existence cannot be established by naturalistically sanctioned methods, or at least by methods which are incompatible with naturalistically sanctioned methods. Some naturalists, such as Maddy, have also argued that methodological questions are primary in the practice (Maddy 1997, 2007): debates in the mathematical community concern what count as good definitions of given notions, acceptable axioms, or acceptable methods of proof, rather than questions about mathematical ontology. While ontological questions are sometimes debated by the mathematical community, methodological disputes occupy a more central role in the practice. For example, we do not find number theorists wondering whether prime numbers really exist, nor algebraists wondering whether groups really exist, nor do we find natural scientists wondering whether the existence of prime numbers or groups is confirmed or disconfirmed when they occur in scientific theories and explanations. On the other hand, we do find mathematicians discussing whether Mochizuki's alleged proof of the *abc conjecture* can be considered as verified, or whether the classification of finite simple groups really captures all finite simple groups.[18]

Methodological naturalism can be distinguished from *epistemological naturalism*. The former is a thesis on legitimate methods of enquiry, while the latter is a thesis that directly concerns epistemological notions such as knowledge or justification. For example, mathematical epistemological naturalism may include the the thesis that mathematical knowledge is not a priori. However, since methodological naturalism is a thesis about the methods of enquiry that produce knowledge, for the purposes of this paper these two kinds of naturalism will not be distinguished.

Brouwer's intuitionism still qualifies as a methodologically (and epistemologically) revisionist account of mathematics. Naturalists have widely criticised founding mathematical knowledge on a faculty of intuition whose mysterious powers to grant us access to abstract mathematical objects does not seem liable to scientific investigation (see e.g. Kitcher 1985 and Maddy 2000). Intuitionism's distinctive epistemology sanctions a radical revision of the methods of most practicing mathematicians in ways incompatible with classical methods of proof.

It is important to note here that many mathematical naturalists subscribe to the thesis that a (naturalistic) philosophy of mathematics requires an *interpretation* of the practice. For example, Shapiro (1997, pp. 6–8 and 32–34) argues that providing a philosophical account of mathematics is an intrinsically interpretive enterprise. It is not possible to simply "read off" from mathematical practice the answers to the distinctively philosophical questions that a philosophical account of mathematics aims at answering, such as what is the logical form of mathematical statements,

[18] See e.g. Rittberg (2021) for a discussion of the former, and Habgood-Coote and Tanswell (2021) for the latter.

what are the valid rules of inference, what is the best semantics for mathematical language, whether mathematical activity describes a realm of abstract objects given independently of our description, or what is mathematics' true ontology. Each one of these questions requires an interpretation of mathematical language in order to be answered. For example, reading published mathematical papers will not by itself suffice to reveal which rules of inference are used in each proof.[19] This interpretation will likely require the introduction of notions that are not already present in mathematical discourse. For example, Heyting's formalisation of intuitionistic logic required the definition of the intuitionistic interpretation of the connectives and of the notion of intuitionistic proof, neither of which was already part of intuitionistic informal mathematics. This is even more striking in the case of the interpretation of historical practices, which were often cast in vague terms and were sometimes even inconsistent; no precise, let alone formal, account of any such practice could be a perfect reconstruction of it (see e.g. Urquhart 2008, pp. 424–425). In a similar way we cannot, starting from a philosophical account of the underlying form of mathematical statements, their semantics and ontology, and their associated rules of inference, "read off" how mathematics is done (or ought to be done) in practice. Engaging in such an exercise would surely result in a number of different accounts of mathematical practice, with serious doubts that any of them would even resemble actual mathematical practice.

From this perspective, the mathematical naturalist's commitment to anti-revisionism involves a commitment to engage with mathematical practice without preconceived first principles concerning what mathematics should be, and how it should be done properly (see Shapiro 1997, p. 32 and Maddy 2007, Part IV).[20] The naturalist anti-revisionist does not start their interpretation of the practice from the critical observation that mathematics as practiced is incoherent, and that therefore it must be provided with a corrected, coherent interpretation that can place it on a more secure foundation, as many foundational programmes have typically done. Rather, the grounds for any criticism and interpretation of the practice should be the practice itself: mathematics as it is practiced should be taken as a starting point, and it should be subject to ordinary mathematical criticism. The anti-revisionist naturalist is in the position to recognise that every practice (and every practitioner) is fallible, and thus that no practice (or practitioner) is immune from correction:

---

[19] See Antonutti Marfori (2010) for a discussion of this point.

[20] The mathematical naturalist can perhaps be seen as engaging in a *hermeneutic*, rather than *revolutionary*, account of the practice (Burgess 1983 and Burgess and Rosen 1997, p. 6). I agree with Weir's and Shapiro's reservations on this philosophical move: philosophy of mathematics, even if naturalistic and anti-revisionist, cannot be construed as a purely descriptive enterprise insofar as it deals in *correct* mathematical reasoning, rather than any reasoning carried out by someone employed in a mathematics department (see e.g. Shapiro 1997, p. 30 and Weir 2005; see also Antonutti Marfori 2020). However, cfr. Burgess' distinction between *prescriptive* and *descriptive* criticism of classical logic in (Burgess 2005, p. 729), where the former is the thesis that classical logic correctly captures the types of inference made in mathematical practice, which are however incorrect, whereas the latter is the thesis that classical logic ratifies incorrect principles that are not used in the practice.

mistakes, even systematic ones, occur in the practice. This is compatible with the naturalistic recognition that mathematics as a practice is hugely successful and occupies a central place in our best science (see e.g. Maddy 1998, Shapiro 2000, pp. 14–15).

The tasks of the mathematical naturalist who embraces anti-revisionism then include the analysis of the goals of historical and contemporary mathematical practices and the investigation of how those goals were pursued by the mathematical community. Especially when substantive disagreements arise concerning what axioms or definitions should be adopted in a certain domain of mathematics, and what methods of proof are rigorous in reasoning about that domain, a naturalist account of how such controversies were adjudicated by practitioners can reveal important aspects of the rationality of the practice. This critical activity may differ in perspective from that of the practitioners, but should not differ in its evidential standards. The philosophical account of the standards for what counts as a reasonable assumption or definition, and a reasonable form of argument, must be rooted in an understanding of the norms that govern mathematical practice—norms that are already at work in the practice. Similar standards presumably apply to formal accounts of the practice: consider, for example, Heyting's formalisation of the logic implicit in informal intuitionistic mathematics mentioned above, or Feferman's proof-theoretic results concerning the extent of predicative mathematics mentioned in Sect. 9.4 which formally captures the notion of predicatively acceptable proof as a proof that does not make use of impredicative definitions.

If this is correct, then mathematical naturalism is compatible with some degree of pluralism concerning accounts of the practice: nothing prevents the mathematical naturalist from taking different accounts of the goals implicit in the practice, and the means-ends activities employed to reach them, as bringing out different aspects of the same practice. After all, a philosophical investigation of a practice that does not start from preconceived conceptions of it will reveal that mathematics, like any other scientific practice, is not unitary. While there are collectively recognised standards at work in the mathematical community–for example, those applied by mathematical journal editors and referees–we need to be cautious about the idealisations embedded in representing mathematical practice as an integrated, concerted endeavour. The mathematical community at a given time is composed of a number of different groups and communities, which are in turn composed of highly specialised practitioners whose assumptions, methods, mathematical culture and mathematical style may differ in important ways from those of other groups of practitioners working even in the same subfields.

If we concede that every philosophical account of mathematics, including a naturalistic account, needs to engage in this kind of interpretive effort, it is hard to imagine that there can be an account of mathematics that is completely immune from criticism with respect to its ability to capture salient aspects of the practice. To the extent that an account purports to capture the underlying goals of a practice, the logical form of mathematical statements, the valid rules of inference, the correct semantics for mathematical language, rather than the ontology of mathematics or some other aspect of the practice, it is liable to be criticised for holding a revisionist attitude about that particular aspect of the practice, or other aspects of the practice that may be misrepresented or even eliminated. So, if we push the definition of naturalistic anti-revisionism about mathematical practice to the limit, we may end up with no viable philosophical account of the practice, or some aspects thereof, that lives up to the standard of avoiding *any* type of revisionism whatsoever.

It would be too hasty, however, to conclude that anti-revisionist mathematical naturalists are looking for something unattainable due to a tension between the interpretive aspect of philosophical accounts of mathematics, and the naturalistic anti-revisionist requirement. Based on the considerations so far, the mathematical naturalist should not regard all types of revisionism as being equally anti-naturalist. The recognition of the centrality of methodological questions in mathematical practice indicates that a third anti-revisionist criterion can be formulated in the context of mathematical naturalism as follows: the illegitimate revisions to mathematics are those that (i) require the abandonment of currently accepted theorems, and (ii) require the abandonment of established mathematical methodology.

The revised naturalistic criterion captures what matters to the mathematical naturalist—namely, the centrality of methodological concerns—which is missing from the second criterion, which merely prescribes not to abandon classical theorems. Since it is the distinctive methodology of mathematics that makes it a successful discipline, and this methodology is preserved in a cumulative way through the history of mathematics, it should be particularly important for the naturalist that recommended revisions to the practice do not prescribe practitioners to abandon established proving practices. Formulated in this way, the new criterion clearly implies the second criterion for anti-revisionism discussed in Sect. 9.4, i.e. that we should not revise the extent of ordinary mathematics, and it seems therefore a good candidate criterion for the naturalist anti-revisionist. In the next section, the new criterion will be tested again the desiderata discussed so far.

## 9.6 Assessing the Revised Criterion

This criterion fares better than the first naturalistic criterion considered, according to which only revisions to mathematical practice prescribed on extra-mathematical grounds are illegitimate. According to the revised criterion presented in the previous section, the approaches which are typically considered revisionist—e.g. intuitionism—are still identified as such, and the approaches that are not typically considered revisionist are correctly not identified as such. Consider for example the case of Hilbert's introduction of merely existential proofs: the introduction of merely existential proofs broke with previously established proving practices, but not in a way that required mathematicians to abandon any of them, or to abandon any theorem provable by use of previously established methods of proof.

For what concerns capturing the naturalistically illegitimate cases of revisions, since this third criterion for mathematical anti-revisionism implies the second one (a revisionist account of mathematics is one that requires us to abandon accepted theorems), we may wonder how the strengthened version of the criterion is to be motivated. The reply to this objection consists of two parts. Firstly, the analysis above argued that the naturalist has reasons to adopt a distinctively naturalistic version of the second criterion for revisionism that go beyond merely capturing the intended cases of naturalistically unacceptable revisions to the practice. Secondly, the amended naturalist criterion for antirevisionism is more fine-grained than the second in that it discriminates against certain philosophical accounts of mathematics that could be considered non-revisionist on the basis of the second criterion, but would arguably be considered revisionist on the basis of the third. Consider for example the case of predicativism. It could be argued that predicativism is not really a revisionist programme because it recovers all mathematics that is necessary for the purposes of science, i.e. all of that portion of mathematics that is an ineliminable component of our scientific worldview, and thus accounts for the central place that mathematics occupies in our best science.[21] If we accept this correction to the second criterion, according to which a revisionist account of mathematics would be one that requires us to abandon accepted theorems that are important to science, then predicativism can be considered as succeeding in recovering classical mathematics. It then follows that predicativism does not constitute a revisionist account of mathematics because it does not prescribe the abandonment of previously accepted classical theorems (that are important to do science). On the third criterion, on the other hand, predicativism would count as a revisionist programme. Interpreted as an account of mathematics, it can be seen as showing that ordinary mathematical practice can forego the use of impredicative definitions. This could generate a prescription to change mathematical practice to eliminate the use of impredicative definitions altogether. However, the predicativist is not bound to issue such a prescription for radical methodological change in the practice. The predicativist is in a position to acknowledge that the abandonment of impredicative definitions would overly complicate everyday mathematical activity, and argue that while pragmatically useful and thus widely used in ordinary mathematics for convenience, impredicative definitions are provably unnecessary to mathematical practice.

On the proposed revised criterion, this move would still qualify as a revisionist proposal. Naturalistic attention to the practice of science tells us that it is the methodology of a successful scientific practice that is at the heart of the success of the discipline. If mathematicians did not take the results of the predicativist account of mathematics as a sufficient reason to dispense with the use of impredicative definitions in everyday mathematical practice, we should take this to show that the use of impredicative definitions is indeed part of those established proving practices that make mathematics successful, contrary to what the philosophical and formal outcomes of the predicativist programme seem to show. Hence, from a mathematical naturalist point of view, predicativism can be construed as prescribing a revision to established proving practices even in the version of the programme which does not

[21] For considerations suggesting that predicative mathematics does suffice to recover scientifically indispensable mathematics, see chapter 14 of Feferman (1998).

prescribe a revision to mathematical practice. It follows that predicativism would be identified as a revisionist approach on the basis of the the revised naturalist criterion, even though it would arguably not be considered revisionist on the basis of the second criterion.

The discussion of predicativism shows that the revised criterion does capture certain types of revision which the mathematical naturalist should care about, even though the non-naturalistic anti-revisionist might not have any reason to. In conclusion, I will argue that the revised criterion is not too restrictive in two different senses. Firstly, it leaves room to allow for certain methodological revisions to the practice that do not lead to the abandonment of accepted theorems—in other words, it does not require the mathematical naturalist to maintain a dogmatic reverence towards established mathematical methodology. Secondly, it tolerates revisions to aspects of mathematical practice that do not lead to the abandonment of accepted theorems, provided that they do not betray established mathematical practice.

In order to illustrate why the revised criterion leaves space for criticisms of established mathematical methodology it is useful to consider another case study, namely Bolzano's criticism of his contemporary fellow mathematicians. Bolzano's view on the proper methods of proof is inspired by a substantive philosophical conception of the notion of explanatory proof and grounding, which is embedded in the context of a larger axiomatic conception of science, the proper order of truths, and proper scientific method. In particular, Bolzano criticised his contemporaries for giving proofs that, while establishing the certainty of the result, fail to provide the grounds for the truth in question, or rather reverse the correct order of ground and consequence. The proper order of exposition of knowledge is one that respects the structure of science by ordering truths according to the objective ground-consequence structure: the direction of explanation is from the more complex (*explanandum*) to the simpler (*explanans*), according to a specific analysis of generality and simplicity (for more details, see Kitcher 1975, Mancosu 1999, and Betti 2010). Bolzano extensively discusses the case of the intermediate value theorem, an analytic theorem that is most often proved by appealing to geometrical facts. This practice is flawed because analytical truths are more general than geometrical truths, and thus such proofs reverse the proper order of ground and consequence by proving more general truths by appeal to less general truths (Mancosu 1999, p. 438, Detlefsen 2008, pp. 183–186).[22]

It should be pointed out that Bolzano is not considered to be a revisionist about mathematical practice: he certainly never advocated the abandonment of any accepted theorems. Bolzano's attitude towards the methodological practice of his contemporaries, however, seemingly qualifies as a revisionist stance according to the revised criterion proposed in Sect. 9.5. At a closer inspection though, Bolzano is not prescribing the abandonment of any established method of proof on the

[22] Mancosu points out that mathematical practice more generally was influenced by views of explanatory vs. non-explanatory proofs, and that this often led advocates of specific conceptions to criticise methods of proof that were common practice at the time (Mancosu 2008, 141–142).

grounds that such method, albeit commonly used by mathematicians, is incorrect. What Bolzano is advocating is a more rigorous use of geometrical methods in mathematical proving practice, not an abandonment of geometrical methods. There is therefore an asymmetry with the predicativism case study discussed above, where the abandonment of impredicative definitions is prescribed on the grounds that they are incorrect because viciously circular. We could make more precise sense of Bolzano's attitude as a revisionist version of "methodological perfectionism" by echoing Burgess's terminology (Burgess 2005, pp. 738–739). Mathematical practice abounds with geometrical proofs of analytical theorems, but in principle it ought not to do so. This is a perfectionist view of mathematical methodology that does not call for the rejection of any previously accepted theorem nor of any previously sanctioned method of proof, but merely advocates the superiority, for example, of purely analytical proofs of analytical theorems over geometrical proofs of analytical theorems. This case study shows that the revised criterion allows the mathematical naturalist to make room for calls to change existing methodology while doing justice to the cumulative character of the development of mathematical knowledge. Geometrical methods of proof sit squarely among the sanctioned methods of proof in mathematical practice; nonetheless, according to Bolzano, the rigorisation of analysis is better served by using those methods in the context of proving geometrical, rather than analytic theorems.

Finally, I want to argue that the revised criterion is liberal with respect to revisions that do not fall under the two kinds specified (abandonment of established methodology, abandonment of accepted theorems). As discussed in Sect. 9.4, there are many aspects of mathematical practice that one can be revisionist about, and as argued in Sect. 9.5, for the mathematical naturalist established methodology and ordinary mathematical theorems are not two of those. What is the mathematical naturalist's attitude with respect to other kinds of revisionism? The answer will have to be evaluated on a case-by-case basis: recall that the naturalist does not look favourably upon preconceived philosophical attitudes that do not take mathematical practice as the starting point.

The case I want to briefly examine here is that of some contemporary antirealist programmes, which include different kinds of mathematical nominalism. The approaches in this camp provide accounts of ordinary mathematics that are often explicitly motivated by naturalistic and anti-revisionist considerations. Consider for example mathematical fictionalism. In some versions (for example, the one proposed by Leng 2010), it holds that our best science does not warrant belief in the existence of the mathematical objects seemingly posited by ordinary mathematical theories. Mathematical practice is viewed as an investigation into the consequences of certain assumptions. Whether these assumptions are true, and whether we believe that they are, does not make a difference to the outcome of this investigation. In fact, neither mathematical practice nor scientific practice require the literal truth of mathematical statements or our belief in their truth; the successful applications of mathematics in science can be explained by assuming that the mathematical objects that figure in scientific theories are merely useful fictions. In order to hold that only some uses of "there is" are ontologically committing, the fictionalist needs to provide a nominalistic reinterpretation of all the statements involving apparent reference to mathematical objects. This view is explicitly motivated by a naturalistic trust in the methods of our best scientific theories as being the best methods we have to obtain knowledge of the world.

Since it denies the existence of mathematical objects and it claims that mathematical statement should not be taken at face-value to refer to mathematical objects, fictionalism appears to be a revisionist position about ontology and semantics. Some naturalists have explicitly argued along these lines:

> Having studied Euclid's Theorem, we are prepared to say that there exist infinitely many prime numbers. Moreover, when we say so, we say so without conscious mental reservations or purpose of evasion. [...] For those of us for whom something like this is the starting-point, any form of nominalism will have to be revisionary, and any revision demands motivation. (Burgess and Rosen 1997, pp. 10–11)

Along different lines but in a somewhat similar spirit, Parsons (2008, p. xv, p. 2) argues that we do not have sufficiently compelling reasons for abandoning current mathematics or for denying the truth of mathematical axioms and theorems. Giving an account of mathematics as engaged in "speaking of remarkable configurations of objects that just do not exist [...] would leave the existence of mathematics and its importance to science and practical life a much greater mystery than its objects are at the outset." How can these charges of revisionism be reconciled with the anti-revisionist tenet at the core of mathematical naturalism?

Through the lens of the revised naturalistic criterion for anti-revisionism, a view such as Leng's fictionalism is not revisionist about the aspects of the practice that matter the most to the naturalist: mathematical fictionalism does not require the abandonment of accepted theorems nor the rejection of established methodology. In other words, fictionalism leaves ordinary mathematics as it is: no revision is prescribed to the way in which mathematics is done. This assessment fits with the prevailing attitude in the naturalistic debates. Despite some charges of revisionism, nominalistic accounts of various kind occupy an important place among contemporary naturalistic accounts of mathematics, all of which take anti-revisionism to be a central component of their view. This considerations do not, of course, go anywhere near answering the question whether naturalistic accounts of mathematics should adopt some form of realism about mathematics, which ultimately rests on more complex considerations than those which can be discussed here.

From this perspective, we can appreciate how mathematical naturalists are strong critics of revisionist foundational programmes, but less critical of philosophical accounts of the practice that revise e.g. the ontology and semantics of mathematics. From the point of view of recovering classical theorems (the second anti-revisionist criterion discussed in Sect. 9.4), foundational programmes like predicativism fare reasonably well, almost as well as approaches that do not prescribe any revisions to mathematical practice at all. From the mathematical naturalist point of view, on the other hand, approaches such as fictionalism, which imply a radical revision of both the ontology and semantics of mathematics, fare much better than programmes like predicativism, which imply a far less radical revision of mathematical ontology, but prescribe the abandonment of an established method of proof in classical mathematics. In this sense, even the revised criterion for anti-revisionism leaves considerable room for both realist and anti-realist naturalistic accounts of mathematical practice.

## References


Antonutti Marfori, M. (2010). Informal proofs and mathematical Rigour. *Studia Logica, 96*(2), 261–272.

Antonutti Marfori, M. (2012). Naturalising mathematics: A critical look at the Quine-Maddy debate. *Disputatio, IV*(32), 323–342.

Antonutti Marfori, M. (2020). A new look at Quine on set theory. In F. Janssen-Lauret (Ed.), *Quine, structure, and ontology*. Mind association occasional series (pp. 253–279). Oxford University Press.

Betti, A. (2010). Explanation in metaphysics and Bolzano's theory of ground and consequence. *Logique et Analyse, 53*(211), 281–316.

Burgess, J. P. (1983). Why i am not a nominalist. *Notre Dame Journal of Formal Logic, 24*(1), 93–105.

Burgess, J. P. (2005). No requirement of relevance. In Shapiro (2005) (pp. 727–750).

Burgess, J. P., & Rosen, G. (1997). *A subject with no object: Strategies for nominalistic interpretation of mathematics*. Oxford University Press.

Clark, K. J. (Ed.) (2016). *The Blackwell companion to naturalism*. Blackwell.

Daly, C., & Liggins, D. (2011). Deferentialism. *Philosophical Studies, 156*, 321–337.

Detlefsen, M. (2008). Purity as an ideal of proof. In Mancosu (2008) (pp. 179–197).

Dieterle, J. M. (1999). Mathematical, astrological, and theological naturalism. *Philosophia Mathematica, 7*(2), 129–135.

Dreben, B., & Kanamori, A. (1997). Hilbert and set theory. *Synthese, 110*(1), 77–125.

Ewald, W. B., & Sieg, W. (Ed.) (2013). *David Hilbert's lectures on the foundations of arithmetic and logic 1917–1933*. Springer.

Feferman, S. (1964). Systems of predicative analysis. *The Journal of Symbolic Logic, 29*(1), 1–30.

Feferman, S. (1998). *In the light of logic*. Logic and computation in philosophy. Oxford University Press.

Fogelin, R. (1997). Quine's limited naturalism. *The Journal of Philosophy, 94*(11), 543–563.

Gödel, K. (1944). Russell's mathematical logic. In P. A. Schilpp (Ed.), *The philosophy of Bertrand Russell* (pp. 125–153). Northwestern University.

Habgood-Coote, J., & Tanswell, F. S. (2021). Group knowledge and mathematical collaboration: A philosophical examination of the *classification of finite simple groups. Episteme, 20*(2), 281–307.

Hansson, S. O. (2021). Science and pseudo-science. In E. N. Zalta (Ed.), *The Stanford encyclopedia of philosophy* (Fall 2021 ed.). Metaphysics Research Lab, Stanford University.

Hilbert, D. (1890). Ueber die Theorie der algebraischen Formen. *Mathematische Annalen, 36*(4), 473–534.

Hilbert, D. (1928). Die Grundlagen der Mathematik. *Abhandlungen aus dem Seminar der Hamburgischen Universität, 6*, 65–85. Reprinted in Ewald and Sieg (2013, pp. 917–942). English translation in van Heijenoort (1967, pp. 464–479).

Iemhoff, R. (2020). Intuitionism in the philosophy of mathematics. In E. N. Zalta (Ed.), *The Stanford encyclopedia of philosophy* (Fall 2020 ed.). Metaphysics Research Lab, Stanford University.

Jerkert, J. (2013). Why alternative medicine can be scientifically evaluated. countering the evasions of pseudoscience. In M. Pigliucci, & M. Boudry (Eds.), *Philosophy of pseudoscience. reconsidering the demarcation problem* (pp. 305–320). Chicago University Press.

Kitcher, P. (1975). Bolzano's ideal of algebraic analysis. *Studies in History and Philosophy of Science Part A, 6*(3), 229–269.

Kitcher, P. (1985). *The nature of mathematical knowledge*. Oxford University Press.

Klein, F. (1926). *Vorlesungen Über die Entwicklung der Mathematik im 19. Jahrhundert. Die Grundlehren der mathematischen Wissenschaften in Einzeldarstellungen mit besonderer Berücksichtigung der Anwendungsgebiete* (Vol. I). Springer.

Leng, M. (2010). *Mathematics and reality*. Oxford University Press.

Maddy, P. (1997). *Naturalism in mathematics*. Oxford University Press.

Maddy, P. (1998). Naturalizing mathematical methodology. In M. Schirn (Ed.), *The philosophy of mathematics today* (pp. 175–193). Clarendon Press.

Maddy, P. (2000). Naturalism and the a priori. In P. Boghossian, & C. Peacocke (Eds.), *New essays on the a priori* (pp. 92–116). Clarendon Press.

Maddy, P. (2005). Three forms of naturalism. In Shapiro (2005) (pp. 437–459).

Maddy, P. (2007). *Second philosophy: A naturalistic method*. Oxford University Press.

Maddy, P. (2011). *Defending the axioms*. Oxford University Press.

Mancosu, P. (1999). Bolzano and Cournot on mathematical explanation. *Revue d'histoire des Sciences, 52*(3–4), 429–456.

Mancosu, P. (Ed.) (2008). *The philosophy of mathematical practice*. Oxford University Press.

McLarty, C. (2012). Theology and its discontents: David Hilbert's foundation myth for modern mathematics. In A. Doxiadis, & B. Mazur (Eds.), *Circles disturbed: The interplay of mathematics and narrative* (pp. 105–129). Princeton University Press.

Moore, G. H. (1982). Zermelo's axiom of choice: Its origins, development, and influence. Springer-Verlag.

Noether, M. (1914). Paul Gordan. *Mathematische Annalen, 75*(1), 1–41.

Parsons, C. (2008). *Mathematical thought and its objects*. Cambridge University Press.

Posy, C. (2005). Intuitionism and philosophy. In Shapiro (2005) (pp. 318–355).

Quine, W. V. (1981). *Theories and things*. Harvard University Press.

Rathjen, M., & Weiermann, A. (1993). Proof-theoretic investigations on Kruskal's theorem. *Annals of Pure and Applied Logic, 60*(1), 49–88.

Resnik, M. D. (2005). Quine and the web of belief. In Shapiro (2005) (pp. 412–436).

Rittberg, C. J. (2021). Intellectual humility in mathematics. *Synthese, 199*, 5571–5601.

Roland, J. W. (2016). Naturalism and mathematics: Some problems. In Clark (2016) (pp. 289–304).

Schütte, K. (1964). Eine Grenze für die Beweisbarkeit der transfiniten Induktion in der verzweigten Typenlogik. *Archiv fur Mathematische Logik und Grundlagenforschung, 7*, 45–60.

Schütte, K. (1965). Predicative well-orderings. In J. N. Crossley, & M. Dummett (Eds.), *Formal systems and recursive functions: Proceedings of the eighth logic colloquium, Oxford, July 1963. Studies in logic and the foundations of mathematics* (Vol. 40, pp. 280–303). North-Holland.

Shapiro, S. (1997). *Philosophy of mathematics: Structure and ontology*. Oxford University Press.

Shapiro, S. (2000). *Thinking about mathematics. The philosophy of mathematics*. Oxford University Press.

Shapiro, S. (Ed.) (2005). *The Oxford handbook of philosophy of mathematics and logic*. Oxford University Press.

Simpson, S. G. (1985). Nonprovability of certain combinatorial principles of finite trees. In L. A. Harrington, M. D. Morley, A. Scedrov, & S. G. Simpson (Eds.), *The philosophy of mathematical practice. Studies in logic and the foundations of mathematics* (Vol. 117, pp. 87–117). North-Holland.

Smith, K. (2012). Homeopathy is unscientific and unethical. *Bioethics, 26*(9), 508–512.

Urquhart, A. (2008). Mathematics and physics: Strategies of assimilation. In Mancosu (2008) (pp. 417–440)

van Atten, M. (2017). The development of intuitionistic logic. In E. N. Zalta (Ed.), *The Stanford encyclopedia of philosophy* (Spring 2017 ed.). Metaphysics Research Lab, Stanford University.

van Heijenoort, J. (Ed.) (1967). *From Frege to Gödel: A source book in mathematical logic, 1879–1931*. Harvard University Press.

Weir, A. (2005). Naturalism reconsidered. In Shapiro (2005) (pp. 460–482).
Weir, A. (2014). Quine's naturalism. In G. Harman, & E. Lepore (Eds.), *A Companion to W. V. O. Quine. Blackwell companions to philosophy* (pp. 114–147). Blackwell.
Weyl, H. (1918). *Das Kontinuum: Kritische Untersuchungen über die Grundlagen der Analysis*. Veit.
Weyl, H. (1994). *The continuum: A critical examination of the foundation of analysis*. Dover.
Whitehead, A. N., & Russell, B. (1910). *Principia mathematica* (Vol. I, 1st ed.). Cambridge University Press.
Zach, R. (2019). Hilbert's program. In E. N. Zalta (Ed.), *The Stanford encyclopedia of philosophy* (Fall 2019 ed.). Metaphysics Research Lab, Stanford University.

# Chapter 10
# Perspectives on the Universe

**Graham Priest**

**Abstract** Joel Hamkins has advanced a well known view to the effect that there is no unique universe of sets. There is simply a plurality of such universes. We have, then, a pluriverse. A natural objection to this view is that there is still a single universe: the totality, $V$, in which all the members of the pluriverse find themselves. In this paper I consider a reply to the objection, to the effect that there is no such thing as $V$ in itself. Rather, each member of the pluriverse simply gives a different perspective on what $V$ is like. This view is then generalised in the light of mathematical pluralism. What emerges is a vastly expanded and logic-neutral view of the pluriverse.

**Keywords** Joel Hamkins · Set theoretic pluriverse · Perspectivalism · LP · Mathematical pluralism · Relativism

*Dedication* It gives me enormous pleasure to dedicate this essay to Alan Weir. Though we have never spent a lot of time together, our paths have crossed on many occasions, both in person and in print. Alan's work is always imaginative and insightful, and he is never afraid to beat paths in novel directions. As well as being a great philosopher, when it comes to giving talks, he is undoubtedly the funniest person I know. It is impossible to leave a talk by Alan without a smile on one's face. I have no doubt that he would have some words of wry humour about the following thoughts!

---

G. Priest (✉)
Departments of Philosophy, CUNY Graduate Center, New York, NY, USA

University of Melbourne, Parkville, VIC, Australia

Ruhr University of Bochum, Bochum, Germany

 
A. Rieger, S. Leuenberger (eds.), *Themes from Weir: A Celebration of the Philosophy of Alan Weir*, Synthese Library 484,
https://doi.org/10.1007/978-3-031-54557-3_10

## 10.1 Introduction

Perspectivalism about some situation is, roughly, the view that there is no fact of the matter about it, *simpliciter*. There are just the views afforded by different perspectives (literal or metaphorical), all of them equally correct. It is common enough to endorse perspectivalism about whether a joke is funny or a dish is tasty. Such things are the case only with respect to a perspective afforded by a sense of humour or a gastronomic palate, respectively. It is less common to apply it to areas where we would standardly take it that there is indeed an objective fact of the matter. The point of this essay is to do just this. In what follows, I will apply the view to certain aspects of mathematics—an area normally taken to be a paradigm of objectivity.

The essay falls into two (related) parts. The first concerns the matter of whether there is a single universe of sets or a plurality thereof. The discussion here will naturally take us into the second half of the essay, where the issue will be subjected to concerns drawn from mathematical pluralism, the view that there are different kinds of mathematics which are all equally correct.

## 10.2 The Pluriverse

### *10.2.1 The Set-Theoretic Universe*

A fairly orthodox view—at least until the last couple of decades—is that there is a unique universe of sets, $V$, and that this is characterised, at least partially, by the axioms of Zermelo Fraenkel set theory (with Choice), $ZFC$. (Of course, $V$ is not itself a set in the totality, but in some undeniable sense, it is the totality of all sets.) The axioms of $ZFC$ are insufficient to settle fundamental questions about sets, such as the Continuum Hypotheses ($CH$); but the answers to these questions are determinately true or false, none the less.

In the last dozen years or so, a rival view has emerged. There is no unique such universe. Instead, there is a plurality of universes. Each of them may satisfy the axioms of $ZFC$, but the $CH$ (say) is true in some and false in others. As Joel Hamkins, one of the main defenders of this picture, puts it:[1]

> There are diverse distinct concepts of set, each instantiated in a corresponding set-theoretic universe, which exhibit diverse set-theoretic truths. Each such universe exists independently in the same Platonic sense that proponents of the universe regard their universe to exist.

[1] Hamkins (2012, p. 416 f). See this essay for a defence of the view.

Hamkins calls this totality of universes the *multiverse*.[2] I prefer to use the term *pluriverse*. (As we will see, it's going to get much bigger!)

The pluriverse view is driven by the fact that attempts to determine the answers to the questions left open by $ZFC$ in any satisfactory way have proven largely unsuccessful—at least as judged by any substantial consensus. On the other hand, set-theorists have a large experience of constructing and investigating different models of $ZFC$ and the relationships between them. It is natural enough, then, to think of each such model as delivering its own universe; and each seems to be an equally valid realisation of the structure of sets. Hamkins calls these different *concept(ion)s* of sets. It might be better to say that we have but a single conception of set here (the $ZFC$ conception), and that these are different *realisations* of it—in the same way that different mathematical groups deliver different realisation of one and the same notion of *group*. However, I will return to the topic of different conceptions of sethood in due course.

Natural as this picture of sets may seem in the light of developments in set-theory, it appears to face an obvious objection. The different denizens of the pluriverse would seem to inhabit the totality of *all* such universes, which, therefore, has a claim to be *the* universe of sets. A defender of the more traditional view will point out that all the models of $ZFC$ that have been constructed appear to live within this, the cumulative hierarchy, $V$, as usually conceived.[3] Indeed, when the set-theorist is constructing and establishing the relations between the models of $ZFC$, they actually appear to be working *within* this (hyper-)universe. The situation, then, would appear to be this:

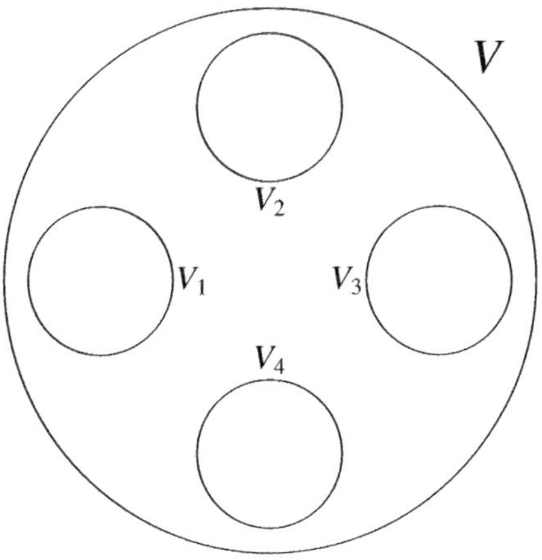


[2] As is clear, Hamkins subscribes to a Platonism concerning the universes. However, the pluriverse view is not, in itself, committed to platonism, and can be endorsed by various kinds of non-platonists as well.

[3] Though of course one cannot *prove* that there are such models in $ZFC$ without adding extra axioms, such as large cardinal axioms.

$V$ is the unique (hyper-)universe. Each of the $V_i$s is simply some sub-collection which happens to validate all the axioms of $ZFC$. The pluriverse view, then presupposes that there is a unique universe, and so seems to be self-refuting.

### *10.2.2 Enter Perspectivalism*

There is, however, a radical but intriguing reply. There is no (hyper-)universe as such: each member of the pluriverse has a universe, and these all provide equally valid perspectives on what the totality of all sets is like. This is perspectivalism.[4]

Perspectivalism about some matter is the view that there is no perspective-independent truth about it; we simply have a bunch of different perspectives, each of which is equally legitimate. As I observed, perspectivalism is the natural view about the funniness of a joke or the tastiness of a food. There is no truth *simpliciter* about such a matter. Truth is relative to a perspective. So the claim 'such and such a joke is funny' makes no more sense that 'this is to the right of that'. The latter makes sense only with respect to a visual perspective; the former only with respect to a sense of humour. What *are* true or false *simpliciter* are things of the form 'with respect to such and such a spatial position this is to the right of that' and 'with respect to a certain sense of humour this a is funny joke'. Note that such perspectival truths are not subjective, at least in one sense, since different people can occupy the same spatial position (at different times), and different people can have the same sense of humour.

Perspectivalism about spatial orientation and humour are a forms of local perspectivalism. They apply only to certain properties of certain things. But global perspectivalism, according to which *every* claim is true or false only with respect to a perspective, is also well known in philosophy. It is a view often attributed to Nietzsche.[5] Interpreting Nietzsche is always a contentious matter. However, it is not contentious that Indian Jain philosophers held a global perspectivalism. For the Jains, reality was like a cut diamond with many facets. Each facet encodes a different aspect of reality, all of equal reality, and any claim is true or false only with respect to one facet.[6]

---

[4] The view was suggested to me by Joel Hamkins in conversation. As far as I know, he has not endorsed it explicitly in print, though he gestures towards it here (Hamkins 2012, p. 419): 'On the universe view, of course, forcing extensions of $V$ are deemed illusory, for $V$ is already everything, while the multiverse perspective regards $V$ as a relative concept, referring to whichever universe is currently under consideration, without there being any absolute background universe. On the multiverse view, the use of the symbol $V$ to mean "the universe" is something like an introduced constant that might refer to any of the universes in the multiverse, and for each of these the corresponding forcing extensions $V[G]$ are fully real.'

[5] See Anderson (2017, esp. §6.2).

[6] See, e.g., Priest (2008).

Of course, given perspectivalism about something, it is a fair question to ask what it is that the perspectival properties are properties *of*. For a local perspectivalism about, e.g., jokes, it is easy enough to answer this. They are all perspectives of the story told, the picture shown, etc. In the case of global perspectivalism, things are more problematic. To say anything about the object in question is but to provide a perspective of it. One can say nothing of the thing itself. It thus becomes a *ding an sich*, a "something I know not what".

A further question concerns the relationship of perspectivalism to fictionalism. Fictionalist accounts are of many kinds.[7] However, the basic idea of fictionalism about some topic is that there are no truths *simpliciter* about it. All truths are of the form 'According to fiction $F$, such and such'. Clearly, 'According to fiction $F$, such and such' and 'According to perspective $P$, such and such' at least have a similar syntactic form. There may, in fact, be no formal difference between the two operators. However, the intents of fictionalism and perspectivalism are quite distinct. Fictionalism is an anti-realist view: the objects in question are, after all, *fictions*. Thus, if one is a fictionalist about numbers, it is not true that $1 + 1 = 2$. What is true is that in the "fiction" about numbers $1 + 1 = 2$. And that delivers no commitment to the existence the number 1, any more than Conan Doyle's stories deliver a commitment to the existence of Sherlock Holmes. By contrast, perspectivalism is not an anti-realist view. If a joke is funny from a perspective, or an apple is tasty from a perspective, there is no suggestion that the joke or the apple does not exist—quite the contrary.

### *10.2.3 Perspectivalism and the Pluriverse*

Let us apply a perspectivalist view to the multiverse. All claims about $V$ are relative to a certain perspective, and each perspective is perfectly legitimate. In particular, each member of the pluriverse, $i$, has its own take on the universe of sets, $V_i$, and each of these provides an equally correct perspective on the object of the perspective. Of course, this raises the question of what that thing is. Matters are not so straightforward as in the case of a humour. But since this is not a global perspectivalism, we are not reduced to taking it to be an ineffable *ding an sich*. We may take the thing to be that mathematical object which is the totality of all collections, though this has no intrinsic structure. Supposing it had such a structure would be to take us back to a determinate-universe view, which is exactly what perspectivalism was invoked to avoid.[8]

---

[7] See Eklund (2019).

[8] Another possibility is to adopt a conceptualist position, bringing matters of representation to bear. The object in question is the *concept* of the totality of all collections. One may take this to be vague, in that it can be precisified in many different ways, one for each perspective. But, again, it is crucial that the concept does not deliver any determinate structure. Such must remain a matter of perspective.

Let us write the target of the perspectives as $\mathfrak{V}$. The $V$ of each member of the pluriverse gives a perspective on this. The situation may be depicted as follows:

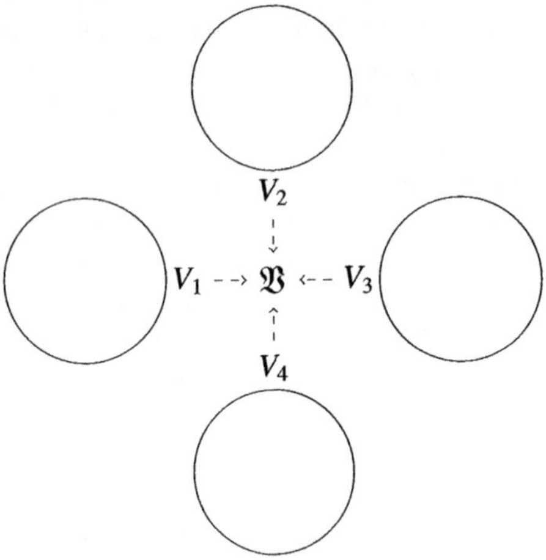


The different perspectives may well agree on many of the properties of $V$. For example, according to each of them, it satisfies the axioms of $ZFC$. Thus, when a set-theorist proves things using the axioms of $ZFC$, they establish something common to all perspectives. But other properties, such as whether the $CH$ holds, may vary from perspective to perspective—the perspectives afforded by each of member of the pluriverse, as encoded in its own take on matters. Each of these perspectives is equally legitimate: there is no hyper-universe as such—in particular, no hyper-universe which contains all the others.

The differences in perspective can be more radical than so far indicated. Let us call something *absolute* if it holds in all universes/perspectives. Then that the totality of sets is uncountable is absolute. But suppose, as can happen, that we have two members of the pluriverse, $V_1$ and $V_2$, such that $V_1$ is a substructure of $V_2$. $V_1$ takes itself to be uncountable. But according to $V_2$, $V_1$ may be countable. We might depict this as follows:

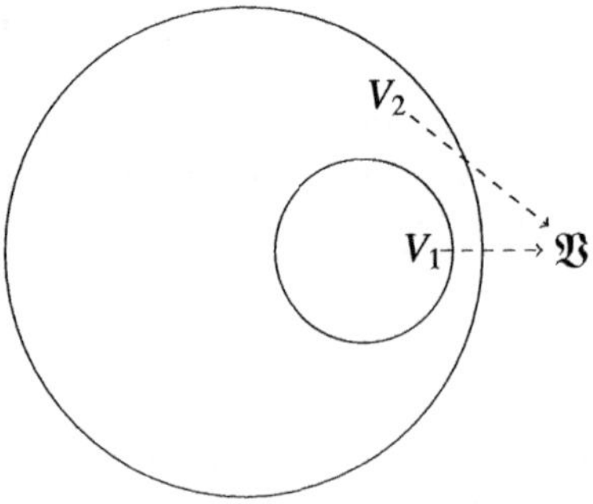


From the perspective of $V_2$, $V_1$'s perspective of $\mathfrak{V}$ will be but partial.

Such, then is a perspectivalist view about the universe of sets. If one endorses the pluriverse, and if perspectivalism does provide a solid reply to the obvious objection, this will speak in its favour. However, the aim of this paper is not to defend perspectivalism about the cosmos of sets as such. The aim of the present essays is much more modest: simply to explore the idea, and see where else it may take us.

### 10.2.4 *Changing the Logic, Not the Subject*

One further step is natural and obvious to those familiar with paraconsistent set theory.

Given the pluriverse understanding of set-theory, there is nothing sacrosanct about $ZFC$. We can take the pluriverse to contain things that do not validate all of the axioms of $ZFC$ as Hamkins notes:[9]

> The background idea of the multiverse, of course, is that there should be a large collection of universes, each a model of (some kind of) set theory. There seems to be no reason to restrict inclusion only to $ZFC$ models, as we can include models of weaker theories $ZF$, $KP^-$, $KP$, and so on, perhaps even down to second-order number theory, as this is set-theoretic in a sense. At the same time, there is no reason to consider all universes in the multiverse equally, and we may be simply more interested in the parts of the multiverse consisting of universes satisfying very strong theories, such as $ZF$ plus large cardinals.

We can indeed consider stronger theories as well, and not just those of the kind that Hamkins notes. Using various model-theoretic techniques, one can construct set-theoretic models which validate all the theorems of $ZFC$, together with the naive comprehension principle ($CP$):[10]

- $\exists x \forall y (y \in x \equiv A(y))$

In such models, in particular, the universal set, $V$, will actually be a member of the totality. That is, $V \in V$. Obviously, the usual models of $ZFC$ do not verify this. The reason, of course, is that $CP$ is inconsistent with $ZFC$. Indeed, the set of things true in each of the models in question *is* inconsistent, though not trivial. This is because the underlying logic of these models is not classical logic, but the paraconsistent logic, $LP$.—Actually, the trivial model is one of these; but, as Hamkins notes, there is no reason why we have to find all members of the pluriverse (equally) interesting.[11]

---

[9] Hamkins (2012, p. 436).

[10] For these models, see Priest (2006, §18.4), and Priest (2017, §11).

[11] Naturally, one may ask in what theory these models are constructed. The answer is that they are constructed in $ZFC$. Clearly, there is an element of boot-strapping here. But boot-strapping is everywhere in set theory. Thus, for example, to prove that the Gödel's constructible hierarchy, $L$, is a model of the axioms of $ZFC$ requires the very axioms of $ZFC$.

Moving to this paraconsistent logic actually *expands* the pluriverse, since all classical models are $LP$ models. It still contains all the classical universes verifying $ZFC$, but it now also contains all the $LP$ $ZFC$ universes. Indeed, we may throw in as well, for good measure, all the $LP$ models of naive set theory which do not verify the theorems of $ZFC$. As Hamkins notes, we do not have to fetishize $ZFC$.

Nor is there any reason why the pluriverse *should* be constrained by the straight-jacket of classical logic. The non-classical constructions deliver models of set theory which have clear mathematical interest.[12] And each provides a perspective on the totality of sets—indeed, an interesting and novel kind of perspective!

Of course, it might well be suggested that these paraconsistent perspectives on $V$ are not legitimate because they use the wrong logic. We need not go into this here, however, since the matter is finessed by a consideration of mathematical pluralism, as we will now see.

## 10.3 ...and Mathematical Pluralism

### *10.3.1 Mathematics (Pl.)*

A common enough assumption—again until relatively recently—is that there is one correct kind of mathematics, and that this is based on so called classical logic—the logic developed by Frege, Russell, and others around the turn of the twentieth Century to do justice to the reasoning in the mathematics of their time.

The classical pluriverse does not, as such, threaten the picture. Set theory has to be reconceptualised as a theory like ring theory or topology, rather than like the theory of the natural numbers. The notion of a ring or a topological space is not meant to be categorical: it can have multiple realisations. By contrast, the theory of the natural numbers is standardly taken to have a unique intended realisation.[13]

Nor need the move to the paraconsistent logic $LP$ threaten the claim that there is a uniquely correct logic for mathematics. This may not be classical logic, but is more general. (Classical logic can be obtained from $LP$ by adding the principle of Explosion; and this may be taken to be a "contingent" feature of some mathematical structures.)

However, one thing that recent developments in mathematics has taught us is that interesting mathematical structures may be formed on the basis of a number of different non-classical logics. Thus, there are not only the various paraconsistent theories, such as set theory and topology, but intuitionist theories such as smooth

[12] The relationships between these and the classical models is also a topic which requires much further investigation.

[13] However, because of the supposed foundational role of set theory in mathematics, the plurality of set-theoretic universes challenges the supposed categorical nature of other structures, such as the natural numbers, as Hamkins (2012, §5), notes. See also Gitman and Hamkins (2010).

infinitesimal analysis and Heyting Arithmetic plus Church's Thesis, which would collapse into triviality if classical logic were used. Hence, we have seen a number of authors defending the notion of mathematical pluralism.[14] There are many kinds of pure mathematics. In particular, there are those based on various non-classical logics. This is not to say that all such mathematics are equally interesting, rich, applicable, etc—just that they are all equally valid pure mathematical structures.[15]

It should be stressed that what is at issue here is *pure mathematics*. All of these pure mathematical structures are equally legitimate. Matters are different when it comes to applied mathematics. When a piece of mathematics is applied for some purpose (in physics, economics, linguistics, or whatever), one must find the mathematical domain whose structure is isomorphic (or near enough isomorphic) to the "real world" system in question, so that establishing results about the mathematical structure tells us about this as well.[16] The structure of the real-world system, hence, imposes constraints on the choice of the pure mathematical structure in question; all are not, therefore, equally legitimate.

### *10.3.2 The Real Pluriverse*

Given mathematical pluralism, and the fact that we have already expanded the classical pluriverse once to obtain one in which there are universes whose underlying logic is—of necessity—one particular non-classical logic, it is natural to expand it further to universes of sets whose underlying logics are other non-classical logics.

Thus, one might have a universe of predicatively definable sets; a universe of sets with the set theoretic axioms of $ZFC$, but the underlying logic of which is intuitionist logic; or a set-theory based on the intuitionistic notion of a spread; or a fuzzy set-theory based on, say, Łukasiewicz continuum-valued logic; or a set-theory based on a relevant logic; or one based on quantum logic.[17] Alan Weir, of course, has his own version of set theory with the naive comprehension principle.[18] This would fit snugly into this framework—though to what extent he would accept the picture presented here, I leave it to him to say.

---

[14] See, e.g., Shapiro (2014), Priest (2013), and Priest (2019). As the last of these argues, the pluralism seems to dispose of any last vestiges of foundationalism in mathematics—in the sense that there is some ur-mathematics, which encompasses all of its branches—and so of set-theoretic reductionism.

[15] This does not, note, imply logical pluralism. The logical structures, and so the mathematical structures built on them, have clear mathematical interest, even if the logics are not the "one true logic". See Priest (2021).

[16] See Priest (2016, §7.8).

[17] See, e.g., Feferman (1964), Crosilla (2019), Van Atten (2017, §3), Hájek and Haniková (2003), Weber (2012), Takeuti (1981), respectively.

[18] See Weir (1998). The underlying logic has been something of a work in progress. For the most recent account of which I am aware, see Weir (2015).

I think it fair to say that at least some of these theories of sets *are* different conceptions of sethood. But once one is a mathematical pluralist, there would appear to be no reason why the set-theoretic pluriverse should not contain all of them. Indeed, once one has given up the thought that there is a unique universe of sets, the very spirit of the enterprise is one of pluralism; and one should not put arbitrary limits on the plurality.

Of course, the pluriverse is now amazingly diverse, and in practice one might expect mathematicians to be interested in only fragments of it, e.g., those universes based on classical logic, or those universes verifying the $CP$. But no mathematician is interested in *all* of mathematics.

### 10.3.3 The Structure of the Pluriverse

So we now have a large plurality of universes. Each universe, $\mathfrak{U}$, will have its own logic, $L_{\mathfrak{U}}$. Different things may hold in each $\mathfrak{U}$; so let us write $\mathfrak{U} \Vdash A$ to mean that the sentence $A$ holds in $\mathfrak{U}$.[19] Then the set, $\Sigma_{\mathfrak{U}} = \{A : \mathfrak{U} \Vdash A\}$ will be closed under $L_{\mathfrak{U}}$. Set theorists may investigate each $\Sigma_{\mathfrak{U}}$; but of course, they will want to do more than that: they will want to investigate the relationships between the different universes.[20] For such an investigation one needs an account of the the properties of, and relations between, different universes. In other words, one needs a set theoretic structure in which the universes are themselves objects. The question is: which one?

Hamkins himself notes that his multiverse view provides a natural take on this matter. He says:[21]

> The multiverse perspective ultimately provides what I view as an enlargement of the theory/metatheory distinction. There are not merely two sides for this distinction, the object theory and the metatheory, but rather there is a vast hierarchy of metatheories. Every set-theoretic context, after all, provides in effect a meta-theoretic background for the models and theories that exist in that context, a model theory for the models and theories one finds there. Every model of set theory provides an interpretation of second-order logic, for example, using the sets and predicates existing there. Yet, a given model of set theory $M$ may itself be a model inside a larger model of set theory $N$, and so what was previously the absolute set-theoretic background, for the people living inside $M$, becomes just one of the possible models of set theory, from the perspective of the larger model $N$. Each meta-theoretic context becomes just another model at the higher level. In this way, we have theory, metatheory, metametatheory and so on, a vast hierarchy of possible set-theoretic backgrounds.

---

[19] If this is a statement of set theory, and $A$ itself is such a statement, this is really of the form $\mathfrak{U} \Vdash \langle A \rangle$, where angle brackets are whatever form of quotation is deployed in the set theory. As is standard practice in logic, I simply abuse notation here.

[20] As noted, e. g., by Shapiro (2014, p. 171).

[21] Hamkins (2020, ch. 8).

Of course, this view is not committed, *per se*, to the view that all universes are equally good for the purpose at hand. Thus, one might suppose, e.g., that the bigger the universe the better.

However, perspectivalism endorses just this thought. We may take any universe which includes the universes to be investigated, and investigate what happens within this structure. Note that there is no reason why the universes to be investigated must have the same internal logic as the universe in which they are embedded. Thus, a classical universe (one whose truths are closed under classical logic) may appear in a paraconsistent universe. Or an intuitionistic universe (one whose truths are closed only under intuitionist logic) may occur in a classical universe. As Hamkins notes, each hyper-universe with its internal logic provides, in effect, a metatheory for investigating the different universes. Classical logic, note, has no difficulty in making sense of a Kripke model for an intuitionist theory; and this structure may be investigated in a classical metalanguage.

The different hyper-universes in which the universes are embedded may well establish somewhat different relationships between them. Indeed, they may well deliver different properties of the relation $\mathfrak{U} \Vdash A$. Thus, Kripke models for intutitionist logic are complete, given a classical metalanguage; but not, given an intuitionist metalanguage.[22] So the different hyper-universes are liable to deliver quite different perspectives on matters.[23] But given perspectivalism, all perspectives are equally legitimate.

In his *Varieties of Logic*, Stewart Shapiro discusses the question of whether one should suppose there to be a unique foundation for mathematics; that is, whether one should suppose there to be a unique mathematical theory—and so logic—which can be used to formulate and compare all mathematical theories or structures.[24] This is somewhat more general than the matter I have been discussing. A hyper-universe provides, in effect, a very specific kind of metatheory for its sub-universes. (There can be other kinds of metatheories—for example, category theoretic ones.) However, Shapiro provides a pluralist answer to the question; and it is much the same as the one at which we have arrived above. He says:[25]

> To put the matter in general terms, if we become interested in relations among particular theories or systems, $S_1, S_2, \ldots, S_n$, then we need a perspective—theory, structure—from which we can refer to $S_1, S_2, ..., S_n$, at once, so to speak, and we need some account of the relations between theories of systems $S_1, S_2, ..., S_n$. But we do not need a single foundational theory that accommodates every possible mathematical theory, or every possible logic...

Indeed we do not. Nor should one expect one, given the pluriverse and mathematical pluralism.

---

[22] See, McCarty (1991). For further discussion of this sort of variability, see Shapiro (2014, ch. 7).

[23] I note also that not all universes have to occur in all meta-universes. As I have already noted, a perspective can be partial.

[24] Shapiro (2014, pp. 177–81).

[25] Shapiro (2014, p. 180 f).

### 10.3.4 Relativism

There remains one issue to be discussed: relativism. The truth of many claims is relative. Whether something is funny is relative to one's sense of humour. Whether something is surprising is relative to what one knows (or believes). This is not problematic. A form of relativism is problematic when its articulation appears to presuppose the non-relativeness of a claim that falls within the scope of the relativism in question. This certainly happens with global relativism: the claim that the truth of any claim is relative. The articulation of this appears to be a non-relative claim. One way to bring this out is to imagine (as does Plato in the *Theaetetus*, 161c, ff.) a confrontation between a relativist of this kind and a non-relativist. On pain of self-refutation, the relativist must admit that the view of the non-relativist, relative to their own view, is just as correct as that of the relativist, relative to their view. Hence, they can no longer defeat their opponent in debate.[26]

Now, mathematical pluralism itself is not self-defeating. It claims that all mathematical structures, based on whatever logic, are equally valid. But this is a claim of philosophy, not of mathematics, and so does not fall within its own scope; any problematic self-reference does not, then, arise.[27]

However, we are not concerned here with mathematical pluralism as such, but with the pluriverse in particular. As we noted, this does appear to be self-refuting, since it appears to presuppose a unique hyperverse. Perspectivalism was exactly a move to avoid this problem. However, perspectivalism may have relativist problems of its own. Certainly the Jain global perspectivalism does, since it is committed to the thought that the Jain view, the Hindu view, the Buddhist view, etc, are all equally correct relative to the appropriate perspective. Hence, a Jain can no longer claim that their view is superior to a rival's. Let us not pursue this matter here, however.[28] The question is whether a perspectival view of $V$ has similar problems.

It might well be thought to do so, since, one might suppose, the claim that there in no absolute universe of sets, but only different perspectives thereupon, appears to be giving a non-relative perspective on the matter. However, the perspectivalism in question is to the effect that any mathematical claim about $V$ must be interpreted as relative to some universe. Since any claim in the language of set theory has its quantifiers relativised to $V$ (at least implicitly) this is equivalent to the thought that any claim of set theory must be relativised to some universe. It is hence akin to the thought that any claim about a group is true or false only with respect to some group or other. As such, it is a claim *about* mathematics, not *of* mathematics. Hence, it does not fall within its own scope, and so is no more self-refuting than mathematical pluralism itself.

---

[26] For a discussion of relativism in general, and Plato in particular, see Baghramian and Carter (2015).

[27] As Shapiro (2014, p. 183), notes.

[28] It is pursued in Priest (2008, §9).

## 10.4 Conclusion: A Perspective on this Paper

This short paper has been concerned with the pluriverse of sets. We have seen that there are good reasons to generalise Hamkin's classical pluriverse to a pluriverse containing universes based on non-classical logics. We have also seen that taking each universe to have its own perspective on the totality of sets, each of which is equally legitimate, resolves the most pressing philosophical objection to the pluriverse. We have seen, moreover, that such perspectivalism provides an answer to the question of what metalogic should be used to investigate the totality of universes itself. Each member of the pluriverse gives us a perspective on the matter, and that perspective will be articulated with the internal logic of that universe. Finally, as we saw, this does not engender a pernicious relativism.

The pluriverse of sets in vast and diverse. Each universe in the pluriverse has a take on what it is like, and there is no sense in which any of them gets it objectively right. As far as the universe of sets goes, perspectives are all we have.[29]

## References

Anderson, R. L. (2017). Friedrich Nietzsche. In E. Zalta (Ed.), *Stanford encyclopedia of philosophy*. https://plato.stanford.edu/entries/nietzsche/.

Baghramian, M., & Carter, J. A. (2015). Relativism. In E. Zalta (Ed.), *Stanford encyclopedia of philosophy*. https://plato.stanford.edu/entries/relativism/.

Crosilla, L. (2019). Set theory: Constructive and intuitionist $ZF$. In E. Zalta (Ed.), *Stanford encyclopedia of philosophy*. https://plato.stanford.edu/entries/set-theory-constructive/.

Eklund, M. (2019). Fictionalism. In E. Zalta (Ed.), *Stanford encyclopedia of philosophy*. https://plato.stanford.edu/entries/fictionalism/.

Feferman, S. (1964). Systems of predicative analysis. *Journal of Symbolic Logic, 29*, 1–30.

Gitman, V., & Hamkins, J. (2010). A natural model of the mutiverse axioms. *Notre Dame Journal of Formal Logic, 51*, 475–484.

Hájek, P., & Haniková, Z. (2003). A development of set theory in fuzzy logic. In M. Fitting, & E. Orłowska (Eds.), *Beyond two: Theory and applications of multiple-valued logic* (pp. 273–285). Springer.

Hamkins, J. (2012). The set-theoretic multiverse. *Review of Symbolic Logic 5*, 416–449.

Hamkins, J. (2020). *Lectures on the philosophy of mathematics*. MIT Press.

McCarty, D. (1991). Incompleteness in intuitionist mathematics. *Notre Dame Journal of Formal Logic, 32*, 323–358.

Priest, G. (2006). *In contradiction* (2nd ed.). Oxford University Press.

Priest, G. (2008). Jaina logic: a contemporary perspective. *History and Philosophy of Logic, 29*, 263–278.

Priest, G. (2013). Mathematical pluralism. *Logic Journal of the IGPL, 21*, 4–14.

Priest, G. (2016), *Towards non-being* (2nd ed.). Oxford University Press.

[29] For helpful comments on an earlier draft of this paper, I'm grateful to Hartry Field, Joel Hamkins, Stewart Shapiro, and Alan Weir himself. The material was presented to a couple of seminars in Hartry Field's course on the philosophy of mathematics in the spring semester of 2020. I'm also grateful to a number of students in the class for their comments.

Priest, G. (2017). What if? The exploration of an idea. *Australasian Journal of Logic, 14*(1), article 4. https://ojs.victoria.ac.nz/ajl/article/view/4028/3574.

Priest, G. (2019). From the foundations of mathematics to mathematical pluralism. In S. Centrone, D. Kant, & D. Sarikaya (Eds.), *Reflections on the foundations of mathematics: univalent foundations, set theory, and general thoughts* (pp. 363–380). Springer.

Priest, G. (2021). A note on mathematical puralism and logical pluralism. *Synthese, 198*, 4937–4946.

Shapiro, S. (2014), *Varieties of logic*. Oxford University Press.

Takeuti, G. (1981). Quantum set theory. In E. Beltrametti, & B. van Frassen (Eds.), *Current issues in quantum logic* (pp. 303–322). Plenum.

Van Atten, M. (2017). The development of intuitionist logic. In E. Zalta (Ed.), *Stanford encyclopedia of philosophy*. https://plato.stanford.edu/entries/intuitionistic-logic-development/.

Weber, Z. (2012). Transfinite cardinals in paraconsistent set theory. *Review of Symbolic Logic, 5*, 269–293.

Weir, A. (1998). Naive set theory is innocent! *Mind, 107*, 763–798.

Weir, A. (2015). A robust non-transitive logic. *Topoi, 34*, 99–107.

# Chapter 11
# Elements of an Alternative Philosophy of Set Theory

**Marcus Giaquinto**

**Abstract** The two favoured solutions of the class paradoxes have been the 'limitation of size' view of sets and the iterative view of sets. A rival is offered: a view of sets based on an idea of definite pluralities, augmented to allow for sets of just one member and an empty set. An advantage of this alternative is that it flows from an ordinary non-mathematical idea, but with a slightly refined understanding of indefiniteness.

Some consequences for axioms of set theory are presented. The axioms of Separation and Replacement are natural outcomes of this view. Another consequence is that the Foundation axiom is not universally true. So, we should take seriously certain non-well-founded set theories, and let the context of investigation determine when to use a set theory with an anti-foundation axiom and when to stay with well-founded set theory.

**Keywords** Set · Definite plurality · Class paradox · Separation · Replacement · Foundation

One of Alan's intellectual virtues is a reluctance to run with the crowd in abandoning a natural view or prevalent practice in order to get round a blockage. An example is Alan's interest in the possibility of saving naïve set theory despite the class paradoxes.[1] One motivation for this might be dissatisfaction with the past and current orthodox rationales for the most widely accepted solutions to the class

I am most grateful for incisive comments on earlier drafts from Neil Barton and Alan Weir.

[1] See for example Weir (1998).

M. Giaquinto (✉)
University College London, London, UK
e-mail: m.giaquinto@ucl.ac.uk



A. Rieger, S. Leuenberger (eds.), *Themes from Weir: A Celebration of the Philosophy of Alan Weir*, Synthese Library 484,
https://doi.org/10.1007/978-3-031-54557-3_11

paradoxes. An early and long-lasting orthodoxy was the idea that some pluralities are too big to be members of a set, sometimes known as the 'limitation of size' theory.[2] That was superseded by the idea that the members of any set are in some sense 'prior' to the set itself; a set is formed by collecting prior entities and a universe of sets is formed by iterating this set-forming operation.[3] This is known as the 'iterative conception' of sets and is the basis of the currently orthodox solution to the class paradoxes. Though I am unable to see a way of saving naïve set theory, I am rebel enough to reject those widely accepted views. One of my aims in this contribution is to show that there is an alternative to them, an alternative which flows from an ordinary non-mathematical idea, that of a definite plurality.

I have two other aims. First, I hope to show that the idea of a definite plurality can sometimes guide us when deciding whether or when to adopt a proposition as an axiom. I consider only three axioms (Separation, Replacement and Foundation) so as not to stretch the patience of readers, and for the same reason my discussion of these axioms is far from comprehensive.[4] Second, I aim to support the claim that there are several different kinds of axiom when categorised on the basis of reasons for adopting them, kinds which are not often distinguished in discussions of the nature of mathematical axioms.[5]

## 11.1 From Definite Pluralities to Sets

The idea of *a definite plurality* of items is a major pre-theoretical source of our concept of a set. It is well known that we have no operational criterion for definiteness.[6] But there are some pluralities we know to be definite and others we know to be indefinite. For example, the plurality of English technical words is indefinite, whereas the English words in the second sentence of this paragraph constitute a definite plurality. Moreover, I hope to show that, despite the lack of an

---

[2] The title 'limitation of size theory' is given also to other related doctrines discussed in Hallett (1984).

[3] An early formulation of the iterative conception can be found in Gödel (1947). A couple of decades later it was enshrined in textbooks. See, for example, section 1 of chapter 9 in Shoenfield (1967).

[4] I also omit discussion of the *historical* origins of set theory and its axioms. For this see Ferreirós (1999) and Fraenkel et al. (1973).

[5] For an overall framework for considering axioms in mathematics, I strongly recommend Schlimm (2013). That there are different kinds of axiom is widely accepted. In discussions of axioms for set theory one often finds a bare two-fold distinction between those with 'intrinsic' justification and those with only 'extrinsic' justification, as reported in Maddy (1988).

[6] This lack motivated Skolem's revision of Zermelo's (1908) Axiom of Separation, as the latter used the condition that a statement A($x$) is 'definite' for every member of a set M. See the discussion of the Axiom schema of Subsets in Ch.2 §3 of Fraenkel et al. (1973).

operational criterion, the distinction between definite and indefinite pluralities can in some cases help to guide evaluation of candidate axioms.

The items of a definite plurality can be of any metaphysical character, as long as each of them is a single entity. Those items collectively constitute the plurality, which is itself a single item. I will say, of each of the items which together constitute a plurality, that is *in* or *of* or *a member of* the plurality. Any single item, even if it is itself a plurality, can be a member of a definite plurality. This view arises naturally from the cognitive capacity to view a definite plurality of items which are not pluralities, such as the plurality of books now in a pile on my desk, as a single entity while simultaneously viewing those books as single entities. This is in the primary case a perceptual ability.[7] But it is not restricted to perception; it is also a feature of thinking which persists at an abstract level into elementary arithmetic, as when one says "two threes make six", referring to any two non-overlapping pluralities of three items each. We quite naturally take the two triples as single items, keeping in mind that each is collectively constituted by three single items, and regard the two triples as the items which collectively constitute another definite plurality.

The definite pluralities that we meet in experience are typically distinct from their members and the members typically have some describable common property or relation which distinguishes them from non-members.[8] But neither of these features of experienced definite pluralities is guaranteed by the very idea of a definite plurality. What is guaranteed is that, as the members of a definite plurality constitute the plurality, the identity of the plurality, in the sense of *which* plurality it is, is entirely fixed by those items. This will be familiar. It is the principle of extensionality for definite pluralities:

> If the members of definite plurality A are exactly those of definite plurality B, A is the same definite plurality as B.

Here is part of my main proposal:

> Any definite plurality is a plural set and any plural set is a definite plurality.

The only serious objection to this proposal is that there are some apparently definite pluralities which cannot be sets. I will defend the proposal against this objection in the section after next, by arguing that the pluralities in question are not in fact definite. Before that I will make a case for allowing for sets of fewer than two members.

---

[7] See section 2 of chapter 2 of Maddy (1992) for a defence of this claim.

[8] Set aside the property of being a member of that plurality and properties dependent on that property.

## 11.2 Sets of Fewer Than Two Members

Why sets of fewer than two members? The answer depends on the answers to prior questions: Why do we want a mathematical theory of sets? What purposes do we want it to serve? One important motivation is that we want a systematic general theory for dealing with applications, which is efficient for this purpose, hence without needless complications.

Before introducing an example to illustrate the point, I should make explicit a feature of our grasp of definite pluralities which are not perceived. When it comes to reasoning about applications, any such plurality under consideration is likely to be given by a *condition* which all and only the items in the plurality satisfy. An example might be the plurality of irrational points in the unit interval of the real line. The *extension* of such a condition is a kind of association: its members are just whatever satisfies the condition.

Consider a simple example of how we might want to apply a systematic general theory of sets. Suppose we are doing tests of a certain medical treatment on patients A to E as a pilot study; the patients have been chosen to represent a relevantly diverse section of the population. We have devised criteria for a positive outcome, so that for each patient the outcome is positive or non-positive, *tertium non datur*. We are interested in the outcome possibilities, which collectively we call the *event space* (also known as the *sample space* in probability texts). These possibilities are in one-to-one correspondence with the possible extensions of the condition "$x$ is a patient among A, B, C, D and E for whom the outcome of the test is positive." There are 32 possibilities. Clearly there are 5 possibilities in which the outcome is positive for just one patient. So, using sets to represent the possibilities of the event space, there are 5 sets with just one member. Just as clearly, there is 1 possibility in which no outcome is positive. So there is a set without a member.

This way of dealing with an event space is extremely useful. It simplifies combinatorics and it is the starting point for probability theory. Any way of dealing with event spaces which *forbids* sets with fewer than two members will have restricted application or be needlessly complicated or both. So here is a pragmatic stipulation: there are sets with just one member and there is a set with no member.

The character of this proposal is in one way like some of Euclid's geometric postulates. Euclid's first postulate, for example, stipulates that a straight line can be drawn from any point to any other point. It is a stipulation or a demand: take it or leave it.[9] This serves to ward off objections of the following sort. "A line cannot be drawn from a point on the surface of the earth to a point on the surface of the moon, or from a point on one bank of a river to a point on the opposite bank." The objector fails to appreciate that simplifying assumptions are needed for a general geometrical theory which is sufficiently uncomplicated to be usable for both new

[9] Euclid (1956). The noun translated as "postulate" is aitēma (αιτημα). This meant *a request, requirement* or *demand*, not anything like *a self-evident truth*. Here is a direct translation: "Let it be postulated: to draw a straight line from any point to any point."

theoretical discoveries and practical applications. In the case of set theory, it is a simplifying move to adopt a notion of sets that allows for single-membered sets and an empty set as well as plural sets, both for applications of set theory and for the theory itself. Moreover, a notion of sets relaxed enough to allow for non-plural sets is easily grasped, as well as pragmatically useful.

## 11.3 Some Basic Principles

Accordingly, I propose the following account of what a set is.

> [SET] An entity is a *set* iff it is a definite plurality or the extension of a condition satisfied by exactly one item or by nothing at all.[10]

Extensionality is trivially true for extensions. So there is only one empty set, namely, the extension of any condition satisfied by nothing at all; and for each item *e*, there is only one set with sole member *e*, namely, the extension of the condition of being *e*.[11]

Given Extensionality for definite pluralities, this entails Extensionality for sets.

> [EXT] If the members of set A are exactly the same as the members of set B, A is the same set as B.

Closely related is the following principle: If the members of a set A are exactly the same as the members of a plurality B, then B is a set. Why? As A is a set, it is a definite plurality; as A's members are *exactly* the same as the members of B, B too is a definite plurality; so, by [SET] B is a set. This is an instance of the following principle:

> [CORR] If there is an exact 1-to-1 correspondence of the members of a set A with the members of a plurality B, then B is a set.

The idea underlying [CORR] is that if the plurality B were not a set, there would be some indefiniteness of the membership B, which would be transferred back via the exact 1-to-1 correspondence to the membership of A; so A's membership would be indefinite; so by [SET] A would not be a set.

[10] This is close to Linnebo's proposition Collapse in (Linnebo, 2010). As I understand it, Collapse is the claim that for any zero, one or more things *xx*, there is a set *y* such that any entity *u* is one of *xx* if and only if *u* is a member of *y*. Linnebo's use "*xx*" does not allow it to range over any *indefinite* plurality.

[11] This is the condition expressed by the predicate "$x = e$" where '$e$' names $e$.

## 11.4 The Class Paradoxes and Proposed Solutions

The extension of a condition may fail to be definite, as for example the condition of being *a difficult problem*. The extension of a condition is a set if and only if it is definite (i.e., satisfied by a definite plurality of items or by exactly one item or by nothing at all). But a specification of a condition, that is, a predicate true of exactly the satisfiers of the condition, need not by itself reveal whether the extension of the condition is definite. One way of reducing the risk of inserting indefinite pluralities into our domain of discourse is to exclude the use of predicates which we know might express conditions with indefinite extensions, such as "is a difficult problem". One idea that surfaced in the late nineteenth century was to permit only predicates used in scientific language. But science uses predicates for conditions with borderline, fuzzy or unstable cases, such as "is a species" and "is a pandemic". A better idea was to start with a clear, restricted mathematical language and extend it carefully as needed for applications.

This approach foundered on the class paradoxes: Burali-Forti's paradox, Cantor's paradox, and the Zermelo-Russell paradox.[12] The predicates involved were clearly mathematical, two of them central to talk about sets and set-membership: "is an ordinal", "is a class" (or "is a set"—no distinction was drawn between 'classes' and 'sets' at the time) and "is not a member of itself". These paradoxes are not merely paradoxical: while examination of the arguments reveals no clear fallacy, the conclusions are clear contradictions.[13]

When the dust settled two approaches to the paradoxes remained. One of them distinguishes between two kinds of class, *sets* and *proper classes*, the difference being that while a set can be a member of a class (and is in fact a member of many classes), a proper class cannot be a member of any class. Proper classes are what Cantor called absolutely infinite multiplicities, in contrast to merely transfinite multiplicities; examples of proper classes are the class of all sets, the class of all ordinals, and any class of things that stand in one-to-one correspondence with the ordinals. To illustrate how this distinction blocks the paradoxes, consider Cantor's paradox about the class of all classes. If there were such a class it would have to have all proper classes as members; this is impossible, as it is stipulated that no proper class is a member of a class. So there is no class of all classes; hence the premiss of the argument of Cantor's paradox is false.

---

[12] Here are the discovery data. Burali-Forti's paradox: Cantor 1895, Burali-Forti 1897. Cantor's paradox: Cantor 1899. The Zermelo-Russell paradox: Zermelo 1899 or 1900 (see Rang & Thomas, 1981), Russell 1901.

[13] A paradox, in this context, is an argument from apparently true premisses by apparently valid steps to an apparently false conclusion. Some paradoxes have true conclusions, such as Cantor's sound argument that the number of rational numbers is no greater than the number of integers, despite the fact that the integers can be embedded in the rationals and between any two integers there are infinitely many rationals.

But what is stopping a proper class from being a member of a set or proper class? If a proper class P is a real entity, why cannot its unit set {P} also be a real entity? The answer given was that a proper class is *too big* to be a member of a set or proper class. This is known as the *limitation of size* theory. No clear and convincing reason has been given for the claim that, in this respect, size matters. Though the limitation-of-size theory came to be widely accepted, some reflective people do not accept it, as they find it lacks a cogent rationale. I agree with them, and refer readers to previous discussion.[14]

The second approach to the class paradoxes makes no distinction between sets and classes; it avoids positing a realm of proper classes. Instead, its first tenet is that all members of a set are in some sense *prior to* the set itself. This entails that no set is a member of itself, as no set is prior to itself. So there is no universal set. So, also, there is no set of all non-self-membered sets.

The second tenet of this approach is that every set belongs to a cumulative hierarchy of sets, which is a well-ordered succession of stages $V\alpha$, each indexed by an ordinal $\alpha$. One set S is prior to another set T iff the first stage to which S belongs is earlier than the first stage to which T belongs.

The first stage $V_0$ (which may be empty) is the set of atoms, single items which are not sets and which have no sets as components; atoms are prior to sets. The next stage after any given stage $V\alpha$ is the result of adding to $V\alpha$ all of its subsets: $V\alpha + 1 = V\alpha \cup \mathbf{P}V\alpha$. Each limit stage is the union of all earlier stages: $V\lambda = \cup \kappa < \lambda\ V\kappa$. If $V\alpha$ is the first stage to which all M's members belong, $V\alpha + 1$ is the first stage to which M belongs, and M belongs to all later stages. The first stage to which an ordinal $\alpha$ belongs is $V\alpha + 1$. So there is no set of all ordinals.

This gives a pleasing picture of a universe of sets. But how do we know that every set belongs to a universe of this description, a cumulative hierarchy of well-ordered stages? The answer given is known as the *iterative conception* of sets. The mathematical logician Joseph Shoenfield put it this way: according to "the intuitive notion of a set ... a set A is formed by gathering together certain objects to form a single object, which is the set A. Thus before the set A is formed, we must have available all of the objects which are to be members of A." Shoenfield continues:

> We are thus led to the following description of the construction of sets. We start with certain objects which are not sets and do not involve sets in their construction. We call these objects *urelements*. We then form sets in successive stages. At each stage we have available the urelements and the sets from earlier stages; we form into sets all collections of these objects. A collection is to be a set only if it is formed at some stage in this construction.[15]

The iterative conception has seemed to offer a more convincing solution to the paradoxes than the limitation-of-size idea.

---

[14] In fact there are distinct versions of the limitation-of-size idea. For the purposes of this paper "the-limitation-of-size theory" refers to the view that some classes are too big to be members of a class. For excellent discussion of the various versions of the limitation-of-size idea see Ch.5 of Hallett (1984), and Ch.13 §5 of Potter (2004).

[15] P. 238 in Shoenfield (1967).

The main problem for this approach results from the fact that accounts of the iterative conception are clearly metaphorical, as proponents of it have conceded.[16] This would not matter if one could give an acceptable strict and literal statement of the conception. But this is no easy task and to my mind has not been achieved.[17] An interesting and penetrating discussion of the difficulties has been given by Michael Potter, a defender of the claim that every set belongs to a cumulative hierarchy. His own defence is an argument that every descending membership chain terminates after finitely many components or, in other words, there is no infinite sequence of sets such that each set after the first is a member of the preceding one. To put it in symbols, there is no infinite sequence of sets $\{S_k: k \in N\}$ such that

$$\ldots S_{n+1} \in S_n \in \cdots \in S_2 \in S_1 \in S_0.$$

If Potter's argument were sound, it would suffice to establish that every set belongs to a cumulative hierarchy of the sort described a couple of paragraphs back, because that is provable (using the Axiom of Choice) from the non-existence of any infinite descending membership chain. But Potter's argument is questionable. Here is what I think the argument is, using the phraseology in the text, but extracted from the context.[18]

> Mathematics is part of our attempt to represent the world. Any conceptual scheme which genuinely represents a world cannot contain infinite backwards chains of meaning, and so collections which mirror such chains could only be idle wheels in such a scheme. So it is a constraint on set theory, as part of our attempt to represent the world, that no set can lie at the head of an infinite descending membership chain.

Let us accept that set theory is part of a genuine representation of the world. Are chains of meaning in set theory and set membership chains really related in such a way that the existence of an infinite descending set membership chain would be as useless as an infinite backwards chain of meaning (such as a circle of definitions)? Why take sets and their membership relations to be elements of a conceptual scheme? Potter acknowledges this point of dialectical weakness in his argument, saying that it rests on an 'internalist' premiss which readers may not find compelling.[19]

I will return to the question of infinite descending membership chains later in this paper, in connection with the Axiom of Foundation. My conclusion at this point is that we still lack a cogent argument for believing either of the past or present diagnoses of the paradoxes, the limitation of size theory and the iterative conception of sets. So, it is not unreasonable to consider the possibility of an alternative diagnosis of the class paradoxes.

---

[16] Boolos (1971); Potter (2004).

[17] Rieger (2011) surveys the literature and makes a good case against its achievability.

[18] See Potter (2004), p. 40. The context of Potter's argument is section 3.3, headed "Metaphysical dependence", p.38–40.

[19] Potter (2004) p. 53.

## 11.5 An Alternative Diagnosis

My alternative is that none of the predicates involved in the class paradoxes expresses a condition with a definite extension. (From now on I will use 'the extension of a predicate' for the extension of the condition expressed by the predicate, and 'the satisfiers of the predicate' for the satisfiers of that condition.) This is not initially obvious, because these predicates do not seem to have 'border problems', that is, a case of indeterminate membership. The presence of a border problem is a sufficient condition for indefiniteness of extension:

> [BORDER INDEF] A predicate has indefinite extension if there is at least one entity *e* such that the question "Is the predicate true of *e*?" lacks a correct yes-or-no answer.

This gives a sufficient condition for indefiniteness, but not a necessary condition: "if" but not "only if". The indefiniteness of familiar examples used in discussions of the Sorites paradox is due to border problems. In contrast, the source of indefiniteness in the case of the class paradoxes is the unboundedness of the predicate-extensions, in the following sense:

> [UNBOUND INDEF] A predicate $\Phi(x)$ has indefinite extension if for any set of satisfiers of $\Phi(x)$ there is at least one satisfier of $\Phi(x)$ not in that set.[20]

The extension of each of the paradox-generating predicates ($x$ is an ordinal / is a class / is non-self-membered) is indefinite, because unbounded in this sense. For example, for any set S of non-self-membered sets, there is a non-self-membered set not in S, namely S itself.

Instances of this kind of indefiniteness can be importantly different from instances of border indefiniteness. For a case of border indefiniteness, consider a crowd of people with a close-packed core, gradually thinning out with distance from the core, having no boundary separating outlying members of the crowd from non-members in the vicinity. Despite its indefiniteness the crowd may be treated as if it were a single thing and an estimate of its approximate size by lower and upper limits may be possible and informative. For contrast, consider the predicate "$x$ is an ordinal". Its extension is not fixed but is domain-dependent, where a domain is a universe of set theoretic discourse fixed in a particular context. As there is no most inclusive domain of sets, no single thing is the extension of "$x$ is an ordinal". In many contexts of set theoretic talk, no domain is specified at all and so the talk is schematic, applying in *any* domain of sets in which the chosen axioms are satisfied; the ordinals could be those in a domain without a strongly inaccessible[21] cardinal, for example, or those in a domain such that for every cardinal there is a greater

[20] Michael Dummett's various formulations of 'indefinite extensibility' have some similarity to this kind of indefiniteness, but Dummett's notion is somewhat obscure. See Oliver (1998) and Linnebo (2018).

[21] A cardinal $\kappa$ is *strongly inaccessible* iff (i) it is not the sum of fewer than $\kappa$ cardinals less than $\kappa$ and (ii) it is greater than 2 to the power of $\mu$, for every cardinal $\mu$ less than $\kappa$.

strongly inaccessible cardinal. The corresponding extensions of "$x$ is an ordinal" would be vastly different things.

The significance of this for the class paradoxes is that when the range of the variable in the predicate of the paradox is unrestricted, the paradoxical argument gives us a proof by contradiction that no set is the predicate's extension, but when the range of the predicate's variable is restricted to a set, the paradoxes vanish. To illustrate for the Zermelo-Russell paradox: when the range of "is non-self-membered" is restricted to a set M, that is, when $S = \{x: x \notin x\}$ is replaced by $S' = \{x \in M: x \notin x\}$, we find $S'$ cannot be a member of $S'$; hence $S'$ is non-self-membered and beyond M.

This approach to the class paradoxes rests on the claim that for any set there is something not in it, that is, that there is no universal set, otherwise none of the 'paradoxical' predicates would have indefiniteness of extension due to unboundedness. This rejection of a universal set is not an unjustified dogma.[22] It is based on other beliefs. One is the belief that every set has a cardinal number. This together with the belief that there is no greatest cardinal number is a problem for the existence of a universal set: if there were such a thing it would have to have the greatest cardinal number, as every set has a cardinal number and everything, including every cardinal number, would be in a universal set. Another belief is that for every set there is a set of all its subsets; Cantor's argument shows that this precludes a universal set. These are some of the beliefs that would have to be rejected in order to maintain consistently that there is a universal set. So a choice has to be made. I believe that considerations of plausibility and pragmatic costs weigh against the existence of a universal set, but I will not attempt to justify this belief here.

Earlier I undertook to reply to an objection to the claim that any definite plurality is a set. If the objection were correct, my identification of plural sets with definite pluralities, hence my central claim [SET], would be false. The objection is that the paradoxical pluralities, those which cannot be sets on pain of contradiction, such as the plurality of all ordinals, are definite. But this is wrong, even though "is an ordinal", for example, has no border problem. The plurality of ordinals is indefinite because it is unbounded, in the sense that for any set of ordinals there is an ordinal not in that plurality. Similarly for the other paradoxical pluralities. In the absence of any other objection, the identification of plural sets with definite pluralities and the view of sets as expressed in [SET] remain tenable.

[22] Whoever said that it *is* an unjustified dogma? Answer: Tim Button, in the abstract of his talk "Loving the Universe", Philosophy of Mathematics Seminar, 27 January 2020, University of Oxford.

## 11.6 From Comprehension to Separation

The practical implication of all this is to move away from the principle of Comprehension to the principle of Separation. Comprehension for a language L is this:

> For any predicate $\Phi(x)$ of L there is a set which has as members exactly the items of which $\Phi(x)$ is true.

Here L is a suitable language—more about that shortly—and "$\Phi(x)$" is a variable for predicates applicable only to terms referring to single items. The Comprehension principle runs straight into the class paradoxes, hence into contradictions, as the predicates involved belong to the language of set theory; and that language, when syntactically regimented and recursively defined, surely counts as a suitable language. The Separation principle, proposed by Zermelo (1908), is a way of restricting the range of the variable of the predicate to a set. As was shown in the case of the Zermelo-Russell paradox, this transforms the class paradoxes into unobjectionable arguments with innocuous conclusions. Here is the Separation principle:

> For any predicate $\Phi(x)$ of L and any set M, there is a set which has as members exactly the members of M of which $\Phi(x)$ is true.

This is a metalinguistic statement, as it is about the predicates of the language L of the theory; for an axiomatization of the theory we need to express the essential content of the principle without recourse to a metalanguage. We can do this by means of a statement schema, (as long as L is syntactically regimented and recursively defined, because only then is there an effective procedure for telling whether a given sentence of L is an instance of the schema). Using "$\Phi(x)$" as a *schematic* variable for predicates of L, we can express the principle thus:

> [SEP] For any set M there is a set K such that for all $x$, $x$ is in K iff $x$ is in M and $\Phi(x)$.[23]

The danger of introducing a set term which in fact refers to an indefinite plurality cannot be totally ruled out simply by abandoning Comprehension in favour of Separation. This is because the language L might allow predicates expressing conditions for which there are border problems. Consider, for example, "$x$ is in N and any $x$-day old person is a child", where we assume that the natural numbers constitute a set N. This is pertinent even if we are confident that the standard languages for set theory do not have predicates susceptible to border problems. Why? Because we want to be able to apply set theory to non-mathematical domains, and for this we need to use languages which extend beyond the language of set theory.[24] The extended language needs to be carefully chosen and properly

[23] There is a checkable side condition, namely that "K" does not occur free in the instances of $\Phi$. The principle also holds for $n+1$-place predicates with $n$ parameters.

[24] Neil Barton (personal communication) points out that mathematical discourse *about* set theory can have predicates with border problems and suggests the following example: "$x$ is the G-code of

specified to reduce the risk of troublesome predicates. But it is quite conceivable that unobvious cases lurk, even when we are careful with the language.

This has significance for the status of Separation as a principle for generating axioms, which would be the instances of the axiom-schema for a given language L. A proposition that has strong intrinsic plausibility is that *if* the only possible sources of indefiniteness of a predicate-extension are unboundedness and border problems, and *if* no predicate of the language L has border problems, *then* the members of a set which satisfy a predicate of L is also a set. Alternatively put, if those two antecedent conditions are fulfilled, the Separation principle holds for L. But if those conditions are not both fulfilled, strong confidence in the Separation principle for L is misplaced. What, then, is the basis of our belief in the antecedent conditions, where L is any one of the familiar languages of set theory (LST)? I strongly suspect that we believe them solely on the basis of our collective experience of set theory in practice: we have not found a novel kind of indefiniteness, nor have we found any border problem for a predicate in LST when its variable is restricted to a set, nor have we found any reason to suspect that there is such a predicate. If that is right, our belief in the antecedent conditions is an empirical generalization and so our belief in the Separation principle for LST rests on that empirical generalization. And if *that* is right, [SEP] for LST is neither a conceptual truth, nor a pragmatic stipulation; it is a *hypothesis*, in the sense of that word as used to refer to universal claims in empirical sciences. But [SEP] for LST is not just a guess. It is a hypothesis supported by over a century of mathematical experience.

## 11.7 Set Existence

The next axiom to be discussed is the Axiom of Replacement. Before that I will table axioms which I am going to rely on later but not discuss, and I will use them to rehearse briefly a familiar account of how in set theory the existence of ever more sets can be proved.

To establish in the theory that there are any sets, we first need to know that there is something in our domain of discourse. This is usually built into first order logic, but it does not have to be.[25] When it is not, we can just stipulate that there is something in our domain of discourse. If our intention is to apply set theory to a given domain of things outside set theory, those things will be in our domain of discourse *ab initio* as atoms. Having experienced small positive numbers of items as definite pluralities we find it obvious that for any pair of items there is a set of precisely those items. This is expressed by the Pairing Axiom [PAIR], which is formulated so as to guarantee also the existence of the unit set $\{x\}$ for any item $x$ in the domain. As

a large cardinal axiom" where G is a fixed coding (Gödel-numbering) for a fixed language of set theory.

[25] Logic which allows for empty domains is known as *inclusive logic*, a subspecies of free logic.

we have stipulated that there is to be something in our domain of discourse, there is a set S. So, by [SEP] there is a set of all items $x$ in S such that $x \neq x$. By [EXT] there is only such set, the empty set ∅. Then by repeated application of [PAIR], there is a set {∅} and a set {∅, {∅}}.

The Union Set Axiom, [UNION], enables us to prove the existence of more sets.[26] Following custom, we write "B ∪ C" for the union set of {B, C}. We use [UNION] to define the Von Neumann successor of a set.

$$\mathrm{M}^{+} =_{\mathrm{df}} \mathrm{M} \cup \{\mathrm{M}\}.$$

This delivers an unending sequence of sets. The sequence starts with ∅ and continues by repeated application of the successor operation: {∅}, {∅, {∅}}, {∅, {∅}, {∅, {∅}} and so on. These are the customary representatives of the natural numbers in set theory. Most of us have a strong inclination to regard this sequence as a definite plurality, reflecting our confidence that the Arabic decimal place numerals for the natural numbers collectively constitute a definite plurality. We show that the members of this sequence do indeed form a set by using [SEP] and the Axiom of Infinity, [INF], which says that there is an *inductive set*, that is, a set which has ∅ as a member and which is closed under the successor operation.[27]

For any infinite set S already proven to exist, we can prove the existence of sets which are much larger than S by means of the Power set Axiom, [POW].[28] [POW] and [UNION] together enable us to prove the existence of ever larger sets. We can get even more power by adding the Axiom of Replacement.

## 11.8 The Axiom of Replacement: The Problem

Like the Axiom of Separation, the Axiom of Replacement enables one to deduce the existence of a set from a set already proven to exist. Replacement is like Separation also in that it is usually formulated as a schema. To state it succinctly, let us abbreviate thus:

> A binary predicate $\Phi(x, y)$ is function-like on a set M iff for any member $c$ of M there is at most one item $d$ such that $\Phi(c, d)$.

[26] [UNION]: For any set M there is a set with members exactly the members of members of M. By [EXT] there is only one such set, the union set of M.

[27] By [INF] there is an inductive set A. Put NN($x$) for the property of being a member of every inductive set. By [SEP] there is a set B such that $x \in$ B iff $x \in$ A & NN($x$). By [EXT] there is only one such set.

[28] [POW]: For any set M there is a set with members exactly the subsets of M. By [EXT] there is only one such set, the power set of M.

Then the Axiom of Replacement, formulated as a schema, is this:

> [REP]: For any set M, if $\Phi(x, y)$ is function-like on M, there is a set B with members exactly the items $d$ such that, for some $c$ in M, $\Phi(c, d)$.

Why should we accept this? The reasons given are not always compelling. Here is what the influential philosopher and mathematical logician George Boolos once said about it. "The reason for adopting the axioms of replacement is quite simple: they have many desirable consequences and (apparently) no undesirable ones."[29] *Really*? How is this claim, as Boolos states it, an acceptable reason for adopting a proposition as an axiom?[30]

I am not objecting to looking to the consequences of a candidate axiom for reason to adopt it. Gödel famously suggested that a candidate axiom could get inductive support from its *independently verifiable* consequences. In fact, for inductive support Gödel wanted more:

> consequences demonstrable without the new axiom, whose proofs with the help of the new axiom, however, are considerably simpler and easier to discover, and make it possible to contract into one proof many different proofs.[31]

The contrast with the 'desirability' rationale could hardly be greater. The desirable consequences of Replacement such as the existence of sets with cardinality $\aleph\omega$, and those mentioned by Boolos, are not demonstrable without Replacement (from Zermelo's axioms), and so do not give it any inductive support whatsoever. Gödel considers the possibility of a candidate axiom which has *independently provable* consequences and which, therefore, can get inductive support (in the absence of refutable consequences, of course).[32]

Gödel's comment shows how a candidate axiom might get not merely *inductive* support but also *explanatory* support, that is, the support that a good explanation gets from the phenomena it explains. He considers the possibility that the candidate axiom provides its independently provable consequences with proofs which are

---

[29] Boolos (1971), p. 27 in Boolos (1998). In that article Boolos goes on to say what he took the desirable consequences to be, apparently taking the desirability rationale for adopting Replacement quite seriously. Boolos later changed his mind about it. On a visit to MIT I started to query him about Replacement, but before I could finish my sentence he handed me a typescript in which he took a different line. The typescript was a pre-publication version of "Iteration again", in which he pointed out that Replacement was derivable in an axiom system, FN, which he himself devised. Boolos (1989), pp. 98–103 in Boolos (1998).

[30] Think of the same principle used in a different context: "In a couple of years global warming will start to reverse, with no change in human activity, due to factors that climate scientists have yet to discover; there will be a gradual cooling over the next few decades stabilising at global temperatures around mid-twentieth-century levels. The reason for taking this view is quite simple: it has many desirable consequences and (apparently) no undesirable ones."

[31] Gödel (1947, 1964).

[32] Russell too, decades before Gödel, had argued that, just as in empirical sciences, mathematical propositions could be inductively supported by their independently known consequences. (Russell, 1907)

shorter, more accessible and *unifying*,[33] by contracting many proofs into one proof.[34]

Gödel continued his account of conceivable grounds for adopting as an axiom a proposition which is not intrinsically compelling[35]:

> There might exist [new] axioms so abundant in their verifiable consequences, shedding so much light on a whole field, and yielding such powerful methods for solving problems ... that, no matter whether or not they are intrinsically necessary, they would have to be accepted at least in the same sense as any well-established physical theory.[36]

What if a candidate axiom is not only unifying, but also 'abundant in [its] verifiable consequences, shedding light on a whole field'? Then the candidate axiom would be an explanatory hypothesis which, in the absence of an equally good or better rival hypothesis, could be part of the best explanation of its consequences within the field it illuminates. Adopting the candidate axiom in that circumstance would be an inference to the best explanation.[37] In this way the basis for adopting a candidate axiom could be of the same kind and strength as the basis on which many physical theories come to be established.

While this is a conceivable rationale for adopting a proposition as an axiom, Gödel did not suggest that it applies to Replacement or to any other of the already established axioms of set theory. Though Replacement continued to appear in textbooks on axiomatic set theory as an axiom (or axiom schema), skepticism about it remained. Here is how the philosopher and mathematical logician Hilary Putnam expressed it:

> Quite frankly, I see no intuitive basis at all for ... the Axiom of Replacement. Better put, I do not see that a *notion* of set on which that axiom is clearly true has ever been *explained*.[38]

---

[33] Philosophers of science have taken it to be an explanatory virtue of a hypothesis that it has a unifying property. Friedman (1974); Kitcher (1981).

[34] It is not clear to me exactly what Gödel had in mind by 'contracting'. Perhaps the idea is that the candidate axiom enables one to construct one or more highly general proofs each having *as specific cases* the independent proofs of many of the candidate axiom's consequences.

[35] As Gödel is using the expression, a proposition is *intrinsically* necessary just when it is implied by the concepts involved in the proposition. A paradigm example is the Axiom of Extensionality.

[36] Gödel (1947, 1964).

[37] Inference to the best explanation in empirical science has been a staple of late twentieth century philosophy of science: Harman (1965); Lipton (1991). For defence of the view that 'explanatory considerations are the fundamental driver in set-theoretic [axiom] justification' see Barton et al. (2020). This is much stronger than (my construal of) Gödel's claim, which is only that there *could be* an explanatory justification of a proposed axiom.

[38] Putnam (2000), p. 24. Quoted in Potter (2004), p. 298.

## 11.9 The Axiom of Replacement: The Solution

The notion of set I will use to show why Replacement is true is the one presented in the first three sections of this paper. Components of this notion that I will use are these:

[SET] A *set* is a definite plurality or the extension of a condition satisfied by exactly one item or by nothing at all.
[EXT] If the members of set A are exactly the same as the members of set B, A is the same set as B.
[CORR] If there is an exact one-to-one correspondence of the members of a set A with the members of a plurality B, then B is a set.

The root idea of Replacement is already captured by [CORR]. We could restate it by saying that if all the members of a set A can be replaced by the members of B, one for one using every member of B, then B too is a set. The Axiom of Replacement has a weaker antecedent condition, allowing the association between A and B to be one-to-one *or* many-to-one. (This weakening facilitates applications). To get from [CORR] to the Axiom of Replacement I will use three other axioms, Extensionality, Separation and Power Set.

Let two-place predicate $\Phi(x, y)$ be function-like on set M.[39] The task is to show that there is a set B whose members are exactly those items *d* such that, for some *c* in M, $\Phi(c, d)$. In short, we need to show that there is a set $B = \{d\text{: } \exists c \in M\ \Phi(c, d)\}$. If there is no item *d* such that for some *c* in M $\Phi(c, d)$, the empty set $\varnothing$ is the relevant set B. If there is just one such item *d*, the unit set $\{d\}$ is the relevant set B. If there is more than one such item *d*, we need to be convinced that those items form a definite plurality. To do this in an easily digestible way, I will introduce this symbol, "$\Phi$[M]", as an abbreviation for "the plurality of *d*s such that, for some *c* in M, $\Phi(c, d)$." We can read "$\Phi$[M]" as "the range of $\Phi$ restricted to M", as long we keep in mind that this 'range' may be indefinite for all that has been shown so far.

The argument strategy is simple: (a) describe a plurality, show that it is a set and (b) show that its members stand in one-to-one correspondence with the members of $\Phi$[M]. Then it follows from the correspondence principle [CORR] that $\Phi$[M] is a set, which is the hoped for set B in the formulation of Replacement:

[REP]: For any set M, if $\Phi(x, y)$ is function-like on M, there is a set B with members exactly the items *d* such that, for some *c* in M, $\Phi(c, d)$.

For brevity, let $\Phi[M] = \{d\text{: for some } c \text{ in } M, \Phi(c, d)\}$. For ease of reading, I will use an expression from function-talk: for each *d* in $\Phi$[M], let us call the collection of members *c* of M for which $\Phi(c, d)$, *the pre-image of d*. Figure 11.1 and its explanation may help.

Here is the argument.

---

[39] A reminder: For any set M, a binary predicate $\Phi(x, y)$ is function-like on M if and only if for any member *c* of M there is at most one item *d* such that $\Phi(c, d)$.

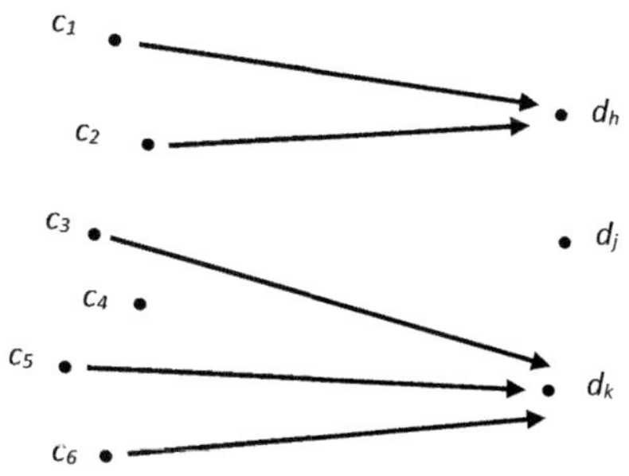


**Fig. 11.1** The dots labelled with *c*s are members of M. An arrow from a dot $c_x$ to a dot $d_z$ represents that $\Phi(c_x, d_z)$. Dots $d_h$ and $d_k$ are in $\Phi$[M] but $d_j$ is not. The pre-image of $d_h$ is $\{c_1, c_2\}$. The pre-image of $d_k$ is $\{c_3, c_5, c_6\}$

Consider the plurality of pre-images of members of $\Phi$[M]. In terms of the diagram, a pre-image is a collection of *c*s which have arrows to the same *d*. We need to show that the plurality of these pre-images is a set. A member of such a pre-image is a member of M, and M is a set (given). Hence by [SEP] each such pre-image is a set. For example, the pre-image of $d_k$ in Fig. 11.1 is the set of *c*s in M such that $\Phi(c, d_k)$.

Moreover, each such pre-image is a subset of M; so it is a member of the power set of M. Hence the plurality of pre-images of members of $\Phi$[M] is a subcollection of the power set of M. Hence by [SEP] the plurality of pre-images of members of $\Phi$[M] is a set.

Step (a) is done. Now for step (b): show that the pre-images of members of $\Phi$[M] stand in one-to-one correspondence with the members of $\Phi$[M]. Here is the correspondence: For each *d* in $\Phi$[M], *d* corresponds to its pre-image, $\{c \in \text{M}: \Phi(c, d)\}$.

(i) For each *d* in $\Phi$[M] *exactly one* set of this form is *d*'s pre-image, by [EXT].

(ii) For each pre-image S of some *d* in $\Phi$[M], S is the pre-image of *exactly one d* in $\Phi$[M].

Why? Suppose S were the pre-image of *d* and of *d'* when $d \neq d'$. As *d* and *d'* are in $\Phi$[M], S must be non-empty. Then we would have, for some *c* in S, both that $\Phi(c, d)$ *and* that $\Phi(c, d')$, contradicting the given condition that $\Phi(x, y)$ is function-like on M. By (i) and (ii) the correspondence is one-to-one.

I have spelled this out in more detail than will be necessary for most readers, only to assure other readers that there is no funny business here. The crux of the matter is the one-to-one correspondence principle [CORR], which flows from the idea of sets given by [SET], as argued in Sect. 11.3. Having presented this notion of sets in the first three sections of this paper, the foregoing is my response to Putnam's challenge to show that, according to a notion of set, [REP] is clearly true.[40]

[40] Replacement is deducible from Zermelo's axioms plus the Axiom of Foundation plus a proposition known as the Reflection Principle. Some people, assuming Foundation and Reflection to be justified, take this as their justification for accepting Replacement. The disadvantages of that justification are that (i) it applies only to domains restricted to well-founded sets, whereas the justification in terms of [CORR] applies without this restriction and (ii) it depends on some justification of Reflection, and that in turn will depend on a belief about *universes* of sets. The

However, the status of [REP] as an axiom schema is subject to the same considerations as the status of [SEP]. [SEP] is in fact derivable directly from [REP] by putting "$\Psi(x)$ & $x = y$" in place of "$\Phi(x, y)$" in [REP], noting that any predicate of the form "$\Psi(x)$ & $x = y$" is function-like on any set. So if the language we are using has a one-place predicate with a border problem, it also has a binary predicate with a border problem, and this opens up the danger of introducing indefiniteness by [REP]. To adapt the example used in discussing [SEP], the binary predicate "any $x$-day old person is a child and $x = y$" is function-like on N. But the plurality of entities $y$ equal to some $x$ in N such that all $x$-day olds are children is indefinite. So our belief in [REP] for LST, the language of set theory, depends on our belief that none of its predicates have a border problem; and that belief rests at least partly on our collective experience in using and investigating set theory. So the credibility of [REP] for LST rests on an empirical generalization. For this reason [REP], just like [SEP], is a hypothesis. But again like [SEP], it is a hypothesis that is supported by considerable mathematical experience.

## 11.10 The Axiom of Foundation

Might there be a set S which is a member of itself, or a member of a member of itself? Such cases would entail that there are infinite backwards (or 'descending') membership sequences:

$$\cdots \in \mathrm{S} \in \mathrm{S} \in \mathrm{S} \in \mathrm{S} \in \mathrm{S}$$

$$\cdots \in \mathrm{S} \in \mathrm{T} \in \mathrm{S} \in \mathrm{T} \in \mathrm{S}$$

There is an obvious generalisation of the question. Define $\in_n$ ($n$ in $\mathrm{N}^+$, the set of positive integers) as follows:

$$x \in_1 y \equiv_{\mathrm{df}} x \in y \text{ and } x \in_{n+1} y \equiv_{\mathrm{df}} \exists w\ [x \in_n w \,\&\, w \in y]\,.$$

Might there be a set S such that S $\in_k$ S for some $k$ in $\mathrm{N}^+$?

If there is such a set there would be membership loops, and this may seem puzzling. We are not aware of any such sets from our perceptual experience of definite pluralities, nor from the standard parts of mathematics for which set theory is used as a framework. Why not accept the simplifying hypothesis that there are no

truth of that belief is far less obvious than [CORR], which falls out of the basic notion of *sets* given in the early part of this paper. Reflection, however, is a true principle about the well-founded sets, being deducible from Replacement along with Zermelo's axioms. (Lévy, 1960).

such things, and adopt as an axiom a proposition that ensures this? The (set version of the) Axiom of Foundation does the job:

> [FOUND] Any non-empty set M has a member S which has no member in common with M.

Consider the possibility that there is a set A which *is* a member of itself. Then putting M = {A}, we find that the *only* member of M is A, which *does* have a member in common with M, namely A itself. This contradicts [FOUND], which therefore rules out self-membered sets. Again, suppose that we have two sets A and B, each a member of the other. Putting M = {A, B}, it follows that A has a member in common with M, namely B, and B has a member in common with M, namely A. As M has no members other than A and B, every member of M has a member in common with M, again contradicting [FOUND]. This kind of argument is easily generalised to show that [FOUND] rules out the possibility of an unending sequence of sets each having the next one as a member:

$$\cdots \in S_{n+1} \in S_n \in \cdots \in S_3 \in S_2 \in S_1$$

The argument takes the set M to be $\{S_n: n \in N^+\}$ and proceeds as before: all M's members have a member in common with M, contradicting [FOUND].

The *membership tree* of a non-empty set S is a directed graph with nodes representing S and all members of S and all *their* members and all ***their*** members and so on, where an arrow from a node representing $x$ to a node representing $y$ represents that $y$ is a member of $x$. Figure 11.2 presents a membership tree.

It follows from what has been said that, given [FOUND], every branch of the membership tree of a non-empty set terminates after finitely many nodes, each terminal node representing the empty set or an atom, (otherwise we can take M to be the set of sets represented by the nodes in the infinitely descending branch and derive a contradiction with [FOUND].) A set is said to be *well-founded* when

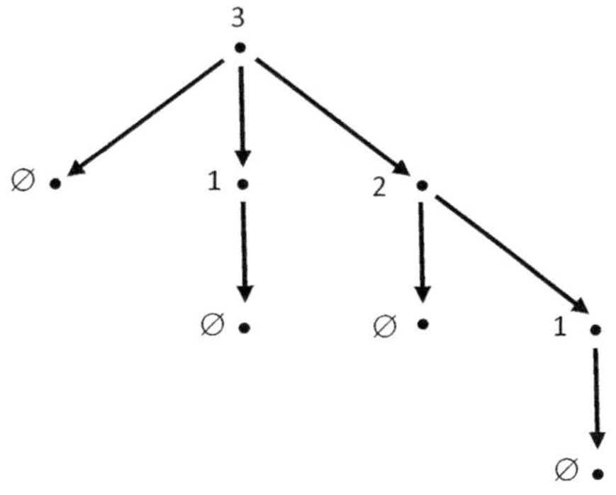


**Fig. 11.2** The membership tree for Von Neumann $3 = \{\varnothing, \{\varnothing\}, \{\varnothing, \{\varnothing\}\}\}$

its membership tree has no infinite descending branch. The Axiom of Foundation entails that every set is well-founded.[41]

But is the Axiom of Foundation true? To see that it is *not* true, on the simple view of sets spelled out earlier, consider these two sentences, named "S" and "T" respectively.

[S] The author of *Middlemarch* did not study at a university.
[T] The set of denotations of definite descriptions in S or T is finite.

The definite descriptions in S or T are:

"The author of *Middlemarch*"
"The set of denotations of definite descriptions in S or T"

The denotations of these definite descriptions are, respectively, George Eliot and the set of denotations of definite descriptions in S or T. For ease of reading let "M" stand for that set. Then M = {George Eliot, M}. M is a plurality which has exactly two members. It is a definite plurality, hence a set. The unit set {M} is a counter-instance to the Axiom of Foundation.

"Objection! Not every definite description has a denotation. Why assume that there is such a set as the set of denotations of definite descriptions in S or T?" Here is why. Suppose for the sake of *reductio* that there is no such set. Then "The set of denotations of definition descriptions in S or T" has no denotation. So the only denotation of a definite description in S or T is the denotation of "The author of *Middlemarch*", that is, George Eliot. In that case the set of denotations of definite descriptions in S or T is the unit set {George Eliot}; hence there is such a set as the set of denotations of definite descriptions in S or T, contradicting the supposition.

M is clearly not well-founded. It has non-empty intersection with one of its members. Figure 11.3 indicates how its membership tree has an infinitely descending branch.

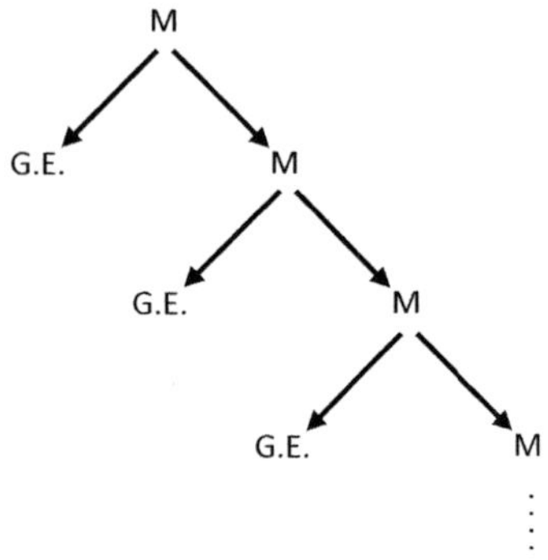


**Fig. 11.3** Initial fragment of the membership tree of M

41 The converse can be proved using the Axiom of Choice (or the Axiom of Dependent Choices).

The Axiom of Foundation is not true, at least not if [SET] is true. But the strangeness of non-well-founded sets makes one pause. Here is how George Boolos expressed it:

> It is important to realize how odd the idea of something's containing itself is. Of course a set can and must *include* itself (as a subset). But *contain* itself? Whatever tenuous hold on the concepts of *set* and *member* that were given one by Cantor's definitions of "set" and one's ordinary understanding of "element", "set", "collection" etc. is altogether lost if one is to suppose that some sets are members of themselves. The idea is paradoxical not in the sense that it is contradictory ... but [in the sense that] if one understands "∈" as meaning "is a member of", it is very, very peculiar to suppose it true.[42]

The understanding of "set" presented in the first section of this contribution *is*, I reckon, ordinary, and it is this understanding which compels me to believe that there are self-membered sets, despite their oddness. Mathematical phenomena that we find odd or 'very, very peculiar' may in fact abound. Think of continuous nowhere differentiable functions on a real interval and space-filling curves.

Those things are counter-intuitive, but we have come to accept that our intuitions are sometimes misleading in the context of real analysis. Self-membered sets, in contrast, have not always seemed counter-intuitive. Consider the idea of a universal set, which would obviously have to be self-membered if it existed. We could not have got a *paradox* from Cantor's theorem (that there is no surjection from a set to its power set) unless it was assumed that there *is* a universal set, hence a self-membered set. The fact that we feel non-well-founded sets to be odd may be mostly because, *up to the present* we have come across them rarely, or not at all, in our mathematical or non-mathematical experience.[43]

## 11.11 When and Why We Should Keep the Axiom of Foundation

What is the significance of all this for the question whether we should continue to treat the proposition that all sets are well-founded as an axiom? The matter is certainly not settled by the fact that some sets are not well-founded, because there could be perfectly good mathematical reasons for ignoring non-well-founded sets.

One of the principal reasons for developing an axiomatic theory of sets was to provide a consistent 'foundation' for classical mathematics[44] by interpreting it in the theory. This can be done smoothly in the well-founded sets. But what

[42] Boolos (1971), pp.17–18 in Boolos (1998).

[43] Another, and perhaps better, adjective than "odd" or "peculiar" for non-well-founded sets is Mirimanoff's adjective "extraordinary" (Mirimanoff, 1917). He is said to have been the first person to investigate the well-founded sets, having drawn the distinction between well-found and non-well-founded sets.

[44] From the theory of the integers up to and including complex analysis, at least.

about consistency? If we adopt the principle of foundation as an axiom, thereby confining the universe of the theory to well-founded sets, the arguments of the class paradoxes do not go through, as each of those arguments assumes the existence of a self-membered set. While this is no guarantee that set theory with the Axiom of Foundation is consistent, it certainly reduces the worry that it might not be.[45] Moreover, a universe of well-founded sets has the structure of a cumulative hierarchy which often simplifies deductive work and has proved useful for constructing models (e.g. for establishing relative consistency results). If, in addition, non-well-founded phenomena are isolated curiosities which do not merit systematic investigation, rejecting the Axiom of Foundation could hinder mathematical progress.

However, it is not true that non-well-founded phenomena are isolated curiosities unworthy of investigation. Set theories allowing for non-well-founded sets have already been employed in such investigations. Semantics is one source of non-well-founded phenomena to which a non-well-founded set theory has been applied.[46] Another is an area of computer science involving concurrent processing.[47] Yet another concerns foundations of non-standard analysis.[48] This is not the place to evaluate these applications of a non-well-founded set theory, but it is clearly possible, given what has already been done, that versions of non-well-founded set theory will turn out to have fruitful applications.

What about consistency? The version of non-well-founded set theory presented by Aczel (and its modification by Barwise) is consistent relative to standard set theory (ZFC). Aczel's theory is ZFC with an 'anti-foundation' axiom dubbed AFA instead of [FOUND].[49] AFA does not exclude well-founded sets; rather it ensures that for every graph of the type which could be the membership tree of a set, including those with a non-terminating path, there is a set with that membership tree. Working within ZFC without [FOUND] or AFA, Aczel showed that a universe of well-founded pure sets can be embedded in a universe of sets satisfying Barwise-Aczel non-well founded set theory,[50] in the same sense in which the real numbers can be embedded in the complex numbers.[51] Barwise called the sets in the universe of this theory "hypersets." So the well-founded sets stand to the hypersets as the real

[45] This relative confidence is clearly expressed by Gödel in "What is Cantor's Continuum Hypothesis?" towards the end of the 2nd paragraph of p. 180 in Gödel (1990) and his footnote 14.

[46] Barwise (1986); Barwise and Etchemendy (1987).

[47] Aczel (1988). Concurrent processing, as distinct from parallel processing, involves two or more processing 'threads' using the same processor over the same time period, involving communication between threads.

[48] Ballard and Hrbacek (1992).

[49] Aczel (1988). AFA is one among several 'anti-foundation' axioms that have been proposed.

[50] Reported in Barwise and Etchemendy (1987), but I have not been able to find the publication they refer to.

[51] A structure H can be embedded in a structure K if and only if there is an isomorphism from H to a substructure of K.

numbers stand to the complex numbers. This entails that the foundational goal of interpreting classical mathematics in set theory can be achieved in Barwise-Aczel non-well-founded set theory just as in ZFC with [FOUND].

What are the practical implications of all this for keeping or dropping [FOUND]? The answer is context-dependent, the relevant contextual factor being the problems to be solved. For the purpose of solving problems arising from non-well-founded phenomena, we should not be inhibited from considering non-well-founded set theory[52] as a means of investigation. That entails dropping [FOUND] and replacing it with an anti-foundation axiom. There is no good reason to ban non-well-founded sets in such a context. As there is a model of non-well-founded set theory (given that there is a model of ZFC), banning non-well-founded sets has no more justification than banning complex numbers with a non-zero imaginary part.

In other contexts of applying or investigating set theory, our aims do not require us to consider non-well-founded sets. In those contexts, there is no reason against keeping [FOUND] as an axiom. On the contrary, it could be a totally reasonable pragmatic stipulation, on a par with the stipulation that $V = L$ when investigating the constructible sets. Having [FOUND] as an axiom is a good way of making explicit the intended area of investigation, but a mistake if meant as a true assertion about all sets whatever.

In short, [FOUND] as an axiom is a *context-dependent* pragmatic stipulation. As long as contexts requiring non-well-founded sets are exceptional, [FOUND] will and should continue to be presented as an axiom in introductory textbooks on axiomatic set theory. What must go are the claims that [FOUND] should be kept as an axiom of set theory in all contexts, and that [FOUND] is true on any ordinary understanding of "set".

## 11.12 Conclusion

The idea of a definite plurality is the heart of a natural idea of sets. Pragmatic considerations lead us to allow that, in addition to definite pluralities, there are sets of just one member and an empty set. Accordingly, the view of sets I have proposed is this:

> [SET] An entity is a set iff it is a definite plurality or the extension of a condition satisfied by exactly one item or by nothing at all.

While we have no operational criterion for telling whether a plurality is definite, we are not completely in the dark. There are two circumstances in which a plurality

[52] A note of caution. Several distinct theories are counted as non-well-founded set theories. The principle of extensionality [EXT] does not hold in one of them: Finsler (2012). So Finsler's theory is not a theory of *sets*, strictly speaking. But what Rieger (2000) calls "Finsler-Aczel set theory" is extensional. Aczel's theory does not allow for atoms (urelemente) and so has restricted application; Barwise's modification of Aczel's theory removes this restriction.

given as the extension of a predicate $\Phi(x)$ will be indefinite; these are (1) when $\Phi(x)$ has a border problem, and (2) when $\Phi(x)$ has unbounded extension:

[INDEF BORDER] A predicate has indefinite extension if there is at least one entity $e$ such that the question "Is the predicate true of $e$?" lacks a correct yes-or-no answer.
[INDEF UNBOUND] A predicate has indefinite extension if, for any set of its satisfiers, it has at least one satisfier not in that set.

I have argued that this second kind of indefiniteness is the source of the class paradoxes. We avoid the paradoxes by restricting the troublesome predicates to a set, replacing "$\Phi(x)$" by "$x \in M$ & $\Phi(x)$". This is the effect of replacing Naïve Comprehension by the Separation Axiom. The idea that some definite pluralities are too big to be members of a set is not needed. Nor do we need the idea that every set is formed by 'gathering up previously available' entities.

From the view of sets I have proposed, namely [SET], and principles that issue from that view, namely [EXT] and [CORR], I argued that the Axiom of Replacement is justified, using a couple of relatively uncontroversial axioms, [SEP] and [POW]. The heart of the matter is [CORR], which is intrinsically plausible given [SET].

Although axioms for unit sets and an empty set are not usually stated, other axioms, namely [PAIR] and [SEP], are formulated in a way which quickly guarantees their existence. Sometimes an empty-set axiom is stated.[53] This is a *pragmatic stipulation*, a prescription not a description[54]; its function is to help specify what kind of domain is intended. Pragmatic stipulation is one of four *kinds* of axiom—there may be others—that need to be distinguished in assessing the acceptability of candidate axioms:

Pragmatic stipulation
Intrinsically plausible proposition
Hypothesis
Context-restricted stipulation

Intrinsically plausible propositions are implications or articulations of the concept or concepts we use to determine the subject matter, in our case concepts of *set* and *member of a set*.[55] Perhaps [EXT] is the clearest case of an intrinsically plausible proposition. I have argued that [SEP] and [REP] have intrinsic plausibility *given* a couple of empirical assumptions. As our confidence in these assumptions

---

[53] See for example Drake (1974) p. 9, and Enderton (1977) p. 18.

[54] I have taken *prescription* versus *description* as a way of drawing a major distinction between kinds of axiom from Dirk Schlimm (2013).

[55] This surely needs amplification and refinement. One problem, the so-called paradox of analysis, concerns the notion of 'implication or articulation of a concept'. Another problem arises from the fact that what seems to be the same mathematical subject matter can be thought of under radically different concepts in different ages. For example, the objects of the number theory in books VII to IX of Euclid's *Elements* were conceived of as definite pluralities (arithmoi) and single items or units, not as cardinals, ordinals or positions in the natural number structure; the geometry in books I and II of the *Elements* was thought to be about ideal graphical points, lines and figures in or on a plane surface, whereas in the seventeenth century plane geometry came to be thought of as elements or parts of a spatial plane.

rests on our collective experience, [SEP] and [REP] are *hypotheses*. This is a heterodox view, but not threatening: were either assumption found false, we would almost certainly change the language of the theory and retain the schemas adjusted to the new language. Finally, [FOUND] is not an outright truth about sets. There are non-well-founded-sets. But in ordinary contexts, where it is safe to ignore non-well-founded sets, it is fine to adopt [FOUND] as an axiom. In that circumstance [FOUND] is a *context-restricted stipulation*. In other contexts, AFA (or one of the other anti-foundation axioms) in place of [FOUND] is again an axiom of this kind.

Although this taxonomy of axiom kinds probably needs to be refined and extended, some distinctions among axiom kinds roughly along the lines I have suggested are essential for a proper understanding of what can be involved in arriving at an axiomatisation of a mathematical subject area.

## References


Aczel, P. (1988). *Non-well-founded sets* (CSLI lecture notes 14). Stanford.

Ballard, D., & Hrbacek, K. (1992). Standard foundations for non-standard analysis. *Journal of Symbolic Logic, 57*(2), 741–748.

Barton, N., Ternullo, C., & Venturi, G. (2020). On forms of justification in set theory. *The Australasian Journal of Logic, 17*, 158–200.

Barwise, J. (1986). Situations, sets, and the axiom of foundation. In Paris, Wilkie, & Wilmers (Eds.), *Logic colloquium '84*.

Barwise, J., & Etchemendy, J. (1987). *The liar. An essay on truth and circularity*. Oxford University Press.

Boolos, G. (1971). The iterative conception of set. *Journal of Philosophy, 68*, 215–232. Reprinted in Boolos (1998).

Boolos, G. (1989). Iteration again. *Philosophical Topics, 42*, 5–21. Reprinted in Boolos (1998).

Boolos, G. (1998). *Logic, logic and logic*. Harvard University Press.

Drake, F. (1974). Set theory. An Introduction to Large Cardinals. .

Enderton, H. (1977). *Elements of set theory*. Academic.

Euclid. (1956). *The thirteen books of Euclid's elements* (trans. T. L. Heath). Dover Publications.

Ferreirós, J. (1999). *The labyrinth of thought. A history of set theory and its role in modern mathematics*. Birkhäuser.

Finsler, P. (2012). *Finsler set theory: Platonism and circularity*. Translation of Paul Finsler's papers on set theory with introductory comments by D. Booth and R. Ziegler. Birkhäuser.

Fraenkel, A., Bar-Hillel, Y., & Levy, A. (1973). *Foundations of set theory* (2nd Rev. Edn.).

Friedman, M. (1974). Explanation and scientific understanding. *Journal of Philosophy, 71*, 5–19.

Gödel, K. (1947, 1964). *What is Cantor's continuum problem?* Both versions reprinted in Gödel (1990).

Gödel, K. (1990). *Kurt Gödel. Collected works. Volume II. Publications 1938–1974*. S. Feferman (Ed.). Oxford University Press.

Hallett, M. (1984). *Cantorian set theory and limitation of size*. Oxford University Press.

Harman, G. (1965). The inference to the best explanation. *Philosophical Review, 74*, 88–95.

Kitcher, P. (1981). Explanatory unification. *Philosophy of Science, 48*, 507–531.

Lévy, A. (1960). Principles of reflection in axiomatic set theory. *Fundamenta Mathematicae, 49*, 1–10.

Linnebo, Ø. (2010). Pluralities and sets. *The Journal of Philosophy, 107*(3), 144–164.

Linnebo, Ø. (2018). Dummett on indefinite extensibility. *Philosophical, 28*, 196–220.

Lipton, P. (1991). *Inference to the best explanation*. Routledge.
Maddy, P. (1988). Believing the axioms 1. *Journal of Symbolic Logic, 53*, 481–511.
Maddy, P. (1992). *Realism in mathematics*. Oxford University Press.
Mirimanoff, D. (1917). Les antinomies de Russell et de Burali-Forti et le problème fundamental de la théorie des ensembles. *L'Enseignement Mathematique, 19*, 37–52.
Oliver, A. (1998). Hazy totalities and indefinitely extensible concepts. *Grazer Philosophische Studien, 55*, 25–50.
Potter, M. (2004). *Set theory and its philosophy*. Oxford University Press.
Putnam, H. (2000). Paradox revisited II: Sets. In G. Sher & R. Tieszen (Eds.), *(2000) between logic and intuition: Essays in honor of Charles parsons*. Cambridge University Press.
Rang, B., & Thomas, J. (1981). Zermelo's discovery of the 'Russell paradox'. *Historia Mathematica, 8*, 15–22.
Rieger, A. (2000). An argument for Finsler-Aczel set-theory. *Mind, 109*, 241–253.
Rieger, A. (2011). Paradox, ZF, and the axiom of foundation. In P. Clark, M. Hallett, & D. DeVidi (Eds.), *Logic, mathematics, philosophy, vintage enthusiasms: Essays in honour of John L. Bell* (pp. 171–187). Springer.
Russell, B. (1907). The regressive method of discovering Premisses in mathematics. In D. Lackey (Ed.), *Essays in Analysis. Bertrand Russell*. George Allen & Unwin. 1973.
Schlimm, D. (2013). Axioms in mathematical practice. *Philosophia Mathematica (III), 21*, 37–92.
Shoenfield, J. (1967). *Mathematical logic*. Addison-Wesley.
Weir, A. (1998). Naïve set theory is innocent! *Mind, 107*, 763–798.
Zermelo, E. (1908). Untersuchungen über die Grundlagen der Mengenlehre I. *Mathematische Annalen, 65*, 261–281. Translated as "Investigations in the foundations of set theory I" in *From Frege to Gödel*, ed. J. van Heijenoort. Cambridge, MA., Harvard University Press 1967.

# Chapter 12
# Eliminativism About Phenomenology

**Graham Peebles**

**Abstract** Illusionism about consciousness holds that consciousness merely seems or appears to have a phenomenology, although this appearance is incorrect. On this view, the notion of phenomenology is sourced directly from introspection. It is therefore not merely introduced as a result of a conceptual error in theorizing about consciousness. This paper raises some difficulties for illusionism and argues for the latter eliminativist view. The characterization of conscious states as having a phenomenology is not one whose conceptual basis is sourced directly from introspection. Rather, it is introduced theoretically after reflection on what introspection tells us. This introduction is mistaken and the notion of phenomenology should therefore be eliminated.

**Keywords** Consciousness · Phenomenology · Illusionism · Eliminativism

## 12.1 Introduction

I would like to defend eliminativism about phenomenology as follows.[1] The characterization of conscious states as having a phenomenology is not one whose conceptual basis is sourced directly from introspection. Rather, it is introduced theoretically after reflection on what introspection tells us. This introduction is

[1] I can quite vividly remember first outlining some of the arguments in this paper while attempting to convince Alan of the plausibility of the view that there is no phenomenology. Thinking about this topic over the years has often raised this and other fond memories of the thoroughly enjoyable and rewarding PhD supervision sessions from which I learned a great deal about philosophy and how it is best done. I am therefore delighted to be able to contribute this paper to this much deserved *festschrift* in Alan's honour.

G. Peebles (✉)
Center for Global Education, Tokyo University, Tokyo, Japan
e-mail: gpeebles@g.ecc.u-tokyo.ac.jp

 
A. Rieger, S. Leuenberger (eds.), *Themes from Weir: A Celebration of the Philosophy of Alan Weir*, Synthese Library 484,
https://doi.org/10.1007/978-3-031-54557-3_12

mistaken and the notion of phenomenology should therefore be eliminated. This type of view, in various forms, has had some prominent defenders, the most well-known of which is perhaps Dennett (1988, 1991) who argues that the motivating "intuition pumps" adduced in favour of qualia, or phenomenological properties, are unsound. Similar views are advanced by other "Type A" materialists (e.g., Armstrong, 1968a, b, Dretske, 1995, Lewis, 1999, Ryle, 1949, Wittgenstein, 1953). Nevertheless, such a view remains unpopular and is regularly objected to on the grounds that it clashes with ordinary common sense (e.g., Balog, 2016; Chalmers, 1996; Nida-Rümelin, 2016; Strawson, 1994).

Recently, however, the theory of "illusionism" has revived a type of eliminativism about phenomenology in a different way (e.g., Frankish, 2012, 2017; Graziano, 2016; Humphrey, 2011; Kammerer, 2019, 2021; Pereboom, 2011). Illusionists accept the phenomenological realist's introspective claim that experiences seem to have a phenomenology, but they hold that this is merely an "illusion". According to the illusionist, even though no phenomenological properties are exemplified by any conscious states, it still seems—in an illusory sense to be specified but which is not due to a theoretical error made about experience—as though some conscious states do have a phenomenology. Illusionists argue that it is introspection itself that is mistaken in this respect, not that we are mistaken about what we introspect. The view, were it successful, could therefore provide an answer to the widespread intuitive scepticism regarding eliminativism while still eliminating the notion of phenomenology in any substantial sense.

I will raise some difficulties for illusionism and argue that eliminativism is the better view. After outlining the background in Sect. 12.2, in Sect. 12.3 I will discuss how to frame illusionism and argue against two formulations of the view according to which either: (a) introspection directly presents experience as having qualia as phenomenological realists also hold, only this is illusory in a sense modeled on perpetual illusion where non-exemplified properties are merely projected onto experiences in the sense of Locke's secondary qualities; or (b) introspection does not project qualia in this way but involves or triggers a disposition or other similar psychological mechanism that directly causes us to conceive of experiences as having qualia without support or interjection from theoretical argument. Contrary to these two views, however, introspection in fact provides no grounds for the introduction of the notion of phenomenology in either of the senses described in (a) and (b). The notion must therefore be introduced theoretically. In Sect. 12.4 I will argue that this introduction is fallacious and suggest a common, but fallacious, way in which the introduction is based on a comparison between perception and thought.

## 12.2 Background

Here is a traditional view with which everyone is familiar. There are two kinds of conscious states: those which have a phenomenology, i.e. those states which used to be called "sensations" and are now called "phenomenal" or "qualitative" states which have a "phenomenal character" that is accounted for by their "qualia" or "phenomenal properties", and those which do not. We can call the former "experiences" and the latter "thoughts". In Nagel's (1974) ubiquitous terms, there is "something it is like" to be undergo an experience but not to have a thought. To introduce more standard terminology from Block (1995), seeing a red square, for example, is "phenomenally conscious", whereas judging that there is a red square in front of one is merely access conscious in that the judgement puts the subject in a position to immediately report, act on, and reason about its content. Unlike perceptual experience, thought has no qualitative, phenomenological nature. This phenomenological distinction between experiences and thoughts has long been common currency. It is assumed, for example, in Lewis' (1929) introduction of the notion of qualia, in Moore's (1903) discussions of transparency and sense-data, Russell's (1912) further distinction between acquaintance with sense-data in perception and universals in thought, and in Smart's (1959) identity theory.[2] In the contemporary debate, the underlying general notion is no less taken to be a pre-theoretical and neutral notion. Famously, in Block's case, for example: "You ask: What is it that philosophers have called qualitative states? I answer, only half in jest: As Louis Armstrong said when asked what jazz is, 'If you got to ask, you ain't never gonna get to know.'" (Block, 1978: 281).[3]

A central debate in the philosophy of mind is between substantive and reductive realists about phenomenology. In contrast to substantive realists, reductive realists

[2] Smart, for example, holds that while sensations pose a problem for materialism, thoughts do not as they have no phenomenology and can therefore be handled behaviouristically. Moore contrasts perception with thought, which for Moore is the act of understanding and does not present its objects in the distinctive phenomenological way that perception presents its sense-data. This distinction is more vividly drawn by Russell. For Russell, the sense-data of perceptual acquaintance exemplify universals and therefore have colour, shape, and extension, thus bestowing a phenomenology on experiences of them. As thought, in contrast, acquaints us with universals and not their instances, they do not have such a phenomenology. Lewis describes the qualia given to us in experience as properties of experience which, although never presented without concurrent interpretation, are metaphysically independent from the conceptualization that is found in the concurrent thought. Lewis' intrinsic qualia are forerunners of the contemporary conception of qualia—irreducible, intrinsic, monadic, ineffable, qualitative properties of experience—although contemporary qualia are generally held to be experienced uninterpreted.

[3] This traditional view has recently been denied by some philosophers (e.g., Pitt, 2004). Those who take this contrary position argue that all conscious states have a phenomenology. My reason for objecting to this view will become clear in Sect. 12.4. This view suffers from an extension of the fallacious introduction of phenomenology employed in characterization of experience: first the notion is fallaciously introduced through a comparison between perception and thought, and then once it has gained theoretical traction it is further mistakenly used to also characterize thought.

hold that, while experiences do have a phenomenology, these properties can be reduced to more fundamental (physical-functional) properties. While there is a significant metaphysical disagreement, therefore, on the nature of phenomenological properties, the reductive realist is in agreement with the substantive realist on the explanandum. Here, for example, is the opening passage from Tye (2002), a much-cited paper defending a functional-representational theory of experiences, where Tye introduces "phenomenal character" as the explanandum.

> Representationalism is a thesis about the phenomenal character of experiences, about their immediate subjective 'feel'. ... Strong or pure representationalism goes further. It aims to tell us what phenomenal character *is*. According to the theory developed in Tye (1995), phenomenal character is one and the same as representational content that meets further conditions (Tye, 2002: 137).

There can be no doubt that such an intuitive acceptance of experiences as having a phenomenology is the standard view. And illusionists, unlike eliminativists, agree with this starting position. They are in agreement with both the substantive and reductive realist, and in opposition to the eliminativist, that experiences do seem to have a phenomenology. However, in contrast to the reductive and substantive realist, they argue that this is as far as we need to go: experiences *only seem* to have a phenomenology, when in fact they do not. This seeming is merely an illusion and there is therefore no question as to the nature of the phenomenological properties that experiences merely seem to have.

Frankish, for example, begins his paper "Illusionism as a Theory of Consciousness" by outlining illusionism as follows:

> introspection delivers a partial, distorted view of our experiences, misrepresenting complex physical features as simple phenomenal ones. Sensory states have complex chemical and biological properties, representational content, and cognitive, motivational, and emotional effects. We can introspectively recognize these states when they occur in us, but introspection doesn't represent all their detail. Rather, it bundles them all together, representing it as a simple, intrinsic phenomenal feel. (2017: 17)

This shifts the focus from the nature of the properties which account for the phenomenology of experience, the question tackled by the substantive and reductive realists, to the "the illusion problem", which is the main problem that illusionists seek to solve:

> Illusionists may hold that introspection issues directly in dispositions to make phenomenal judgments—judgments about the phenomenal character of particular experiences and about phenomenal consciousness in general. Or they may hold that introspection generates intermediate representations of sensory states, perhaps of a quasi-perceptual kind, which ground our phenomenal judgments. Whatever details, they must explain the content of the relevant states in broadly functional terms, and the challenge is to provide an account that explains how real and vivid phenomenal consciousness seems. This is the illusion problem. (2017: 27)

Frankish is careful to contrast this version of illusionism with "eliminativism", which is the view that:

> belief in phenomenal consciousness is simply a theoretical error, that rejection of phenomenal consciousness is part of a wider rejection of folk psychology, and that there is no role at all for talk of phenomenal properties—claims that are not essential to the approach. (2017: 14)

The illusionist therefore stakes out the following position.[4] They are in agreement with substantive and reductive realists, and in opposition to eliminativists, that experiences do genuinely seem to have a phenomenology. That is, they agree that the notion of phenomenology is genuinely sourced directly from introspection of experience. But in opposition to realists, they hold that introspection is mistaken in this respect about experience. However, they do not agree with eliminativists that the error involved is "simply a theoretical error", one that we make about what we introspect. Rather, the error is a pre-cognitive error made by introspection itself. Hence the "illusion problem" of explaining this illusion in non-cognitive, non-theoretical, terms. Illusionism therefore offers an attractive threading of the needle: acceptance of the intuitive characterization of experience as having a phenomenology, but without the attendant metaphysical difficulties involved in giving an account of phenomenological properties. However, as I will now argue, the "illusion problem" poses significant difficulties and eliminativism is the preferable view.

## 12.3 Two Types of Illusionism

Frankish (2012) asks whether there is a "theory-neutral explanandum" regarding the phenomenology of experience which would be acceptable to all parties (2012: 667). He proposes three different notions: classic qualia, diet qualia, and zero qualia. Classic qualia are the straightforward qualia that everyone knows well: intrinsic, ineffable, purely qualitative. Diet qualia are merely the "phenomenal characters (subjective feels, what-it-is-likeness, etc.) of experience" (2012: 668). Frankish argues that once diet qualia, presented as an intuitive theory-neutral or pre-theoretical conception of the phenomenology of experience, are stripped of the attributes that characterize classic qualia we are left only with "zero qualia": "the properties of experiences that dispose us to judge that experiences have introspectible qualitative properties that are intrinsic, ineffable, and subjective" (2012: 669). As zero qualia are defined in terms of their cognitive effects and functional role, we can explain zero qualia and therefore describe how it is to

[4] Frankish's characterization of illusionism and the illusion problem and its contrast with eliminativism is shared by other illusionists such as Graziano, Humphrey, Kammerer, and Pereboom, even if they offer different answers to the illusion problem.

undergo an experience with zero qualia without recourse to notions such as the qualitative, ineffable, intrinsic etc.—i.e. in non-phenomenological terms.

A robust notion of diet qualia would need to be, on the one hand, "weak enough to distinguish it from that of classic qualia, so that functional or representational theories of consciousness are not ruled out a priori. On the other hand, it needs to be strong enough to distinguish it from the concept of zero qualia, so that belief in diet qualia counts as realism about phenomenal consciousness" (2012: 669). Frankish considers some strategies—diet qualia as appearing to be classic qualia, appeals to conceivability, direct ostension of diet qualia, recognitional concepts, diet qualia as given—and shows that they all fail to steer a stable middle course between classic and zero qualia. If there are no classic qualia, then, as there is no robust notion of diet qualia, there are only zero qualia. Thus, "the chief moral is that the fundamental dispute about consciousness should be over the nature of the explanandum, not the explanans" (2012: 575). If the illusion of qualia, therefore, can be explained in terms of zero-qualia, then the illusionist can thread the needle: their theory will account for the appearance of phenomenology without needing to offer an account of phenomenology which exceeds that of zero qualia.

So, illusionists argue that what needs to be explained are really zero qualia, which are properties which merely dispose us to judge, incorrectly, that experiences have classic qualia (which I will refer to as just "qualia" from now on). Once zero qualia have been explained, there is no need to explain the qualia that experiences merely seem to have as a result of their zero qualia.

However, there are two possible construals of illusionism, and it can be unclear in the literature as to which is being proposed. Views on introspection and qualia will presumably divide illusionist theories into one of two camps in which either: (a) introspection directly presents experience as having qualia, only this is illusory in more or less the same sense as perceptual illusions are illusory and non-exemplified qualia are projected onto experiences in a way analogous to Locke's secondary qualities; or (b) introspection does not project qualia in this way but involves or triggers a disposition or other psychological mechanism that directly causes us to conceive of experiences as having qualia without support or interjection from theoretical argument. Illusionists who offer the explanation in (a) would hold that introspection does present qualia as this is traditionally understood, but that introspection is mistaken in this respect. With Pereboom (2011), for example, one could draw an analogy with illusory theories of secondary qualities, and then hold that introspection projects qualia on to experiences. Illusionists who would deny that introspection projects qualia in this way would offer an explanation of the type outlined in (b). On this view, it would be incorrect to hold that experiences introspectively "seem" to have qualia, if this is understood in a perception-like sense, but introspection nonetheless involves or triggers some automatic mechanism which disposes or causes us to believe that experiences have qualia.

Proponents of (b) will need to explain why there is such a psychological mechanism and how it works. Proponents of (a) will need to explain why and how introspection has evolved to project qualia. These are both significant explanatory burdens (see Frankish, 2017 for some discussion). The eliminativist has no such burden, as all that needs to be explained is the error through which the notion of phenomenology is introduced. Although some will argue that the explanatory burdens on (a) or (b) can be met, I will argue, in Sects. 12.3.1 and 12.3.2 respectively, against both characterizations of the central aspect of the view as outlined in (a) and (b). Illusionism as outlined in (a) falls to transparency. And there is no such psychological mechanism as outlined in (b). Experience does not even seem to have a phenomenology. In Frankish's terms, not only are there no classic qualia, there are not even zero qualia.

### *12.3.1 Transparency and qualia*

I believe that experience is transparent and introspection does not present qualia as properties of experience. The construal of illusionism in (a) should therefore be rejected. An enormous amount has been written, and continues to be written, about transparency since Harman (1990). Many philosophers defend transparency and it is a well-known view. It would therefore be possible, noting that many oppose transparency, to take an adherence to transparency off the shelf and move on. However, I would like to make the following point about the strength of the transparency claim, one which will be relevant to the discussion in Sect. 12.4.

The core of transparency was outlined by Moore (1903). Moore's central contention was that introspection reveals experience to be a relation to its objects, not itself an object which has the properties that are in fact properties of the objects which are its relata. Moore seems to have viewed the experiential relation, if not its objects, in the same way as contemporary direct realists. According to this view, when *s* perceptually experiences *x*, *s* stands in a primitive, dyadic relation, *A*, to *x*. *A* is common to all experiences, regardless of what *x* is. The central point that we can extract from Moore is that upon introspection, all of the properties with which we are familiar in experience are properties of *x* (see Hellie, 2007 for a full discussion of Moore's argument). Although we can, according to Moore, become aware of *A*, *A* is not an object of one's experience or introspection as *x* is in *A*, nor does *A* have any properties of the kind that are exemplified by *x*. What we discover in introspection is that *A* is the relation through which we are aware of *x*. This relation is, as Moore famously put it, "diaphanous" in the sense that *A* does not come before the mind in introspection in the way that *x* does in *A*. There is something of a tension in this view, as it is not immediately clear what the awareness that *A* is a relation amounts to. However, present purposes require only the negative point that introspection makes one aware of there being an experience of *x* through *A* without this resulting from a direct awareness of any features of *A* in the way that *A* makes one directly aware of features of *x*—i.e. *A* makes one aware of intrinsic features of *x*, but introspection of *A* does not make one aware of any intrinsic features of *A*, (nor of any intrinsic features of any accompaniments to *A* which are distinct from *x*). One common way in which this is put is to contrast experience *of x* with a cognitive awareness *that* one is undergoing an experience of *x* (e.g., Dretske, 1995, 1999; Tye, 2014).

Proponents of contemporary direct realism essentially agree with proponents of sense-data that the experiential relation is dyadic, primitive, and unanalyzable, as well as transparent. Proponents of contemporary functional-representational theories also agree with this transparency claim but hold that representation provides structure to experience (e.g., Dretske, Harman, Tye). Representational content determines the objects experienced as what it is to experience, say, $x$ as $F$, is for the experience to have $x$ being $F$ as part of its representational content. To experience a cup on the table, for example, is not to be related through a dyadic relation to the cup on the table (or some parts such as the facing surfaces) but to be in a mental state which represents a cup on the table and therefore brings the cup (or that there is a cup) before perceptual consciousness.

Great efforts have been exerted to raise counterexamples to the representationalist transparency thesis and thus to attribute qualia to experience. Transparency is committed to at least the supervenience of the subjective or qualitative aspects of experience on its representational content, and so if a case can be found where this supervenience fails, then there must be some qualitative aspect to experience which is non-representational. However, the transparency thesis is extremely difficult to refute, the representationalist version especially so.

Consider trying to refute transparency in the face of a determined sense-data theorist. Any qualitative aspect to experience will be attributed to the sense-data. The "blur" in blurry vision, for example, will be attributed to the sense-data. One could argue that there is in some sense a "qualitative" difference between myopic vision and 20/20 vision of a blurry painting which duplicates the "painterly eye" as cast over a myopic experience. However, the sense-data theorist will argue that what explains any such difference between the cases is non-perceptual knowledge. The pattern of attributing differing qualitative aspects of experience either to properties of the sense-data or to an awareness of the circumstances of the experience will repeat itself in response to each proposed counterexample. The anti-transparentist is in something a bind here. For the fewer properties that they allow to sense-data, for example limiting them only to colour, shape and extension in two dimensions, the less scope there is to argue that there are cases where experiences with identical sense-data differ in a strictly experiential, and non-cognitive, aspect. However, should they increase the range of properties that they allow sense-data to have, such as being three-dimensional or exemplifying higher level properties, then the sense-data theorist is presented with a greater range of properties of sense-data which they will argue can change across the two cases. More expansive sense-data could likely account in this way for other classic counter examples such as Peacocke's (1983) trees, in which a tree twice the size and twice the distance away as another still "looks bigger" despite the two (allegedly) taking up identical sections of the visual field and thus "looking the same size", aspect switching cases in which, for example, the duck-rabbit "looks identical" yet in one case appears to be a rabbit and in another a duck, or cases where the elements of a scene can be held fixed but different aspects given visual "prominence" (Macpherson, 2006).

I find it difficult to see how to refute the transparency aspect of the sense-data theory, as the proponent of sense-data can pack any genuine change of experienced properties into the sense-data. The functional-representational theory duplicates the sense-data theory's strategy, but with the addition that it is the content that determines the objects of experience and how they are experienced. So, changes in the way that things are seen are understood not in the first place as changes in the objects of experience but as changes in the contents of experience which in turn determine the ways in which the objects are experienced. This is a superior, more flexible, and more plausible version of the same tactic. Thus, blur, for example, becomes a decrease in the determinacy or accuracy of representation, and aspects-switches are explained by different representations of some aspect of the scene such as visual depth, size, etc. This version of transparency, then, seems to me to be at least as difficult to refute as the sense-data version.

Transparency is a matter of significant controversy and the foregoing will, of course, no doubt fail to convince its opponents. However, I do believe that experience is transparent, and that introspection therefore reveals no qualia.

### *12.3.2 Introspection and Phenomenology*

If experience is transparent, then introspection does not project qualia onto experiences, either in an illusory or accurate way. Illusionism is therefore not best understood in the way described in (a). This leaves (b): that introspection involves or triggers a disposition or other psychological mechanism that directly causes us to conceive of experiences as qualitative without cognitive interjection and support from theoretical argument. But the lack of qualia projection which rules out the version of the theory described in (a) also raises a problem for (b).

Consider transparency in respect of the illusionist commitment that introspection represents experience as having a "simple, intrinsic phenomenal feel" (Frankish, 2017: 17). If this is correct, then introspection both represents experience as possessing properties which give it a "simple, intrinsic phenomenal feel" but does not make us directly aware of these properties in the way that introspection of qualia would were they projected onto experience on the model of Lockean secondary qualities as in (a). There is at least a tension here: introspection involves no projection and thus no direct awareness of qualia, but nonetheless represents experiences as having such properties—properties which are traditionally understood as being necessarily objects of experience and as revealing their nature in totality when they are experienced. The proponent of this theory, though, may argue that this air of tension depends on a particular conception of introspection as a perception-like state. If—as, note, transparency theorists likely hold—introspection is not perception-like in at least some aspects, then the lack of direct introspective awareness of qualia could be accompanied by a representation of a "simple, intrinsic phenomenal feel" even in the absence of the projection of these properties. Indeed, in eliminating illusionism understood as in (a) the proponent of transparency perhaps presents the proponent of (b) with exactly this theoretical distinction in their account of introspection. Nevertheless, even if we can make sense of this claim, I do not believe that introspection contains or triggers the mechanism described in (b).

I do also doubt, though, that transparency can be comfortably squared with this characterization of introspection. Without qualia as the direct objects of introspection, the view postulates a significant mismatch between the overall content output by introspection and the direct objects of introspection. A similar mismatch could be claimed for the transparency theorist's account of introspection as, for example, making one indirectly aware that one is undergoing an experience while making one directly aware only of the objects of that experience, but the illusionist's mismatch is more significant. Even if introspection makes one aware that one is undergoing an experience "indirectly", or without making one aware of the experience as an object in the way that one is aware of the objects of experience, there actually is an experience and introspection is not mistakenly making one aware of any properties of the experience that it does not have. And it is not a great introspective leap from undergoing an experience to being aware that one is undergoing that experience through introspection, even if one is not directly aware of the experience. This is a relatively modest and easy to grasp theoretical advance over the contents of experience: one can easily scrutinize experience and introspection and see how the objects of which one is directly aware in introspection are related to the experience that one must be in when one experiences them. And it is also easy to see why we possess such an introspective faculty. The same cannot be said for the illusionist's claim. Here, we are claimed to be aware not just that we are undergoing an experience in addition to awareness of the objects of experience, but also that this experience has significant and mysterious properties whose nature is usually held to be revealed in totality only by direct awareness of them, but of which we are not directly aware and which, in fact, experience does not have.

There is a further problem for (b). The eliminativist view is that the notion of phenomenology is introduced theoretically through cognitive reflection on the results of introspection. In order to avoid collapsing illusionism into a view more amenable to this interpretation, introspection itself must characterize experience as phenomenological prior to any independent cognitive reflection. The proponent of (b) must therefore claim that a characterization of experience in terms which are not reducible to non-phenomenological terms is either the direct causal output of introspection or the result of such a mechanism that is triggered by introspection and without the aid of broader theoretical reflection. But the alternative explanation according to which the characterization of experience as having a phenomenology is introduced as a consequence of theoretical reflection is significantly more plausible.

In a pair of recent papers, Kammerer (2019, 2021) identifies both the main motivation for this view of (b), which involves a problem that he calls the "illusion meta-problem", and clarifies the different senses of "illusion", "rich" and "sparse" illusions, that could be employed by proponents of (b):

> [the illusion meta-problem is] the problem of explaining not only why phenomenal consciousness seems to exist even though it does not (why we have an illusion of phenomenality), but also why it seems so strongly to exist (why the illusion of phenomenality is so peculiarly strong); notably, and that's the crucial point, why it is so hard for us, not only

> to believe that phenomenality is an illusion, but even to seriously entertain the possibility that it is. (2021: 5)
>
> "By "rich illusion", I mean an illusion in which an object is positively presented in an incorrect way by a cognitively impenetrable representational process; by "sparse illusion" I mean an illusion in which the object is not positively presented in an incorrect way by a cognitively impenetrable representational process—but presented in a merely partial or incomplete way, which in turn leads us to infer incorrect beliefs about the object (2019: 1–2).

The final claim in the first passage motivates the idea that it is introspection itself, and not a consequence of theoretical reflection on introspection, which is directly responsible for the illusory characterization of experience. If this characterization is a causal output of introspection, then this would explain the alleged strength of the illusion and illusionists therefore look for a mechanism which outputs this characterization. But I do not believe that there is good reason to posit such mechanism as the "illusion" does not have the features which would support this. As a consequence, it is not a rich illusion. Indeed, it is not really a sparse illusion either if this is understood to involve introspection's misrepresention of experience leading us to believe in phenomenology in a way which does not involve inference. There are, to again make use of Frankish's distinction, not only no classic qualia, there are not even zero qualia.

After the passage quoted above, Kammerer (2021) goes on to argue that the strength of the illusion is not the result of a mere tendency or disposition to believe, only one which just happens to be particularly strong. He compares visual illusions, which, even though they do not disappear, we can disavow in light of countervailing evidence, and argues that the strength of the illusion about phenomenology is different in kind, not just degree. It is, he argues, not just that it is difficult to come to believe that there is no phenomenal consciousness, but that it is difficult to even entertain or conceive of this. This claim seems to me to be too strong. I for one have no difficulty in conceiving of this, and I imagine that the same will be true of many readers. Note that I am not saying that I can conceive of or entertain not being in a conscious state when introspecting one (although it is also not immediately obvious that this should be rejected), but that I can conceive of a conscious state that I introspect as not having a phenomenology—not having qualia, being phenomenally conscious, etc. Indeed, it seems quite easy for me to conceive of this and entertain it.

I believe that we should therefore read the claim made by the proponent of (b) as a weaker one which does not require it to be difficult or even impossible to conceive of experiences which lack phenomenology. This, however, likely does not significantly alter the dialectic, for either way the illusionist will still look to "hard-wired features of our introspective mechanisms" (Kammerer, 2021: 5) as an explanation for the allegedly stubborn nature of this illusory belief. The question to consider, then, is whether it is correct to couple the belief in experience as having a phenomenology so tightly to introspection as to require such an explanation. Again, I believe not.

In order for the illusionist to defend a theory on which experiences do genuinely seem to have a phenomenology, it is not enough for introspection merely to present experiences as merely conscious states or as having any cluster of properties which, no matter how complicated, do not entail the genuine notion of phenomenology captured by qualia. The characterization of experience as phenomenological must come to be believed directly as an output of introspection and unaided by independent reflection about the contents of introspection which makes use of premises whose content exceeds that which can be derived from introspection of experience. But it seems to me as though the psychological facts do not match this description of introspection and phenomenological beliefs. Here, I would like to echo objections I made to Chalmers' (2018) discussion of the meta-problem of consciousness in Peebles (2020).

Chalmers' meta-problem of consciousness is very closely related to the illusionist meta-problem. Frankish and Kammerer ask for an explanation as to why our belief in the phenomenology of experience is so resistant to revision. Chalmers asks for an explanation of the source of the hard problem of consciousness, our allegedly stubborn belief or intuition that consciousness cannot be physical. As the hard problem follows from the nature of the alleged phenomenological properties of experience leading to problems with physical reduction, the illusion meta-problem and the meta-problem of consciousness are closely aligned. And Chalmers, like the proponent of (b), couples our belief in the hard problem very tightly to introspection:

> Problem intuitions can result from inferences. So that judgments that result from philosophical arguments will count as problem intuitions. Still, it is plausible that, in solving the meta-problem, the most important problem intuitions will be non-inferential judgments that arise prior to philosophical argument. (Chalmers, 2018: 11–12)

Applied not just to the hard problem, but to the underlying belief in the phenomenology of experience from which the hard problem follows, the following is the view that I believe to be more plausible. We introspect, scrutinize experience and its objects, reflect upon it in relation to other conscious states and our background beliefs and theories, and then we draw philosophical conclusions. One of these conclusions is often that experiences have a phenomenology, another that there is therefore the hard problem of consciousness. But this is not an automatic, causal result of introspection. I raised two points in objection to Chalmers' characterization of the meta-problem of consciousness, both of which also apply, I think, to the illusion meta-problem. First, that this description of introspection and its results is not true to the psychological facts. Second, that there is a significant content mismatch between the content of introspection and the content of the belief that introspection is alleged to support.

I ask the reader to consider when the idea that there is such a thing as phenomenology first struck them. Not the general notion of consciousness, the mind, the subject, or the notions of a mental state or experience, but a notion of phenomenology. I submit that this characterization, if it exceeds the former group of notions, does not automatically first strike one merely as a result of introspection. Yet that is the claim that the illusionist must make. In order to avoid collapsing the view into eliminativism in which there is theoretical reasoning between introspection and belief in phenomenology, the illusionist must hold that either introspection directly causes the belief in phenomenology or disposes us to automatically form this belief, perhaps subject to some further conditions which cannot include any theoretical reflection or independent support. As with the claim that introspection so directly discloses the hard problem, this does not seem to me to an accurate reporting of the psychological facts. Readers will be familiar with explaining such concepts to undergraduate students and will no doubt recall that it can sometimes be difficult for students to grasp even such seemingly—to philosophers—obvious ideas as that of experience having a phenomenology. Sometimes extended discussion of, for example, Mary's room is required to sufficiently pump the students' intuitions, and sometimes these intuitions remain stubbornly unforthcoming. This simple fact seems to me to raise a significant problem for illusionism understood as the thesis that mere introspection issues directly in a belief in phenomenology. If there is such a mechanism, one which causally outputs belief in phenomenology, or at the very least disposes one to automatically believe in phenomenology, then why does this mechanism so often fail?

A possible rejoinder to this argument is that although it may take some work to come to realize that experiences have a phenomenology, this can be done solely by unpacking the contents of introspection and without the aid of any (non-trivial) conceptual resources sourced from outwith introspection.[5] As with the meta-problem of consciousness, this does not seem to me to be plausible for the illusion meta-problem. Here, this point I raised against non-inferential views of the meta-problem of consciousness is perhaps stronger than when applied to the illusion meta-problem, but it still has traction.

It is difficult to see how plausible contents of introspection alone could suffice for the content of beliefs in the hard problem. The hard problem is a problem of multifaceted complexity involving such notions as the physical, causal, intentional, modal, fundamental, etc. It is hard to see how such notions can be derived or unpacked from the contents of introspection without the aid of independent premises sourced from outwith introspection. Introspection alone does not lead to belief in the hard problem. Inference which draws in content from outside introspection and applies it to the content of introspection is required. The formation of this belief is a broader, rational, inferential process. As the hard problem of consciousness is in a sense more complex than the notion of phenomenology as it relies on it, there will perhaps be less distance from the contents of introspection to the notion of phenomenology than there is to the hard problem. However, there still is a significant level of conceptual complexity involved in even the simplest characterization of phenomenology as, for example, intrinsic, ineffable, qualitative, etc. These are highly complex notions, not those of everyday discourse—and not notions that are plausibly output by introspection either directly and plainly or that can be unpacked without significant theoretical reflection on what introspection tells us

[5] This seems to be Chalmers' (2020: 17–18) response to this objection to non-inferentialist views of the meta-problem.

about experience. And the point I made against illusionism understood as in (a) is again relevant for it removes any scrutable grounds for the notion.

I believe, then, that we ought not to characterize our belief in experience as having a phenomenology as, in Kammerer's terms, a rich illusion, whereby introspection directly misattributes properties to experience that it does not have. And I also believe it not quite right to characterize introspection as a sparse illusion if this means that introspection also presents experience in a way which directly leads to a belief in phenomenology in the way suggested in (b). Kammerer (2019) discusses Graziano's attention schema theory in these terms, for example. However, if Graziano's view is to count as a genuine type of illusionism of the type described in (b), then this schematic representation will have to directly lead to belief in phenomenology without inferential support from independent premises and a broader theoretical reflection on the schematic representation and its structure, consequences, etc. Otherwise, and this is perhaps a better characterization of such views, introspection can lead one into making mistakes about experience when one reflects on it. But this is not illusionism as described in (b). Rather, it is the eliminativist view that our belief in phenomenology arises from (mistaken) theoretical argument, not directly from introspection.

So, it seems to me that as experience is transparent, illusionism should not be understood as in (a). Introspection does not project qualia onto experience. But we should not understand it as in (b) either, for the belief in phenomenology is not so tightly coupled to introspection as to support the mechanism described in (b). Introspection does not cause, dispose, or otherwise automatically lead to beliefs in phenomenology. There are, then, not only no classic qualia but no zero qualia either.

## 12.4 A Fallacy Regarding a Comparison Between Perception and Thought

This points the illusionist back towards the view that the introduction of phenomenology is a mistake that we make about what introspection tells us, not a mistake that we inherit from introspection. If not sourced directly from introspection, the introduction of the notion of phenomenology must lie in our reasoning about experience. To close, I would like to briefly discuss what I believe to be a central, perhaps foundational, example where reasoning about experience can introduce the notion of phenomenology, but where this reasoning is strictly fallacious. My point is, of course, not that this entirely explains the general belief in phenomenology, but I do believe that it is a pervasive and recurring feature of philosophical theory, one that plays a, if not the, central role in the introduction and solidification of the notion of phenomenology.[6]

At the beginning of Sect. 12.2, I discussed the traditional view that divides conscious states into sensations, such as perceptual experiences, pains, and the like, and thoughts. If it is to exceed such notions as those of the mental or conscious, the notion of phenomenology must be fundamentally contrastive in this way. And indeed it standardly is introduced by pointing to the difference between, say, seeing a red square and thinking about a red square. The perception of the red square is "vivid", or "phenomenal", or "sensational", and so on, in a way that the mere thought of the red square is not—i.e. the perception of the red square has a phenomenology that the thought of the red square does not. But the difference between perception and thought is, in fact, only epistemological. The introduction of any non-merely epistemological metaphysical notion of phenomenology to characterize any introspective distinction between the two groups is therefore strictly fallacious, no matter how intuitive its pull may be.

Consider again the discussion of transparency in Sect. 12.3.1. Introspection does not reveal anything more about experience than that it is a relation to its objects. When one experiences $x$, one is in a relation $A$ to $x$ which reveals some features of $x$. Upon introspection, one becomes aware that one is in a relation to $x$; however, one is not aware of any features of $A$ in the way that one is aware of features of $x$ in $A$. There are thus no intrinsic properties of $A$ which are introspected. But consider now whether the same is true of thought. It is the same in this respect. If one thinks about $x$, one is aware of $x$ in whatever way one thinks about it., e.g. as the drawer in which I left my keys. If one perceives $x$, one is aware of it in whichever way it is that one perceives it, e.g. as (to vastly simplify) the upper left drawer in the brown desk over by the window. The very same transparency thus holds of thought as holds of perception. Upon introspection, one becomes aware that one is thinking about $x$, not aware of any properties of the thought other than that it is about $x$ in the way that one is thinking about it—i.e. its content.

These introspective differences between perception and thought are differences only in the features that we are aware of, or in the way that we are aware of those features. To take two vivid examples, perception makes us aware of colours in a way that thought does not, and perception locates objects egocentrically in both space and time from the origin of both axes in a way that thought does not. But these differences are merely epistemological differences. Experience makes us aware of what colours are, for example, and makes us aware of objects in the similarly most fundamental way—knowing the egocentric location from one's own relative position at the centre of the axes of time and space of an object through perception is the epistemically most fundamental, rock-bottom way that one can understand, conceive of, or know some external object. Thought does not make us directly

[6] Similar points have, of course, been made in the literature, both historically (e.g., Dennett, 1988; Armstrong, 1968b) and in contemporary responses to illusionism (e.g., Garfield, 2016; Mandik, 2016)

aware of these features of objects and their perceptible properties. In thought, we remember objects that we have previously egocentrically identified, identify them as objects described by another, conceive of objects that we have not egocentrically identified, and so on, but we do not immediately identify objects egocentrically as we do in perception. Nor are their colours, for example, revealed to us. But, again, these are epistemological differences, differences in the contents and in the causal, functional, inferential, or psychological roles that these two types of conscious states play in our psychology.

This distinction leads to much metaphysics of the mind—acquaintance, conceptual and non-conceptual content, singular or Russellian propositions, analog and digital contents, in past times sense-data, etc.—but whatever merits these have as metaphysical theories of aspects of the mind they are not phenomenological. One cannot derive such notions as the qualitative, intrinsic, sensational, subjective etc., from these epistemological differences. To characterize aspects of these distinctions, for example non-conceptual contents of experience, or acquaintance relations in experience, as such is to fallaciously introduce some metaphysics which the underlying epistemological distinctions can do without and therefore do not support. Although it may be difficult to resist describing the underlying epistemological distinctions in terms which lead to phenomenological characterization, this does seem to me to be either redundant loose talk, or technically fallacious in that it makes new metaphysics out of epistemological distinctions which do not license it.

And we can easily explain the source of this difference between experience and thought. Perceptual experiences (as types) result from the workings of the sensory system, whereas thoughts are not the direct causal outputs of the sensory system (this is overly simplistic, but the explanatory point holds). Both their functional role and the properties of their vehicles are thus quite different: experiences directly encode and report the output of the sensory system which directly represents the immediate environment and have the representational capacity required to represent these contents in significant detail; thoughts are at least a step removed and do not encode information in the same detailed way. Indeed, this seems to me to the only difference between perception and thought. Perceptual experiences are thoughts, or judgements, which are directly output from the sensory system (see Armstrong, 1968a, b and Pitcher, 1971 for famous defences of this type of view). It may be difficult to avoid invoking such notions as "sensations", "vividness", "phenomenal character", "something it is like", and so on—i.e. notions of phenomenology—when we think about the difference between perception and thought, but these notions cannot be read-off either from experiences themselves nor from introspection, and upon examination their grounds reveal themselves to be merely epistemological in nature and thus not grounds for the introduction of any novel metaphysics of phenomenology.[7]

[7] As I mentioned in Sect. 12.2, some philosophers hold that all conscious states have a phenomenology. Once the notion of phenomenology—"what it is like", etc., has taken hold—it is easy to see

## 12.5 Conclusion

I have made the case for the eliminativism as opposed to illusionism as follows. Introspection does not reveal qualia either as mistaken projections onto experience on the model of Lockean secondary qualities, or as a characterization of experience in these terms which is a direct output from introspection. There are thus no classic qualia and no zero qualia. The notion of phenomenology is therefore not introduced directly from or by introspection, but theoretically. In order to get some purchase beyond such notions as the conscious, or the mental, the theoretical introduction is fundamentally contrastive of thoughts and experiences. However, the contrast between perception and thought which is the basis for this introduction only brings out epistemological differences which do not license the introduction of any notion of phenomenology. The notion should therefore be eliminated.

## References

Armstrong, D. M. (1968a). *A materialist theory of the mind*. Routledge.
Armstrong, D. M. (1968b). The headless woman illusion and the defence of materialism. *Analysis, 29*, 48–49.
Balog, K. (2016). Illusionism's discontent. *Journal of Consciousness Studies, 23*(11–12), 40–51.
Block, N. (1978). Troubles with functionalism. *Minnesota Studies in the Philosophy of Science, 9*, 261–325.
Block, N. (1995). On a confusion about a function of consciousness. *Brain and Behavioral Science, 18*(2), 227–247.
Chalmers, D. J. (1996). *The conscious mind: In search of a fundamental theory*. Oxford University Press.
Chalmers, D. J. (2018). The meta-problem of consciousness. *Journal of Consciousness Studies, 25*(9–10), 6–61.
Chalmers, D. J. (2020). Is the hard problem of consciousness universal? *Journal of Consciousness Studies, 27*(5-6), 227–257.
Dennett, D. C. (1988). Quining qualia. In A. J. Marcel & E. Bisiach (Eds.), *Consciousness in contemporary science* (pp. 42–77). Oxford University Press.
Dennett, D. C. (1991). *Consciousness explained*. Little, Brown and Company.
Dretske, F. (1995). *Naturalizing the mind*. MIT Press.
Dretske, F. (1999). The mind's awareness of itself. *Philosophical Studies, 95*(1), 103–124.
Frankish, K. (2012). Quining diet qualia. *Consciousness and Cognition, 21*(2), 667–676.
Frankish, K. (2017). Illusionism as a theory of consciousness. *Journal of Consciousness Studies, 23*(11–12), 11–39.
Garfield, J. L. (2016). Ilusionism and givenness. *Journal of Consciousness Studies, 23*(11–12), 73–82.
Graziano, M. (2016). Consciousness engineered. *Journal of Consciousness Studies, 23*(11–12), 98–115.
Hacker, P. M. S. (2002). Is there anything it is like to be a bat? *Philosophy, 77*(300), 157–174.

---

how this can naturally be extended to cover all conscious states. Especially when one also considers how language can lead us astray in such circumstances (see, e.g., Hacker, 2002).

Harman, G. (1990). The intrinsic quality of experience. *Philosophical Perspectives, 4*, 31–52.
Hellie, B. (2007). That which makes the sensation of blue a mental fact: Moore on phenomenal relationism. *European Journal of Philosophy, 15*(3), 334–366.
Humphrey, N. (2011). *Soul dust: The magic of consciousness*. Princeton University Press.
Kammerer, F. (2019). How rich is the illusion of consciousness. *Erkenntnis*.
Kammerer, F. (2021). The illusion of conscious experience. *Synthese, 198*, 845–866.
Lewis, C. I. (1929). *Mind and the world order: Outline of a theory of knowledge*. Charles Scribners.
Lewis, D. (1999). What experience teaches. In *Papers in metaphysics and epistemology* (pp. 262–290). Cambridge University Press.
Macpherson, F. (2006). Ambiguous figures and the content of experience. *Noûs, 40*, 82–117.
Mandik, P. (2016). Meta-illusionism and qualia quietism. *Journal of Consciousness Studies, 23*(11–12), 140–148.
Moore, G. E. (1903). The refutation of idealism. *Mind, 12*, 433–453.
Nagel, T. (1974). What is it like to be a bat? *Philosophical Review, LXXXIII*(4), 435–450.
Nida-Rümelin, M. (2016). The illusion of illusionism. *Journal of Consciousness Studies, 23*(11–12), 160–171.
Peacocke, C. (1983). *Sense and content*. Oxford University Press.
Peebles, G. (2020). Two problems for non-inferentialist views of the meta-problem. *Journal of Consciousness Studies, 27*(5–6), 156–165.
Pereboom, D. (2011). *Consciousness and the prospects of physicalism*. Oxford University Press.
Pitcher, G. (1971). *A theory of perception*. Princeton University Press.
Pitt, D. (2004). The phenomenology of cognition, or, what is it like to think that P? *Philosophy and Phenomenological Research, 69*(1), 1–36.
Russell, B. (1912). *Problems of philosophy*. Williams and Norgate.
Ryle, G. (1949). *The concept of mind*. Hutchinson and Co.
Smart, J. J. C. (1959). Sensations and brain processes. *Philosophical Review, 68*, 141–156.
Strawson, G. (1994). *Mental reality*. MIT Press.
Tye, M. (1995). *Ten problems of consciousness*. MIT Press.
Tye, M. (2002). Representationalism and the transparency of experience. *Noûs, 36*(1), 137–151.
Tye, M. (2014). Transparency, qualia realism and representationalism. *Philosophical Studies, 170*(1), 39–57.
Wittgenstein, L. (1953). *Philosophical investigations* (G. E. M. Anscombe, Trans.). Basil Blackwell.

# Chapter 13
# What Is the Sceptical Solution?

**Alexander Miller**

**Abstract** In chapter 3 of *Wittgenstein on Rules and Private Language*, Kripke's Wittgenstein offers a "sceptical solution" to the sceptical paradox about meaning developed in chapter 2 (according to which there are no facts in virtue of which ascriptions of meaning such as "Jones means addition by '+'" can be true). Although many commentators have taken the sceptical solution to be broadly analogous to non-factualist theories in other domains, such as non-cognitivism or expressivism in metaethics, the nature of the sceptical solution has not been well understood. The main aim of this paper is to advance our understanding of the nature of the non-factualism about meaning proposed in the sceptical solution. It attempts to outline some desiderata that should be respected by interpretations of the sceptical solution and considers two objections raised against it in Barry Stroud's paper "Wittgenstein on Meaning, Understanding and Community". It attempts to correct misconstruals of the sceptical solution that have been promulgated by Davidson and some of his followers and suggests that the sceptical solution developed by Kripke's Wittgenstein is best viewed as a form of quasi-realism about meaning. It ends by outlining what it takes to be the most pressing challenges facing the sceptical solution.

**Keywords** Kripke · Wittgenstein · Meaning · Rule-following

Alan Weir is famously anti-Wittgenstein, often referring in discussion to the author of the *Tractatus Logico-Philosophicus* and *Philosophical Investigations* as "Loopy Ludwig". It may seem slightly odd, then, to contribute a paper on Kripke's Wittgenstein's "Sceptical Solution" to a Festschrift volume in his honour. On reflection, though, it is not inappropriate. Kripke himself remarks on the "points of contact" between the sceptical argument suggested to him by Wittgenstein's discussions

---

A. Miller (✉)
University of Otago, Dunedin, New Zealand
e-mail: alex.miller@otago.ac.nz

 
A. Rieger, S. Leuenberger (eds.), *Themes from Weir: A Celebration of the Philosophy of Alan Weir*, Synthese Library 484,
https://doi.org/10.1007/978-3-031-54557-3_13

of rule-following, and W.V. Quine's famous arguments for the indeterminacy of translation (see e.g. Kripke, 1982, 14); Alan's doctoral dissertation (Weir, 1984) was on Quine's Philosophy of Language, while Quine and Quinean concerns have figured prominently in his work (see e.g. Weir, 2006a, b, 2007, 2014). Alan has also published on the determinacy of content (Weir, 2003). Hopefully, then, the present chapter will give Alan an opportunity to revisit and refine his views on the relevant issues.[1]

## 13.1 Introduction

Suppose that a truth-conditional account of meaning is appropriate for moral sentences. Suppose, also, that moral facts, if they exist, would have to be, as Mackie (1977) puts it, objective and categorically prescriptive: they would have to provide categorical reasons for action, reasons capable of motivating any rational agent independent of facts about that agent's contingent desires. So we have:

(1) "It is right for Jones to assist the injured child" is true if and only if there is a categorical reason for Jones to assist the injured child.

Suppose, too, that Mackie's "argument from queerness" succeeds in establishing that there are no categorical reasons for action, so that we have:

(2) There are no categorical reasons for action.

It apparently follows from (1) and (2) that:

(3) "It is right for Jones to assist the injured child" and other atomic, positive moral sentences are false.

One way of attempting to avoid this apparently unpalatable conclusion would involve giving up the truth-conditional account of the meanings of moral sentences: to embrace a form of non-factualism about moral thought and talk. The emotivism of Ayer and Stevenson, Hare's prescriptivism, and the expressivism of Blackburn and

[1] I first encountered Alan when (as a Masters student) I attended the paper on the philosophy of mathematics that he gave at the Logic and Language conference at the University of Leeds in 1989, and I can still remember the inspiration I felt on hearing such a fine paper delivered by a fellow Glaswegian.* I next heard Alan delivering a paper at the Putnam conference in Dublin in 1992, and I was happy to have an opportunity to talk about philosophy with him the following summer in one of Glasgow's most famous venues for philosophical discussion, The Horseshoe Bar, in Drury Street. Alan has been an inspiration and a great friend ever since, and it is a great honour to be able to contribute to this volume celebrating his career in philosophy. * See the entry on Barrhead at www.urbandictionary.com (accessed 27/2/2020): "Town in Glasgow, Scotland ... full of junkies and what-not ... commonly referred to as 'Borrheed' ... best avoided".

Gibbard, could all be cited as instances of this approach to avoiding an error-theory of moral judgement.[2]

Now suppose that a truth-conditional account of meaning is appropriate for ascriptions of meaning themselves. Suppose, also, that semantic facts, if they exist, would have to be both correctness determining and normative. So we have:

(4) "Jones means *addition* by '+'" is true if and only if there is a fact about Jones or his speech community which (a) makes it the case that "x + y = z" is correct iff z is the *sum* of x and y and (b) tells Jones how he ought to use "+".[3]

Suppose, too, that the sceptical argument in chapter 2 of Kripke (1982) succeeds in establishing that even idealised access to facts about Jones or his speech community fails to turn up facts that are both correctness determining and normative, so that:

(5) There are no facts about Jones or his speech community capable of satisfying both (a) and (b).

Then, as in the moral case, we are apparently pushed towards an error theory:

(6) "Jones means *addition* by '+'" and other atomic, positive ascriptions of meaning are false.

This is the predicament we are left in given the argument of the sceptic in chapter 2 of Kripke (1982): it appears that "all language is meaningless" and "the entire idea of meaning vanishes into thin air" (Kripke, 1982, 71, 22). What I'm concerned with in this paper is the response that Kripke's Wittgenstein (as opposed to Kripke's Sceptic) offers in response to this predicament in chapter 3 of Kripke (1982), where he develops his "sceptical solution" to the sceptical argument of chapter 2. Although I will sketch tentative responses to two lines of objection developed by Barry Stroud against the sceptical solution, my primary aim in this paper is not to defend it, but simply to advance our understanding of its nature. Inter alia, I will attempt to correct some misunderstandings about the sceptical solution that have been promulgated by Davidson and some of his his followers. I'll end by outlining what I take to be the two most serious challenges facing the sceptical solution.

---

[2] See e.g. Ayer (1946), chapter 6, Stevenson (1937), Hare (1952), Blackburn (1984a, 1993b, 1998a, b), Gibbard (1990, 2003). See Blackburn (2011, 8) and Blackburn (2013a, 273) for especially clear illustrations of how this type of view attempts to sidestep the error theory.

[3] I'm vastly oversimplifying here, especially with respect to (b), since none of the complications bear on the main points in the paper. For an account of some of the complications, see Hattiangadi (2017).

## 13.2 The Sceptical Solution

Kripke describes the conclusion of the sceptic's argument as "insane and intolerable" (1982, 60) and "incredible and self-defeating" (1982, 71), and distinguishes between two sorts of solution that might be developed in response. *Straight solutions* would attempt to deny (5) (or a defend an account of meaning facts less demanding than that involved in (4)),[4] while sceptical solutions would *concede* (5) to the sceptic but avoid the error theory by giving up on a truth-conditional account of the meanings of ascriptions of meaning.[5] Citing Michael Dummett's famous review of Wittgenstein's *Remarks on the Foundations of Mathematics*, Kripke suggests that in the *Philosophical Investigations* Wittgenstein rejects the truth-conditional account of meaning advocated in the *Tractatus* and proposes to replace it with an account of meaning in terms of assertibility conditions:

> As Dummett says, "the *Investigations* contains implicitly a rejection of the classical (realist) Frege-*Tractatus* view that the general form of explanation of meaning is a statement of the truth-conditions". In place of this view, Wittgenstein proposes an alternative rough general picture. ... Wittgenstein replaces the question, "What must be the case for this sentence to be true?" by two others: first, "Under what conditions may this form of words be appropriately asserted (or denied)?"; second, given an answer to the first question, "What is the role and utility, in our lives of our practice of asserting (or denying) the form of words under these conditions?". (Kripke, 1982, 73, quoting Dummett, 1959)

Thus, to account for the meaning of a given sentence we should describe the conditions in which there is a license to assert it, and show that the practice of licensing its assertion in these conditions has a certain utility in our lives. The central idea in Kripke's Wittgenstein's sceptical solution is that we can avoid the "insane

---

[4] An example of attempting to defend an account of meaning facts less demanding than that in (4) is the view developed in Fodor (1990). Fodor first of all rejects the idea that meaning facts have to be normative in any sense stronger than that associated with the idea that they determine correctness conditions, so that (b) is effectively deleted from (4); he then proceeds to identify facts capable of satisfying condition (a) using the notion of an "asymmetric dependency base". There is of course a whole literature about the precise nature of condition (b) and its analogues, but as noted above we can put this aside for the purposes of the present paper.

[5] The distinction between straight and sceptical solutions could be deployed in the moral case also. Ayer, Stevenson et al could be regarded as proffering sceptical solutions. McDowell (1998) could be regarded as giving a straight solution hospitable to the idea of moral facts as characterised by (1), while Railton (1986), Brink (1989) and the like attempt to defend straight solutions by weakening the account of moral facts in play in (1) (by denying that we need to think of moral facts as intrinsically reason-giving). In viewing the sceptical solution in the meaning case as promoting a form of non-factualism I am agreeing with Wright (1984), McGinn (1984) and Boghossian (1989) and disagreeing with some recent commentators who view the sceptical solution as factualist. See e.g. Wilson (1994), Byrne (1996), Davies (1998), Kusch (2006). I am not going to argue for the non-factualist aspect of the interpretation in the present paper as I have done so elsewhere: for a full defence of the non-factualist interpretation and a critique of some of the main forms of factualism, see Miller (2009).

and intolerable" error theory by applying this model of meaning to ascriptions of meaning themselves:

> We have to see under what circumstances attributions of meaning are made and what role these attributions play in our lives. Following Wittgenstein's exhortation not to think but to look, we will not reason *a priori* about the role such statements *ought* to play; rather we will find out what circumstances *actually* license such assertions and what role this license *actually* plays. (Kripke, 1982, 86–7)

Kripke argues that when we do this, it transpires that meaning is possible only in a communal context,[6] that there could not be a practice of ascribing meanings in the case of an individual solitary from birth, or "considered in isolation from any community".[7]

Consider the case of Jones, a lifelong solitary. If we initially think of the assertibility conditions of a sentence as conditions in which the participants in the practice of using S regard the utterance of S as justified, we can say that the assertibility conditions of

(6) "Jones means *addition* by '+'"

are conditions in which Jones is confident that he knows "how to go on" in responding e.g. to arithmetical queries featuring "+". Now, the assertibility conditions of

(7) "It seems to Jones that he means *addition* by '+'"

will be identical to those of (6): the utterance of (7) will be regarded as justified by Jones when he is confident that he knows "how to go on".

Likewise, in this context the assertibility conditions of

(8) "'125' is the correct response to '68 + 57 = ?'"

---

[6] We'll take a community to consist of two or more people capable of linguistic interaction at a given time, and who share "innate similarity responses"(see §4 below). This means that a cross-temporal sequence of time slices of an individual – in the manner of Blackburn, 1984b – doesn't count as a community for our purposes here. For further discussion of some of the issues here, see Wee (2015).

[7] The phrase is from Kripke (1982, 110). Kripke writes: "The falsity of the private model does not mean that a physically isolated individual cannot be said to follow rules; rather that an individual, considered in isolation (whether or not he is physically isolated), cannot be said to do so". It is not obvious to me that this holds in the case of a lifelong solitary individual: from the fact that a lifelong solitary individual could be said to be following rules if we imagined him to be a member of our community, it does not obviously follow that a lifelong solitary individual could follow rules (for some relevant commentary see Loar (1985, 279–280)). The issues are delicate and require a fuller discussion than I can attempt here, but I don't think any of my substantive claims or arguments turn on them. So for our present purposes I simply take "an individual considered in isolation from any community" to refer to a lifelong solitary individual and focus on the question whether such an individual could be a genuine rule follower. (For a useful discussion of some of the options here, see Ahmed (2007, 155–163)).

are conditions in which Jones is confidently inclined to respond '125', and the assertibility conditions of

(9) "It seems to Jones that '125' is the correct response to '68 + 57 = ?'"

will again be identical to those of (8), namely conditions in which Jones is confidently inclined to respond "125".

The key point for Kripke's Wittgenstein is that in both cases the "seems right/is right" distinction necessary for the propriety of meaning-talk collapses. Matters stand differently when Jones is a member of a community. The assertibility conditions of

(10) "Jones means *addition* by '+'"

will now be conditions in which Jones is inclined to give the same answers as the community when it is confident that it knows "how to go on", whereas the assertibility conditions of

(11) "It seems to Jones that he means *addition* by '+'"

will be conditions in which Jones is confident that he knows "how to go on".

Likewise, the assertibility conditions of

(12) "'125' is the correct response to '68 + 57 = ?'"

will be conditions in which the community is confidently inclined to respond "125", while the assertibility conditions of

(13) "It seems to Jones that '125' is the correct response to '68 + 57 = ?'"

will be conditions in which Jones is confidently inclined to respond "125".

In the communal context the assertibility conditions of "is right" sentences (such as (10) and (12)) can come apart from the assertibility conditions of "seems right" sentences (such as (11) and (13)) when Jones's inclinations clash with those of the community, so the "seems right/is right" distinction is preserved. And the utility of the practice of ascribing meanings under the relevant conditions is that in uttering e.g. "Jones means *addition* by '+'" we can admit Jones into our community or signal our confidence that he can be trusted to answer as we do. (Clearly, this is useful: if Jones means quaddition rather than addition one should be careful when lending him money!). None of this applies in the case of the lifelong solitary individual, so the practice of ascribing meanings cannot get a grip in that context.

## 13.3 Three Desiderata

In this section I'll outline three desiderata that we should attempt to respect when giving an exposition of the sceptical solution.

### 13.3.1 Communitarian Straight Solutions and Communitarian Sceptical Solutions

First, we should respect the fact that Kripke explicitly distinguishes between communitarian straight solutions and communitarian sceptical solutions to the sceptical paradox. Kripke wants to reject the former, but argues in favour of the latter. This is clearest in the following passage:

> Wittgenstein's theory should not be confused with a theory that, for any 'm' and 'n', the value of the function we mean by 'plus', is (by definition) the value that (nearly) all the linguistic community would give as the answer. Such a theory would be a theory of the truth-conditions of such assertions as "By 'plus' we mean such-and-such a function" ... this would be a social, or community-wide, version of the dispositional theory, and would be open to at least some of the same criticisms as the original form. (Kripke, 1982, 111)

We can illustrate this point by considering a recent interpretation of the sceptical solution which arguably fails to respect it. In a recent book, Claudine Verheggen writes:

> Kripke's answer [to the sceptical paradox] is that ascriptions of meaning to a speaker's utterances have assertibility conditions rather than truth-conditions. They are justified if the speaker's applications of her words agree consistently with those made by competent members of her community. (They are also justified in terms of the role and utility they have in peoples lives.) Thus, in effect, the meanings of a speaker's words are determined by what the members of her community mean by those words. Correct applications are applications that conform to the conditions of correctness at play in her community. (Verheggen, 2016, 90)

Verheggen here appears to ascribe a communitarian *straight* solution to Kripke's Wittgenstein, thereby missing the crucial "sceptical" nature of the solution that he offers. As above, let the assertibility conditions of a sentence S be the conditions under which speakers are regarded by participants in the relevant practice as justified in uttering S, and let the truth conditions of S be the conditions whose obtaining is necessary and sufficient for S's truth. On a realist "truth-conditional" picture of meaning, the fact that S has the assertibility conditions it has will be explained in terms of its having the truth-condition that it has. Thus, that I am regarded as justified in uttering "It is currently raining in Roslyn" by the fact that there is moisture streaming down my window and on the paving stones outside can be explained by the relationship between that fact and the sentence's truth-condition: there is (we can suppose) a causal relationship between the fact that it is raining and the presence of moisture on windows and paving stones. And something like this appears to be what is suggested by Verheggen on behalf of Kripke's Wittgenstein in the passage above: that I agree with the members of my community determines that I mean addition by "+", and this fact about agreement with communal meaning explains why the assertion of "Alex means addition by '+'" is regarded as justified when my answers to arithmetical queries involving "+" are along the lines of those that members of my community are inclined to give. Recall, however, that at this point in the development of the sceptical solution Kripke's Wittgenstein has jettisoned the realist, truth-conditional picture of meaning, and replaced it with a view on which the practice of asserting the relevant sentence under the relevant conditions is justified, not by any relationship between the obtaining of those assertibility conditions and a truth-condition, but by the utility of the practice that allows the assertion of the sentence under those conditions. This is all that is available in the meaning case, given that the sceptical argument of chapter 2 could just as easily be utilised by the sceptic to show that there is no fact capable of making it the case that the *community* means one thing rather than another by the expressions of its language.[8]

### *13.3.2 Ethical Expressivism and the Sceptical Solution*

Second, we should note a difference between Kripke's Wittgenstein's response to the threat of an error theory about meaning and broadly expressivist responses to the parallel threat in the moral case. In the latter, a non-factualist approach to meaning is advocated in a limited number of special cases such as moral sentences, but not in general: the emotivists and expressivists have no wish to reject truth-conditional accounts of the meanings of e.g. sentences attributing physical characteristics to medium-sized physical objects. In the case of the sceptical solution, in contrast, we have a rejection of truth-conditional accounts of meaning *across the board*, so that the rejection of such an account in the case of ascriptions of meaning themselves is an instance of a general strategy that includes sentences from any area of thought and talk within its scope.[9]

[8] Paul Horwich appears to make a similar mistake to Verheggen in his critical notice of Kripke's book. According to Horwich's construal of the sceptical solution "[W]hat someone means by 'plus' is not, as assumed in the skeptical argument, an intrinsic property of that person. Rather, his meaning is determined by the meaning of "plus" within a community language which he speaks. Such a community is the external source of the normative standards that will permit some of a speaker's dispositions to qualify as mistaken" (1984, 166). The problem for this suggestion (and the point Kripke is emphasising at (1982, 111)) is that the sceptic can run his argument against even *extrinsic* candidate meaning-constituting facts.

[9] See e.g. Kripke (1982, 71–75). On Kripke's reading, the switch from the *Tractarian* account of language in terms of truth-conditions to that put forward in the *Philosophical Investigations* in terms of assertibility conditions does not apply only to ascriptions of meaning, but to sentences of any sort. In noting this I'm signalling my agreement with a point that George Wilson makes against the accounts of the sceptical solution adumbrated by Crispin Wright (1984) and Paul Boghossian (1989): contrary to Wright and Boghossian, there is no "gap" in the sceptical solution between the rejection of truth-conditional accounts of ascriptions of meaning and the rejection of truth-conditional accounts generally, hence no need for Kripke's Wittgenstein to construct a "globalisation argument" taking us from the specific to the general. However, on the issue of whether the sceptical solution is factualist or non-factualist, I side with Wright and Boghossian (who take it to be non-factualist) against Wilson (who takes it to be factualist). See Miller (2009).

### 13.3.3 The Nature of Assertibility Conditions

Third, although Kripke introduces the distinction between assertibility conditional and truth-conditional accounts of meaning in the context of a reference to Michael Dummett's piece on Wittgenstein's philosophy of mathematics, it is important to note that the notion of an assertibility condition that figures in the sceptical solution is completely different from the notion that appears in Dummett's writings on realism and anti-realism. An assertibility condition, in this latter, more familiar sense, is a condition in which there is evidence that the truth-condition of a sentence obtains. There are at least two reasons why this cannot be the notion that Kripke's Wittgenstein is deploying in the sceptical solution. First, in order for an ascription of meaning to have an assertibility condition in this sense it would need to *have* a truth-condition. But the sceptical argument showed – didn't it? – that if we think of ascriptions of meaning as having truth-conditions we will have to view them as systematically false. So deploying this notion would appear not to save us from the clutches of the sceptical paradox. We can reinforce this point by considering the change in Dummett's conception of semantic anti-realism signalled in the preface to his *Truth and Other Enigmas* (1978, xxii)). There, Dummett suggests that the proper way to express semantic anti-realism is not in terms of an adherence to an assertibility conditional theory of meaning, but rather in terms of adherence to a truth-conditional account of meaning in which the notion of truth is epistemically constrained (i.e. where we reject the idea that the applicability of the principle of bivalence is not restricted by epistemic considerations). Since the role of assertibility conditions in Dummett's earlier writings is (post-1978) played by epistemically constrained truth-conditions, we can surmise that Dummett's assertibility conditions play the same role in the anti-realist theory of meaning as epistemically unconstrained truth-conditions play in the realist theory of meaning or epistemically constrained truth conditions play in the post-1978 anti-realist theory of meaning. Since the account of the meanings of ascriptions of meaning given by each of these latter two positions would straightforwardly succumb to the argument of Kripke's sceptic, we can infer that the sceptical solution cannot be viewed as utilising the Dummetian notion of an assertibility condition on pain of leaving itself exposed to the sceptical paradox.[10]

[10] Note that restricting the search for truth-conditions for ascriptions of meaning to states of affairs that are not potentially evidence-transcendent actually makes it harder to find a straight solution in response to the sceptic's argument. Whatever the sceptical solution is, it is not a form of Dummettian semantic anti-realism. It seems to me that the argument of Wright (1984, 105–6) – to the effect that the sceptical paradox can be re-deployed to establish an indeterminacy result against the sceptical solution – depends on construing the notion of an assertibility condition in the sceptical solution along Dummettian lines. In an interesting discussion, Daniel Boyd (2017) argues that Wright's argument fails even given the usual notion of an assertibility condition. He distinguishes between e.g. "Jones associates assertibility condition X with expression Y" and e.g. "Expression Y is warranted in condition X" and suggests that Wright's argument fails because we could accept a non-factualist construal of the former whilst retaining a factualist construal of

So, the assertibility condition of an ascription of meaning in the sceptical solution is not a situation in which there is evidence warranting the assertion that its truth-condition obtains – at the point at which the sceptical solution takes off we have given up the idea that accounting for the meaning of a sentence of any sort is a matter of identifying its truth-conditions. I think that this is what Kripke is emphasising when he writes:

> It is important to realize that we are *not* looking for necessary and sufficient conditions (truth conditions) for following a rule, or an analysis of what such rule-following 'consists in'. Indeed such conditions would constitute a "straight" solution to the sceptical problem, and have been rejected. (Kripke, 1982, 87)

Overall, then, any account of the sceptical solution must respect the distinction between straight and sceptical forms of communitarianism, the fact that the scope of the sceptical solution is general and not just restricted to ascriptions of meaning, and the fact that Kripke's Wittgenstein's assertibility conditions are not assertibility conditions in the sense familiar from the literature on semantic anti-realism.[11]

## 13.4 Stroud's "Nothing for Us to Assert" Objection

So what *are* Kripke's Wittgenstein's assertibility conditions? We can work towards an answer to this question by considering an objection to the sceptical solution developed by Barry Stroud. Stroud writes:

> [Kripke's Wittgenstein's] suggestion that "assertibility-conditions" for ascriptions of meaning and understanding are all we can find or even hope for can make no sense of *what* we are said to be justified in asserting when the assertibility conditions are fulfilled ... [I]f what is to be asserted has no "truth-conditions", there would seem to be nothing for us to assert. (Stroud, 1990, 89)

So far, this really only amounts to the observation made above, to the effect that Kripke's Wittgenstein's assertibility conditions can't be viewed as "Dummettian" assertibility conditions. However, Stroud goes on to make a useful suggestion about

---

the latter, which is what Kripke's Wittgenstein needs to make the sceptical solution workable. However, if we think of assertibility conditions as akin to rules governing the use of expressions, Boyd's suggestion amounts to the idea that there could be facts about rule *compliance* (i.e. accord and failure to accord with a rule), even if there were no facts about rules being *followed* by agents. This is highly implausible, and it is hard to see Kripke's Wittgenstein accepting it: the idea of a rule with determinate compliance conditions but which it is in principle impossible to follow is merely paradoxical, and Kripke (1982, 55) moves seamlessly from the absence of rule following to the absence of accord/conflict with rules.

[11] The full significance of this last point will not become apparent until the end of the paper.

how the notion of an assertibility condition in the sceptical solution might be understood:

> [Perhaps] it would be enough simply to say that there is an expression we would be justified in using assertively in certain circumstances, without our having to identify anything that could be called the content of that expression, or saying what it means. (Stroud, 1990, 90)

Perhaps we can build on this suggestion on behalf of Kripke's Wittgenstein. If, like Quine (1960, 29–30) with his notion of "assent", or Davidson with his notion of "holding true" (1984, 135), we can identify what it is for a sentence to be used assertively independently of an assumption about the meaning of that sentence, this can be pressed into service in the manner described by Stroud.[12]

However, Stroud himself thinks that the suggestion he contemplates fails in the case of ascriptions of meaning:

> [T]he suggestion ... fails because using assertively the expression "He means plus by that sign" is not the same thing as asserting that the person means plus by that sign. What assertion you are making in using a certain expression assertively depends in part on what the expression you are using means. If you use an expression in order to ascribe a specific, determinate meaning or understanding to a community or a person, the expression you use must itself have a determinate meaning. But the notion of "assertibility conditions" and the suggestion which speaks only of the assertive use of certain mentioned expressions was introduced precisely because it held that there was no such fact as the fact of an expression's having a determinate meaning, or someone's meaning some particular thing by an expression he uses. (Stroud, 1990, 90)

Arguably, though, this objection is unconvincing. *If* you think of meaning in terms of truth-conditions, then in assuming that ascriptions of meaning don't have truth-conditions, you are thinking of them as lacking meaning: in which case, as Stroud points out, there is no sense to the idea of using them assertively. But if, like the purveyor of the sceptical solution, you have given up thinking of meaning in terms of truth conditions, the objection lapses. To put it another way, the objection only works if we have to think of meaning in terms of truth-conditions – so since the sceptical solution has rejected that idea, the objection simply begs the question against the sceptical solution.

We can see this even more clearly by recalling Kripke's comment about the language game of arithmetic:

> Wittgenstein asks that we discard any *a priori* conceptions and *look* ("Don't think, look!") at the circumstances under which numerical assertions are actually uttered, and at the roles such assertions play in our lives. (1982, 75)

Kripke's Wittgenstein's idea is that we *look* at the conditions in which there is assertoric use by the community of ascriptions of meaning and tell a story about the pragmatic utility of that practice. To be sure, in identifying those conditions we will not be identifying the *content* of ascriptions of meaning or materials that could be

---

[12] I'm working here on the assumption that if a notion is acceptable to Quine, invoking it in the course of constructing the sceptical solution is not unacceptable. See Lepore and Ludwig (2005, 177) and Glüer (2011, 73) for the Quine-Davidson link.

used in an analysis of them – but as we saw (see again (1982, 87), quoted above) Kripke has already explicitly put that idea to one side. So Stroud has given us no reason to think that Kripke's Wittgenstein's conception of assertibility conditions is problematic, other than the observation that on it ascriptions of meaning are not assigned truth-conditions. In the context, in the absence of further argument, Stroud is simply begging the question against the alternative proposed in the sceptical solution by assuming that ascriptions of meaning could not be regarded as meaningful if they are not assumed to have truth-conditions.

## 13.5 Stroud's "Specificity" Objection

In fact, Stroud has a prior worry about the assertibility conditions provided by the sceptical solution. In this section, I will outline this objection and sketch a response. Then, in the next section I will use this response to clear up some confusions about the nature of the sceptical solution on the part of Davidson and some of his followers.

Stroud outlines what we might call his "specificity" worry as follows:

> [E]ven restricting ourselves to the conditions under which [ascriptions of meaning] are justifiably asserted, the mere facts of general consensus in the community or the conformity of individual speakers could not give us what we want. Knowing only that all members of a community use the word "otiose" in the same way does not justify us in asserting that what "otiose" means is functionless. And knowing only that someone uses a certain sign he puts between two numerals in the same way as everyone else in the community uses that sign does not in itself justify us in asserting that what he means or understands by that sign is plus. Nor is that justified by my knowing only that he uses or responds to that sign in the same way that I do. We are never justified in making such specific ascriptions on such general and indiscriminate grounds. (1990, 89)

And he continues:

> [M]ere conformity to community practice does not justify us in uttering "He means plus by that sign" rather than, say, "He means times by it" or "He means larger than", as long as his use of the signs for those other functions also conforms to the community's use of them. ... There is conformity in each case. If conformity to community practice is the only evidence, nothing in the evidence shows why we could use only the one expression assertively and not the others. (Stroud, 1990, 90)

On Kripke's Wittgenstein's account – Stroud argues – the assertibility conditions of "Jones means times by 'x'" are circumstances in which Jones's responses to arithmetical queries involving the 'x' sign are the same as those of the community. But the assertibility conditions of "Jones means plus by '+'" are also circumstances in which Jones's responses to arithmetical queries involving the "+" sign are the same as those of the community. In each case, since all that is mentioned is conformity with communal use, the relationship between Jones's use and community use is the same, so according to Stroud the sceptical solution assigns the same assertibility conditions to "Jones means addition by '+'" and "Jones means times by 'x'", so that it is mysterious why "Jones means addition by '+'" is assertible rather than "Jones means times by '+'". He concludes that the assertibility conditions provided by the sceptical solution are simply too "general and indiscriminate" to capture the differences between specific ascriptions of meaning.

If the sceptical solution were indeed unable to discriminate between the assertibility conditions of e.g. "Jones means addition by '+'" and "Jones means times by '+'" this would constitute a serious worry for the account: it is surely a requirement that it pin down assertibility conditions that are suitably specific to "Jones means addition by '+'". We can work towards remedying this apparent defect in the sceptical solution by considering the assertibility conditions we provided earlier for

(10) "Jones means *addition* by '+'".

These were conditions in which Jones gives the same answers as those the community is inclined to give when it is confident that it knows "how to go on". This appears to be susceptible to the worry adumbrated by Stroud, since "Jones means times by '+'" too will be assertible when Jones gives the same answers as those the community is inclined to give when it is confident that it knows "how to go on". But we can start to add the requisite specific detail by including a reference to communal inclination to judgement: the assertibility condition of (10) will now be the circumstance in which Jones gives the same answers as the community when *it* is inclined to judge that "+" means *addition*. Clearly this is different from the circumstance in which Jones gives the same answers as the community when it is inclined to judge that "+" means times. (For example, the first circumstance would be one in which Jones answered "125" to the query "68 + 57 = ?" while the second circumstance would be one on which Jones answered "3876"). However, the reference to communal judgement in these assertibility conditions is illegitimate – the use of the that-clause is not allowed given that the sceptical solution has disavowed the attribution of truth-conditions to ascriptions of meaning. We can overcome this, though, by replacing the reference to communal judgement by a reference to communal dispositions to assertive use, deploying the notion of uttering a sentence assertively that we introduced in the previous section.[13] In other words, we can say that the assertibility conditions of (10) are circumstances in which Jones gives the same answers as the community when it is disposed to utter assertively "'+' means *addition*". This solves the specificity worry. Giving the same answers as the community when it is disposed to utter assertively "'+' means *addition*" would involve answering "125", whereas giving the same answers as the community when it is disposed to utter assertively "'+' means *times*" would involve answering "3876".

[13] Note that this is not intended as an *analysis* of the relevant communal judgement – such an analysis is the preserve of straight solutions to the paradox. Rather, we are cashing out the assertibility conditions by using communal dispositions to use a sentence assertively in order to avoid importing any resources off limits to sceptical – as opposed to straight – solutions.

Overall, then, we can take the assertibility condition of a sentence to be conditions in which the participants in the relevant practice are disposed to utter it assertively. In the kind of communal context envisaged in the sceptical solution, the community will be disposed to utter assertively "Jones means addition by '+'" in conditions in which Jones gives the same answers the community is inclined to give when it is disposed to utter assertively "'+' means addition".[14] Given the question-begging nature of the "nothing to assert" objection considered in the previous section, we can tentatively conclude that on the account offered here the sceptical solution succumbs to neither of Stroud's objections.

## 13.6 Strong Communitarian and Interpersonalist Conceptions of Meaning

We can build on the account sketched above to get clear on the respect in which the sceptical solution involves a broadly social (as opposed to individualistic) conception of meaning. Claudine Verheggen draws a distinction between two ways in which an account of meaning can be social. According to communitarian views "[H]aving a (first) language essentially depends on meaning by one's words what members of some community mean by them" (Verheggen, 2016, 84). I'll refer to this type of view as *strong communitarianism.* In contrast, interpersonalist views hold that "[T]he possession of language and thought essentially depends on having [interacted] linguistically with others ... [but] does not require that [those involved in the interaction] assign the same meaning to the same words" (Verheggen, 2016, 84).[15]

Verheggen takes the sceptical solution to be putting forward a strong communitarian view in this sense. Recall her comment that according to Kripke's Wittgenstein "[T]he meanings of a speaker's words are determined by what the

[14] To be fair, I should say that in replying to the specificity objection in this way, I'm going beyond anything that Kripke says explicitly in the text. I hope, though, that the proposal is at least a plausible extension of his development of the sceptical solution so that it can count as "constructive" exegesis. There may well be other ways of dealing with the specificity objection, perhaps by deploying the disquotational rule that we can always assert that we use "+" to mean +. (This, together with the fact that Jones's use of "+" accords with ours, would license the assertion that he means + by "+"). If this other way of dealing with the problem works, that's fine by me. (I'm grateful here to an anonymous referee).

[15] I've changed the characterisation very slightly. Where Verheggen speaks of "triangulation", I have spoken simply of "interaction". For Verheggen, Davidson's view – which is couched in terms of triangulation – is the paradigm example of an interpersonalist view, but we should at least leave open the possibility that there are ways other than triangulating in which language users can interact as part of a story about the constitution of meaning. Note too that Verheggen speaks of communitarianism and interpersonalism whereas I speak of strong communitarianism and interpersonalism. This leaves open the possibility of describing interpersonalist views as "weak communitarianism". Nothing hangs on this terminological preference.

members of her community mean by those words" (2016, 90). Other writers view the sceptical solution in the same way. For example, Åsa Wikforss and Kathrin Glüer take Kripke's Wittgenstein to be proposing a view on which the meanings of linguistic expressions are determined by "conventions adopted by whole speech communities"(2018, §2.2.2). Further, Davidson himself views the sceptical solution in this way. He takes Kripke's Wittgenstein to be proposing a view on which "to mean something in speaking, one must mean the same thing by the same words as others do" (1992, 114), and he characterises "Kripke's criterion for speaking a language" as involving the idea that "speaking a language ... depends ... on speaking as someone else does (or as many others do)" (1992, 115). In the middle of a critical discussion of Kripke's Wittgenstein, Davidson contrasts his own view with one on which being understood "involves shared rules or conventions" (1992, 114), and elsewhere he construes Kripke's Wittgenstein as holding that "communication depends on speakers using the same words to express the same thoughts" (1991, 209 n.1).

All of these views of Kripke's Wittgenstein's sceptical solution seem to me to make two mistakes. First, and most obviously, they mistakenly construe Kripke's Wittgenstein as proposing a *straight* solution according to which what an individual means by an expression is determined by facts about rules or conventions adopted by his community. Of course, this cannot be right: it is easy for the sceptic to repeat at the level of the community as a whole the arguments he deployed at the level of the individual. What facts about the community make it the case that it is *these* conventions rather than *those* that govern the linguistic activity of its members? Recall (from Sect. 13.3 above) that Kripke himself makes it clear that his Wittgenstein is not proposing a communitarian straight solution to the sceptical paradox: "a social, or community-wide, version of the dispositional theory ... would be open to at least some of the same criticisms as the original [i.e. individualistic] form" (Kripke, 1982, 111).[16]

Second, and less obviously, the sceptical solution is consistent with Davidson's interpersonalism and weak communitarianism. We can see this by asking: is there space within the framework provided by the sceptical solution for it to be assertible that an individual speaker means something different from what the rest of the community means by a given expression? It seems to me that the answer is yes: this will be assertible when the individual concerned uses the expression in the way the community *would be* inclined to use it *were* they to judge that they mean what he says he means by it (where, as above, the reference to communal judgement is eliminated and replaced by a reference to communal dispositions to utter sentences assertively). For example, suppose that Jones is inclined to utter "I mean hexagonal by 'green'". The assertibility condition of "Jones means hexagonal by 'green'" will

[16] Davidson himself writes "I have ignored a very important aspect of Kripke's discussion, his claim that Wittgenstein's 'solution' to the problem is 'skeptical'" (1992, 113 n.7). But the "sceptical" nature of Kripke's Wittgenstein's position is not just a "very important aspect" that can be downplayed if the discussion seems to demand this, but an absolutely essential aspect that cannot be suppressed without seriously distorting the position.

be that Jones uses "green" in a way that agrees with how the community *would* use it *were* they to be disposed to utter assertively "'green' means hexagonal" – and, crucially, this can be so even if the community is in fact disposed to utter assertively "'green' means green".[17] Or, again, "Jones means *quaddition* by '+'" would be assertible in circumstances in which Jones gives the same answers as the community would have given if it had been disposed to utter assertively "'+' means *quaddition*" – and this is consistent with the community being in fact disposed to utter assertively "'+' means addition". So it is consistent with the sceptical solution that "Jones means quaddition by '+'" is assertible even though it is assertible that the rest of the community means addition.

Thus, it is consistent with the sceptical solution that agreement with such *counterfactual* inclinations of the community will license the assertion that an individual means something by an expression, but something different from what the community takes itself to mean. And even though this is not agreement with the way the community actually uses the word, it will often be enough in practice to provide them with a basis for interpreting and understanding the individual. The utility of meaning ascriptions taken up by the sceptical solution is preserved: by saying that Jones means such and such by an expression we can signal to our fellow community members that we can expect him to act as we would were we to judge that we mean such and such by the expression, and this expectation can form the basis for interpretation and understanding even if we do not in fact judge that we mean such and such by the relevant expression ourselves.[18] Thus, it is consistent with the sceptical solution that an individual can mean something by his words – and be understood accordingly – even if he does not mean the same thing by his words as his interpreters. So the sceptical solution as such is consistent with the Davidsonian interpersonalist view favoured by Verheggen: in our terminology, the sceptical solution is consistent with weak communitarianism.

## 13.7 Four Notes

We can make four notes following on from the discussion in the previous section. Firstly, it is true that the argument in section 6 to the effect that the sceptical solution is consistent with an interpersonalist or weak communitarian account of

[17] An anonymous referee has helpfully pointed out that there may be a simpler way – than referring to the counterfactual inclinations of the community – to spell out the assertibility condition for "Jones means hexagonal by 'green'". We could instead say that this is assertible when Jones's use of "green" agrees with our (the community's) inclinations to use "hexagonal". I've held off from replacing the counterfactual analysis with that suggested by the referee because I'm not sure that the simpler formulation is conservative with respect to my reply to the second objection I try to address in §7 below. It would be fine by me if that were so.

[18] As above, strictly speaking the references to judgements about meaning here should be replaced by references to dispositions to use the relevant sentences assertively.

meaning rests in part on the solution to Stroud's "specificity" objection as outlined in Sect. 13.5. Note, though, that the sceptical solution faces the "specificity" worry even on a *strong communitarian* interpretation, so that the possibility of an interpersonalist/weak communitarian reading is opened up by a solution to a problem also faced by the sceptical solution on strong communitarian readings. It follows that anyone proposing a strong communitarian reading cannot object to our interpersonalist or weak communitarian reading unless they provide a different way of dealing with Stroud's "specificity" problem – in particular, a way that doesn't open up the possibility of an interpersonalist construal of the sceptical solution.

Secondly, it might be argued that the "interpersonalist" version of the sceptical solution sketched in section 6 is still appreciably stronger than Davidson's interpersonalist view, insofar as for an individual, Jones, to mean something by an expression it has to be something that his community *could have* meant by the same expression i.e. something that the community means by the expression in some other possible world.

In reply, however, it can be argued that the condition imposed by our reading of the sceptical solution is so weak that even Davidson has to accept it. Consider the sort of example sketched in Davidson (1986). Mrs Malaprop uses the word "derangement" to mean arrangement, and the word "epitaph" to mean epithet, in an exchange with Smith. Now, even though Smith does not mean these himself by "derangement" and "epitaph" (we can suppose that he means *madness* and *short text honouring a dead person*), he is able, on the basis of the relevant clues, to understand what Mrs Malaprop is saying when she utters "A nice derangement of epitaphs". In particular, he can form reasonable expectations about her future use of the words (perhaps that she will say "That is not a nice derangement of epitaphs" on looking at such and such a text). The key point is that Smith would not be able to do this if it was not possible for him to have meant arrangement by "derangement" or epithet by "epitaph". So it appears that the story Davidson tells about Mrs Malaprop involves an interlocutor satisfying the condition imposed on the community in the type of interpersonalist sceptical solution sketched in the previous section (Mrs Malaprop doesn't mean the same as Smith by "derangement" and "epitaph", but there are possible worlds in which Smith means what she means in the actual world).

Thirdly, in Miller (2017) I suggested that Davidson and Kripke's Wittgenstein impose conditions of equal strength on the possibility of meaning and communication: that there be creatures who share the same "innate similarity responses".[19] Olivia Sultanescu and Claudine Verheggen (2019) question this claim. They point out that, in the sceptical solution "what is necessary is that there be agreement with respect to the ways in which expressions are applied in particular contexts" (22, n.30), and, furthermore, that "the agreement that is necessary on Davidson's picture concerns 'facts about salience, attention and tendencies to generalize in some ways rather than others'(Davidson, 1990, 61), which may not result in agreement with respect to the ways in which expressions are applied"(ibid.). This is because for

[19] This phrase is taken from Davidson (1992, 120).

Davidsonian triangulation to be possible we need to be able “to discriminate, pre-semantically, green things from things that are not green”(ibid.). But this does not require that we also “agree in the particular applications we make to green objects or in the ways we categorize those objects”(ibid.). So, Sultanescu and Verheggen conclude, the pre-conditions for the possibility of meaning imposed by Davidson are in fact weaker than those imposed by Kripke’s Wittgenstein in the sceptical solution.

In reply to this, I concede that the sort of agreement required by the sceptical solution (as I sketched it above) requires agreement vis-à-vis the application of linguistic expressions. However, since it is consistent with the sceptical solution that the relevant speakers do not mean the same thing by the same words, they do not have to apply e.g. “black” to the same class of objects. For example, in order for “John means green by ‘black’” to be assertible, John’s linguistic inclinations have to be the same as the linguistic inclinations the community would have if they were disposed to utter assertively “‘black’ means green”. And, crucially, this will be so only if the community and John have similar innate similarity responses vis a vis green things – but, for all that, the community may in fact be disposed to utter assertively “‘black’ means black”, so that while it applies “black” to the bottom half of the anarchist flag, John does not. So it turns out, again, that the condition on the possibility of meaning imposed by Kripke’s Wittgenstein is no stronger than that imposed by Davidson.

Finally, in discussion Timothy Williamson raised the worry that there is a damaging circularity in the account of the assertibility conditions of e.g. “Jones means quaddition by ‘+’” used in the previous section to argue that the sceptical solution is consistent with Davidsonian interpersonalism. It will easiest to outline this using the case of Mrs Malaprop. The account offered in our construal of the sceptical solution contains claims like the following:

(Assertibility) The assertibility condition of “Mrs Malaprop means epithet by ‘epitaph’” = circumstances in which Mrs Malaprop’s use of “epitaph” is the same as that the community would have displayed if it had been inclined to utter assertively “‘epitaph’ means epithet”.

We have to assume that in the set of possible worlds relevant to the evaluation of the counterfactual that appears on the right-hand side of (Assertibility), “epithet” *means the same* as it means on the left-hand side, otherwise the identity will be false. But an assumption like this is a substantial assumption about meaning the making of which would import a damaging circularity into the sceptical solution. (To see why the assumption is necessary, imagine a world in which by “epithet” the community means *epiphany*. In such a world, the community’s use of “epitaph” will be quite different from Mrs Malaprop’s use of it in the actual world: the community will apply it to e.g. moments of revelation experienced by characters in Joyce’s stories, while Mrs Malaprop will refrain from doing so).

I’ll make two comments in response. First, it is not obvious that such an assumption induces a fatal circularity, given that (Assertibility) is not put forward as an *analysis* or *explanation*, but rather as shorthand for the empirical claim that the community is disposed to utter assertively "Mrs Malaprop means epithet by 'epitaph'" when and only when Mrs Malaprop's use of "epitaph" is the same as that the community would have displayed if it had been inclined to utter assertively "'epitaph' means epithet".

But second, and even putting that point to one side, we can deal with the putative problem without making any substantial assumptions about meaning. Suppose that the left-hand side of (Assertibility) is true. Then note that the right-hand side is the following counterfactual:

C) If the community had been inclined to utter assertively "'epitaph' means epithet", the community's use of "epitaph" would be the same as Mrs Malaprop's use of the expression in the actual world.

What is required for the truth of C) is that in the closest worlds to the actual world in which the community is disposed to utter assertively "'epitaph' means epithet", the community's use of "epitaph" would be the same as Mrs Malaprop's in the actual world. We could stipulate that only worlds in which the community means what it means by it in the actual world count as close, but this raises the worry about circularity voiced above. Instead, let's stipulate that for the purposes of confirming C), the only worlds to count as close are those in which the community's non-semantically and non-intentionally characterised use of "epithet" corresponds to its non-semantically and non-intentionally characterised use in the actual world. This rules out the kind of "epiphany" world mentioned in the objection: in such a world the community's non-semantically and non-intentionally characterised use of "epithet" will be different from that in the actual world. Hence, the claim that (Assertibility) is true needn't introduce a damaging circularity into the sceptical solution.

## 13.8 The Sceptical Solution and Quasi-Realism

In Sect. 13.3 above, I noted some dissimilarities between the non-factualist position about meaning advocated in the sceptical solution and the sort of non-factualism exemplified in broadly expressivist and non-cognitivist theories of moral judgement. In this section, I will try to shed further light on the nature of the sceptical solution by discerning some *similarities* between the two, taking as my paradigm expressivist view the sort of quasi-realism developed by Simon Blackburn.[20] I'll then outline what I take to be the two most serious challenges facing this position.

In Blackburn's work, the fundamental task of metaethics is explaining the nature of moral or ethical judgement.[21] A realist about moral judgement is someone who

[20] See in particular Blackburn (1984a), chapters 5 and 6.

[21] See e.g. Blackburn (1993a, 368): "[M]y instinct is to start, with Kant, as far back as possible: how *can there be* such a thing as moral judgement?"

starts off the attempt to explain the nature of moral judgement by locating distinctively moral states of affairs. Once identified, we can explain moral judgement as the expression of belief about the obtaining or otherwise of these states of affairs. To cut a long story short, the realist account falters because of a dilemma that opens up at the outset of the realist's attempt to deploy this strategy. If we think of the distinctively moral states of affairs as natural,[22] we run afoul of (something like) Moore's open-question argument: we fail to capture the action-guiding nature of moral judgement.[23] On the other hand, if we think of the distinctively moral states of affairs as non-natural, we face the metaphysical and epistemological worries routinely taken to undermine the non-naturalism of Moore and the intuitionists who followed him in the first third of the twentieth century.[24] The moral that Blackburn draws from this dilemma is that we should not begin our attempt to explain the nature of moral judgement from the starting point favoured by the realist, i.e. by isolating states of affairs that we deem to be distinctively moral. Rather, we should "in the first instance, think of moral judgements as expressing states of mind whose function is not to represent anything about the world" (Blackburn, 1988a, 184), and otherwise restrict ourselves to a methodologically naturalist toolkit which includes only: natural states of affairs (none of which we try to identify with moral states of affairs), beliefs about natural states of affairs, and desires/sentiments towards natural states of affairs.[25] In Blackburn's quasi-realist project, we use these materials to construct a notion of moral truth, inter alia solving the Frege-Geach problem and the other familiar problems that beset classical emotivism.[26] A "first approximation" might be that an expression of a moral attitude is true if and only it belongs to the set of attitudes "which would result from taking all opportunities for improvement of attitude" (Blackburn, 1984a, b, 198). If we pull off this task, we can happily conclude that e.g.

(14) "Torture is wrong" is true if and only if torture is wrong

and, indeed, that the judgement that torture is wrong is true. In concluding thus, though, we will not have "sold out" to moral realism: moral states of affairs and beliefs about them were not part of the methodologically naturalist toolkit we deployed in giving the account, so we have not made the assumptions that

---

[22] Where natural properties are (something like) those that are part of the natural and social sciences (a conception of natural that derives from Moore (1903)).

[23] For a good account of this, see the opening sections of Darwall et al. (1992).

[24] See Miller (2013), chapters 2 and 3 for a summary.

[25] For "toolkit", see e.g. Blackburn (1993a, b, 378). I say "methodologically" naturalist, because although Blackburn is attempting to give a naturalistically respectable account of moral judgement, there is no attempt to specify purely naturalistic truth-conditions for moral judgements, or naturalistic semantic values for moral predicates.

[26] The Frege-Geach problem is the problem of accounting for the role of e.g. moral statements, as construed by the emotivist, in the antecedents of conditionals such as "If lying is wrong, getting little brother to lie is wrong". For an overview, see Miller (2013) chapters 3 and 4.

generated the damaging dilemma for the realist.[27] The states of affairs designated on the right-hand sides of (14) and the like are not natural (wrongness is not a property attributed by any of the natural or social sciences), and the role given to desire and sentiment in the explanation of moral judgement ensure that the view we have reached is immune to worries relating to the open-question argument and the motivational characteristics of moral judgement. But, in addition, because there were no irreducibly moral states of affairs in our methodologically naturalist tool-kit, the metaphysical and epistemological worries that afflicted Moorean non-naturalism don't arise. Since the state of affairs designated on the right hand of (14) is not assigned a role in the quasi-realist explanation of the nature of moral judgement, there is no need for us to give a metaphysical account of its nature, or an epistemological story about how we access it, over and above a story about the nature of, and our access to, states of affairs the description of which utilises materials from the methodologically naturalist toolkit used in the quasi-realist construction of moral truth.

I think that we can view the sceptical solution as having a similar overarching structure to the position of Blackburn's just outlined. In the general case, if we were giving a realist account of the judgement that snow is white we would do this via a truth-conditional account of the meaning of e.g. "snow is white". We would attempt to identify a state of affairs whose obtaining is necessary and sufficient for its truth: a state of affairs, in other words, that *makes it the case* that snow is white. In the case of meaning itself, if we were aiming to give a realist account of e.g. the judgement that Jones means addition by "+", we would do this via a truth-conditional account of the meaning of the sentence "Jones means addition by '+'" and attempt to identify a state of affairs whose obtaining makes it the case that Jones means addition by "+". Kripke's Wittgenstein's sceptic's argument shows, we are supposing, that the two constraints (4(a) and 4(b) on page 295 above) that have to be met in order for such an account to be successful cannot in fact be met. In a quasi-realist style of reconstruction of the sceptical solution there would be no requirement to identify such a state of affairs, since we have self-consciously given up the aspiration to explain the nature of the judgement that Jones means addition by "+" in terms of

[27] It might look like I'm misconstruing quasi-realism here, insofar as I'm viewing it as providing truth-conditions for moral claims. However, commenting on the nature of quasi-realism, Blackburn himself writes: "*Initially, an expressive theory stands in stark contrast to one giving moral remarks truth-conditions*. But if we sympathise with the pressures I have described, we come to appreciate why it should be natural to treat expressions of attitude as if they were similar to ordinary judgements. We come to need a predicate, whose behaviour is like that of others. Why not regard ourselves as having *constructed* a notion of moral truth? If we have done so, we can happily say that moral judgements are true or false, only not think that we have sold out to realism when we do so" (1984a, b, 196, emphasis on first sentence added). Blackburn's point is that although *initially*, an expressive theory contrasts sharply with those theories which give moral remarks truth-conditions, *once the quasi-realist project has been executed* we can happily view moral remarks as having truth-conditions without thereby "selling out" to realism. Ironically, among commentators this aspect of Blackburn's view has been best appreciated by one of his most trenchant opponents, John McDowell. See §2 of McDowell (1987).

a state of affairs. So the sceptical argument of chapter 2 of Kripke (1982) cannot get a grip. Instead of attempting to account for ascriptions of meaning in terms of their truth-conditions, in the first instance we think of them as expressing states of mind whose function is not to represent the world, and then move from that starting point to construct a notion of truth genuinely applicable to them. Ascriptions of meaning, such as "Jones means addition by '+'", can in the first instance be viewed as marking our acceptance of Jones as a member of our linguistic community and expressing our trust in him to respond to arithmetical queries involving "+" just as we would (Kripke, 1982, 93, 95). We then identify the assertibility conditions of such ascriptions and tell a story about the utility of the relevant practice. Finally, we use the assertibility conditions identified in the sceptical solution to construct a notion of truth in semantics in a manner analogous to Blackburn's quasi-realist construction of moral truth. Then, if all went well, we would end up with e.g.

(15) "Jones means addition by '+'" is true iff Jones means addition by "+".

Since we are not trying to explain the meaning of the ascription of meaning in terms of a condition whose obtaining is necessary and sufficient for its truth, we are under no obligation to tell a story about what makes it the case that that state of affairs obtains. And the epistemological challenge that beset non-reductionism about meaning – to account for the first-person authority and non-inferential character of knowledge of meaning – lapses also.[28] On a realist picture, knowledge of the meaning of an ascription of meaning involves knowledge of the state of affairs that makes it the case that you mean what you mean, and it was difficult to see how this picture could adequately capture the first-person epistemology of meaning.[29] On the alternative, quasi-realist, story, the first person epistemology simply falls out of the assertibility conditions that figure in the sceptical solution:

> The criterion by which others judge whether a person is obeying a rule in a given instance cannot simply be his sincere inclination to say that he is; otherwise there would be no distinction between his thinking he is obeying the rule and his really obeying it (§202), and whatever he thinks is right will be right (§258). However, *after* the community judges (based on the original criteria) that he has mastered the appropriate rule, the community may (for certain rules) take the subject's sincere claim to follow it in this instance as in itself a new criterion for the correctness of his claim, without applying the original criteria. (Kripke, 1982, 101 n.82)

[28] See e.g. Wright (1989a, 177–78). As Wright sees it, the epistemological challenge is to account for the non-inferential, first-person authoritative character of knowledge of meaning while accommodating its "disposition-like theoreticity" (Wright, 2001, 87).

[29] The realist runs into trouble because he has to see the knowledge as a matter of cognizing or tracking independent states of affairs, but has nothing useful to say about how this alleged "tracking" accomplishment underwrites the first-person authority of meaning. See Wright's critique of McGinn on this point (Wright, 1989b, 147–150). The issues here are probably best framed in terms of the cartography of debates about realism developed in Wright (1992). For a useful discussion, see Butler (2019).

The "normativity" of meaning is also easily captured. If Jones desires to tell the truth and means addition by "+", then he ought to answer "125" in response to "68+57 =?" – and there is no need to capture this normative fact on the basis of a state of affairs necessary and sufficient for Jones's meaning addition by "+".[30]

## 13.9 Questions and Challenges for the Sceptical Solution

I'll finish by noting some questions and challenges for the sceptical solution construed as we have construed it.

How, in detail, do the assertibility conditions in the sceptical solution feed into a broadly quasi-realist construction of truth applicable to ascriptions of meaning? In the moral case, this involves dealing with the Frege-Geach problem, earning the right to the idea of improvement and deterioration of ethical sensibilities, then characterising truth for moral commitments in terms of membership of the set of attitudes that remains after all opportunities for improvement have been taken (see again Blackburn, 1984a, b, 198) – all on the basis of the "methodologically naturalist" toolkit outlined in section 8. One possible disanalogy with Blackburn's approach as outlined above is that Kripke's Wittgenstein appears to shy away from any attempt to confront the Frege-Geach problem and related issues. Kripke writes:

> Like many others, Wittgenstein accepts the 'redundancy' theory of truth: to affirm that a statement is true (or, presumably, to precede it with 'It is a fact that ...') is simply to affirm the statement itself, and to say it is not true is to deny it: ('p' is true = p). However, one might object: (a) that only utterances of certain forms are called 'true' or 'false' – questions, for example, are not – and these are so called precisely because they purport to state facts; (b) that precisely the sentences that 'state facts' can occur as components of truth-functional compounds and their meaning is hard to explain in terms of assertibility conditions alone. Wittgenstein's way with this is also short. We *call* something a proposition, and hence true or false, when in our language we apply the calculus of truth conditions to it. That is, it is just a primitive part of our language game, not susceptible of deeper explanation, that truth-functions are applied to certain sentences. (1982, 86)[31]

This appears to militate against our construal of the sceptical solution as a kind of quasi-realism insofar as that must include an attempt to solve the Frege-Geach problem. However, we should note that so far in this paper we have been construing quasi-realism as what Blackburn calls "slow-track" quasi-realism. Speaking of

[30] Kripke's remarks on "inversion of conditionals" (1982, 93–5) are relevant here. Note that I'm glossing over the issue, central to discussions of the alleged normativity of meaning (i.e. condition 4(b) on page 295 above), as to whether the "ought" here would survive the removal of the background desire.

[31] Note that Kripke actually misquotes Wittgenstein in this passage: in *Philosophical Investigations* §136 Wittgenstein does not italicise "call" but rather "in our language".

the quasi-realist attempt to explain why e.g. ethical sentences "sustain a fully propositional role", Blackburn writes:

> [W]e should distinguish between a slow track and a fast track to this result. The slow track involves patiently construing each propositional context as it comes along. This is the line I took in trying to meet Geach's problem. (1988, 185)

And he continues:

> Fast-track quasi-realism would get there in better style. It would make sufficient remarks about truth to suggest that we need a comparable notion to regulate evaluative discourse (even though this is nonrepresentational) and then say that our adherence to propositional forms needs no further explanation than that. (1988, 185)

Perhaps Kripke's Wittgenstein's disavowal of the need to account for how ascriptions of meaning "can occur as components of truth-functional compounds" is consistent with his embracing a position similar to the fast track variant of quasi-realism? I won't attempt to answer this question here, but will finish by pointing out what I take to be the two main challenges facing the sceptical solution.[32]

The first challenge concerns the scope of the quasi-realism advanced by Kripke's Wittgenstein's solution to the paradox. As noted in section 3 above, expressivist accounts such as Blackburn's only apply to specific areas of thought and talk, such as morals (or causation or modals), and not across the board. Blackburn is a projectivist quasi-realist about morals, for example, but not about medium sized dry goods. In contrast, the quasi-realism of the sceptical solution would appear to be global: recall that in Kripke's Wittgenstein's account the truth-conditional account of meaning is rejected, not just for ascriptions of meaning, but for all areas of thought and talk. Perhaps, ultimately, this will prove to be its undoing: does its global nature mean that in any given attempt to carry out its explanatory project the "tool-kit" of admissible materials that can figure in its explanations turns out to be empty? If so, the sceptical solution, construed as we have construed it, appears to be in serious trouble.[33]

Secondly, although the task facing the quasi-realist in e.g. the case of morality is daunting, requiring as it does a solution to the Frege-Geach problem, the task facing the quasi-realist sceptical solution to Kripke's Wittgenstein's challenge is even more forbidding. Recall that the quasi-realist attempts to construct a notion of

---

[32] That is to say, challenges *over and above* those faced by non-factualist positions about any area of thought and talk.

[33] Perhaps this shouldn't come as a surprise, given that other non-factualist and projectivist construals of the sceptical solution point to its tendency to "globalise" as a potential source of worry. See e.g. Wright (1984) and Boghossian (1989). Blackburn himself considers a worry of the sort outlined in the text, and tentatively floats the idea that although forms of quasi-realism akin to what he calls "foundational pragmatism" won't globalise without leaving us prone to an "empty toolkit" worry, less hegemonic forms of quasi-realism akin to what he calls "rolling pragmatism" may be able to do so (see Blackburn (2013b, 78-81)). Although Kripke's Wittgenstein does not loom large in them, the kind of sceptical solution discussed in the text may be a relative of the "global expressivism" or "global anthropological pragmatism" discussed in a series of fascinating works by Huw Price. See in particular McArthur and Price (2007) and Price (2015).

moral truth whose "first approximation" (Blackburn, 1984a, b, 198) goes roughly as follows: a remark expressing a moral attitude is true if the attitude belongs to the set of attitudes that results from taking all opportunities for improvement of moral attitude. We can view this as an attempt to construct a notion of moral truth on the basis of a notion of justification that (according to the quasi-realist) can itself be explicated independently of the notion of moral truth: for example, an attitude can be regarded as justified insofar as it is the product of an admirable moral sensibility.[34] In effect, this is to attempt to construct a notion of moral truth out of the notion of assertibility: a moral statement is assertible to the extent that the attitude it expresses is the product of an admirable moral sensibility. On the face of it, it looks like the quasi-realist in the case of meaning faces an analogous task. In the case of meaning, we take the assertibility conditions of ascriptions of meaning as the basis for constructing a notion of truth applicable to them, and this notion of assertibility is held to be explicable independently of the notion of truth in semantics: roughly, an ascription of meaning is assertible if the relevant speaker's use of the relevant expression agrees with that of the community. Despite this similarity, though, there is a key difference. In the case of morals, for example, the quasi-realist is attempting to construct a notion of truth out of a notion of assertibility, where this is a notion of *justification*; or, as we may put it, a notion of *correctness* applicable to moral attitudes that meet a particular *standard*. The moral quasi-realist is thus attempting to construct one notion of correctness (truth) out of another (assertibility construed as a form of justification). Although not readily apparent, the quasi-realist position about meaning, although it too is attempting to construct a notion of truth out of the notion of assertibility, is faced with an altogether more radical task. This can be seen by reflecting on an insufficiently appreciated remark made by Paul Boghossian about a comment on the rule-following considerations contained in a classic paper by Simon Blackburn. Blackburn wrote:

> I intend no particular theoretical implications by talking of rules here. The topic is that there is such a thing as the correct and incorrect application of a term, and to say that there is such a thing is no more than to say that there is truth and falsity. I shall talk indifferently of there being correctness and incorrectness, of words being rule-governed, and of their obeying principles of application. Whatever this is, it is the fact that distinguishes the production of a term from mere noise, and turns utterance into assertion – into the making of judgment (Blackburn, 1984a, b, 28–29).

Boghossian comments:

> My only disagreement with this passage concerns its identification of correctness conditions with truth conditions. Truth conditions are simply one species of a correctness condition; proof conditions or justification conditions supply further instances (Boghossian, 1989, 152 n.18).

[34] For our purposes here, a moral sensibility can be thought of as a disposition to have moral attitudes in response to naturalistically characterised states of affairs. See McNaughton (1988, 183) for a clear explanation of the notion as it is deployed by Blackburn.

Although Boghossian himself does not draw any conclusions from this in his own subsequent discussion of the sceptical solution and semantic non-factualism, we can see that it has potentially devastating consequences for Kripke's Wittgenstein's reconstructive project. Since the notion of assertibility condition at play in the sceptical solution is *entirely* divorced from *any* notion of correctness, from *any* notion of a *standard* that utterances of ascriptions or meaning might or might not meet, the quasi-realist about meaning (unlike his cousin in the moral case) is not attempting to construct one notion of correctness (truth) out of another (justification), but rather attempting to construct a notion of correctness out of materials which are to be described with no invocation of any notion of correctness whatsoever. That is to say, although the quasi-realist in the moral case is barred from simply assuming that moral sentences have truth-conditions, he is permitted to view moral discourse as regulated by standards that can be explicated without recourse to the notion of a moral truth-condition; the quasi-realist about meaning, however, can have no recourse to the notion of a standard of any sort, irrespective of whether that standard can be explicated without recourse to the idea of a meaning-constituting state of affairs. Whereas the quasi-realist in the moral case may (perhaps unfairly) be accused of trying to have his cake and eat it, the proponent of the analogous position in the case of meaning appears to be attempting to conjure something out of nothing.[35] I suggest that dispelling this impression is the most pressing task facing any future development of Kripke's Wittgenstein's sceptical solution.[36, 37]

---

[35] In footnote 27 (page 313 above), I noted that quasi-realists are happy to view moral remarks as having truth conditions. It may be objected that although this is the case, quasi-realists don't have to view themselves as *specifying* the truth conditions of the target sentences, so that the objection in the text lapses (since the quasi-realist in the meaning case is no longer expected to specify determinate truth conditions for ascriptions of meaning on the basis of assertibility conditions). In reply, however, we can say that a quasi-realist who sees themselves as engaged in constructing a notion of truth applicable to e.g. moral sentences *is* committed to specifying their truth conditions. To see this, recall from §8 above that, in the moral case, "a first approximation" to the notion of moral truth involves the idea that an expression of a moral attitude is true if and only if it belongs to the set of attitudes "which would result from taking all opportunities for improvement of attitude" (Blackburn, 1984a, b, 198). In any given case, the application of this strategy *cannot but* issue in a specification of a truth condition for the relevant sentence, the details of which will depend on the details of the debate at the level of normative ethics about what can and cannot be regarded as an improvement in a moral sensibility. So there is no escape route here for a quasi-realist about meaning, since the task of constructing a notion of truth applicable to semantic discourse and the task of specifying truth conditions for ascriptions of meaning go hand-in-hand, and cannot be separated in the manner envisaged in the objection. (I'm grateful here to an anonymous referee).

[36] Given the broad structural similarities between Blackburn's position and the sorts of position developed in Gibbard (1990) and (2003), it seems to me highly likely that worries of the sort outlined in this section will arise for the account of ascriptions of meaning proposed in Gibbard (2012). I pursue this issue in work-in-progress.

[37] For some interesting reflections on quasi-realism, see Weir (1994).

**Acknowledgements** A slightly abridged version of this paper was published in the *Journal for the History of Analytical Philosophy*, Volume 8 Number 2 (February 2020). Earlier versions have been delivered at the University of Otago Philosophy Seminar in September 2018 and the *Themes From Weir* conference at the University of Glasgow in December 2018. I'm grateful to Adam Rieger and Stephan Leuenberger for organizing the conference and for the invitation to participate in it. For comments and/or discussion, I'm grateful to Finn Butler, Ali Hossein Khani, Christoph Lernpass, James Levine, Olivia Sultanescu, Claudine Verheggen, Cristovao Atilio Viero, Alan Weir, Timothy Williamson, and two anonymous referees for the *Journal for the History of Analytical Philosophy*.

## References

Ahmed, A. (2007). *Saul Kripke*. Continuum.

Ayer, A. (1946). *Language, truth and logic* (2nd ed.). Dover Books.

Blackburn, S. (1984a). *Spreading the word*. Clarendon Press.

Blackburn, S. (1984b). The individual strikes back. Reprinted in Miller and Wright (2002), pp. 28–44.

Blackburn, S. (1993a). Realism, quasi, or queasy? In J. Haldane & C. Wright (Eds.), *Reality, representation and projection* (pp. 365–383). Oxford University Press.

Blackburn, S. (1993b). *Essays in quasi-realism*. Oxford University Press.

Blackburn, S. (1998a). Attitudes and contents. Reprinted in Blackburn (1993b), pp. 182–197.

Blackburn, S. (1998b). *Ruling passions*. Oxford University Press.

Blackburn, S. (2011). Pragmatism in philosophy: The hidden alternative. *Philosophic Exchange, 41*, 1–14.

Blackburn, S. (2013a). Deflationism, pluralism, expressivism, pragmatism. In N. Pedersen & C. Wright (Eds.), *Truth and pluralism: Current debates* (pp. 263–277). Oxford University Press.

Blackburn, S. (2013b). Pragmatism, all or some? In H. Price (Ed.), *Expressivism, pragmatism and representationalism* (pp. 67–84). Cambridge University Press.

Boghossian, P. (1989). The rule-following considerations. Reprinted Miller and Wright (2002), pp. 141–187.

Boyd, D. (2017). Semantic non-factualism in Kripke's Wittgenstein. *Journal for the History of Analytic Philosophy, 5*(9), 1–12.

Brink, D. (1989). *Moral realism and the foundations of ethics*. Cambridge University Press.

Butler, F. (2019). *The tenability of meaning irrealism*. MA dissertation, University of Otago.

Byrne, A. (1996). On misinterpreting Kripke's Wittgenstein. *Philosophy and Phenomenological Research, 61*, 339–343.

Darwall, S., Gibbard, A., & Railton, P. (1992). Towards Fin de Siecle ethics: Some trends. *Philosophical Review, 101*, 115–189.

Davidson, D. (1984). *Inquiries into truth and interpretation*. Oxford University Press.

Davidson, D. (1986). A nice derangement of epitaphs. Reprinted in his *Truth, language and history* (pp. 89–107). Oxford University Press.

Davidson, D. (1990). Meaning, truth and evidence. Reprinted in his *Truth, language and history* (pp. 47–62). Oxford University Press.

Davidson, D. (1991). Three varieties of knowledge. Reprinted in his *Subjective, intersubjective, objective* (pp. 205–237). Oxford University Press, 2001.

Davidson, D. (1992). The second person. Reprinted in his *Subjective, intersubjective, objective* (pp. 107–122). Oxford University Press, 2001.

Davies, D. (1998). How sceptical is Kripke's 'sceptical solution'? *Philosophia, 26*, 119–140.

Dummett, M. (1959). Wittgenstein's philosophy of mathematics. *Philosophical Review, 68*, 324–348.

Dummett, M. (1978). *Truth and other enigmas*. Duckworth.

Fodor, J. (1990). *A theory of content and other essays*. MIT Press.

Gibbard, A. (1990). *Wise choices, apt feelings*. Harvard University Press.
Gibbard, A. (2003). *Thinking how to live*. Harvard University Press.
Gibbard, A. (2012). *Meaning and normativity*. Oxford University Press.
Glüer, K. (2011). *Donald Davidson: A short introduction*. Oxford University Press.
Glüer, K., & Wikforss, Å. (2018). The normativity of meaning and content. In *Stanford encyclopedia of philosophy*. https://plato.stanford.edu/entries/meaning-normativity/. Accessed 18 Dec 2019.
Hare, R. M. (1952). *The Language of Morals*. Oxford University Press.
Hattiangadi, A. (2017). The normativity of meaning. In Hale, Miller, & Wright (Eds.), *A companion to the philosophy of language* (2nd ed., pp. 649–669). Wiley-Blackwell.
Horwich, P. (1984). Critical Notice of *Wittgenstein on rules and private language*. *Philosophy of Science, 51*, 163–171.
Kripke, S. (1982). *Wittgenstein on rules and private language*. Harvard University Press.
Kusch, M. (2006). *A sceptical guide to meaning and rules*. Acumen.
Lepore, E., & Ludwig, K. (2005). *Donald Davidson: Meaning, truth, language and reality*. Oxford University Press.
Loar, B. (1985). Review of *Wittgenstein on rules and private language*. *Noûs, 19*(2), 273–280.
Mackie, J. L. (1977). *Ethics*. Penguin.
McArthur, D., & Price, H. (2007). Pragmatism, quasi-realism and the global challenge. In C. Misak (Ed.), *The new pragmatists* (pp. 91–121). Oxford University Press.
McDowell, J. (1987). Projectivism and truth in ethics. Reprinted in McDowell (1998), pp. 151–166.
McDowell, J. (1998). *Mind, value and reality*. Harvard University Press.
McGinn, C. (1984). *Wittgenstein on meaning*. Basil Blackwell.
McNaughton, D. (1988). *Moral vision*. Blackwell.
Miller, A. (2009). Kripke's Wittgenstein, factualism and meaning. In D. Whiting (Ed.), *The later Wittgenstein on language* (pp. 167–190). Basingstoke.
Miller, A. (2013). *Contemporary metaethics: An introduction*. Polity Press.
Miller, A. (2017). Postscript: Radical interpretation. In B. Hale, A. Miller, & C. Wright (Eds.), *A companion to the philosophy of language* (2nd ed., pp. 317–323). Wiley-Blackwell.
Miller, A., & Wright, C. (Eds.). (2002). *Rule-following and meaning*. Acumen.
Moore, G. E. (1903). *Principia ethica*. Cambridge University Press.
Price, H. (2015). From quasi-realism to global expressivism – And back again? In R. Johnson & M. Smith (Eds.), *Passions and projections: Themes from the philosophy of Simon Blackburn* (pp. 134–152). Oxford University Press.
Quine, W. V. (1960). *Word and object*. MIT Press.
Railton, P. (1986). Moral realism. *Philosophical Review, 95*, 163–207.
Stevenson, C. (1937). The emotive meaning of ethical terms. *Mind, 16*, 14–31.
Stroud, B. (1990). Meaning, understanding and community. Reprinted in his *Meaning, understanding and practice* (pp. 80–94). Oxford University Press, 2001.
Sultanescu, O., & Verheggen, C. (2019). Davidson's answer to Kripke's sceptic. *Journal for the History of Analytical Philosophy, 7*(2), 1–28.
Verheggen, C. (2016). Language, Thought and knowledge. Part I of C. Verheggen and R. Myers *Donald Davidson's triangulation argument: A philosophical inquiry*. Routledge.
Wee Ming Kho, D. (2015). *Wittgenstein, objectivity and rule-following: Towards resolving the communitarian vs. individualist debate*. PhD dissertation, University of Otago.
Weir, A. (1984). *The philosophy of language of W.V. Quine*. DPhil dissertation, University of Oxford.
Weir, A. (1994). Review of Blackburn *Essays in Quasi-Realism*. *International Journal of Philosophical Studies, 2*, 345–347.
Weir, A. (2003). Objective content. *Proceedings of the Aristotelian Society Supplementary, 77*, 47–72.
Weir, A. (2006a). Review of Roger. F. Gibson Jnr (ed.) *The Cambridge Companion to Quine*. *Theoria, 72*, 240–247.

Weir, A. (2006b). Indeterminacy of translation. In E. LePore & B. Smith (Eds.), *The Oxford handbook of the philosophy of language* (pp. 233–249). Oxford University Press.
Weir, A. (2007). Naturalism reconsidered. In S. Shapiro (Ed.), *The Oxford handbook of philosophy of mathematics and logic* (pp. 460–482). Oxford University Press.
Weir, A. (2014). Quine's naturalism. In G. Harman & E. LePore (Eds.), *A companion to W.V.O. Quine* (pp. 114–147). Wiley-Blackwell.
Wilson, G. (1994). Kripke's normativity argument. Reprinted in Miller and Wright (2002), pp. 234–259.
Wright, C. (1984). Kripke's account of the argument against private language. Reprinted in Wright (2001), pp. 91–115.
Wright, C. (1989a). Wittgenstein's rule-following considerations and the central project of theoretical linguistics. Reprinted in Wright (2001), pp. 170–213.
Wright, C. (1989b). Excerpts from a critical study of Colin McGinn's *Wittgenstein on meaning*. Reprinted in Wright (2001), pp. 143–169.
Wright, C. (1992). *Truth and objectivity*. Harvard University Press.
Wright, C. (2001). *Rails to infinity*. Harvard University Press.

# Chapter 14
# Truthmaker Equivalence

**Stephan Krämer**

**Abstract** Truthmaker semantics provides a powerful framework for the study of hyperintensional operators: operators that distinguish between some pairs of intensionally equivalent contents. Even under a truthmaker based account though, an operator will not distinguish between *truthmaker equivalent* contents. This paper determines the propositional logics of truthmaker equivalence for a range of related conceptions of truthmaking, supplementing previous results on this topic obtained by Kit Fine and Fabrice Correia.

**Keywords** Truthmakers · Truthmaker semantics · Hyperintensionality · Logic · Equivalence

I first met Alan Weir in 2017. I was a new lecturer in Glasgow, and he a regular presence again at the department's Senior Seminar, as well as the pub afterwards, his health thankfully having improved enough to permit as much. Due to the nature and context of these highly enjoyable and memorable encounters, I probably learned even more about his views on Scottish Independence than about his views on logic. Even though I am inclined to regard the latter as less likely correct than the former, both, to my mind, provide excellent illustrations of his powerful mind and admirably fearless and uncompromising intellectual character. I am delighted to dedicate the present essay to the celebration of Alan's philosophy. I hope you will forgive me one more—undoubtedly doomed—attempt to convince you of the case for hyperintensionalism, Alan!

S. Krämer (✉)
University of Hamburg, Hamburg, Germany
e-mail: stephan.kraemer@uni-hamburg.de

 
A. Rieger, S. Leuenberger (eds.), *Themes from Weir: A Celebration of the Philosophy of Alan Weir*, Synthese Library 484,
https://doi.org/10.1007/978-3-031-54557-3_14

## 14.1 Introduction

Truthmaker semantics is a powerful formal framework for the study of *hyperintensional* operators. To give just some examples, it has been used with considerable success to develop hyperintensional accounts of the counterfactual conditional (Fine 2012a), the notion of partial content (Fine 2016, 2017a), the notion of ground (Fine 2012c,b, 2017b), imperatives (Fine 2018a,b), the notions of permission and obligation (Fine 2018b; Anglberger et al. 2016), the notion of evidential relevance (Krämer 2017), belief revision (Krämer 2022), and the notion of a whole truth (Krämer 2023a,b). The distinctive feature of hyperintensional operators is that necessarily equivalent sentences are not in general substitutable *salva veritate* within their argument places. Thus, suppose $\varphi(\ldots)$ is a one-place, sentence-forming operator on sentences. Then if $\varphi$ is hyperintensional, even if $A$ and $B$ are necessarily equivalent sentences, $\varphi(A)$ may be true and yet $\varphi(B)$ false, or the other way around. If so, the semantics of $\varphi$ cannot be captured within the intensional framework of standard possible world semantics. In that framework, necessarily equivalent sentences are assigned the same semantic content—viz., the same set of possible worlds—and so there is no way for the semantics to distinguish between $\varphi(A)$ and $\varphi(B)$, and thus for them to receive different truth-values.

Truthmaker semantics provides a more fine-grained account of the content of sentences than possible world semantics. As a result, within this framework, necessarily and even logically equivalent sentences are often assigned different contents. So the mere necessary equivalence of $A$ and $B$ no longer prevents us from distinguishing within our semantics between $\varphi(A)$ and $\varphi(B)$, or from assigning different truth-values to them. But although necessary equivalence does not imply semantic indistinguishability within truthmaker semantics, there is a different (non-trivial[1]) equivalence relation which does—call it *truthmaker equivalence*: If two sentences $A$ and $B$ are truthmaker equivalent, they are assigned the same content even within truthmaker semantics, and any truthmaker based account of $\varphi$ will have the consequence that $\varphi(A)$ and $\varphi(B)$ must have the same truth-value. In order to identify the predictions of a proposed truthmaker semantics for an operator $\varphi$, it is therefore important to know when two sentences are truthmaker equivalent.

Generally speaking, the considerations relevant to answering this question may vary with the subject matter of the specific sentences we are looking at. For instance, in order to say whether '2+2=4' and '8/2=4' are truthmaker equivalent, we may need to determine what the truthmakers of arithmetical truths are, and how they are individuated. But in some cases, general considerations about the *logical* structure of a pair of sentences suffice to show that the sentences are truthmaker equivalent.

[1] Identity is an equivalence relation which will trivially ensure interchangeability *salva veritate* of equivalent sentences. But truthmaker equivalence, as we shall see, is weaker than identity.

The focus of this paper will be on the question when the *propositional logical form* of two sentences guarantees their truthmaker equivalence.[2]

As it turns out, there are a number of subtly different interpretations we can give to the notion of truthmaker equivalence. Which of them is most relevant, and most tightly connected to the matter of intersubstitutability, depends on the specific application of truthmaker semantics, i.e. on the specific account of a given hyperintensional operator that is under consideration. For some of these notions, the logics of equivalence have already been determined, but not for some others.[3] In particular, the logic for what is arguably the most central, and most discriminating notion of truthmaker equivalence has not been established in the existing literature.[4] My aim in this paper is, firstly, to close this gap, and secondly, to offer a single, comprehensive point of reference on the logic of truthmaker equivalence, describing the logics for all the different conceptions of truthmaker equivalence and discussing their interrelations.

The plan for the paper is as follows. After introducing the framework of truthmaker semantics (14.2), I explain the relevant different conceptions of truthmaking and truthmaker equivalence, and describe applications of truthmaker semantics for which they may plausibly be regarded as pertinent (Sect. 14.3). I then specify, for each of these different conceptions, a deductive system for deriving the logically true equivalences between sentences of a propositional language and prove soundness and independence results (Sect. 14.4), before establishing completeness (Sects. 14.5–14.7).

## 14.2 Truthmaker Semantics

The best way to introduce truthmaker semantics is to describe how it deviates from the familiar framework of possible world semantics.[5] Firstly, instead of a space of all the possible worlds, we work with a space of *states*. Like worlds, states are *specific*, non-disjunctive entities. Unlike worlds, they need not be *complete*, so they need not decide the truth-value of every proposition. They also need not be *possible*, so they can make true contradictions and other necessary falsehoods. Secondly, while the notion of a possible world's making true a proposition may be understood in modal

[2] While proposals for a truthmaker semantics for first-order logic exist, the question of how best to deal with quantifiers within truthmaker semantics is considerably harder and more contentious than the treatment of the truth-functional operators. More importantly, the question of the logic of the truthmaker equivalence under the existing proposals is very difficult.

[3] The relevant papers are Fine (2016) and Correia (2016). I will be more specific about what is established in these paper in due course.

[4] (Fine 2018a, p. 625) claims of a certain deductive system that it is sound and complete for this notion of equivalence, but as I show below (in Footnote 19), this is a mistake, as the system in question is actually not complete.

[5] For a more in-depth presentation, the best source is Fine (2017a).

terms—that the world *necessitates* the truth of the proposition—we will work with a notion of a state's *exactly making true*, or *exactly verifying* a proposition that is much narrower than necessitation. Exact verification is subject to the requirement that the state be *relevant as a whole* to the proposition in question. Thus, the state of grass being green does not verify the proposition that snow is white or not, because it is wholly irrelevant to the truth of that proposition. And while the state of it being sunny exactly verifies the proposition that it is sunny, the state of it being sunny and hot does not, because it is not relevant *as a whole*: it contains as an irrelevant part the state of it being hot. So we can see that exact verification is non-monotonic: a state may contain an exact verifier without itself being an exact verifier. In addition, the sense in which the obtaining of a state is required to be *sufficient* for the truth of a proposition it exactly verifies is more demanding than necessitation, and akin to the characteristic way in which the truth of *full grounds* are sufficient for the truth of what they ground. For instance, the state of Socrates being wise does not exactly verify that Socrates is wise and snow is white or not, even though it is both wholly relevant and such that it necessitates the truth of the proposition: it lacks a part that is sufficient in the right way for snow's being white or not.

As can be inferred from the previous remarks, states are taken to enter into *part-whole* relations. We write $s \sqsubseteq t$ to say that state $s$ is part of state $t$. We assume that for any set of states $T = \{t_1, t_2, \ldots\}$ there exists their *fusion* $\bigsqcup T = t_1 \sqcup t_2 \sqcup \ldots$, which we take to be the smallest state to contain all of $t_1, t_2, \ldots$ as part. More precisely, the basic structure in truthmaker semantics is that of a *state-space*, which is defined as follows.

**Definition 14.1** A *state-space* is a pair $(S, \sqsubseteq)$ such that

(1) $S$ is a non-empty set,
(2) $\sqsubseteq$ is a partial order[6] on $S$ such that for any subset $T$ of $S$, there exists the least upper bound[7] $\bigsqcup T$ of $T$ with respect to $\sqsubseteq$

Relative to a state-space, we then define the notion of a proposition.

**Definition 14.2** A *unilateral proposition* on a state-space $(S, \sqsubseteq)$ is any subset of $S$.

Under this definition, a proposition is thus identified with the set of its verifiers. Note that the definition reflects the conception of exact verification as non-monotonic: there is no requirement that if $P$ is a proposition and $s \in P$, then any state $t$ with $s \sqsubseteq t$ should also be a member of $P$.

[6] Recall that $\sqsubseteq$ is a partial order on $S$ iff $\sqsubseteq$ is a binary relation on $S$ that is reflexive ($s \sqsubseteq s$ for all $s \in S$), transitive ($s \sqsubseteq u$ whenever $s \sqsubseteq t$ and $t \sqsubseteq u$) and anti-symmetric ($s = t$ whenever $s \sqsubseteq t$ and $t \sqsubseteq s$).

[7] A state $s$ is an *upper bound* of $T \subseteq S$ iff $t \sqsubseteq s$ for all $t \in T$, and $s$ is a least upper bound of $T$ iff $s$ is an upper bound of $T$ and $s \sqsubseteq u$ whenever $u$ is an upper bound of $T$. Least upper bounds are easily seen to be unique if they exist.

One might ask, however, whether some weaker closure principles should be imposed. In particular, one might consider the following closure constraints (cf. e.g. Fine 2017a, p. 628f):[8]

Convexity $t$ verifies $P$ whenever $s \sqsubseteq t \sqsubseteq u$ and $s$ and $u$ verify P
Closure under Fusion $s \sqcup t \sqcup \ldots$ verifies $P$ whenever all of $s, t, \ldots$ verify $P$

For reasons I shall give below, I think that exact verification does not in general satisfy either of these constraints. At the same time, as we shall see, it is sometimes useful to focus not on exact verification itself, but on derivative, more inclusive notions of verification which by definition satisfy these constraints.

**Definition 14.3** For $s$ a state and $P$ a unilateral proposition, say that

- $s$ *convexity-inclusively* (short: c-inclusively) verifies $P$ iff $t \sqsubseteq s \sqsubseteq u$ for some states $t$ and $u$ verifying $P$
- $s$ *fusion-inclusively* (short: f-inclusively) verifies $P$ iff $s$ is the fusion of some verifiers of $P$
- $s$ *regularly* verifies $P$ iff $t \sqsubseteq s \sqsubseteq \bigsqcup P$ for some state $t$ verifying $P$.

Thus, the c-inclusive verifiers are exactly the states that lie *between* some verifiers, the f-inclusive verifiers are the states that may be obtained by *fusing* verifiers, and the regular verifiers are the states that lie between a verifier and the fusion of all verifiers. Note that both every f-inclusive verifier and every c-inclusive verifier is a regular verifier, so regular verification is the least discriminating among the four conceptions of verification. However, some c-inclusive verifiers are not f-inclusive verifiers and some f-inclusive verifiers are not c-inclusive verifiers. For let $p, q, r$ be pairwise distinct and consider the state-space $(\wp(\{p, q, r\}), \subseteq)$. Then $\{p, q\}$ is a c- but not f-inclusive verifier of $\{\{p\}, \{p, q, r\}\}$, and $\{p, q\}$ is an f- but not c-inclusive verifier of $\{\{p\}, \{q\}\}$. So in terms of fineness of grain of the four conceptions of verification, we have the following structure with maximal fineness of grain at the top node, corresponding to exact verification (Fig. 14.1).[9]

Given the view of propositions as sets of their exact verifiers, we may define operations of conjunction and disjunction as follows:

**Definition 14.4** For $P, Q$ unilateral propositions on some state-space,

- $P \wedge Q = \{s \sqcup t : s \in P \text{ and } t \in Q\}$
- $P \vee Q = P \cup Q$

[8] Another constraint Fine discusses that one might wish to impose is that of *verifiability* (ibid.), so that only non-empty sets of states are taken to represent propositions. For our purposes, it makes no difference whether the constraint is imposed.

[9] Note that the term 'exact verification' is sometimes applied to any form of non-monotonic verification. Correspondingly, Fine calls a state an *inexact* verifier of a proposition iff it contains an exact verifier (cf. e.g. 2017c, p. 558). So I should emphasize that although none of c-inclusive, f-inclusive, and regular verification are inexact forms of verification, in my terminology they are distinct from what I call exact verification.

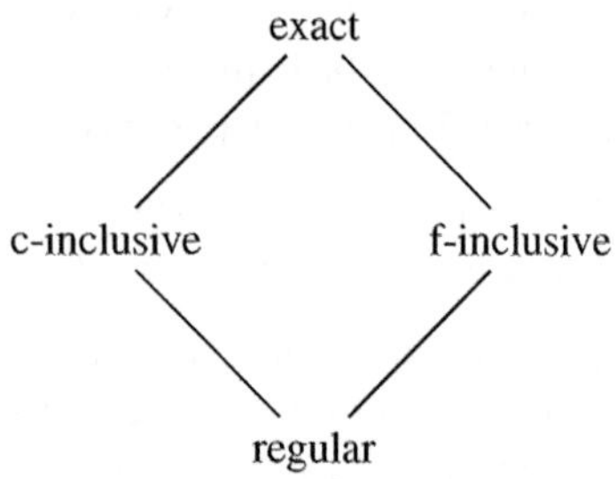


**Fig. 14.1** Varieties of verification

Note how the definition of disjunction reflects the conception of states as non-disjunctive: the only way for a state to get to verify a disjunction is by verifying one of its disjuncts. Again, this is just as for worlds within possible worlds semantics. The definition of conjunction in turn reflects the requirement that a verifier be *wholly relevant* to what it verifies. For note that in contrast to its possible worlds counterpart, the clause is not that a state verifies a conjunction iff it verifies both conjuncts. The reason is that even for conjuncts that are perfectly compatible with one another—like the propositions that snow is white and that grass is green, say—there will not in general be a state that is both sufficient for and wholly relevant to both conjuncts. So instead, a state is counted as a verifier of a conjunction if it may be obtained by fusing a verifier of one conjunct with a verifier of the other.

Negation is somewhat trickier to handle than conjunction and disjunction. Since states are not subject to a completeness requirement, any proposition divides the states into three typically non-empty sets: those that verify it, those that verify its negation, and those that do neither. As a result, we cannot in any straightforward way read off from the set of verifiers of a proposition the set of states verifying its negation.

We may deal with this difficulty by adopting a *bilateral* conception of propositions, specifying separately both the set of verifiers and the set of falsifiers, and taking the verifiers (falsifiers) of the negation of a proposition $P$ to be the falsifiers (verifiers) of $P$. For the falsifiers of conjunctions and disjunctions, we adopt the obvious 'dual' clauses, so that a state falsifies a conjunction iff it falsifies either conjunct, and it falsifies a disjunction iff it is the fusion of a falsifier of one disjunct with a falsifier of the other disjunct.

**Definition 14.5** A *bilateral proposition* $\mathbf{P}$ on a state-space $(S, \sqsubseteq)$ is any pair of unilateral propositions on $(S, \sqsubseteq)$. If $\mathbf{P}$ is a bilateral proposition, $\mathbf{P}^+$ ($\mathbf{P}^-$) is its first (second) coordinate, and will be called the positive (negative) component or content of $\mathbf{P}$.

**Definition 14.6** For $\mathbf{P}$, $\mathbf{Q}$ bilateral propositions on some state-space,

- $\mathbf{P} \wedge \mathbf{Q} = (\mathbf{P}^+ \wedge \mathbf{Q}^+, \mathbf{P}^- \vee \mathbf{Q}^-)$
- $\mathbf{P} \vee \mathbf{Q} = (\mathbf{P}^+ \vee \mathbf{Q}^+, \mathbf{P}^- \wedge \mathbf{Q}^-)$
- $\neg\mathbf{P} = (\mathbf{P}^-, \mathbf{P}^+)$

This completes my exposition of the basic framework of truthmaker semantics. It is easy to see that, due to the move from possible worlds to (possible or impossible) states, and from necessitation to exact verification, the framework allows us to make much more fine-grained distinctions between propositions than are available within the standard framework of possible worlds semantics. For example, the logically true proposition that snow is white or not is no longer identified with the logically true proposition that grass is green or not: the state of snow's being white verifies the former but not the latter. The contradictory proposition that snow is both white and not white is no longer identified with the logically false proposition that grass is both green and not green: the fusion of the state of snow's being white and the state of snow's being red verifies the former but not the latter.

Note that these two examples do not even depend on the non-monotonicity of exact verification. For an example that does, consider the proposition that snow is white and the proposition that snow is white or (snow is white and grass is green). They are distinguished within truthmaker semantics because the latter, but not the former is verified by the state that snow is white and grass is green. But since the former is verified by the state that snow is white, its failure to be verified by the state that snow is white and grass is green is an instance of exact verification's non-monotonicity. The example is of some significance because it illustrates how truthmaker equivalence is more demanding than equivalence in familiar relevant logics, such as the logic of first-degree entailment, for these do render pairs of the form $A$ and $A \vee (A \wedge B)$ equivalent.[10]

## 14.3 Varieties of Truthmaker Equivalence

Within the general framework of truthmaker semantics just outlined, we may distinguish a variety of equivalence relations between propositions. The strongest, most demanding one is simply numerical identity, requiring sameness both of verifiers and falsifiers of a bilateral proposition. Then there are two ways in which this requirement may be weakened. The first is to disregard one of the two components in our bilateral propositions, and require only sameness of verifiers, or only sameness of falsifiers. We may call equivalence relations considering both components, only the positive component, or only the negative component of a bilateral proposition *full*, *positive*, and *negative* equivalence relations, respectively.

These relations differ in how they bear on the question of intersubstitutability. Suppose we are considering some propositional language under a bilateral truthmaker semantics. Obviously, then, any two sentences $A$ and $B$ that are fully equivalent will be freely interchangeable. Not only will the truth-value of $C(A)$ and

[10] Indeed, it is known that the logic of *inexact truthmaker* equivalence is the logic of mutual first-degree entailment, where a state inexactly verifies a proposition iff it contains an exact verifier of the proposition; cf. Fine (2016, sec. 10).

$C(B)$ be the same for any context $C$, but $C(A)$ and $C(B)$ themselves will also have the same verifiers and falsifiers, i.e. be fully equivalent.

However, a certain degree of interchangeability may be guaranteed already by positive or negative equivalence. For instance, in many cases, the semantics for an operator $\varphi$ will be such that the positive content of $\varphi(A)$ depends only on the positive content of $A$. This holds, for example, for conjunction and disjunction, but also for many other operators for which truthmaker semantical accounts have been proposed.[11] In such a case, the positive equivalence of $A$ and $B$ will guarantee sameness of positive content of $\varphi(A)$ and $\varphi(B)$. And since the truth-value of a sentence will typically be fixed by the positive content—a sentence being true iff one of its truthmakers obtains—the positive equivalence of two sentences will often be enough to tell us that a certain substitution of one for the other is bound to preserve truth-value.

It is therefore of considerable interest to know when two sentences are bound by their logical form to be positively equivalent, to have the same truthmakers, even if their falsitymakers differ. Fortunately, once we have found the answer to this question, we can easily extend it to also answer the corresponding questions for negative and full equivalence. For $A$ and $B$ are bound by their logical form to be negatively equivalent iff $\neg A$ and $\neg B$ are bound by their logical form to be positively equivalent, and they are bound to be fully equivalent iff bound to be both positively and negatively equivalent. In what follows, we shall therefore focus on positive rather than negative or full equivalence.

The second way to weaken the requirements for equivalence is to appeal to the more coarse-grained conceptions of c-inclusive, f-inclusive, or regular truthmaking introduced before, and define corresponding relations of equivalence as sameness of c-inclusive, f-inclusive, or regular verifiers and/or falsifiers. For any unilateral proposition $P$, let $P^C$, $P^F$, and $P^R$ denote the set of c-inclusive, f-inclusive, and regular verifiers of $P$, respectively.[12]

**Definition 14.7** For $P$, $Q$ unilateral propositions, say that

- $P$ is *exactly equivalent* to $Q$ ($P \approx_E Q$) iff $P = Q$
- $P$ is *c-equivalent* to $Q$ ($P \approx_C Q$) iff $P^C = Q^C$
- $P$ is *f-equivalent* to $Q$ ($P \approx_F Q$) iff $P^F = Q^F$
- $P$ is *r-equivalent* to $Q$ ($P \approx_R Q$) iff $P^R = Q^R$

---

[11] This is not to say, of course, that operators that do not have this feature would have to be peculiar in some way. Most obviously, negation does not have this feature, since it flips verifiers and falsifiers.

[12] It is worth noting that $P^C$, $P^F$, and $P^R$ have an equivalent characterization in terms of corresponding closure properties on propositions. Thus, say that $P$ is *convex* iff $t \in P$ whenever $s \sqsubseteq t \sqsubseteq u$ and $s, u \in P$, *closed under fusion* iff $\bigsqcup T \in P$ whenever $\emptyset \subset T \subseteq P$, and regular iff convex and closed under fusion. It is routine to verify that $P^C$, $P^F$, and $P^R$ are the smallest supersets of $P$ that are convex, closed under fusion, and regular, respectively. We may also note that $P^R = (P^F)^C$, though not in general $P^R = (P^C)^F$. For example, if $s$ and $t \sqcup u$ are states neither of which is part of the other, then $(\{s, t \sqcup u\}^C)^F = \{s, t \sqcup u\}^F = \{s, t \sqcup u, s \sqcup t \sqcup u\} \neq \{s, t \sqcup u, s \sqcup t, s \sqcup u, s \sqcup t \sqcup u\} = \{s, t \sqcup u\}^R\}$.

Which of these relations is most important, and most intimately connected to the question of intersubstitutability, depends on the application of truthmaker semantics under consideration, and whether it depends on our ability to distinguish between r-equivalent, c-equivalent, and/or f-equivalent propositions. For example, it is plausible that for the purposes of a theory of *partial content*, we need not distinguish between r-equivalent propositions. Fine (2016, 2017a, sec. 5) has defended an explication of the notion of one (unilateral[13]) proposition being part of another by the condition that (i) every verifier of the first is part of a verifier of the second, and (ii) every verifier of the second has a verifier of the first as a part. It can be shown that r-equivalent propositions enter into the same relations of parthood, so understood. Relatedly, parthood is better behaved as a relation on the set of regular propositions rather than an on the set of all propositions, since it is guaranteed to be anti-symmetric only on the set of regular propositions. So for the study of partial content, it is appropriate to work with the concept of regular verification, and the most important concept of equivalence is that of r-equivalence.[14] Similar points apply to the theory of partial truth in Fine (202x) and the theory of verisimilitude in Fine (2021).

But for other applications, we need to work with a more fine-grained conception of verification, and thus of equivalence. One such application is the theory of *metaphysical grounding*. As Fine has shown (2012c; 2012b; 2017b), we can give a plausible explication of a worldly notion of ground within the truthmaker framework. Briefly, Fine proposes to take some propositions $P_1, P_2, \ldots$ to *weakly, non-factively* ground a proposition $Q$ iff $s_1 \sqcup s_2 \sqcup \ldots$ verifies $Q$ whenever $s_1, s_2, \ldots$ verify $P_1, P_2, \ldots$, respectively. We then take $P_1, P_2, \ldots$ to *strictly, non-factively* ground $Q$ iff there are no propositions together with which $Q$ weakly grounds any of $P_1, P_2, \ldots$. The ordinary factive notion of ground is then obtained by demanding that all the $P_i$—and hence $Q$—be true.

Now we obtain subtly different accounts of ground depending on whether verification is here understood as exact, f-inclusive, c-inclusive, or regular. It has been argued, independently but for the same reasons, both in Krämer and Roski (2015) and Correia (2016), that adopting the regular or c-inclusive conception of truthmaking leads to implausible results. To see why, consider the truth $X$ that Frege is an American philosopher or Quine is an American philosopher. If verification is at least f-inclusive, the claim that Frege is an American philosopher and Quine is an American philosopher comes out a non-factive ground of $X$, prevented from being a ground only by its falsity. If verification is also c-inclusive, however, the claim that Frege is a philosopher and Quine is an American philosopher—since it 'lies between' the claim that Quine is an American philosopher and the claim that Frege and Quine are both American philosophers—is also a non-factive ground. And since

[13] For bilateral propositions, it may also be required that every falsifier of the second proposition is also a falsifier of the first; cf. (Fine 2017a, p. 643).

[14] It is for this reason that Fine (2016) focuses on the logic of what I have called r-equivalence; cf. (Fine 2016, sec. 4).

this claim is true, it comes out an actual ground of $X$, and so it follows that the truth that Frege is a philosopher helps ground the truth that either Frege or Quine is an American philosopher. But this seems implausible. For it seems that Frege's being a philosopher could only play a role in bringing it about that Frege or Quine is an American philosopher in conjunction with Frege's being American, which he is not. Since this point applies also with respect to the truth that Frege is an American philosopher, or Quine is an American philosopher, or Frege and Quine are both American philosophers, dropping f-inclusiveness will not help, and so we need to drop c-inclusiveness, and work with either the exact or the f-inclusive conception of verification.

The logics of regular and f-equivalence are known; they are described in Fine (2016) and Correia (2016), respectively. But there are also applications for which even f-inclusiveness needs to be dropped. One plausible example is the application to what may be called the *logic of totality*. In Krämer (2023a,b), I argue that truthmaker semantics allows us to give a more satisfactory account of the so-called totality operator '…and that's it' than is available within intensional frameworks. As I point out there, it seems that this operator is sensitive to the exact truthmakers of its argument, and not just its f-inclusive ones. For there is a natural reading of the totality operator under which the statements

(1) I had a dram of Lagavulin or a dram of Ardbeg, and that's it.
(2) I had a dram of Lagavulin, or a dram of Ardbeg, *or a dram of each*, and that's it.

have different truth-conditions, with only the latter being compatible with me having had both a dram of Lagavulin and a dram of Ardbeg. But in order to assign such truth-conditions, we need to allow that the embedded sentence in (1) is verified by the state of me having a dram of Lagavulin, and by the state of me having a dram of Ardbeg, but not by the fusion of these two states.

Similarly, there is a natural reading of

(3) I had coffee, or coffee and eggs and bacon, and that's it.
(4) I had coffee, or coffee and eggs, and coffee and eggs and bacon, and that's it.

under which they have different truth-conditions, with only the latter allowing for a scenario in which I had eggs but no bacon. But in order to assign such truth-conditions, we need to allow that the embedded sentence in (3) is verified by the state $c$ of me having coffee, and by the state $c \sqcup e \sqcup b$ of me having coffee, eggs, and bacon, but not by the state $c \sqcup e$ in between those two. So to capture these readings of the totality operator, we need to work with the exact notion of truthmaking, assuming neither c- nor f-inclusiveness.

Similar examples support the claim that truthmaker based theories of imperatives and of deontic operators of permission and obligation need to distinguish between f- and/or c-equivalent propositions (cf. Fine 2018a,b). The permission to have ice-cream or chocolate is not automatically a permission to have both. By shutting the door and closing the window, one may fail to comply with the imperative to shut the door or close the window. As a result, as Fine points out (2018a, sec. 6), the propositional logic of exact equivalence constitutes an essential component of a fully developed truthmaker account of these notions.

So in addition to regular and f-equivalence, we have strong reasons to be interested in the logic of exact equivalence as well. I am not aware of a compelling example of an operator that does not distinguish between c-equivalent propositions but that does distinguish between r-equivalent propositions. From a technical point of view, however, it is a very natural further question in the present context what the logic of c-equivalence is, and how it relates to the others.[15]

For exact truthmakers, it is easily seen that since disjunction is defined in terms of set-theoretic union, we can express the subset-relation between propositions in terms of disjunction and exact truthmaker equivalence: $P \subseteq Q$ iff $P \vee Q \approx_E Q$. We shall later have occasion to make use of the fact that this relationship extends to f-inclusive, c-inclusive, and regular truthmakers:

**Proposition 14.1** *For $P$, $Q$ unilateral propositions:*

*(1)* $P^C \subseteq Q^C$ *iff* $P \vee Q \approx_C Q$
*(2)* $P^F \subseteq Q^F$ *iff* $P \vee Q \approx_F Q$
*(3)* $P^R \subseteq Q^R$ *iff* $P \vee Q \approx_R Q$

***Proof*** For (1), left-to-right: Suppose $P^C \subseteq Q^C$. We need to show that $(P \cup Q)^C = Q^C$. Suppose first that $t \in (P \cup Q)^C$. Let $s \sqsubseteq t \sqsubseteq u$ with $s, u \in P \cup Q$. Then both $s$ and $u$ are either in $P$ or in $Q$. But $Q \subseteq Q^C$ and $P \subseteq P^C \subseteq Q^C$ by assumption, so $s$ and $u$ are both in $Q^C$, and hence so is $t$. Suppose now that $t \in Q^C$. Then there are $s, u \in Q \subseteq P \cup Q$, with $s \sqsubseteq t \sqsubseteq u$, so $t \in (P \cup Q)^C$ as desired. So $(P \cup Q)^C = Q^C$.

Right-to-left: Suppose that $(P \cup Q)^C = Q^C$. We wish to show that $P^C \subseteq Q^C$. Suppose $t \in P^C$, so $s \sqsubseteq t \sqsubseteq u$ for some $s, u \in P$. Then $s, u \in P \cup Q$, hence $t \in (P \cup Q)^C$. By assumption, it follows that $t \in Q^C$.

For (2), left-to-right: Suppose $P^F \subseteq Q^F$. We need to show that $(P \cup Q)^F = Q^F$. Suppose first that $s \in (P \cup Q)^F$. Let $s = \bigsqcup T$ with $T \subseteq P \cup Q$. Let $T_P = T \cap P$ and $T_Q = T \cap Q$. Then $s = \bigsqcup T_P \sqcup \bigsqcup T_Q$. But $\bigsqcup T_P \in P^F$, hence $\bigsqcup T_P \in Q^F$. So for some $W \subseteq Q$, $\bigsqcup T_P = \bigsqcup W$. But then $s = \bigsqcup W \sqcup \bigsqcup T_Q = \bigsqcup (W \cup T_Q)$. And since $W \cup T_Q \subseteq Q$, it follows that $s \in Q^F$. Now suppose $s \in Q^F$. Let $s = \bigsqcup T$ with $T \subseteq Q$. Then $T \subseteq P \cup Q$, hence $s = \bigsqcup T \in (P \cup Q)^F$.

Right-to-left: Suppose $(P \cup Q)^F = Q^F$ and suppose $s \in P^F$. Let $s = \bigsqcup T$ with $T \subseteq P$. Then $T \subseteq P \cup Q$, so $s = \bigsqcup T \in (P \cup Q)^F = Q^F$, as desired.

For (3), left-to-right: Suppose $P^R \subseteq Q^R$. We wish to show that $(P \cup Q)^R = Q^R$. Suppose first that $t \in (P \cup Q)^R$. Let $s \sqsubseteq t \sqsubseteq \bigsqcup (P \cup Q)$ with $s \in P \cup Q$. Then by assumption, $s \in Q^R$. Moreover, by assumption, $\bigsqcup P \sqsubseteq \bigsqcup P^R \sqsubseteq \bigsqcup Q^R$. And since also $\bigsqcup Q \sqsubseteq \bigsqcup Q^R$, $\bigsqcup (P \cup Q) = \bigsqcup P \sqcup \bigsqcup Q \sqsubseteq \bigsqcup Q^R$. So $s \sqsubseteq t \sqsubseteq \bigsqcup Q^R$ and $s \in Q^R$, hence by regularity of $Q^R$, $t \in Q^R$. Suppose now that $t \in Q^R$ and let

[15] In particular, given that a set of states is regular iff both closed under fusion and convex, it is natural to ask whether the logic of r-equivalence can be obtained by combining the logics of f- and c-equivalence. As we shall prove below, this is indeed the case.

$s \sqsubseteq t \sqsubseteq \bigsqcup Q$ with $s \in Q$. So $s \in P \cup Q$, and since $\bigsqcup Q \sqsubseteq \bigsqcup(P \cup Q)$, it follows that $t \in (P \cup Q)^R$.

Right-to-left: Suppose $(P \cup Q)^R = Q^R$. We wish to show that $P^R \subseteq Q^R$. Suppose $t \in P^R$, and let $s \sqsubseteq t \sqsubseteq \bigsqcup P$ with $s \in P$. Then $s \in (P \cup Q)^R$ and thus by assumption $s \in Q^R$ as well as $\bigsqcup P \sqsubseteq \bigsqcup(P \cup Q)$ and $\bigsqcup(P \cup Q) \in Q^R$, so by regularity of $Q^R$, it follows that $t \in Q^R$. □

## 14.4 Deductive Systems, Soundness, and Independence

Let $\mathcal{L}$ be a standard propositional language with connectives $\wedge, \vee, \neg$. We call the formulas of $\mathcal{L}$ the *$\mathcal{L}$-sentences*, and call an *$\mathcal{L}$-equivalence* any expression of the form $A \approx B$ with $A, B$ $\mathcal{L}$-sentences. The *language of equivalence* $\mathcal{L}_\approx$ over $\mathcal{L}$ is the set of all $\mathcal{L}$-equivalences.

**Definition 14.8** An $\mathcal{L}$-model is any triple $\mathcal{M} = (S, \sqsubseteq, [\cdot])$ such that $(S, \sqsubseteq)$ is a state-space, and $[\cdot]$ maps every atomic $\mathcal{L}$-sentence to a bilateral proposition on $(S, \sqsubseteq)$. Given an $\mathcal{L}$-model, $[\cdot]$ is extended to the complex formulas of $\mathcal{L}$ in the obvious way:

- $[\neg A] = \neg[A]$
- $[A \wedge B] = [A] \wedge [B]$
- $[A \vee B] = [A] \vee [B]$

Relative to an $\mathcal{L}$-model, we then define the exact, c-inclusive, f-inclusive, and regular content of an $\mathcal{L}$-sentence $A$:[16]

**Definition 14.9** Given an $\mathcal{L}$-model $\mathcal{M} = (S, \sqsubseteq, [\cdot])$, for any $\mathcal{L}$-sentence $A$, the

- *exact* content $[A]_e$ of $A$ is $[A]$
- *c-inclusive* content $[A]_c$ of $A$ is $(([A]^+)^C, ([A]^-)^C)$
- *f-inclusive* content $[A]_f$ of $A$ is $(([A]^+)^F, ([A]^-)^F)$
- *regular* content $[A]_r$ of $A$ is $(([A]^+)^R, ([A]^-)^R)$

**Definition 14.10** Given an $\mathcal{L}$-model $\mathcal{M} = (S, \sqsubseteq, [\cdot])$, an $\mathcal{L}$-equivalence $A \approx B$ is *true* in $\mathcal{M}$ under the

- *exact* interpretation of $\approx$ iff $[A]^+ \approx_E [B]^+$
- *c-inclusive* interpretation of $\approx$ iff $[A]^+ \approx_C [B]^+$
- *f-inclusive* interpretation of $\approx$ iff $[A]^+ \approx_F [B]^+$
- *regular* interpretation of $\approx$ iff $[A]^+ \approx_R [B]^+$

**Definition 14.11** An $\mathcal{L}$-equivalence $A \approx B$ is *valid* under an interpretation of $\approx$ iff for every $\mathcal{L}$-model $\mathcal{M}$, $A \approx B$ is true in $\mathcal{M}$ under that interpretation of $\approx$. We write

[16] In (Fine 2016, p. 208), the regular content of a formula is called its *replete* content, and the f-inclusive content is called the *complete* content; there is no special term for the c-inclusive content.

$\models_e$ ($\models_c$, $\models_f$, $\models_r$) $A \approx B$ to say that $A \approx B$ is valid under the interpretation of $\approx$ as exact (c-inclusive, f-inclusive, regular) equivalence.

Before moving on, we note that by using disjunction, we can express within $\mathcal{L}_\approx$ not only *identity* but also *subset*-relations between exact, c-inclusive, f-inclusive, or regular content. For it is immediate from Proposition 14.1 above that

**Proposition 14.2** *An $\mathcal{L}$-equivalence $A \vee B \approx B$ is true under*

- *the exact interpretation of $\approx$ iff $[A]_e^+ \subseteq [B]_e^+$*
- *the c-inclusive interpretation of $\approx$ iff $[A]_c^+ \subseteq [B]_c^+$*
- *the f-inclusive interpretation of $\approx$ iff $[A]_f^+ \subseteq [B]_f^+$*
- *the regular interpretation of $\approx$ iff $[A]_r^+ \subseteq [B]_r^+$*

### *14.4.1 Deductive Systems*

We shall now specify deductive systems for the different readings of $\approx$.The *base system* $\mathfrak{D}_b$ consists of the following axioms and rules.

| | |
|---|---|
| Collapse($\vee$) | $A \vee A \approx A$ |
| Commutativity($\vee$) | $A \vee B \approx B \vee A$ |
| Associativity($\vee$) | $A \vee (B \vee C) \approx (A \vee B) \vee C$ |
| Commutativity($\wedge$) | $A \wedge B \approx B \wedge A$ |
| Associativity($\wedge$) | $A \wedge (B \wedge C) \approx (A \wedge B) \wedge C$ |
| Collapse($\neg\neg$) | $\neg\neg A \approx A$ |
| DeMorgan($\neg\vee$) | $\neg(A \vee B) \approx \neg A \wedge \neg B$ |
| DeMorgan($\neg\wedge$) | $\neg(A \wedge B) \approx \neg A \vee \neg B$ |
| Distributivity($\wedge/\vee$) | $A \wedge (B \vee C) \approx (A \wedge B) \vee (A \wedge C)$ |
| Preservation($\vee$) | $A \approx B \ / \ A \vee C \approx B \vee C$ |
| Preservation($\wedge$) | $A \approx B \ / \ A \wedge C \approx B \wedge C$ |
| Symmetry | $A \approx B \ / \ B \approx A$ |
| Transitivity | $A \approx B,\ B \approx C \ / \ A \approx C$ |

For any equivalence $A \approx B$ in $\mathcal{L}_\approx$, we write $\vdash_b A \approx B$ to say that $A \approx B$ can be derived using only the rules and axioms of $\mathfrak{D}_b$, and similarly for the various extensions of $\mathfrak{D}_b$ to be considered below.[17]

The system $\mathfrak{D}_b$ coincides with the system used in (Fine 2016, p. 201) minus Fine's axioms E2 and E11. Fine shows (ibid, p. 202) that a rule of Positive Replacement is an admissible rule in $\mathfrak{D}_b$, in the sense that it preserves theoremhood. We can state the rule as follows. If $A$, $B$, $C$ are sentences of $\mathcal{L}$, let $C(A/B)$ be

[17] Note that there is no Reflexivity axiom $A \approx A$ since it would be redundant: any instance can be derived using, for example, Collapse($\neg\neg$) in conjunction with Symmetry and Transitivity.

the result of replacing the occurrences of $A$ in $C$ by $B$. Then the rule of Positive Replacement is

PR $\quad A \approx B \;/\; C \approx C(A/B)$, if no occurrences of $A$ in $C$ are in the scope of $\neg$

It is easy to verify that if a sentence $D$ may be obtained from $C$ by replacing one or more, but not necessarily all occurrences of $A$ in $C$ by $B$, then as long as the replaced occurrences are not in the scope of $\neg$, $C \approx D$ will also be derivable from $A \approx B$, and I shall often tacitly rely on this fact in applying PR. As becomes clear on inspection of Fine's proof of the admissibility of PR, it depends only on the availability of the commutativity axioms and the preservation rules for $\wedge$ and $\vee$. PR therefore remains admissible under the various extensions of $\mathfrak{D}_b$ to be considered in what follows.

We shall abbreviate formulas of $\mathcal{L}_{\approx}$ of the form $A \vee B \approx B$ as $A \rightarrow B$. By Proposition 14.2, $A \rightarrow B$ will be true under a given interpretation of $\approx$ iff the content of $A$ pertinent to that interpretation is a subset of the relevant content of $B$. The *exact* system $\mathfrak{D}_e$ is then obtained by adding to $\mathfrak{D}_b$ the *exact collapse* axiom for $\wedge$:

ECollapse($\wedge$) $\quad A \rightarrow A \wedge A$

In effect, this axiom registers the fact that since fusing a state with itself yields that same state, every truthmaker of $A$ is also a truthmaker of $A \wedge A$.

For f-inclusive contents, which are closed under fusion, the converse is also true: every truthmaker of $A \wedge A$ is already a truthmaker of $A$. To obtain the *f-inclusive system* $\mathfrak{D}_f$, we therefore add to $\mathfrak{D}_e$ the *f-inclusive collapse* axiom for $\wedge$:

FCollapse($\wedge$) $\quad A \wedge A \rightarrow A$

ECollapse($\wedge$) and FCollapse($\wedge$) are jointly equivalent, in the presence of the other axioms and rules, to the simpler axiom

Collapse($\wedge$) $\quad A \wedge A \approx A$

(This is axiom E2 in Fine 2016.) The latter can be derived from the former by means Commutativity($\vee$), Transitivity, and Symmetry. Both FCollapse($\wedge$) and ECollapse($\wedge$) may be derived from Collapse($\wedge$) by means of Transitivity, Symmetry, Collapse($\vee$) and Preservation($\vee$).

Turning to c-inclusive contents, we note that every c-inclusive truthmaker of $A \wedge B$ is a c-inclusive truthmaker of $A \vee (A \wedge B \wedge C)$. As a result, under the c-inclusive interpretation of $\approx$, the following axiom is valid:

Convexity $\quad (A \wedge B) \rightarrow A \vee (A \wedge (B \wedge C))$

Adding this axiom to $\mathfrak{D}_e$ results in the *c-inclusive system* $\mathfrak{D}_c$.

Finally, the system obtained by adding both FCollapse($\wedge$) and Convexity to $\mathfrak{D}_e$ will be called the *regular system* $\mathfrak{D}_r$. Within $\mathfrak{D}_r$, we can derive the second distributivity rule, or E11 in the nomenclature of (Fine 2016):

Distributivity($\vee/\wedge$) $\quad A \vee (B \wedge C) \approx (A \vee B) \wedge (A \vee C)$

**Proposition 14.3** $\vdash_r A \vee (B \wedge C) \approx (A \vee B) \wedge (A \vee C)$

***Proof*** We first establish

(*) $\vdash_r A \vee B \approx (A \vee B) \vee (A \wedge B)$

by the following derivation:[18]

| | |
|---|---|
| $A \vee B \approx (A \vee B) \wedge (A \vee B)$ | by Collapse($\wedge$) |
| $\approx ((A \vee B) \wedge A) \vee ((A \vee B) \wedge B)$ | by Distributivity($\wedge/\vee$) |
| $\approx (A \wedge A) \vee (A \wedge B) \vee (A \wedge B) \vee (B \wedge B)$ | by Distributivity($\wedge/\vee$) |
| $\approx A \vee B \vee (A \wedge B)$ | by Collapse($\wedge$) |

Then our result may be established as follows:

| | |
|---|---|
| $(A \vee B) \wedge (A \vee C)$ | |
| $\approx (A \wedge A) \vee (A \wedge B) \vee (A \wedge C) \vee (B \wedge C)$ | by rep. appl. of Distributivity($\wedge/\vee$) |
| $\approx A \vee (A \wedge B) \vee (A \wedge C) \vee (B \wedge C)$ | by Collapse($\wedge$) |
| $\approx A \vee (A \wedge B) \vee (A \wedge C) \vee (A \wedge B \wedge C) \vee (B \wedge C)$ | by (*), Collapse($\wedge$) |
| $\approx A \vee (A \wedge B \wedge C) \vee (B \wedge C)$ | by Convexity |
| $\approx A \vee (B \wedge C)$ | by (*) |

This completes the proof. □

Adding just Collapse($\wedge$) and Distributivity($\vee/\wedge$) to $\mathfrak{D}_b$ yields the set of axioms and rules that Fine uses in his 2016. It follows from Fine's completeness result relative to r-equivalence together with the soundness results below that this axiomatization is equivalent to $\mathfrak{D}_r$. As an axiomatization of the logic of r-equivalence, the Finean version is undoubtedly more natural and elegant. The advantage of the present system $\mathfrak{D}_r$ is the straightforward correlation between, on the one hand, the properties of closure under fusion and convexity on contents, and, on the other hand, the characteristic axioms FCollapse($\wedge$) and Convexity to which they give rise.

### 14.4.2 Soundness and Independence

We shall now prove that the systems $\mathfrak{D}_e$, $\mathfrak{D}_c$, $\mathfrak{D}_f$, and $\mathfrak{D}_r$ are *sound* relative to the pertinent interpretations of $\approx$.

[18] Here and in what follows, we often apply the commutativity and associativity rules as well as Positive Replacement tacitly, and we drop brackets that only disambiguate between formulas equivalent by the associativity rules. That is, we write $A \vee B \vee C$ in place of $(A \vee B) \vee C$ or $A \vee (B \vee C)$, and similarly in other cases.

**Theorem 14.1 (Soundness)** *For all equivalences $A \approx B \in \mathcal{L}_{\approx}$,*

*(1) if $\vdash_e A \approx B$ then $\models_e A \approx B$*
*(2) if $\vdash_c A \approx B$ then $\models_c A \approx B$*
*(3) if $\vdash_f A \approx B$ then $\models_f A \approx B$*
*(4) if $\vdash_r A \approx B$ then $\models_r A \approx B$*

***Proof*** Note first that since the f-inclusive, c-inclusive, and regular contents of a formula are functions of its exact content, sameness of exact content implies sameness of f-inclusive, c-inclusive, and regular content, and hence $\models_e A \approx B$ implies $\models_c A \approx B$, $\models_f A \approx B$, and $\models_r A \approx B$.

For (1):

- The validity of the Symmetry and Transitivity rules is immediate from the fact that $\approx$ is interpreted by an equivalence relation.
- The validity of the Preservation rules is immediate from the fact that, by definition of $\wedge$ and $\vee$, the positive content of a disjunction and a conjunction is a function of the positive content of the disjuncts and conjuncts, respectively.
- The validity of Collapse($\vee$), Commutativity($\vee$) and Associativity($\vee$) follows from the idempotence, commutativity and associativity of set-theoretic union.
- The validity of Commutativity($\wedge$) and Associativity($\wedge$) is readily verified by noting that the binary fusion operation on states is commutative and associative.
- For Distributivity($\wedge/\vee$), suppose first that $s \in [A \wedge (B \vee C)]^+$. Then $s = t \sqcup u$ for some $t$, $u$ with $t \in [A]^+$ and $u \in [B \vee C]^+$, hence either (i) $u \in [B]^+$ or (ii) $u \in [C]^+$. If (i), then $t \sqcup u \in [A \wedge B]^+$. If (ii), then $t \sqcup u \in [A \wedge C]^+$. Either way, $s = t \sqcup u \in [A \wedge B]^+ \cup [A \wedge C]^+ = [(A \wedge B) \vee (A \wedge C)]^+$. Suppose now that $s \in [(A \wedge B) \vee (A \wedge C)]^+ = [A \wedge B]^+ \cup [A \wedge C]^+$, hence either (i) $s \in [A \wedge B]^+$ or (ii) $s \in [A \wedge C]^+$. If (i), then $s = t \sqcup u$ for some $t \in [A]^+$ and $u \in [B]^+$. But then $u \in [B \vee C]^+$, hence $s \in [A \wedge (B \vee C)]^+$. If (ii), then $s = t \sqcup u$ for some $t \in [A]^+$ and $u \in [C]^+$. Then again, $u \in [B \vee C]^+$, hence $s \in [A \wedge (B \vee C)]^+$.
- Collapse($\neg\neg$): $[\neg\neg A]^+ = [\neg A]^- = [A]^+$ by the definition of negation.
- DeMorgan($\neg\wedge$): $[\neg(A \wedge B)]^+ = [A \wedge B]^- = [A]^- \vee [B]^- = [\neg A]^+ \vee [\neg B]^+ = [\neg A \vee \neg B]^+$
- DeMorgan($\neg\vee$): $[\neg(A \vee B)]^+ = [A \vee B]^- = [A]^- \wedge [B]^- = [\neg A]^+ \wedge [\neg B]^+ = [\neg A \wedge \neg B]^+$
- For ECollapse($\wedge$), note that by idempotence of state-fusion, $[A]^+ \subseteq [A \wedge A]^+$.

For (2), we need to establish the validity of Convexity relative to the c-inclusive interpretation of $\approx$, i.e. to show that $[A \wedge B]_c^+ \subseteq [A \vee (A \wedge B \wedge C)]_c^+$. So suppose $t \in [A \wedge B]_c^+$, and hence $s \sqsubseteq t \sqsubseteq u$ for some $s$, $u \in [A \wedge B]^+$. Then $s^- \in [A]^+$ and hence $s^- \in [A \vee (A \wedge B \wedge C)]^+$ for some $s^- \sqsubseteq s$. Moreover, $u \sqsubseteq u^+$ for some $u^+ \in [A \vee (A \wedge B \wedge C)]^+$. Since $s^- \sqsubseteq t \sqsubseteq u^+$, it follows that $t \in [A \vee (A \wedge B \wedge C)]_c^+$.

For (3), we also need to establish the validity of FCollapse($\wedge$) relative to the f-inclusive interpretation of $\approx$. It suffices to show that $[A \wedge A]_f^+ \subseteq [A]_f^+$. So suppose $s \in [A \wedge A]_f^+$, and hence $s = \bigsqcup T$ for some $T \subseteq [A \wedge A]^+$. We may associate each $t \in T$ with states $t_1$, $t_2 \in [A]^+$ with $t = t_1 \sqcup t_2$. Let $T^* = \bigcup \{\{t_1, t_2\} : t \in T\}$. Then

$T^* \subseteq [A]^+$, and hence $\bigsqcup T^* \in [A]^+_f$. But by associativity of fusion, $\bigsqcup T = \bigsqcup T^*$, and hence $s = \bigsqcup T^*$, so $s \in [A]^+_f$, as desired.

For (4), we need to establish the validity of FCollapse($\wedge$) and Convexity relative to the regular interpretation of $\approx$. So we need to show, firstly, that $[A \wedge A]^+_r \subseteq [A]^+_r$. So suppose $t \in [A \wedge A]^+_r$, and hence $s \sqsubseteq t \sqsubseteq \bigsqcup [A \wedge A]^+$ for some $s \in [A \wedge A]^+$. Then $s = s_1 \sqcup s_2$ for some $s_1, s_2 \in [A]^+$. By associativity of fusion, $\bigsqcup [A \wedge A]^+ = \bigsqcup [A]^+$, so $s_1 \sqsubseteq t \sqsubseteq \bigsqcup [A]^+$, and since $s_1 \in [A]^+$, it follows that $t \in [A]^+_r$. We need to show, secondly, that $[A \wedge B]^+_r \subseteq [A \vee (A \wedge B \wedge C)]^+_r$. So suppose $t \in [A \wedge B]^+_r$, and hence $s \sqsubseteq t \sqsubseteq \bigsqcup [A \wedge B]^+$ for some $s \in [A \wedge B]^+$. Then some part $s_A$ of $s$ is in $[A]^+$ and hence in $[A \vee (A \wedge B \wedge C)]^+$. Since $\bigsqcup [A \wedge B]^+ \sqsubseteq \bigsqcup [A \vee (A \wedge B \wedge C)]^+$, it follows that $s_A \sqsubseteq s \sqsubseteq \bigsqcup [A \vee (A \wedge B \wedge C)]^+$, and hence that $s \in [A \vee (A \wedge B \wedge C)]^+_r$. □

It is known for the Finean system, which is obtained by adding both Collapse($\wedge$) and Distributivity($\vee/\wedge$) to $\mathfrak{D}_b$, that its axioms and rules are independent of one another (cf. Fine 2016, p. 201n3). We can show that the same is true for $\mathfrak{D}_e$ by proving that ECollapse($\wedge$) is not derivable within $\mathfrak{D}_b$.[19] To do this, we give an alternative interpretation to the language $\mathcal{L}_{\approx}$ under which $\mathfrak{D}_b$ is sound but ECollapse($\wedge$) is not. The interpretation will make crucial use of multi-sets, which are like sets but may contain each member any finite number of times. If $S$ is a multi-set, let $\#(S, x)$ be the number of occurrences of $x$ in $S$. Given multi-sets $S$ and $T$, let their *sum* $S + T$ be the multi-set $U$ such that $\#(U, x) = \#(S, x) + \#(T, x)$ for all $x$. We write $|x, x, \ldots, y, \ldots|$ for the multi-set containing each of $x, y, \ldots$ exactly as many times as it is listed.

**Definition 14.12** A *multi-model* of $\mathcal{L}_{\approx}$ is any pair $(S, [\cdot])$ where $S$ is a non-empty set of multi-sets closed under summation, and $[\cdot]$ maps every sentence letter $p$ of $\mathcal{L}$ to a pair $([p]^+, [p]^-)$ of non-empty subsets of $S$.

Relative to a given multi-model, we extend $[\cdot]$ to all formulas of $\mathcal{L}$ as follows:

- $[A \vee B] = ([A]^+ \cup [B]^+, \{s + t : s \in [A]^- \text{ and } t \in [B]^-\})$
- $[A \wedge B] = (\{s + t : s \in [A]^+ \text{ and } t \in [B]^+\}, [A]^- \cup [B]^-)$
- $[\neg A] = ([A]^-, [A]^+)$

In effect, rather than interpreting conjunction in terms of *fusions* of verifiers, we now interpret it in terms of *sums*, in the multi-set sense just defined, of verifiers.

**Proposition 14.4** *ECollapse($\wedge$) is not derivable within $\mathfrak{D}_b$.*

***Proof*** It is straightforward to verify that $\mathfrak{D}_b$ is sound relative to the class of multi-models of $\mathcal{L}_{\approx}$, when $A \approx B$ is taken as true iff $[A]^+ = [B]^+$. Pick a multi-model including a state $|x|$. Then pick a sentence letter $p$ from $\mathcal{L}$ and consider a multi-model with $[p]^+ = \{|x|\}$. Then $[p \wedge p]^+ = \{|x, x|\}$ whereas $[p \vee (p \wedge p)]^+ =$

[19] Fine (2018a, p. 625) states, in effect, that $\mathfrak{D}_b$ is complete with respect to exact truthmaker equivalence. It follows from Proposition 14.4, together with the soundness Theorem 14.1, that this is a mistake.

$\{|x|, |x, x|\}$, and so $p \vee (p \wedge p) \approx p \wedge p$ is false. Since this is an instance of ECollapse($\wedge$), by soundness of $\mathfrak{D}_b$, it follows that ECollapse($\wedge$) is not derivable within $\mathfrak{D}_b$. □

We may establish parallel independence results for FCollapse($\wedge$) and Convexity.

**Proposition 14.5** *FCollapse($\wedge$) is not derivable within $\mathfrak{D}_e$ and not derivable within $\mathfrak{D}_c$.*

***Proof*** Let $x$ and $y$ be distinct objects and consider the state-space $(\wp(\{x, y\}), \subseteq)$. Pick a sentence letter $p$ and a model with $[p]^+ = \{\{x\}, \{y\}\}$ and thus $[p \wedge p]^+ = \{\{x\}, \{y\}, \{x, y\}\}$. Note that $[p]^+_c = [p]^+$ and likewise $[p \wedge p]^+_c = [p \wedge p]^+$. Since $[p \wedge p]^+$ is not a subset of $[p]^+$, it follows that $p \wedge p \rightarrow p$ is false under the interpretations of $\approx$ as exact or c-inclusive truthmaker equivalence. Together with the soundness theorem, this establishes our claim. □

**Proposition 14.6** *Convexity is not derivable within $\mathfrak{D}_e$ and not derivable within $\mathfrak{D}_f$.*

***Proof*** Let $x, y, z$ be pairwise distinct and consider the state-space $(\wp(\{x, y, z\}), \subseteq)$. Pick distinct sentence letters $p, q, r$ of $\mathcal{L}$ and a model with $[p]^+ = \{\{p\}\}$, $[q]^+ = \{\{q\}\}$ and $[r]^+ = \{\{r\}\}$. Then $[p \wedge q]^+_f = [p \wedge q]^+ = \{\{p, q\}\}$ and $[p \vee (p \wedge q \wedge r)]^+_f = [p \vee (p \wedge q \wedge r)]^+ = \{\{p\}, \{p, q, r\}\}$. Since $[p \wedge q]^+$ is not a subset of $[p \vee (p \wedge q \wedge r)]^+$, $p \wedge q \rightarrow p \vee (p \wedge q \wedge r)$ is false under both the exact and the f-inclusive truthmaker equivalence interpretation of $\approx$. Then by the soundness theorem, our claim follows. □

## 14.5 Disjunctive Normal Forms

Our completeness proofs are adaptations of Fine's (2016) completeness proof for regular truthmaker equivalence to the other interpretations of equivalence. They proceed via disjunctive normal forms. For each of our notions of truthmaker equivalence, we shall identify a particular class of disjunctive normal forms in $\mathcal{L}$ with the following properties. Firstly, any two distinct members of the class fail to be truthmaker equivalent in some model. Secondly, any sentence of $\mathcal{L}$ is provably equivalent to some member of the class. From these results, it is straightforward to establish completeness.[20]

[20] Note that just as in Fine (2016) and Correia (2016), the completeness results established here are weak completeness results: the deductive systems are shown to prove all the logically true equivalences. There is also a natural way to define a notion of logical consequence within our semantics, and a different kind of deductive system would be required in order to derive all the logical consequences from a given set of premises. Although it would certainly be of interest to develop such systems for our notions of equivalence, doing so is beyond the scope of this paper. It may be worth noting that the applications of a logic of equivalence sketched in the introduction to this paper mainly seem to depend on weak completeness.

As usual, we call any sentence letter and its negation a *literal* of $\mathcal{L}$. Any formula obtainable from literals purely by successive applications of $\wedge$ is called a conjunctive form. As a limiting case, a literal is also considered a *conjunctive form*. When I speak of the conjuncts of a conjunctive form, I mean only the literals that occur as conjuncts in the conjunctive form. Any formula obtainable from conjunctive forms purely by successive applications of $\vee$ is called a *disjunctive form*. As a limiting case, a conjunctive form is also considered a disjunctive form. When I speak of the disjuncts of a disjunctive form, I mean only the conjunctive forms that occur as disjuncts in the disjunctive form.

Given a conjunctive form $A$, let $lit(A)$ be the multi-set of literals of $\mathcal{L}$ that includes each literal of $\mathcal{L}$ exactly as many times as it occurs as a conjunct in $A$. (We count a literal as occuring as a conjunct in itself, so $lit(A) = |A|$ if $A$ is a literal.) Note that by the Associativity and Commutativity rules for $\wedge$, in all of our deductive systems, two conjunctive forms $A$ and $B$ are provably equivalent—that is, $A \approx B$ is derivable—if $lit(A) = lit(B)$. We now pick some function $cf$ that maps each multi-set $X$ of literals of $\mathcal{L}$ to a conjunctive form $cf(X)$ with $lit(cf(X)) = X$, and denote the image of the $cf$-function SCF, the set of standard conjunctive forms.

Given a disjunctive form $A$, let $v(A)$ be $\{lit(B) : B \text{ is a disjunct in } A\}$. (We count a conjunctive form as occurring as a disjunct in itself, so $v(A) = \{lit(A)\}$ if $A$ is a conjunctive form.) By the Associativity and Commutativity rules for $\vee$ and $\wedge$, in all of our deductive systems, two disjunctive forms $A$ and $B$ are provably equivalent if $v(A) = v(B)$. We now pick some function $df$ mapping each set $M$ of multi-sets of literals of $\mathcal{L}$ some disjunctive form $df(M)$ such that (i) $v(df(M)) = M$, (ii) no disjunct in $df(M)$ occurs more than once, and (ii) all of the disjuncts in $M$ are standard conjunctive forms. We denote the image of the $df$-function SDF, the set of standard disjunctive forms.

**Lemma 14.1** *Every sentence in $\mathcal{L}$ is provably equivalent to a disjunctive form.*

***Proof*** This is a straightforward corollary of Theorem 15 in (Fine 2016, p. 214). □

We say that sentences $A$ and $B$ are *strictly equivalent* within a given deductive system iff $A$ and $B$ are provably equivalent within the deductive system and $v(A) = v(B)$. Then

**Lemma 14.2** *Every disjunctive form is strictly equivalent to a standard disjunctive form.*

***Proof*** By standard methods using Collapse($\vee$), Commutativity($\vee$), Associativity($\vee$), Commutativity($\wedge$), Associativity($\wedge$) and PR. □

### *14.5.1 Exact Truthmaking*

We now wish to specify a subclass of SDF with the two desired properties mentioned above, that any sentence in $\mathcal{L}$ is provably equivalent, within $\mathfrak{D}_e$, to a member of the class, and that any two distinct members of the class can differ with respect to their exact truthmakers. Note that SDF itself does not have the second property. For let $A$ be some literal and consider the formulas $A \wedge A$ and $A \vee (A \wedge A)$. By the soundness of ECollapse($\wedge$) for exact truthmaker equivalence, we know that these have the same exact truthmakers in every model. But note that $v(A \wedge A) = \{|A,\ A|\}$ and $v(A \vee (A \wedge A)) = \{|A|,\ |A,\ A|\}$. Now consider the standard disjunctive form $B = df(v(A \wedge A))$ and the standard disjunctive form $C = df(v(A \vee (A \wedge A)))$. Since $v(B) = v(A \wedge A)$, $B$ is provably equivalent to $A \wedge A$, and since $v(C) = v(A \vee (A \wedge A))$, $C$ is provably equivalent to $A \vee (A \wedge A)$, so $B$ is provably equivalent to $C$, and hence $B$ and $C$ have the same exact truthmakers in every model. But $B$ and $C$ are distinct members of SDF, since $v(B) = v(A \wedge A) = \{|A,\ A|\} \neq \{|A|,\ |A,\ A|\} = v(A \vee (A \wedge A)) = v(C)$.

So from any subclass of SDF whose members are pairwise related relevantly like $A \wedge A$ and $A \vee (A \wedge A)$ are, we need to pick a unique representative. To this end, we define a notion of *subsumption* between multi-sets of literals.

**Definition 14.13** For $X$ and $Y$ multi-sets of literals of $\mathcal{L}$, say that

- $X$ is subsumed by $Y$ iff (i) the same set underlies $X$ and $Y$, and every literal that occurs in $X$ occurs at least as many times in $Y$,
- $X$ is properly subsumed by $Y$ iff $X$ is subsumed by $Y$ and $Y$ is not subsumed by $X$.

By extension, we say that a conjunctive form $A$ is (properly) subsumed by a conjunctive form $B$ iff $lit(A)$ is (properly) subsumed by $lit(B)$. Next, we define a notion of fullness for sets of multi-sets of literals.

**Definition 14.14** A set $M$ of multi-sets of literals of $\mathcal{L}$ is *full* iff every multi-set of literals properly subsumed by some member of $M$ is itself a member of $M$.

Again, by extension, we say that a disjunctive form $A$ is full iff $v(A)$ is full.

**Lemma 14.3** *Every disjunctive form is provably equivalent within $\mathfrak{D}_e$ to a full disjunctive form.*

***Proof*** Let $F$ be a disjunctive form. Let $F$'s fall-off value $n$ be the number of multi-sets of literals $X$ such that: (i) $X$ is properly subsumed by some member of $v(F)$, and (ii) $X$ is not itself a member of $v(F)$. If $n = 0$, then $F$ is itself a full disjunctive form and we are done. So suppose $n > 0$. We now pick some multi-set of literals $X$ which is not a member of $v(F)$ but which is *immediately* subsumed by some member $Y$ of $v(F)$, i.e. $X$ is subsumed by $Y$, and there is exactly one literal in $Y$ that occurs less often in $X$ than in $Y$, and it occurs exactly one time less in $X$ than in $Y$. Now let $B$ be a disjunct in $F$ with $lit(B) = Y$. Then $B$ and $cf(X)$ are respectively provably equivalent to conjunctive forms of the forms $(C \wedge C) \wedge D$ and $C \wedge D$. But by ECollapse($\wedge$), $C \wedge C \approx C \vee (C \wedge C)$, and so

$$
\begin{aligned}
B &\approx (C \wedge C) \wedge D \\
&\approx (C \vee (C \wedge C)) \wedge D && \text{by ECollapse}(\wedge)\text{, Preservation}(\wedge) \\
&\approx (C \wedge D) \vee ((C \wedge C) \wedge D) && \text{by Distributivity}(\wedge/\vee) \\
&\approx B \vee cf(X) && \text{by PR}
\end{aligned}
$$

By another application of PR, $F$ is provably equivalent to the result $F'$ of replacing $B$ in $F$ by $B \vee cf(X)$. The fall-off value of $F'$ is strictly less than $n$. □

**Lemma 14.4** *Every standard disjunctive form is provably equivalent, within $\mathfrak{D}_e$, to a full standard disjunctive form.*

***Proof*** Let $F$ be a standard disjunctive form. By the Lemma 14.3, $F$ is provably equivalent to a full disjunctive form $F'$. By Lemma 14.2, $F'$ is strictly equivalent to some standard disjunctive form $F^*$. Since $F'$ and $F^*$ are strictly equivalent, $v(F') = v(F^*)$ and so since $F'$ is full, so is $F^*$. □

### 14.5.2 C-Inclusive Truthmaking

The distinctive feature of c-inclusive equivalence is that it does not distinguish between $A \vee (A \wedge B \wedge C)$ and $A \vee (A \wedge B) \vee (A \wedge B \wedge C)$. So for proving completeness for the system $\mathfrak{D}_c$, we need to pick one representative from each class of full standard disjunctive forms that are related in this way. In order to define the relevant relationship, let us say that, for any multi-sets $X$ and $Y$, $X$ is a sub-multiset of $Y$ ($X \subseteq_m Y$) iff every element of $X$ occurs at least as many times in $Y$ as it does in $X$. We then say that a multi-set $Y$ is *between* multi-sets $X$ and $Z$ iff $X \subseteq_m Y \subseteq_m Z$. Finally, a disjunctive form $A$ will be said to be convex iff whenever a multi-set of literals $X$ lies between $lit(B)$ and $lit(C)$ for $B$ and $C$ disjuncts in $A$, some conjunctive form $D$ with $X = lit(D)$ is also a disjunct in $A$. The class of standard disjunctive forms relevant to proving completeness will be the class of full and convex standard disjunctive forms.

**Lemma 14.5** *Every full disjunctive form is provably equivalent, within $\mathfrak{D}_c$, to a full and convex disjunctive form.*

***Proof*** Let $A$ be a full disjunctive form. Let $A$'s fall-off value $n$ be the number of multi-sets $X$ witnessing $A$'s non-convexity, i.e. that lie between $lit(B)$ and $lit(C)$ for some disjuncts $B$ and $C$ of $A$, while there is no disjunct $D$ in $A$ with $lit(D) = X$. If $n = 0$, then A is already full and convex. So suppose $n > 0$ and let $B$ and $C$ be a witnessing pair of disjuncts in A. Note that $B$ is not subsumed by $C$, for otherwise, by fullness of A, there would be a disjunct $D$ with $lit(D) = X$ for every multi-set $X$ between $lit(B)$ and $lit(C)$. So there is some literal $q$ that occurs at least once in $C$ but not at all in $B$.

Thus $B$ and $C$ are provably equivalent, respectively, to conjunctive forms of the forms $E$ and $E \wedge q \wedge F$. Using PR, $A$ may be shown to be provably equivalent to a disjunctive form of the form $E \vee (E \wedge q \wedge F) \vee G$. By the Convexity rule of $\mathfrak{D}_c$, $E \vee (E \wedge q \wedge F)$ is provably equivalent to $E \vee (E \wedge q) \vee (E \wedge q \wedge F)$, and thus $A$ is provably equivalent to $A' = E \vee (E \wedge q) \vee (E \wedge q \wedge F) \vee G$. The fall-off value of $A'$ is then less than $n$.

Now $A'$ may no longer be full, since there may be multi-sets $Y$ subsumed by $lit(E \wedge q)$ for which there is no corresponding disjunct in $A'$. But by the methods used in the proof of Lemma 14.3, $A'$ may be expanded to a full disjunctive form $A^*$ that is provably equivalent to $A'$. It remains to show that the fall-of value of $A^*$ is still less than $n$. To this end, note first that for any disjunct $G$ added in forming $A^*$, $lit(G) \subseteq_m lit(C)$. Now say that a conjunctive form $A$ is *repetition-free* iff no literal occurs more than once as a conjunct in $A$. Then by fullness of $A$, there is a repetition-free counterpart $B^*$ of $B$ that occurs in $A$, and hence $A'$. We then note, secondly, that for any disjunct $G$ added in forming $A^*$, $lit(B^*) \subseteq_m lit(G)$. It follows that any multi-set $X$ witnessing the non-convexity of $A^*$ also witnesses the non-convexity of $A'$. So the fall-off value of $A^*$ is less than or equal to that of $A'$, and hence less than $n$. □

**Lemma 14.6** *Every standard disjunctive form is provably equivalent, within $\mathfrak{D}_c$, to a full and convex standard disjunctive form.*

***Proof*** Let $A$ be a standard disjunctive form. By the previous lemma, $A$ is provably equivalent to a full and convex disjunctive form $A'$. By Lemma 14.2, $A'$ is strictly equivalent to a standard disjunctive form $A^*$. Since $v(A') = v(A^*)$ by strict equivalence, $A^*$ is full and convex. □

### 14.5.3 F-Inclusive Truthmaking

When we turn to the notion of f-inclusive equivalence, two things change in comparison to exact equivalence.[21] Firstly, we no longer need to distinguish between conjunctive forms one of which subsumes the other: $A$ will be equivalent to $A \wedge A$, and similarly in other cases. Secondly, we no longer need to distinguish between disjunctive forms related like $A \vee B$ and $A \vee B \vee (A \wedge B)$, where one contains some extra disjuncts, all of which are however conjunctions of disjuncts present in the other. So let us say firstly, that a disjunctive form is repetition-free iff every one of its disjuncts is. Secondly, let us say that a disjunctive form $A$ is *closed* iff for any two disjuncts $B$ and $C$ in $A$, $A$ includes a disjunct $D$ with $slit(D) = slit(B) \cup slit(C)$, where $slit(X)$ is the set of literals occurring as conjuncts in the conjunctive form $X$. Note that since any disjunctive form has only finitely many disjuncts, closure of $A$ implies that for any collection $\Gamma$ of disjuncts in $A$, $A$ has a disjunct $D$ with $slit(D) = \bigcup\{slit(B) : B \in \Gamma\}$. The subclass of standard disjunctive forms relevant for f-inclusive equivalence will that of the closed and repetition-free standard disjunctive forms.

**Lemma 14.7** *Every disjunctive form is provably equivalent, within $\mathfrak{D}_f$, to a closed disjunctive form.*

[21] The completeness of $\mathfrak{D}_f$ with respect to f-inclusive equivalence has already been established in Correia (2016). Since Correia's proof proceeds in a slightly different way than the ones given here, I'm giving another proof that proceeds in exact parallel to those for exact and c-inclusive equivalence.

***Proof*** We show first that any instance of $A \vee B$ is equivalent to the corresponding instance of $A \vee B \vee (A \wedge B)$.

$$\begin{aligned} A \vee B &\approx (A \vee B) \wedge (A \vee B) && \text{by FCollapse}(\wedge) \\ &\approx ((A \vee B) \wedge A) \vee ((A \vee B) \wedge B) && \text{by Distributivity}(\wedge/\vee) \\ &\approx (A \wedge A) \vee (A \wedge B) \vee (A \wedge B) \vee (B \wedge B) && \text{by Distributivity } (\wedge/\vee) \\ &\approx A \vee B \vee (A \wedge B) \vee (A \wedge B) && \text{by FCollapse}(\wedge) \\ &\approx A \vee B \vee (A \wedge B) && \text{by Collapse}(\vee) \end{aligned}$$

Now let $A$ be a disjunctive form, and let $A$'s fall-off value $n$ be the number of collections $\Gamma$ of disjuncts in $A$ for which there is no disjunct $D$ in $A$ with $slit(D) = \bigcup\{slit(B) : B \in \Gamma\}$. We prove the result by induction on $n$. If $n = 0$, $A$ is already closed. Suppose $n > 0$. Since there are only finitely many disjuncts in $A$, there will be some pair $\{B, C\}$ that satisfies this condition. Using the above equivalence and PR as well as appropriate reshuffling of disjuncts using Associativity($\vee$) and Commutativity($\vee$), we can derive the equivalence of $A$ and $A \vee (B \wedge C)$. The fall-off value of $A \vee (B \wedge C)$ is strictly less than $n$. □

**Lemma 14.8** *Every conjunctive form is provably equivalent, within* $\mathfrak{D}_f$, *to a repetition-free conjunctive form.*

***Proof*** By FCollapse($\wedge$) and PR, using Associativity($\wedge$) and Commutativity($\wedge$) to shuffle around the conjuncts so as to put repeat occurrences of the same literal next to each other. □

**Lemma 14.9** *Every disjunctive form is provably equivalent, within* $\mathfrak{D}_f$, *to a repetition-free disjunctive form.*

***Proof*** By the previous lemma and PR. □

**Lemma 14.10** *Every standard disjunctive form is provably equivalent, within* $\mathfrak{D}_f$, *to a closed and repetition-free standard disjunctive form.*

***Proof*** Let $A$ be a standard disjunctive form. By Lemma 14.7, it is equivalent to closed disjunctive form $A'$. By PR and Lemma 14.8, every disjunct in $A'$ may be replaced by a repetition-free counterpart, yielding a repetition-free disjunctive form $A''$. Note that the replacement preserves closure, since whenever $B'$ is a repetition-free counterpart of $B$, $slit(B') = slit(B)$. So $A''$ is repetition-free and closed. By Lemma 14.2, $A''$ is strictly equivalent to a standard disjunctive form $A'''$. Since both repetition-freedom and closure of a disjunctive form $X$ depend only on $v(X)$, and $v(A'') = v(A''')$ by strict equivalence, $A'''$ is a closed and repetition-free standard disjunctive form, as required. □

### *14.5.4 Regular Truthmaking*

This is the case that is covered in Fine (2016), so I shall merely mention very briefly the main definitions and results; for details I refer the reader to section 7 of Fine's paper. The relevant subclass of standard disjunctive forms comprises those that Fine calls *maximal*, which are defined as follows. A disjunctive form $A$ is maximal iff for every disjunct $D$ in $A$ and every literal $q$ occurring as a conjunct in a disjunct of $A$, $A$ contains a disjunct $D'$ with $slit(D') = slit(D) \cup \{q\}$. Using similar techniques to those we have employed above, Fine shows that every disjunctive form is provably equivalent, within $\mathfrak{D}_r$, to a maximal disjunctive form. The crucial auxiliary result is that the Distributivity($\vee/\wedge$) rule in $\mathfrak{D}_r$ allows us to prove the equivalence of $A \vee (B \wedge C)$ to $A \vee (A \wedge B) \vee (A \wedge C) \vee (B \wedge C)$, the latter being the obvious maximal counterpart to the former. By the usual means, it may then be established that

**Lemma 14.11** *Every standard disjunctive form is provably equivalent, within $\mathfrak{D}_r$, to a maximal standard disjunctive form.*

## 14.6 Canonical Models

In this section, we define, for each of our notions of truthmaker-equivalence, a suitable canonical model to prove completeness, i.e. a model in which no two members of the relevant class of disjunctive normal forms are assigned the same exact, c-inclusive, f-inclusive, or regular content, depending on the notion of equivalence in question.

### *14.6.1 Exact Truthmaking*

The simplest strategy for defining a canonical model is to let each sentence letter $A$ of $\mathcal{L}$ be verified by $A$ and only $A$, and falsified by $\neg A$ and only $\neg A$. But as a canonical model for exact truthmaker equivalence, this will not do. For it is a straightforward consequence of the definition of conjunction that any sentence $A$ with only a single verifier will have the same set of verifiers as its self-conjunction $A \wedge A$. So the simple canonical model will not allow us to invalidate all desired instances of $A \approx A \wedge A$. For example, whenever $A$ is a literal, $A \approx A \wedge A$ will be true in the simple canonical model. To avoid this problem, we may instead associate each sentence letter $A$ with a countable infinity of verifiers, to make sure that the result of conjoining $A$ with itself any finite number of times will always produce some new verifier.

**Definition 14.15** The *exact canonical model* $\mathcal{M}_E$ of $\mathcal{L}$ is the triple $(S, \sqsubseteq, [\cdot])$ with

- $S = \wp(\{(A, i) : A \text{ is a literal of } \mathcal{L} \text{ and } i \in \mathbb{N}\})$
- $\sqsubseteq$ is the restriction of subsethood to $S$
- $[p] = (\{\{(p, i)\} : i \in \mathbb{N}\}, \{\{(\neg p, i)\} : i \in \mathbb{N}\})$

From the fact that $S$ is a powerset and $\sqsubseteq$ the restriction of the subset-relation to that powerset, it is straightforward that $(S, \sqsubseteq)$ is a state-space and hence that the exact canonical model is indeed a model of $\mathcal{L}$.

**Lemma 14.12** *In $\mathcal{M}_E$, for any conjunctive form A and literal q, if q occurs a total of n times as a conjunct in A, then*

*(1) every verifier of A has at most n pairs of the form $(q, i)$ as members, and*
*(2) some verifier of A has n distinct pairs of the form $(q, i)$ as members, and*
*(3) every verifier of A has at least one pair of the form $(q, i)$ as member, provided $n > 0$.*

***Proof*** Part (3) is immediate from the definition of conjunction and the fact that by construction of the model, every verifier of $q$ is of the form $\{(q, i)\}$.

Parts (1) and (2) may be proved by induction on $n$. Case $n = 0$: Part (1) is immediate from the definition of conjunction and the fact that by construction of $\mathcal{M}_E$, no literal except $q$ has a verifier with some pair of the form $(q, i)$ as a member. Part (2) is immediate from part (1) and the fact that $A$ has a verifier, which is clear by construction of the model.

Case $n = m + 1$: Without loss of generality, assume that $A$ is of the form $B \wedge q$. Then $B$ contains a total of $m$ occurrences of $q$, so by IH, every verifier $B$ contains at most $m$ pairs of the form $(q, i)$ as members, and some verifier of $B$ contains $m$ pairs of that form. For part (1), suppose $s$ verifies $A$. Then $s = t \cup u$ for some $t$ verifying $B$ and some $u$ verifying $q$. By construction of the canonical model, $u = \{(q, j)\}$ for some $j$. Since $t$ contains at most $m$ pairs of the form $\{(q, i)\}$, $s$ contains at most $m + 1 = n$ pairs of that form. For part (2), by IH, we may pick a verifier $t$ of $B$ with $m$ pairs of the form $(q, i)$. Let $j = max\{i \in \mathbb{N} : (q, i) \in t\}$. Then $\{(q, j + 1)\}$ verifies $q$, and hence $s = t \cup \{(q, j + 1)\}$ verifies $A$. By definition of $j$, $(q, j + 1)$ is not a member of $t$, so $s$ contains $m + 1 = n$ pairs of the form $(q, i)$. □

**Lemma 14.13** *Let A be a conjunctive form. In $\mathcal{M}_E$, there is some verifier s of A such that every conjunctive form verified by s subsumes A.*

***Proof*** Let $s$ be a verifier of $A$ as per part (2) of the previous lemma. So for any literal $q$ occurring $n \geq 1$ times in $A$, $s$ has $n$ distinct pairs of the form $(q, i)$ as member. Suppose $s$ verifies a conjunctive form $B$. Then by part (1) of the previous lemma, each literal $q$ occurring in $A$ occurs at least as often in $B$. It remains to show that the same set underlies $lit(A)$ and $lit(B)$. We have already seen that every literal occurring in $A$ also occurs in $B$, so it remains only to show that every literal occurring in $B$ occurs at least once in $A$. So suppose $q$ occurs in $B$. Then by part (3) of the previous lemma, every verifier of $B$, and therefore $s$, contains at least one pair of the form $(q, i)$ as a member. But then by part (1) of the previous lemma, $q$ occurs as a conjunct in $A$. □

**Theorem 14.2** *In $\mathcal{M}_E$, if A and B are distinct full standard disjunctive forms of $\mathcal{L}$, then $[A]_e^+ \neq [B]_e^+$.*

***Proof*** Suppose $A$ and $B$ are distinct full standard disjunctive forms. Without loss of generality, assume there is a disjunct $D$ in $A$ which does not occur in $B$. By the previous lemma, there is a verifier $s$ of $D$, and hence of $A$, which verifies only conjunctive forms that subsume $D$. Now suppose for reductio that $s$ verifies $B$. By the definition of disjunction, $s$ verifies some disjunct $E$ of $B$. Since $s$ verifies only conjunctive forms subsuming $D$, $E$ subsumes $D$. Since $B$ is a full standard disjunctive form, $v(B)$ is full. Since $v(B)$ has $lit(E)$ as a member, and $lit(E)$ subsumes $lit(D)$, by fullness, $v(B)$ also has $lit(D)$ as a member. Since $B$ is a standard disjunctive form, $B$ includes $cf(lit(D))$ as a disjunct. But since $D$ is a standard conjunctive form, $D = cf(lit(D))$, so $B$ includes $D$ as a disjunct, contrary to our assumption. So $s$ does not verify $B$, and hence $[A]_e^+ \neq [B]_e^+$. □

Note that for any given pair $A$ and $B$ of distinct full standard disjunctive forms of $\mathcal{L}$, we can also find a *finite* model in which they have different verifiers. For any literal $p$ and formula $F$ of $\mathcal{L}$, let $o(p, F)$ the number of occurrences of $p$ in $F$. Now let $S_0 = \{(p, i) : p$ is a literal that occurs in $A$ or in $B$ and $i \leq max\{o(p, A), o(p, B)\}\}$. Then in place of $\mathcal{M}_E$, we can pick any model with state-space $(\wp(S_0), \subseteq)$ in which any literal $p$ in $A$ and $B$ is interpreted as verified by exactly the states of the form $\{(p, i)\}$. Since only finitely many literals occur in $A$ and $B$, and since they occur only a finite number of times, any such model is finite. For the components of $A$ and $B$, the proofs of Lemmas 14.12 and 14.13, and thus of Theorem 14.2 carry over unchanged to any such model. It follows that the logic of exact equivalence has the finite model property and therefore is decidable.

### 14.6.2 C-Inclusive Truthmaking

The same canonical model that we used in the previous section for exact truthmaker equivalence is also suitable for proving completeness for c-inclusive equivalence.

**Definition 14.16** The *c-inclusive canonical model* $\mathcal{M}_C$ for $\mathcal{L}$ is $\mathcal{M}_E$.

Recall that $\#(X, x)$ is the number of occurrences of $x$ in the multi-set $X$. Now for any state $s$ in $\mathcal{M}_C$, let $X_s$ be that multi-set of literals such that for every literal $q$, $\#(X_s, q)$ equals the number of distinct pairs of the form $(q, i)$ in $s$. Then

**Lemma 14.14** *For every state s and conjunctive form A,*

*(1) if* $lit(A) = X_s$ *then s verifies A*
*(2) if s verifies A, then* $X_s \subseteq_m lit(A)$
*(3) if s verifies A and A is repetition-free, then* $X_s = lit(A)$

***Proof***

Part (1): By induction on the complexity of $A$. If $A$ is a literal, then $X_s = lit(A)$ iff $s = \{(A, i)\}$ for some $i \in \mathbb{N}$. By definition of the canonical model, any such $s$ verifies $A$. Now suppose that $A = B \wedge C$ and that the claim holds for

$B$ and $C$ (IH). Suppose $s$ is such that $X_s = lit(A) = lit(B \wedge C)$. We need to show that $s$ verifies $B \wedge C$. To this end, we aim to construct states $t$ and $u$ with $s = t \sqcup u$ such that $t$ verifies $B$ and $u$ verifies $C$. Since by (IH), the claim holds for $B$ and $C$, it suffices to show that $X_t = lit(B)$ and $X_u = lit(C)$. Now for any literal $q$, $\#(lit(B \wedge C), q) = \#(lit(B), q) + \#(lit(C), q)$. Since $X_s = lit(B \wedge C)$, we know that for any literal $q$ occurring in $lit(B \wedge C)$, $s$ contains $\#(lit(B), q) + \#(lit(C), q)$ distinct pairs of the form $(q, i)$ as members. So we pick $\#(lit(B), q)$ one of them as members of $t$ and the other $\#(lit(C), q)$ as members of $u$. Then $s = t \cup u = t \sqcup u$. Moreover, by construction of $t$ and $u$, $X_t = lit(B)$ and $X_u = lit(C)$, and so by (IH), $t$ verifies $B$ and $u$ verifies $C$, hence $s$ verifies $A = B \wedge C$.

Part (2): By induction on the complexity of $A$. If $A$ is a literal, then by construction of the canonical model, $s = \{(A, i)\}$ for some $i \in \mathbb{N}$, so $X_s = |A| = lit(A)$. Suppose $A = B \wedge C$ and the claim holds for $B$ and $C$. Suppose $s$ verifies $A$. Then $s = t \sqcup u$ for some $t$, $u$ verifying $B$, $C$, respectively. By induction hypothesis, $X_t \subseteq_m lit(B)$ and $X_u \subseteq_m lit(C)$. Then since $s = t \cup u$, for all $q$, $\#(X_s, q) \leq \#(X_t, q) + (X_u, q)$. Since $X_t \subseteq_m lit(B)$, we have $\#(X_t, q) \leq \#(lit(B), q)$ and since $X_u \subseteq_m lit(C)$, we have $\#(X_u, q) \leq \#(lit(C), q)$. So for all $q$, $\#(X_s, q) \leq \#(lit(B), q) + \#(lit(C), q) = \#(lit(B \wedge C), q)$, and therefore $X_s \subseteq_m lit(B \wedge C)$, as desired.

Part (3): It is readily verified that if $s$ verifies a repetition-free conjunctive form $A$, then $s$ contains exactly one member of the form $(q, i)$ for each literal occurring in $A$. So $X_s$ contains each literal in $A$ exactly once. Likewise, since $A$ is repetition-free, $lit(A)$ contains each literal in $A$ exactly once, and so $lit(A) = X_s$.

□

We can now show that for full convex disjunctive forms, the convex truthmakers and the exact truthmakers assigned by the canonical model coincide.

**Lemma 14.15** *In $\mathcal{M}_C$, for any full convex disjunctive form $A$, $[A]^+_c = [A]^+$.*

***Proof*** Let $A$ be a convex disjunctive form. Since $[A]^+_c$ is the convex closure of $[A]^+$, clearly $[A]^+ \subseteq [A]^+_c$. It remains to show that $[A]^+_c \subseteq [A]^+$. So suppose $t \in [A]^+_c$. Then there are states $s$ and $u$ with $s \sqsubseteq t \sqsubseteq u$ and $s \in [A]^+$ and $u \in [A]^+$. By the definition of disjunction and since $A$ is full, $s$ verifies some repetition-free disjunct $A_s$ in $A$ and $u$ verifies some disjunct $A_u$ in $A$. By Lemma 14.14(3), $X_s = lit(A_s)$ and by Lemma 14.14(2), $X_u \subseteq_m lit(A_u)$. Since $s \sqsubseteq t \sqsubseteq u$, we have $lit(A_s) = X_s \subseteq_m X_t \subseteq_m X_u \subseteq_m lit(A_u)$, so by convexity of $A$, there is a disjunct $D$ in $A$ with $lit(D) = X_t$. By Lemma 14.14(1), $t$ verifies $D$ and hence $t \in [A]^+$, as desired. □

**Theorem 14.3** *In $\mathcal{M}_C$, if $A$ and $B$ are distinct, full and convex standard disjunctive forms of $\mathcal{L}$, then $[A]^+_c \neq [B]^+_c$.*

***Proof*** Since $A$ and $B$ are convex, $[A]^+_c = [A]^+$ and $[B]^+_c = [B]^+$. Since $A$ and $B$ are distinct, full standard disjunctive forms, it follows by Theorem 14.2 that $[A]^+ \neq [B]^+$. □

The same method as before for restricting the canonical model to a finite one given any suitable pair $A$ and $B$ can be used to show that the logic of c-equivalence is decidable.

### 14.6.3 F-Inclusive Truthmaking

For this case, since we want to validate $A \approx A \wedge A$ and similar cases, we can work with the simple canonical model.

**Definition 14.17** The *f-inclusive canonical model* $\mathcal{M}_F$ for $\mathcal{L}$ is the triple $(S, \sqsubseteq, [\cdot])$ with

- $S = \wp(\{A : A \text{ is a literal in } \mathcal{L}\})$
- $\sqsubseteq$ is the restriction of subsethood to $S$
- $[p] = (\{p\}, \{\neg p\})$

As before, since $S$ is a powerset and $\sqsubseteq$ the restriction of the subset-relation, $(S, \sqsubseteq)$ is a state-space and hence the f-inclusive canonical model is a model of $\mathcal{L}$.

**Lemma 14.16** *In $\mathcal{M}_F$, for any conjunctive form $A$, $[A]^+ = \{slit(A)\}$.*

***Proof*** If $A$ is a sentence letter, then by definition of the canonical model, $[A]^+ = \{\{A\}\} = slit(A)$. If $A$ is the negation $\neg B$ of a sentence letter $B$, then $[A]^+ = [B]^- = \{\{\neg B\}\} = slit(A)$. Suppose the claim holds for $A$ and $B$. Then $[A \wedge B]^+ = [A]^+ \wedge [B]^+ = \{s \sqcup t : s \in [A]^+ \text{ and } t \in [B]^+\} = \{s \cup t : s \in \{slit(A)\} \text{ and } t \in \{slit(B)\}\} = \{slit(A) \cup slit(B)\} = \{slit(A \wedge B)\}$. It follows that the claim holds for all conjunctive forms. □

**Lemma 14.17** *In $\mathcal{M}_F$, for any closed disjunctive form $A$, $[A]^+_f = [A]^+$.*

***Proof*** Since $[A]^+_f$ is the closure under fusion of $[A]^+$, we have $[A]^+ \subseteq [A]^+_f$. So suppose $s \in [A]^+_f$. Then for some $T \subseteq [A]^+$, $s = \bigsqcup T = \bigcup T$. For each $t \in T$, $t$ verifies some conjunctive form $B_t$ occurring as a disjunct in $A$, so by the previous lemma, $t = slit(B_t)$, so $s = \bigcup\{slit(B_t) : t \in T\}$. Since $A$ is closed, some conjunctive form $B$ with $slit(B) = \bigcup\{slit(B_t) : t \in T\} = s$ occurs as a disjunct in $A$. By the previous lemma, $B$ is verified by $slit(B) = s$, so $s$ verifies $A$ and hence $s \in [A]^+$, as required. □

**Theorem 14.4** *In $\mathcal{M}_F$, if $A$ and $B$ are distinct closed and repetition-free standard disjunctive forms of $\mathcal{L}$, then $[A]^+_f \neq [B]^+_f$.*

***Proof*** Suppose $A$ and $B$ are distinct closed and repetition-free standard disjunctive forms. Without loss of generality, assume there is a disjunct $D$ in $A$ which does not occur in $B$. By Lemma 14.16, $[D]^+ = \{slit(D)\}$, so $slit(D)$ is in $[A]^+$ and hence $[A]^+_f$. Suppose $slit(D)$ is also in $[B]^+_f$ and thus, by the previous lemma, in $[B]^+$. By the clause for disjunction, it follows that $slit(D)$ verifies some standard conjunctive form $E$ that occurs as a disjunct in $B$. But by Lemma 14.16, $[E]^+ = \{slit(E)\}$, so it follows that $slit(D) = slit(E)$. But since both $D$ and $E$ are standard conjunctive forms, it follows that $D = E$, contrary to our assumption. □

Similarly as before, decidability may be established by finding a finite model to do the job of $\mathcal{M}_F$ for any given pair $A$ and $B$, which here requires merely restricting the states to sets of literals occurring in $A$ or $B$.

### 14.6.4 Regular Truthmaking

Again, this case is covered in Fine (2016), so I provide only a very brief summary; for details, readers may consult section 8 in Fine's paper. It turns out that the same, simple canonical model may here be used as was used as the f-inclusive canonical model.

**Definition 14.18** The *regular canonical model* $\mathcal{M}_R$ for $\mathcal{L}$ is $\mathcal{M}_F$.

As shown above, a conjunctive form $A$ has $slit(A)$ as its only verifier, and so a disjunctive form $B$ is verified by all and only the states $slit(A)$ for $A$ a disjunct in $B$. It may then be shown that for any maximal disjunctive form $A$, $[A]^+ = [A]^+_r$. On that basis, we can shown in a similar way as above that

**Theorem 14.5** *In $\mathcal{M}_R$, if $A$ and $B$ are distinct maximal standard disjunctive forms of $\mathcal{L}$, then $[A]^+_r \neq [B]^+_r$.*

and infer decidability as before.

## 14.7 Completeness

Given the results established in the previous sections, the completeness proofs are all easy and follow exactly the same shape.

**Theorem 14.6 (Completeness)** *For all equivalences $A \approx B \in \mathcal{L}_\approx$,*

*(1) if $\models_e A \approx B$ then $\vdash_e A \approx B$,*
*(2) if $\models_c A \approx B$ then $\vdash_c A \approx B$,*
*(3) if $\models_f A \approx B$ then $\vdash_f A \approx B$, and*
*(4) if $\models_r A \approx B$ then $\vdash_r A \approx B$.*

***Proof*** For (1), suppose $A \approx B$ is valid under the exact interpretation of $\approx$. Then $[A]^+_e = [B]^+_e$ in $\mathcal{M}_E$. By the normal form theorems, $A$ and $B$ are provably equivalent within $\mathfrak{D}_e$ to full standard disjunctive forms $A^*$ and $B^*$. By the soundness Theorem 14.1, $[A^*]^+_e = [B^*]^+_e$. By the exact canonical model Theorem 14.2, $A^* = B^*$. So $A^*$ and $B^*$ are provably equivalent, and thus so are $A$ and $B$. The other cases are exactly analogous. □

**Acknowledgments** For helpful discussion and comments on earlier versions of this essay, I am grateful to Stephan Leuenberger, Stefan Roski, Josh Schechter, and Bruno Whittle. Research on this paper was funded by the *Deutsche Forschungsgemeinschaft* as part of the *Relevance* project (grant KR 4526/2-1), and I gratefully acknowledge that support.

## References

Anglberger, A., Korbmacher, J., & Faroldi, F. (2016). An exact truthmaker semantics for permission and obligation. In O. Roy, A. Tamminga, & M. Willer (Eds.), *Deontic logic and normative systems, 13th international conference, DEON 2016* (pp. 16–31). College Publications.

Correia, F. (2016). On the logic of factual equivalence. *The Review of Symbolic Logic*, *9*(1), 103–122.

Fine, K. (2012a). Counterfactuals without possible worlds. *The Journal of Philosophy*, *109*(3), 221–246.

Fine, K. (2012b). Guide to ground. In F. Correia, &B. Schnieder (Eds.), *Metaphysical grounding: Understanding the structure of reality* (pp. 37–80). Cambridge University Press.

Fine, K. (2012c). The pure logic of ground. *The Review of Symbolic Logic*, *5*(1), 1–25.

Fine, K. (2016). Angellic content. *Journal of Philosophical Logic*, *45*(2), 199–226.

Fine, K. (2017a). A theory of truthmaker content I: Conjunction, disjunction and negation. *Journal of Philosophical Logic*, *46*(6), 625–674.

Fine, K. (2017b). A theory of truthmaker content II: Subject matter, common content, remainder and ground. *Journal of Philosophical Logic*, *46*(6), 675–702.

Fine, K. (2017c). Truthmaker semantics. In B. Hale, C. Wright, & A. Miller (Eds.), *A companion to the philosophy of language* (pp. 556–577). John Wiley & Sons Ltd.

Fine, K. (2018a). Compliance and command I—categorical imperatives. *The Review of Symbolic Logic*, *11*(4), 609–633.

Fine, K. (2018b). Compliance and command II—imperatives and deontics. *The Review of Symbolic Logic*, *11*(4), 634–664.

Fine, K. (2021). Verisimilitude and truthmaking. *Erkenntnis*, *86*, 1239–1276.

Fine, K. (202x). A theory of partial truth. Forthcoming in a volume for David Kaplan.

Krämer, S. (2017). A hyperintensional criterion of irrelevance. *Synthese*, *194*, 2917–2930.

Krämer, S. (2022). Mighty belief revision. *Journal of Philosophical Logic*, *51*, 1175–1213.

Krämer, S. (2023a). That's it! Hyperintensional total logic. *Journal of Philosophical Logic*, *52*, 963–1004.

Krämer, S. (2023b). The Whole Truth. In F. L. G. Faroldi, & F. V. D. Putte (Eds.), *Kit Fine on truthmakers, relevance, and non-classical logic*. Springer.

Krämer, S., & Roski, S. (2015). A note on the logic of worldly ground. *Thought: A Journal of Philosophy*, *4*(1), 59–68.

# Chapter 15
# Scope and Probabilistic Reasoning: A Resolution of the Two-Envelope Paradox

**Gary Kemp**

**Abstract** A solution of McGrew et al. is defended and made more precise; it is found to be a simple scope-fallacy. The Paradox: You are shown two envelopes of money, are given one, and informed that the other one contains half the amount in your envelope or double that amount. Elementary reasoning tells you to trade. But the same reasons tell you to trade again...

**Keywords** Fallacy · Probability · Two-envelopes

I first met Alan at Queen's Belfast in 1999, me having been invited to read a paper there. Who was this guy, I thought, so voluble, yet with an astronomically high signal-to-noise ratio? One that remains high whether you're on the look-out for jokes, philosophy, or pungent details of Irish or Scottish history and politics. And this guy shares my interest in Quine. Next time was 6 years later, when he came to interview for a Chair here at Glasgow, with me the head of department, and Fiona Macpherson, David Bain, Mike Brady, Chris Lindsay and Adam Rieger looking on. Of course there was no question, we wanted him (I remember his performance at both the talk and the pub afterwards – with Tim Williamson the external – as being rather restrained, which only served to raise him in our eyes; he knew when to keep his torrent of a personality somewhat under wraps). When he arrived I had gone on sick leave, and then he was shoved almost immediately into the job of Head of Department, during a very difficult time for the, ahem, 'Department', but it was very much on the mend by the time he finished being, ahem, Head of 'Subject'. It is in no small measure that Philosophy at Glasgow owes its recovery and present success to him. He is up there with famous raconteurs in recent philosophy – like say Sidney Morgenbesser – but the depth of the pure philosopher, his reliability

G. Kemp (✉)
University of Glasgow, Glasgow, UK
e-mail: Gary.Kemp@glasgow.ac.uk



A. Rieger, S. Leuenberger (eds.), *Themes from Weir: A Celebration of the Philosophy of Alan Weir*, Synthese Library 484,
https://doi.org/10.1007/978-3-031-54557-3_15

as friend, combined with his being so tough, informed and wise when it comes to political or administrative matters, make for a great man.

1. The resolution offered to Two-Envelope Paradox by McGrew et al. (1997) is substantially correct, but it does not go far enough: it fails to reduce the underlying fallacy to a more general and generally understood mistake. The Paradox is by now well-known. I present it in a slightly fussy way in order to facilitate certain points to be made later on.

There are two envelopes on the table. You are informed that the amount in one envelope is nonzero but twice that in the other. One of these envelopes is selected and presented to you. You are asked, would you like to trade for the other? (Call this the *initial description* of the game.)

You might reason as follows: Suppose I trade. If I lose, I end up with half the amount that I have in my envelope now. But if I win, I end up with double that amount. So winning wins twice what losing loses. So I should trade. (Call that the *informal argument for trading*.)

More exactly (call this the *quasi-formal argument for trading*): Let A be the envelope handed to me, and B be the other envelope. Let n = the amount in A. Let (1) be the case where the amount in B is 2n. Let (2) be the case where the amount in B is 1/2 n. Let V(1) and V(2) be the respective values of trading in case (1) or (2) obtain. With the probability operator **Pr** being a function from propositions to real values m ($0 \leq m \leq 1$), the informal argument might be represented as follows.

(i) The amount in B is either 2n or 1/2n. (Assumption).
(ii) **Pr**(1) = **Pr**(2) = 1/2. (Assumption).
(iii) V(1) = 2n − n = n; V(2) = 1/2n − n = −1/2n. (from i and ii).
Therefore,
(iv) the expected value of trading is **Pr**(1)V(1) + **Pr**(2)V(2) = 1/2n + 1/2 (−1/2n) = 1/4n.

The expected value of trading is positive, so I should trade.

Of course this contradicts one's natural reaction that there is no point in trading; you are simply being asked to choose from two envelopes, in which case preference is surely irrational. Indeed, once you've traded, it seems the same argument could always be repeated, enjoining you to trade back and forth *ad nauseam*. Alternatively it seems we could repeat the above reasoning to 'prove' that you should *not* trade, simply by letting n be the amount in B, so that the amount in A is 2n or 1/2n.

The gist of the diagnosis by McGrew et al. is as follows.

> The real flaw [in the argument for trading] consists in a confusion as to which amount of money – (a) the amount in the selected envelope [A], or (b) the total amount involved in the game – can legitimately be taken as 'fixed' in performing the above probability calculation. Where assumption (a) is appropriate, then the above calculation is correct, but it is not at all paradoxical. This assumption would be appropriate if, e.g., I were handed a sealed envelope containing a certain amount of money and told that a different sum of money, either twice what was in mine or half of it, had been placed in another envelope with the decision as to whether it was twice or half my sum depending on the toss of a fair coin. In these conditions the total amount in the game is not fixed – it is either 3 times the amount in my envelope, or

> 1.5 times that amount, depending on the coin toss. And in these conditions I *should* swap, for precisely the reasons set out in the above calculation ...
>
> In the standard version of the game, however – a version in which I may select either of the two envelopes and the total sum of money is independent of my selection – it is clearly assumption (b) which is appropriate. And on this assumption the amount in the selected envelope cannot be taken as fixed ... contrary to what this presentation assumes, '*n*' cannot be taken to represent the amount in the selected envelope [throughout the argument] ... While it is true (still assuming that the fixed total amount is 3*n*) that I stand to gain *n* by swapping if my envelope contains the lower amount, it is not true that I stand to lose only n/2 if my envelope contains the higher amount. For if it contains the higher amount, then it contains 2n, not n, and what I stand to lose is n, not n/2.

Most of the discussion of the two envelopes has taken a completely different line. Some have pointed out that we cannot soundly calculate expected valuc unless we makc assumptions about the range of n: in particular, that we have to make assumptions about the probable amounts initially in the envelope A. But once we do this, we realize immediately that, as a contingent matter, the probability that an envelope contains an amount m decreases as m increases; at some point the probability that B contains 2n must fall below 1/3, in which case the expected value of trading dips below zero.[1] I shall return to this in my closing remarks. Largely in response to this, others have claimed or assumed that a genuine paradox arises only if it is assumed that the total amount in the two envelopes could be *any* finite amount of money. It is certainly true that if that assumption is made, then the question of whether or not to trade becomes both mathematically and epistemologically more interesting (see Scott & Scott, 1997). But I think that McGrew et al. are right to claim that it would be incredible if that ludicrous factual assumption were somehow implicit in the reasonings of those who are actually puzzled by the game; indeed if that were right, then the Initial Description of the game *itself* is factually impossible.

It significant, then, that McGrew et al. have made a case that the fallacy in the argument for swapping does not involve a mistake in reasoning about infinite (discrete) probability spaces and infinite expected values. But the case needs to be made out more precisely before we can justly conclude these mathematical complexities are not essentially involved in the paradox. In particular, what we need is a much more exact diagnosis of the fallacy in the above reasoning. There are of course various ways in which one might make a certain invalid inference. But the task before us is to identify the fallacy itself: if, for example, some persons might commit the fallacy because they have mishandled a certain more specialized type of mathematical reasoning, that particular cause of their committing the fallacy should not be confused with the fallacy itself.

---

[1] Actually Broome (1995) has pointed out that prior probability distributions can be specified in which the argument for trading goes through irrespective of the initial value. For example (and roughly), if the probability that an envelope contains x decreases 'slowly' enough as x increases, then the expected value of switching is always positive. But of course this works only if we assume that for every finite amount of money there is a greater.

2. First, let us be certain that finite games can be described in which trading is correct, as well as finite games in which it is indifferent, all of which are consistent with the initial description. In each of following games, then, trading will get you either half or double what you've got already in envelope A.

*Game A* An envelope is placed upon the table (call this the 'α-envelope'). Two more envelopes are produced (the 'added pair'). You are informed that one of these contains twice and the other half the amount in the α-envelope; one is selected at random and placed upon the table beside the α-envelope; the other is set aside. The α-envelope is now presented to you, and a trade offered for the envelope still on the table.

*Game B* There are two envelopes, one containing 1, the other 2. You know this. One is randomly selected and presented to you. You are given the opportunity to trade for the other.

*Game C* There are six envelopes in three pairs. The envelopes of the first pair contain, respectively, 1 and 2; those of the second 2 and 4, and those of the third 4 and 8. A pair is selected at random and placed on the table; and now one member of the pair is given to you, again selected at random. You are asked if you want to trade. That is, the possible gains are from 1 to 2, from 2 to 4, and from 4 to 8; the possible losses the reverses of these. In fact this is what we might call an extension of Game B, but it makes it easy to see how the number of pairs could be increased indefinitely without altering the expected value calculation.

It is obvious that the expected value of trading in Games B and C, and in any extension of game B, is zero. Slightly less obvious is that the expected value in Game A is positive, that is, that you ought to trade; in fact, if you are allowed to repeat the game, letting the envelope you selected in the previous round be the new α-envelope, then you ought to keep trading as long as the gamesmaster will allow; for this game is genuinely 'double-or-half', when 'double-or-nothing' is the paradigm even stakes. That is, if you keep playing this game, you end up even if you lose exactly as often as you win; but for any n, winning n times more than you lose earns more than is lost by losing n times more than you win.

3. Let us now formalize the argument for trading more strictly, thereby enabling us to analyze the fallacy. The analysis hinges on the fact that in the fallacious reasoning, the pretended constant 'n' is introduced by a definite description, 'the amount in A'. I write this as a function N(x) from envelopes to amounts of money, but assume Russell's way both with functions and definite descriptions. That is 'N(A) = n' is to mean 'The amount in A = n', which is to mean 'for some unique n, A contains n' (n is always a number of units of currency). The fallacy, it turns out, is a fallacy of scope; in particular, it reduces to the quantifier-fallacy whereby '∀∃' is confused with '∃∀'. I assume it will be granted that this is a common and well-understood fallacy; thus the two-envelope fallacy does not have anything essentially to do with probabilistic reasoning.

First, let us recast the contingent fact which initiates the argument so as explicitly to incorporate what I have just said. We let '$\varphi$' mean 'is an envelope in my hand' and '$\psi$' mean 'is an envelope on the table' (I leave out the associated variables). We could express this fact as

(I) $\exists$n [n = N(The $\varphi$) and (N(The $\psi$) = 1/2n or N(The $\psi$) = 2n)],

or we could write

(II) $\exists$n [n = N(The $\varphi$) and (N(The $\psi$) = 1/2n] or $\exists$n [n = N(The $\varphi$) and (N(The $\psi$) = 2n].

But these are logically equivalent, in view of the equivalence of the forms '$\exists$x(Fx or Gx)' and '$\exists$xFx or $\exists$xGx'. For convenience then I will use (II). Attaching this to the disjuncts of (II), we have

(I-2) **Pr**[$\exists$n(n = N(The $\varphi$) and N(The $\psi$) = 1/2n)] = 1/2
(I-3) **Pr**[$\exists$n(n = N(The $\varphi$) and N(The $\psi$) = 2n)] = 1/2.

There is nothing paradoxical in these, for they tell us that gain and loss are equiprobable, not that the positive value of winning would exceed the negative value of losing. The sense of paradox arises only if we assume that the value of the existentially quantified variable which would satisfy the matrix of (I-2), in a case in which we win, would also satisfy that of (I-3), in a case in which we lose. That is, it arises only if we assume that we could have subjected the disjuncts in (I) directly to the operator **Pr**, yielding

(I*) $\exists$n[n = N(The $\varphi$) and **Pr**(N(The $\psi$) = 1/2n) = 1/2 and **Pr**(N(The $\psi$ = 2n) = 1/2]

But the move from (I) to (I*) leads to truth only in the case of Game A. *Nothing in the initial description tells us that we can do this.*

4. Why not? What is the exact fallacy in the informal or the quasi-formal argument? I am here assuming that **Pr**, like the description operator, is a sentential operator the results of which are terms standing for real numbers. The explanation for why we can derive (I-2) and (I-3) from the initial description, but not (I*), then, should be traceable to the scope-characteristics of '**Pr**' relative to '$\exists$'. The move to (I*) assumes, in effect, that we can drive a probability operator whose scope is an existential quantification to a position inside the scope of the existential quantifier (so long as the term giving the value of the probability function is likewise brought inside): this is the assumption that from '**Pr**($\exists$x ... x ...) = n' we have '$\exists$x(**Pr** ... x ... = n)'. (That is, (I*) implies that operation being applied to (I-2) and (I-3)). This assumption is clearly false.

We can easily explain this fact semantically, by explaining **Pr** as a quantifier over cases (possible worlds), thereby reducing the scope issue to the familiar relationship between '$\exists$' and '$\forall$'. Thus let ***W*** be the set of worlds (possible outcomes); let {w: p} be the set of worlds at which p is true, and let *C* be a cardinality operator. Then we can interpret '**Pr**(p)' as an abbreviation of *C*{w: p})/*C*(***W***). Then of course the translation of **Pr**($\exists$x ... x ...) = n is *C*{w: $\exists$x ... x ...}/*C*(***W***) = n, not $\exists$x [*C*{w: ... x ...}/*C*(***W***) = n]. And it is perfectly obvious why this is: the set of worlds such that something or other has a certain property may well be larger than the set determined by choosing a particular thing, and selecting the worlds at which *it* has that property. To assume these sets identical is to erase the difference between $\forall x \exists y \ldots x..y \ldots$ and $\exists x \forall y \ldots x..y \ldots$

To apply the lesson to the envelopes, then: the fallacy lay in not taking care when generalizing about 'cases' in the informal and quasi-formal arguments. In generalizing about cases we have to look after the ways in which the scope of a definite description interacts with that of the quantifier ranging over cases. If I say that n is the amount in my envelope – whatever it may be – then I cannot assume that it holds constant as I consider the possible cases. Therefore I cannot assume that a given winning amount – '2n' as I might express it – will stand in the ratio 2:1/2 to an arbitrary losing amount – '1/2n', as I might express it. 'n' is here a narrow-scope definite description.

What if we construe the relevant definite descriptions as taking *wide* scope, hence obviating the shift, and accepting the move from (I) to (I*)? Nothing stops you in the case of game A (so long as the referent of the definite description is fixed only after the $\alpha$-envelope is selected). But consider Game B. Suppose, unknown to you, your envelope contains 1 (so the other contains 2). Then 'The amount in my envelope' is semantically equivalent to '1'. Then to say 'Either the other envelope contains twice what mine does, or half', is equivalent to saying 'Either the other envelope contains 1/2 or it contains 2'. That is of course true. But, on the wide-scope reading, you cannot conclude 'the probability that the other envelope contains half what mine contains is 1/2, as is the probability that it contains twice what mine does'. For this would be equivalent to 'The probability that the other envelope contains 1/2 is 1/2, as is the probability that it contains 2', which is false.

5. If the scope-analysis of the envelope-problem is correct, then we should be able to derive a proposition expressing the indifference of trading from (I), in game B and its extensions. First, we need a value function V and an expected value function *EV*. Intuitively, we want V to be a function from propositions to amounts of money, such that V(p) is the value gained (positive or negative) in case p is true. As we will see, that is not quite sufficient.

Let $e_1$ be the proposition in the scope of **Pr** in (I-2), and $e_2$ be the proposition in scope of **Pr** in (I-3). Then we might set

(I-4) $V(e_1) = 1/2N(\text{The } \varphi) - N(\text{The } \varphi) = -1/2N(\text{The } \varphi)$
(I-5) $V(e_2) = 2N(\text{The } \varphi) - N(\text{The } \varphi) = N(\text{The } \varphi)$.

But we cannot then define *EV* simply in terms of N(The $\varphi$) and the probabilities assigned to $e_1$ and $e_2$. We cannot write

(I-6) $EV = 1/2V(e_1) + 1/2V(e_2) = 1/4N(\text{The } \varphi)$.

For N is a narrow-scope operator (more exactly, such is the scope of 'The $\varphi$' and thus 'N(The $\varphi$)'); hence, V is also narrow-scope; we cannot assume that the value of 2N(The $\varphi$) in one world will exceed the value of 1/2N(The $\varphi$) in another. Thus we cannot calculate with $V(e_1)$ and $V(e_2)$ except with respect to particular worlds (ways in which the money may be distributed).

We must relativize V to possible worlds:

(I-4′) (w)[$e_1 \rightarrow (V(e_1) = 1/2N(\text{The } \varphi) - N(\text{The } \varphi) = -1/2N(\text{The } \varphi))$]
(I-5′) (w)[$e_2 \rightarrow (V(e_2) = 2N(\text{The } \varphi) - N(\text{The } \varphi) = N(\text{The } \varphi))$].

The denotation of '$\varphi$' cannot be assumed constant between worlds, owing to the sort of ambiguity illustrated between Game A and Game B. We can build in this ambiguity, expressing *EV* as normal – as the average gain, as the sum of the gains in the winning cases added to the sum of the losses in the losing cases, divided by the number of cases, assumed equiprobable – if we add a certain wrinkle. The lessons of (I-4′) and (I-5′) having been learned, let us for fun employ subscripts on $\varphi$, i.e. '$\varphi_i$' and '$\varphi_j$', to make clear that the envelope in one situation need not be the envelope in another situation. We can strip away explicit use of 'V' etc., and write:

$$EV = \frac{N\left(\text{the } \varphi_i\right)}{2} + \frac{-\frac{1}{2} N\left(\text{the } \varphi_j\right)}{2}$$

Simplified:

$$EV = \frac{N\left(\text{the } \varphi_i\right) - \frac{1}{2} N\left(\text{the } \varphi_j\right)}{2}$$

In B-type games, $N(\text{the } \varphi_i) + N(\text{the } \varphi_j)$ will always be a certain fixed amount. In A-type games, $N(\text{the } \varphi_i) + N(\text{the } \varphi_j)$ will be *either* three times $N(\text{the } \varphi_i)$ *or* one-and-half $N(\text{the } \varphi_i)$. These are in effect lessons in McGrew et al. In B-type games (where $i \neq j$), the losses and gains cancel each other, yielding zero. In Game A, (i.e. $i = j$), they do not; they sum to 1/4 the amount in the $\alpha$-envelope – $1/4N(\text{the } \varphi_i)$: the denotation of 'The $\varphi_i$' does not shift depending on which world we are in, as in B-type games.

6. The conclusion of §1 was that there is an ambiguity in the initial description of the game, between type-A and type-B. The *EV* calculation is of course different depending on which game is being played. What if you know *only* that the initial description is true, that is, if you're told no more that the envelope in your hand has half or double the one on the table, and you're offered a trade? If you know only *that* much, then a certain further variant of Game B would be consistent with the Initial Description of the game: the two possibilities are a gain from x to 2x, a loss from y to y/2; so for all you know x = 1 and y = 100. Clearly *EV* cannot be calculated for that scenario if x and y are independent (bracketing prior probabilities for second). But if you do know a bit more – the game is either game A or game B – then you should trade, since trading is indifferent if the game is of the game B variety but positive if the game is the A variety.

About prior probabilities: If we assume only the initial description and that the definite description 'the amount in my envelope' is understood as taking wide-scope, then the real-world probability of the other envelope's containing twice that amount depends upon the amount in the envelope in one's hand. If your friend hands you an envelope containing money and then shows you another, telling you that it contains either half or double the amount you've just been given, then the probability that it's double must go pretty low as we approach the limits of your friend's bank account. It can change the game materially if you're allowed to look inside your envelope before deciding.

## References

Broome, J. (1995). The two-envelope paradox. *Analysis, 55*, 6–11.

McGrew, T. J., Shier, D., & Silverstein, H. S. (1997). The two-envelope paradox resolved. *Analysis, 57*, 28–33.

Scott, A. D., & Scott, M. (1997). What's in the Two Envelope Paradox? *Analysis, 57*, 34–41.

# Chapter 16
# Representative Range Voting: How to Elect the First President of Scotland

**Adam Rieger**

**Abstract** This paper presents a novel single-winner voting system. After a brief review of existing systems, a variation of range voting is described, in which the winner is the candidate minimizing the total square dislike expressed by the voters. I argue that this method chooses the candidate who is, in a sense made precise, the most representative of the views of the electorate.

**Keywords** Voting methods · Range voting · Score voting · Utilitarian voting · Least-squares

No-one who has spent more than a few minutes with Alan Weir can be unaware of his enthusiastic support for Scottish independence. And if one has spent a bit longer, one knows not to refer to the UK head of state as the monarch 'of England'.[1] More than one unwary visiting speaker, discussing the theory of descriptions or similar, has done so in our departmental seminar, and they were soon made vividly aware of their transgression.

Current SNP policy is to retain the monarchy post independence. But suppose Scotland decides to become a republic? In anticipation, I offer a new method for electing a president.

I will first briefly review existing electoral systems, roughly from worst to best, before describing my own proposal.

[1] See §7 of Alan's *Memoir*, in this volume.

A. Rieger (✉)
School of Humanities, University of Glasgow, Glasgow, UK
e-mail: Adam.Rieger@glasgow.ac.uk

 
A. Rieger, S. Leuenberger (eds.), *Themes from Weir: A Celebration of the Philosophy of Alan Weir*, Synthese Library 484,
https://doi.org/10.1007/978-3-031-54557-3_16

## 16.1 Some Single-Winner Electoral Systems

Our subject here is 'single-winner' electoral systems, as opposed to those that are used to elect a parliament. Given a number of candidates and an electorate with diverse views on their merits, the problem is simply to choose which candidate should be the winner.

### *16.1.1 The Electoral College*

I shall not say much about the USA's famous system, except that I don't know of another democracy where it would be an improvement to change to 'first-past-the-post' (indeed, there currently is a campaign, under the name 'National Popular Vote', to do just that[2]—by the back door mechanism of getting enough states to pledge their electoral college delegates to the candidate with the largest number of votes nationally, thereby avoiding the difficulties of changing the US constitution). A version of the 'block vote' system of the type notorious in the UK in old Labour Party and TUC conferences, it is only down to luck that it has so far produced relatively few clearly unjust results—it is quite possible for such a system to make it virtually certain that one of the parties will win every time, if it has inbuilt majorities in enough large states.[3]

### *16.1.2 The Plurality Method*

The plurality method, or 'first-past-the-post' (FPTP), is the simplest of voting systems: each voter votes for one candidate, and the candidate with the largest number of votes wins. It is familiar as the system used to elect members of legislatures in the UK, the US and other countries, predominantly English-speaking ones. Its main deficiency is also familiar: briefly, that a candidate without a great deal of support can win because the opposition is split.

In 2011, a referendum was held in the UK to determine whether FPTP should be replaced for Parliamentary elections by the Alternative Vote (which I shall discuss next). The Conservatives had been forced to accept such a referendum as the price of forming a coalition with the Liberal Democrats after the inconclusive result of the 2010 election, but they campaigned vigorously (and successfully) to keep FPTP.

---

[2] See https://www.nationalpopularvote.com/.

[3] In fact, demographic trends suggest something of this sort may not be so far off, for example if Florida and Texas become safe Democrat states. See for example https://www.americanprogress.org/article/americas-electoral-future-3/.

The question as to how to elect a parliament is not, as I mentioned, my main subject here. With FPTP one can make a case that its distorting effects—that the proportions in parliament may bear little resemblance to the proportion of votes obtained,[4] and a party may win a comfortable overall majority with a relatively small proportion of the total vote[5]—are a price worth paying for 'strong government'. For the problem at hand, of electing a single winner considered in isolation, it has few supporters amongst social choice theorists.[6] Various politicians, mostly Conservative but also some from the Labour party, did attempt to argue directly for it, generally on grounds either obviously question-begging (that systems that transfer votes allow 'losing' candidates to win), or obviously spurious (e.g. that they are unfair in that they allow some voters more than one vote).[7]

It is very noticeable that parties currently, or recently, in power tend to favour the status quo; a cynic might conclude that having won under the current system, they are reluctant to replace it with a system under which they may do less well. It is notable, however, that neither the Conservatives nor Labour have ever used FPTP to elect their own leaders, using either a run-off system amongst MPs or, more recently, more complex variants of it involving wider party membership in various ways.[8] Recent examples where important political decisions in the UK have been taken by referendum might also have tested the large parties' intellectual attachment to FPTP; for example in 2012, when the arrangements for the Scottish independence referendum of 2014 were being discussed, the prospect was briefly raised of a referendum with three options (perhaps *independence*, *increased devolution* or the *status quo*). Had this been held by FPTP and independence come first but with less than 50% of the vote, David Cameron would have been in a difficult position. Similarly, a three-way referendum was widely suggested as a way of breaking the

---

[4] For example, in 2015, the SNP won 56 seats to UKIP's 1, with about three eighths of UKIP's vote.

[5] In 2005, Tony Blair won his third comfortable majority, with more than 55% of the seats in parliament, with only 35.2% of the votes, less than 3% ahead of the Conservatives.

[6] In the poll of 22 experts taken at a conference in 2010 described in Laslier (2012) it received not a single positive vote under an 'approval voting' system. (For the record, approval voting—discussed briefly below—was the clear winner.)

[7] At the time of writing (2022), the Conservative government is in the process of putting through Parliament a bill which, amongst other things, changes the electoral system for mayors and police commissioners to FPTP; the explanation on the Government website includes the argument that under a transfer system 'a "loser" candidate can win on second preferences'. See https://www.gov.uk/government/news/first-past-the-post-to-be-introduced-for-all-local-mayoral-and-police-and-crime-commissioner-elections.

[8] Rather incredibly, the Conservatives have only formally elected their leader since 1965; before that, the leader 'emerged'.

deadlock over Brexit in 2018; should the UK accept the deal negotiated by Theresa May, leave with no deal, or remain? Brexiteers, typically enthusiastic supporters of FPTP, would no doubt have cried foul if Remain had been declared the winner after topping the poll with 40% of the vote, an outcome suggested by some opinion polls.

Apart from the obvious unfairness of the system in the above cases, we can make a more abstract point against FPTP: it collects very little data from the voters, merely their first preferences. Their views on other candidates are simply not recorded.

### *16.1.3 The Alternative Vote*

The losing option in the 2011 UK referendum was the Alternative Vote (AV), also known as Instant Run-off Voting (IRV). In a simple run-off system, instead of immediately declaring as the winner the candidate with the largest number of votes, if no candidate has more than 50% we eliminate the bottom candidate and hold a new election amongst the remainder; the process is repeated until one candidate has more than 50%. The disadvantage of such a system is the need to hold a potentially large number of ballots. AV eliminates this problem by asking the voters to *rank* the candidates and then using the ranks to simulate how each voter would behave in such a series of run-offs.[9] AV does eliminate the problem of the 'divided opposition', since similar-minded voters split between a number of candidates will tend eventually to coalesce around a single candidate. It also collects more information than FPTP, since we record not just the first preference of each voter, but their ranking of all the candidates.

However, there are good reasons to doubt that AV is an ideal system. Essentially, a good 'compromise' candidate can be eliminated too early, for want of first preferences. Consider, for example, a large field of candidates, of whom one is the second choice of nearly every voter, but has the fewest first preferences. Such a candidate will be eliminated at the beginning, but intuitively has a claim to be the best collective choice.

A real-life example concerning Scottish independence illustrates the problem. In an opinion poll taken after the 2014 independence referendum, voters were asked how they would have voted if an 'increased devolution' option, the 'Smith proposals', had been on the ballot paper. The results are shown below:[10]

[9] Inferior variants exist where only the top two candidates, on first preferences, pass on to a second and final round; such systems are used for presidential elections in France, where the second round is held on a later date, and the 'supplementary vote' used (up till now) for mayoral elections in the UK, where voters are asked to vote for a first and second preference.

[10] Source: Panelbase/Wings over Scotland. Details at https://wingsoverscotland.com/the-unwanted-vow/#more-70961. I have removed 'don't knows' and made simple, but reasonable, assumptions about second preferences; the original poll asked only for first preferences.

| Ranking | Percentage of voters |
|---|---|
| No change > Smith > Ind | 32 |
| Smith > No change > Ind | 20 |
| Smith > Ind > No change | 5 |
| Ind > Smith > No change | 43 |

We can note first that under FPTP, independence wins, with 43% of the vote.

Under AV, Smith is eliminated, and No Change wins 52:48 after transfers.

But plausibly, Smith ought to win! Smith is a bit worse than the others on first preferences, but *much* better on second preferences.

Although AV *gathers* more information from the voters, it doesn't *use* the second and lower preferences of voters whose first preferences survive until the last round—information is 'thrown away', in this case that Smith is a popular second choice amongst voters for the two leading options.

It is not just intuitively plausible that Smith should win: we can produce an *argument* to this conclusion. Note that 57% prefer Smith to Independence, and 68% prefer Smith to No Change. Thus a clear majority of the electorate prefer Smith to each other option. A popular view amongst social choice theorists is that this implies that Smith should win, and hence that we should seek an alternative to AV.

### *16.1.4 Condorcet Methods*

The above example suggests the following:

> A candidate who is preferred by a majority to each of the others in head-to-head contests should win the election.

This idea is associated with the Marquis de Condorcet, a French aristocrat who was a leading figure of the French enlightenment but eventually perished in the turbulent years following the Revolution. A candidate satisfying the above condition is known as the *Condorcet winner* (clearly there can be only one) and a method electing such a winner a *Condorcet method.*

Can we simply recommend electing the Condorcet winner, and leave it at that? Unfortunately not. As Condorcet himself pointed out, the problem is that there may be no Condorcet winner:

| Ranking | Number of voters |
|---|---|
| A > B > C | 40 |
| B > C > A | 25 |
| C > A > B | 35 |

75 of the 100 voters prefer A to B, 65 prefer B to C, and 60 prefer C to A. Thus, there is in this case no Condorcet winner. 'A majority prefer $x$ to $y$' is not in general transitive (and hence, cannot (alone) be used to decide between candidates).

It is worth noting that (under any system that produces a winner at all) one of A, B, C above will win despite the fact that *a majority would have preferred another candidate*—so the italicized cannot in general be a decisive argument against a candidate winning. This has implications for the conduct of business in legislatures and committees.

As mentioned above, many voting theorists believe that a Condorcet winner should be elected if there is one. If there is a cycle, we need a 'tie-breaker', and a lot of ink has been spilled in arguing about the best way to do this. However, no way seems ideal.

Indeed, a large number of pathologies afflict voting systems: various theorems have been proved showing that certain desirable-looking properties of voting systems are incompatible. The most famous is Arrow's theorem [1963], which I won't discuss here; another is the beautiful result of Moulin (1988) that any Condorcet method with more than four candidates has the 'no-show' paradox, that it is possible for a voter to be better off not voting rather than showing up (and voting sincerely). AV has the 'non-monotonicity' property, that you can harm a candidate by ranking them higher. And so on.

As Amartya Sen remarked in his Nobel prize lecture (Sen, 1999, p. 353),

> The force and widespread presence of impossibility results generated a consolidated sense of pessimism, and this became a dominant theme in welfare economics and social choice theory in general.

I shall explain below why I think this pessimism is unjustified.

### *16.1.5 The Borda Method*

Is it really obvious that a Condorcet winner, if there is one, should be elected? Consider the following

| Ranking | Number of voters |
|---|---|
| A > B > C | 10 |
| B > C > A | 9 |

A is the Condorcet winner. But there's something in the thought that B ought to win—from the rankings a bare majority seem to *narrowly* prefer A to B, but an almost equal number seem to *strongly* prefer B to A.

Borda, another Eighteenth Century French voting theorist, proposed giving 1 point for a last-place ranking, 2 for second-last, etc., the winner being the candidate with the highest total score. The intuitive force of the Borda method is that distance apart in the ranking is taken to be indicative of *strength* of preference (to which Condorcet is insensitive).

But actually we can't really tell this; for example, perhaps all the A > B > C voters in the previous example really hate both B and C (but hate C a bit more), whilst the B > C > A voters quite like all of the options. As long as all we have is rankings, we don't have any solid information about strength of preference.

### 16.1.6 Range Voting

To gather information on strength of preferences, we can ask the voters to *score* the candidates on a numerical scale (for example from 0 to 10). The most obvious way to pick the winner is then to choose the candidate with the highest total (or average). This is usually known as *range voting* or *score voting*.

It has at least three advantages over the systems examined so far. Firstly, it gathers more information from the voters than any ranked system (or FPTP), at least if the number of possible scores is reasonably high. Thus, for example, two voters might each rank candidates A, B, C in that order, but the first greatly prefer A to B, which she ranks only just above C, whilst the second voter only narrowly prefers A to B, both of which he greatly prefers to C. A scoring system allows this to be expressed, perhaps by the voters scoring A, B and C as 10, 2, 1 and 10, 9, 1 respectively. In this sense, scoring methods are an advance on ranking methods, just as ranking methods are an advance on a simple plurality method in which a voter is invited only to cast a vote for a single candidate. (The system known as *approval voting* is technically a special case of range voting, but cannot be justified in this way; it asks the voter to give scores of either 0 or 1 to each candidate, the winner being the candidate with the highest total. It is admirably simple (for example, one can use the same ballot paper as with a FPTP election and simply advise the voter they can vote for as many candidates as they like) and certainly an improvement on FPTP, but does not allow a nuanced expression of voters' opinions.)

Secondly, an explicit *justification* for it is available (something one might worry is lacking for other systems, where we have a trade-off of conflicting intutions). If the scores are regarded as *utilities*, then electing the candidate with the highest total score is the utilitarian outcome.

Thirdly, aggregating numerical scores, unlike attempting to aggregate rankings, produces no pathologies in the form of preference cycles, no-show paradoxes, and so on. The theorems which plague the systems discussed earlier are all based on the paradigm in which voters rank the candidates. The vast majority of work on voting theory has taken place in this paradigm.

One might wonder why this is so; why has there been so little interest in systems in which candidates are scored rather than ranked? The reason is a historical accident, in my view. Most of the work on voting theory has been done by economists. In the 1930s, economists became convinced that interpersonal comparisons of utilities were unscientific or meaningless.[11] In contrast, the preferences of individuals show up in behaviour, and are therefore scientifically acceptable. The positivistic philosophy underlying this has long since faded, but, oddly, the paradigm lives on nearly a century later in the economic community.

We can leave economists to decide for themselves what the appropriate boundaries are of their subject. Maybe, even if not meaningless, interpersonal utilities are properly the study of philosophers rather than economists, who, as scientists, must work with what is observable and measurable. But we should not accept that voting theory is part of economics and thereby restricted to the ranked systems. The 'ought' in 'who ought the winner to be?' is one of fairness or justice, and one philosophers should be discussing if the economists won't.

How does range voting compare with the systems discussed earlier? Not only is it not a Condorcet system, it doesn't even guarantee the victory of the majority winner out of two alternatives! (Usually voting theorists say the problems only begin when you have at least 3 candidates.)

Imagine three friends deciding which restaurant to go to: Italian or Indian? The first (call him Alan) really hates curry. The other two narrowly prefer curry to Italian food.

The scores might be

| Voter | Italian score | Curry score |
|---|---|---|
| A | 8 | 0 |
| B | 6 | 7 |
| C | 7 | 8 |
| Average | 7 | 5 |

Thus Italian wins, despite curry having a 2-1 majority. Is this a bug or a feature? I am inclined to think that it is the right result, and indeed what would happen if the three discussed what to do without voting.[12]

The restaurant example is, in my view, one on which utilitarianism gets the right result. Utilitarianism, however, is hardly uncontroversial. Is it safe to rest our choice of voting systems on the assumption that it is correct?

In addition, consideration of a simple example of a *multi*-winner election suggests that utilitarianism should not be the last word in voting theory. Suppose, in a nationwide election contested by two parties, that 51% of the population support the Blue Party, score all their candidates at 10 and all opposing candidates at 0,

[11] See for example Robbins (1938).

[12] Of course, the system can be abused if voters are dishonest; but that is true of nearly all other voting systems too, as we know from the Gibbard-Satterthwaite theorem. The issue of strategic voting is a complex one, partly theoretical and partly empirical, and I will not attempt to address it in this paper. It is of some interest in any case to debate what is the best voting system under the assumption that all voters vote honestly.

whilst the remaining 49% support the Red Party, score all Red candidates at 10, and Blue candidates at 0. If scores are utilities, the utility-maximizing outcome is a parliament of entirely Blue members (since replacing any Blue member by a Red one would reduce total utility from 5100 to 4900 for every 100 voters).

This seems an outrageous result. But why? The answer is that such a parliament would be completely *unrepresentative* of the electorate. My proposal is that, even in the single-winner case, we should seek a winner that is, in some appropriate sense, representative of the voters.

## 16.2 Representative Range Voting

I have suggested that it is natural to think that the winner of an election should be the most representative candidate, that is, the one who, in some appropriate sense, comes closest to summarizing the views of the electorate as a whole. But what exactly does this mean?

One way of making the vague notion of representation more precise is as follows. We can regard voters as being situated at various positions on an ideological axis, from left to right, say. (Such *spatial models* are well-known in the political science literature.) We need enough structure to be able to talk about numerical distances, so we take the axis to be an interval of real numbers. The ideal candidate is then one placed at a *measure of location* (or *central tendency*) of the various positions of the voters. Most naturally, we might take the centroid ('centre of gravity') of the points as this measure and thus the position of the perfect candidate. An actual candidate's suitability may then be judged by their distance from this centroid. Exactly this notion of representativeness was proposed by Achen (1978).

In a more general spatial model, we can consider the positions of the voters and candidates as being situated in a multi-dimensional ideological space. Once again the centroid of the points is a natural measure of location for the ideally representative candidate.

Whatever the theoretical merits of the judging a candidate by their proximity to the ideological centroid, it may seem impractical to apply this idea in an actual election. One might think it necessary to decide first on the number of dimensions for the spatial model, and then, by a series of questionnaires, ascertain the positions of voters and candidates on the various axes.

However, this is not necessary. A simple modification of range voting will, in fact, have the desired result. Suppose we assume that the score a voter awards a particular candidate can be interpreted as the voter's judgment of the proximity of the candidate to the voter in ideological space (call this the *distance assumption*). We can encourage voters to score in a way that validates this assumption by wording on the ballot paper, for example: 'For each candidate, please give a score, on a scale of 0 to 10, as to how closely you agree, overall, with the policies of that candidate, where 0 means "completely disagree" and 10 means "completely agree"'.

A candidate perfectly reflecting the voter's views would therefore be scored at 10, whilst one maximally distant 0. We can thus measure the voter's expressed distance from a given candidate as the difference between the maximum score and the score awarded: in this case, $10 - s$ if the score given is $s$. Call this the voter's *dislike* of the candidate.

Once we interpret the voters' scores in this way, we have a simple measure of ideological distance for each voter-candidate pair. And this can be used to solve the problem above, of locating the position of the ideally representative candidate. It is well known that the centroid of a set of points in Euclidean space is that point which minimizes the sum of the squares of distances to the points in the set. Thus an ideal candidate would have minimal sum of squares of voters' *dislikes*. In practice, of course, such a 'perfect' candidate is unlikely to exist, but to select a winner of the election we simply choose the closest to the ideal, that is, the candidate for whom the sum of the square dislikes is lowest.

More formally, the voting system proposed is as follows. Given (finite) sets of voters $v_i$ and candidates $c_j$,

1. Ask each voter $v_i$ to score each candidate $c_j$ with a score $s_{ij}$, where $0 \leq s_{ij} \leq M$.
2. For each candidate $c_j$, calculate the total square dislike $D_j$, where

$$D_j = \sum_i (M - s_{ij})^2.$$

3. The winning candidate is that $c_j$ for whom $D_j$ is least.

## 16.3 Comparison with Standard Range Voting

It is interesting to compare the current proposal with conventional range voting, where the winning candidate is the one with the highest total (or average) score.

Suppose we have 3 candidates A, B, C, located along a one-dimensional ideological axis, and further suppose that the voting population of 100 is divided into 3 groups, each located at the exact same point as one of the candidates, 51 at A's location, 20 at B's, and 29 at C's (Fig. 16.1):

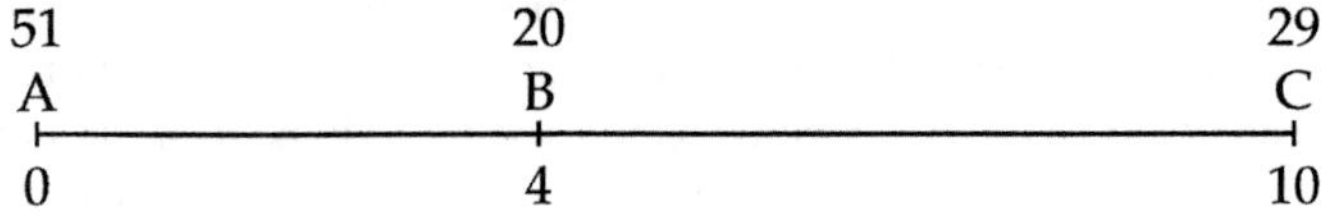


**Fig. 16.1** A voting population

Assume as above that voters are asked to score each candidate from 0 to 10, and that the distance assumption holds. Thus the 51 voters who agree with A all award A a score of 10, B 6 and C 0. In summary, scores will be as follows:

| | A | B | C |
|---|---|---|---|
| 51 voters | 10 | 6 | 0 |
| 20 voters | 6 | 10 | 4 |
| 29 voters | 0 | 4 | 10 |

Calculating the total scores, as recommended by range voting, gives the following:

| Candidate | A | B | C |
|---|---|---|---|
| Total score | 630 | 622 | 370 |

A is therefore the winner of the election.

In contrast, under representative range voting, the dislikes are as follows:

| | A | B | C |
|---|---|---|---|
| 51 voters | 0 | 4 | 10 |
| 20 voters | 4 | 0 | 6 |
| 29 voters | 10 | 6 | 0 |

and the total square dislikes for each candidate are

| Candidate | A | B | C |
|---|---|---|---|
| Total square dislike | 3220 | 1860 | 5820 |

Thus B is elected. Whilst many voting systems would select A—for example, A is the Condorcet winner, indeed the first choice of an absolute majority of voters—nevertheless, the rightness of A's win seems dubious. B, I suggest, seems more closely to 'summarize' the overall views of the voting population. This corresponds to B's being far closer than A to the centroid voter position (easily calculated to be 3.7 in the example).

One can view the choice of winner here, in fact, as a question of choice of measure of location. The ideal candidate under conventional range voting will always be situated at the *median* position (here, candidate A at position 0). This is because maximizing total score is equivalent to minimizing total dislike, and the median is well known to be the value of $m$ which minimizes the sum of distances from each point to $m$. The advantages of the choice of B over A as most representative candidate correspond exactly to the advantages of the mean over the median as a measure of location.

Equally, if we favour the mode as the measure of location, we could implement a voting system where the winner is the candidate with the largest number of maximum scores, a version of the plurality method. The inadequacies of the mode as a measure correspond closely to the inadequacies of the plurality method as a voting system.

Many further measures of location have been proposed, which could be implemented in the current framework. For example, whilst the mean is, as I have argued, the most natural choice, one might think that it is too sensitive to outliers, while the median is not sensitive enough, and seek a compromise between the two. An entire family of new voting systems is thus available. This seems a promising avenue for future exploration.

In the above example, the voters and candidates were assumed to be situated on a single-dimensional ideological scale. But this is in no way essential; given the distance assumption, the proposed system locates the candidate nearest to the centroid, however many dimensions determine the voters' scores. No assumption needs to be made as to how many factors the voters are considering when making their judgments about each candidate.

## 16.4 Further Discussion and Conclusion

As I outlined above, proponents of range voting have generally used three kinds of arguments in its favour: that it extracts more information than ranking methods, that it is justified by utilitarianism, and that it avoids the well-known impossibility results and pathologies which plague ranking methods. My proposal retains the first and third of these, but can no longer straightforwardly be supported by utilitarian principles.

What, in fact, is the relationship between my suggested system and utilitarianism? On one view, we should continue to think of the scores as utilities and the proposal as a variant utilitarianism, which, rather than maximizing total utility, minimizes total square disutility. The justification of this as a moral principle, however, seems likely to be problematic.

More promising, I think, is to cut the link with utility altogether. The multi-winner case—in which, it will be recalled, utility maximization elects a one-party parliament—suggests that representation needs to be centre stage, and the single-winner case is, after all, a limiting case of the multi-winner one. Provided the voters can make sense of what they are being asked—and I suggest that they can—it is reasonable to model their views with the mathematical structure necessary for the calculations.

The system proposed here is on a par with ordinary range voting in terms of what is required from the voters; the only increase in complexity is in calculating the winner. Moreover, it shares with conventional range voting the richness of information obtained from voters and the avoidance of the usual paradoxes. Rather than appealing to a justification via utilitarianism, which I have argued is illusory, it picks as winner the candidate who is, in the sense made precise above, the most representative of the views of the population of voters.

More generally, I hope that even one unconvinced by the details of the current proposal might be persuaded that range voting can be generalized in ways that merit further exploration. Once one has scores from voters, there are other things to do with the numbers than simply add them up.[13,14]

## References

Achen, C. H. (1978). Measuring representation. *American Journal of Political Science, 22*, 475–510.

Arrow, K. J. (1963). *Social choice and individual welfare* (2nd ed.). New York: Wiley.

Balinski, M., & Laraki, R. (2011). *Majority judgment: Measuring, ranking, and electing*. Cambridge: MIT Press.

Laslier, J.-F. (2012). And the loser is... plurality voting. In *Electoral systems* (pp. 327–351). Berlin: Springer.

Moulin, H. (1988). Condorcet's principle implies the no show paradox. *Journal of Economic Theory, 45*(1), 53–64.

Robbins, L. (1938). Interpersonal comparisons of utility: A comment. *Economic Journal, 48*, 635–641.

Sen, A. (1999). The possibility of social choice. *American Economic Review, 89*(3), 349–378.

[13] In this context, I should mention the method of *majority judgment* (Balinski & Laraki, 2011). This uses an ordered but non-numerical scale, and picks as winner the candidate with the highest *median* score, instead of the highest *mean* as in conventional range voting. I do not have the space to discuss it here.

[14] The central ideas of this paper were developed when I held a Leverhulme Fellowship, and I am grateful for the support of the Leverhulme Trust. I also thank audiences at the Universities of Glasgow, Hamburg, Manchester Metropolitan and Melbourne, and at the Society for Social Choice and Welfare meeting in Seoul. I am also grateful to Alan Weir, who read a draft of the paper and made helpful comments.

# Weir Bibliography

## Monograph


| | |
|---|---|
| 2010: | *Truth through Proof: A Formalist Foundation for Mathematics*, (Oxford: Oxford University Press), pp. xi+281. |


## Articles in Reverse Chronological Order


| | |
|---|---|
| 2022: | 'Formalism in the Philosophy of Mathematics', *The Stanford Encyclopedia of Philosophy*, (Spring 2022 Edition), Edward N. Zalta (ed.), forthcoming, URL = <https://plato.stanford.edu/archives/spr2022/entries/formalism-mathematics/>. (Revised Version. First Edition 2011). |
| 2015: | 'Informal proof, Formal proof, Formalism', *Review of Symbolic Logic*, August 2015, doi:10.1017/S1755020315000234; Print Version: *Review of Symbolic Logic* **9**(1), 2016, pp. 23–43. |
| 2013*a*: | 'A Robust Non-Transitive Logic', *Topoi* July 2013 doi: 10.1007/s11245-013-9176-9; Print version: *Topoi*, (2015) **34**, pp. 99–107, 2015. |
| 2013*b*: | 'Metatheoretic Results for a Non-transitive Logic'. Available online at: http://eprints.gla.ac.uk/80590/ |
| 2013*c*: | 'Quine's Naturalism': in Harman, G. and Lepore, E. (eds.) The Blackwell Companion to Quine. Wiley-Blackwell (2014) pp. 114–147; (online version, December 2013). |


 
A. Rieger, S. Leuenberger (eds.), *Themes from Weir: A Celebration of the Philosophy of Alan Weir*, Synthese Library 484,
https://doi.org/10.1007/978-3-031-54557-3

| | |
|---|---|
| 2007*a*: | 'Honest Toil or Sheer Magic?', *Dialectica*, **61** pp. 89–115. |
| 2007*b*: | 'Answers to 'Five questions', in Vincent F. Hendricks and Hannes Leitgeb, (eds.), *Philosophy of Mathematics: Five Questions*, (London/New York: Automatic Press/VIP; 2008 pp. 277–287), Online version 2007, http://www.phil-math.org/) |
| 2006*a*: | 'Is it too much to ask, to ask for everything?', in Agustín Rayo and Gabriel Uzquiano (eds.) *Absolute Generality* (Oxford: Oxford University Press) pp. 333–368. |
| 2006*b*: | 'Indeterminacy of Translation' in Ernest Lepore and Barry C. Smith (eds.) *The Oxford Handbook of Philosophy of Language*, (Oxford: Oxford University Press) Chapter Eleven, pp. 233–249. |
| 2005*a*: | 'Naturalism Reconsidered' in Stewart Shapiro (ed.) *The Oxford Handbook of Philosophy of Mathematics and Logic*, (Oxford: Oxford University Press) Chapter 14, pp. 460–82. |
| 2005*b*: | 'Naïve Truth and Sophisticated Logic', in B.Armour-Garb, J.C. Beall, (eds.) *Deflationism and Paradox*, (Oxford: Oxford University Press) pp. 218–49. |
| 2005*c*: | 'On Kit Fine's *The Limits of Abstraction*', *Philosophical Studies*, **122**, pp. 333–348. |
| 2004*a*: | 'An Ultra-realist Theory of Perception', *The International Journal of Philosophical Studies*, **12**, pp. 105–128. |
| 2004*b*: | 'There are no true contradictions', in G. Priest, J.C.Beall, (eds.) *The Law of Non-Contradiction; New Philosophical Essays*, (Oxford: Oxford University Press) Chapter 22, pp. 385–417. |
| 2003*a*: | 'Objective Content', *Supplementary Proceedings of the Aristotelian Society*, **LXXVII**, pp. 23–48. |
| 2003*b*: | 'Neo-Fregeanism: An Embarrassment of Riches', *The Notre Dame Journal of Formal Logic*, **44**, pp. 13–48.<br>Reprinted in *The Arché Papers on the Mathematics of Abstraction* ed. Roy Cook, (Dordrecht: Springer, 2007) pp. 383–420. |
| 2002: | 'Rejoinder to Laurence Goldstein on the Liar', *Analysis*, **61**, pp. 26–34. |
| 2001: | 'More Trouble for Functionalism', *Proceedings of the Aristotelian Society*, 2000–2001, Part 3, pp. 267–293. |
| 2000*a*: | 'The Force of Reason: why is logic compelling?', *Logica Yearbook 1999* (Academy of Sciences of the Czech Republic) pp. 37–52. |
| 2000*b*: | 'Token Relativism and the Liar', *Analysis*, **60**, pp. 156–170. |
| 2000*c* | '"Neo-logicist" Logic is not Epistemically Innocent', (with Stewart Shapiro), *Philosophia Mathematica*, **8**, pp. 160–189.<br>Reprinted in *The Arché Papers on the Mathematics of Abstraction* ed. Roy Cook, (Dordrecht: Springer, 2007), pp. 119–146.<br>Translated as "La logica 'neologicista' non è epistemicamente innocente" in *Frege E Il Neologicismo* edited and translated by Andrea Pedeferri (Milan: FrancoAngeli, 2005). |
| 1999*a*: | 'New V, ZF and Abstraction' (with Stewart Shapiro), *Philosophia Mathematica*,**7**, pp. 293–321.<br>Reprinted in *The Arché Papers on the Mathematics of Abstraction* ed. Roy Cook, (Dordrecht: Springer, 2007) pp. 275–302. |
| 1999*b*: | 'Naïve Set Theory, Paraconsistency and Indeterminacy II, *Logique et Analyse*,**167-8**, pp. 283–340. |
| 1998*a*: | 'Naïve Set Theory, Paraconsistency and Indeterminacy I', *Logique et Analyse*, **161-163**, pp. 219–266. |
| 1998*b*: | 'Dummett on Impredicativity', *Grazer Philosophische Studien*, **55**, pp. 65–101. |
| 1998*c*: | 'Naïve Set Theory is Innocent!', *Mind***107**, pp. 763–798. |

| | |
|---|---|
| 1998*d*: | 'A Neo-formalist Approach to Mathematical Truth', *World Congress of Philosophy*, Boston 1998; Electronic publication at: http://www.bu.edu/wcp/Papers/Math/MathWeir.htm |
| 1997: | 'Two concepts of rationality?' *Cahiers de Psychologie Cognitive*, **16**, pp. 238–246. |
| 1996*a*: | 'Ultra-Maximalist Minimalism!', *Analysis*, **56**, pp. 10–22. |
| 1996*b*: | 'On an Argument for Irrationalism', *Philosophical Papers*, **XXV**, pp. 95–114. |
| 1995: | 'Gruesome Perceptual Spaces', *Analysis***55**, pp. 27–36. |
| 1993 | 'Putnam, Gödel and Mathematical Realism', *International Journal of Philosophical Studies,***1**, pp. 255–285. |
| 1991: | 'An Instructive Nominalism', Critical Review of *Realism, Mathematics and Modality*, by Hartry Field, *Philosophical Books*, **32**, pp. 17–26. |
| 1988: | 'Rejoinder to George', *Mind*, **97**, pp. 110–112. |
| 1986*a*: | 'Dummett on Meaning and Classical Logic', *Mind*, **95**, pp. 465–477. |
| 1986*b*: | 'Realism and Behaviourism', *Dialectica*, **40**, pp. 167–200. |
| 1986*c*: | 'Classical Harmony', *The Notre Dame Journal of Formal Logic*, **27**, pp. 459–482. |
| 1985*a*: | 'Rejoinder to Tennant', *Analysis*, **45**, pp. 68–72. |
| 1985*b*: | 'Against Holism', *The Philosophical Quarterly*, **35**, pp. 225–244. |
| 1983: | 'Truth Conditions and Truth Values', *Analysis*, **43**, pp. 176–180. |

## Reviews, Dictionary Entries and Other


| | |
|---|---|
| 2014: | 'It's Not Just the Economy, Dunderheid', *The Philosopher's Magazine*, Issue 67, Fourth Quarter, pp. 22–34. |
| 2006: | Review of *The Cambridge Companion to Quine*, (ed. Roger Gibson) *Theoria*, **LXXII**, pp. 240–247. |
| 2004: | Review of *The Limits of Abstraction* by Kit Fine, *Bulletin of Symbolic Logic*, **10**, pp. 554–7. |
| 1999: | Review of *Beyond The Limits of Thought*, by Graham Priest, *The Philosophical Quarterly*, **48**, pp. 121–125. |
| 1996: | Review of *Understanding the Infinite* by Shaughan Lavine, *Philosophical Books*, **37**, pp. 136–139. |
| 1994*a*: | Review of *The Semantic Tradition from Kant to Carnap: To the Vienna Station* by J. Alberto Coffa, *The Philosophical Quarterly*, **44**, pp. 402–406. |
| 1994*b*: | Review of *Essays on Quasi-Realism* by Simon Blackburn, *International Journal of Philosophical Studies*, pp. 345–347. |
| 1993: | Articles on 'Holism', 'Intuitionism' and 'Realism' for *The Encyclopædia of Language and Linguistics* (Oxford: Pergamon Press) December 1993 pp. 1591–1592, 1758–1759 and 3472–3473. Reprinted, with amendments, in *The Concise Encyclopædia of the Philosophy of Language* (Oxford: Pergamon Press) December 1997. |
| 1992: | Articles on *Wittgenstein on Meaning* by C. McGinn and *Spreading the Word* by S. Blackburn for the *Encyclopédie Philosophique Universelle* Vol. III: Les Oeuvres Philosophiques, ed. J-F.Mattei, Presses Universitaires de France, 1992. |
| 1990: | Review of *Relevant Logic*, by Stephen Read, *Philosophical Books*, **31**, pp. 31–33. |
| 1989: | Review of *Anti-Realism and Logic*, by Neil Tennant, *Canadian Philosophical Reviews*, **IX**, pp. 293–296. |
| 1988: | Review of *Ascent to Truth: A critical examination of Quine's Philosophy*, by Paul Gochet, *Mind*, **97**, pp. 496–501. |

Printed by Printforce, the Netherlands